Adolescent Development

Adolescent Development

THIRD EDITION

John Dacey
Boston College

Maureen Kenny
Boston College

Deborah Margolis
Boston College

ISBN: 1-58316-107-4
01-074

3151 Skylane Drive, Suite 102B • Carrollton, TX 75006 • (800) 970-1883

Address all correspondence and order information to the above address.

This book is lovingly dedicated to our wonderful children:

JULIETTE DACEY FAY, JENNIFER DACEY ALLEN, KRISTEN DACEY,
AND
KATHERINE KURGANSKY

and to all the children who inspire us

J. D., M. K., and D.M.

BRIEF CONTENTS

CONTENTS

PART 5
THE END OF
ADOLESCENCE 469

CHAPTER 14
INITIATION INTO ADULTHOOD 470

LIST OF BOXES

PREFACE

Our Purposes for Writing This Book

Our main motivation in writing this book was our belief that in the rapidly changing world of today, major changes—both in the lives of teenagers and in the ways we study them—should be addressed, and have not been. Our book has been designed to be more accessible to a wider variety of readers, who differ in several ways from college students of the past. Today's students represent a much greater cultural diversity. They come to college with more varied experiences, vocabularies, and learning styles. Many of these students are the first in their families to attend college. This is especially so for the community colleges, junior colleges, and state colleges. We have tried very hard, with the help of our many reviewers, to provide a text that is readily accessible to students, in terms of vocabulary and writing style, without sacrificing any of the rigor in describing theory and research that characterizes other books. As we move toward the twenty-first century, it is critical that all individuals should have the opportunity to develop their potential to the fullest. One way of achieving this goal is for us to know as much as possible about adolescent development—how we change from the dependency of childhood to the competence and interdependence of adulthood.

Perhaps the human saga is not written as clearly as we would like. As Thomas Jefferson once noted, although the human condition is not open to complete scrutiny, it is, nevertheless, susceptible to considerable improvement. Clues providing insights into the riddle of adolescent development are beginning to multiply. Genetic discoveries are occurring at a rate that can only be described as breathtaking. For example, the march to identify all the human genes (the Human Genome Project) continues unabated. As the population of our nation continues to change, we have become more sensitive to the influence that culture exerts on development. We recognize that the world's cultures view adolescence from many different perspectives. Never before has it been so clear that development results from the interaction of heredity and environment, and that neglect of either side of our equation can only lead to damaging misinterpretations.

The Organization of This Book

In general, our book is organized along traditional lines; it is in the specifics that we differ. Part One is entitled "Introduction to the Concept of Adolescence." The first chapter offers an explanation to the major questions in the field of adolescent development, and the methods we use to answer these questions. The second chapter provides an overview of the major theories about adolescent development. We return again and again to these foundational ideas throughout the rest of the book.

The title of Part Two is "Developmental Patterns," which refers to those aspects of adolescence that are often considered largely internal, maturational changes. These include chapters describing physical development, cognitive development, morality and spirituality, and the self and identity formation. Part Three, "Interactions with the World," covers important areas in development that are greatly influenced by the person's surroundings. These include chapters that explain family relations, social relations, sexuality, and education and work.

We found it necessary to devote three chapters to Part Four, "Teens Who Have Serious Problems." Although we believe that the great majority of teenagers meet their developmental and environmental challenges successfully, it is necessary to devote considerable coverage to those few who do not, as many of our readers will find themselves dealing with these unfortunate teens. These chapters cover theory, research, and applications in the areas of stress and mental disturbances, substance abuse, and delinquent behavior. They also provide an understanding of the environmental factors that make adolescence an unhappy and dangerous period for some teens. Part Five, labeled "The End of Adolescence," includes the book's final chapter and describes initiation into adulthood in North America. (The sequence in which we have arranged the book is recommended for its use in the classroom, but chapters are relatively independent. Therefore, they may be assigned in any order or not included if so desired.) To further assist readers in their use of these materials, we are including an appendix: "Services Available to Youths in Most Medium-to-Large U.S. Cities, and Their 800 Numbers."

Basic Themes of Adolescent Development

To reflect the exciting changes that are taking place in our knowledge of adolescent development, we have woven our narrative around several integrating themes: the biopsychosocial model, the cultural context of development, the role of gender, multiple levels of development, and applications to daily living. We return to these themes in most chapters as a way to add meaning to a basic knowledge of this field.

THE BIOPSYCHOSOCIAL MODEL

The biopsychosocial model helps you integrate the wealth of information that you will find in the pages to come. If you think of adolescent development as the product of the interaction of biological, psychological, and social forces, you'll better appreciate its complexity. For example, biological influences range from the role of genes in development to adult health concerns; psychological influences include all aspects of cognitive and personality development; social influences refer to such powerful forces as family, school, peers, the media, and other cultural factors. The biopsychosocial model helps explain how the interaction of these forces is the key to understanding human development.

CULTURAL INFLUENCES

Our goal in urging you to adopt a multicultural perspective is to help you develop a greater understanding of those who seem "different." If you adopt this perspective, you will come to realize that different people have different world views that decisively influence their behavior. In this way, you can interact more congenially with others, thus fostering more positive relations in our society. People from different cultures view the world in a variety of different ways and, as we stress, these differences are not deficits. Recognizing how people differ in their thinking and behavior will help you identify variations in how individuals are raised, how they think, and how they become functioning members of their culture. Since we feel so strongly about this matter, you'll find a discussion of cultural issues in each chapter.

THE ROLE OF GENDER

As concern about gender equity receives more publicity, the stereotypes about males and females are slowly eroding. This is because we are becoming more aware that when people are treated according to stereotypical characteristics, their potential is limited. Although gender stereotyping is only one part of the gender story, it illustrates the importance of the relationship between gender and development. For example, children at an early age construct social categories from the world around them, attach certain characteristics to these categories, and then label the category. This process may be positive because it helps to organize the world; it may also be negative if the characteristics associated with the category are limiting: "girls just can't do math"; "boys just can't cook."

MULTIPLE LEVELS OF DEVELOPMENT

We try never to write of "the adolescent" as though there were such a unidimensional stage of life. Rather, where relevant throughout the book, theory and research are presented in terms of early and late adolescence, and in some specific cases, of middle adolescence, a transitional state halfway through the teen years. We also recognize that although all developmental textbooks must make generalizations about people grouped by age, it is vital that we remember the role played by individual differences.

APPLICATIONS TO DAILY LIVING

The study of adolescent development is an exciting, rapidly changing, and highly relevant subject. To help you put the theories and research of this book into a meaningful framework, we have written numerous sections on how the material for each chapter may be applied to real life situations. These range from the role of television to the search for identity, from the problems of the adolescent child of an alcoholic to the appeal of street gangs for some youths.

Third Edition

Many thanks to suggestions from students who used previous editions of our text and to the insightful comments of reviewers.

- The student study guide is incorporated throughout the text. At the end of each chapter, we have provided sets of multiple choice and fill-in-the-blank questions and their answers so that you may review your knowledge of the entire chapter. Professors may well use some of these questions on their tests, so the questions are not only reviews, but may also be previews of actual test questions.

- We have decided to incorporate research on cultural diversity as a regular part of the text but also continue to highlight some of this material in our A Sociocultural View boxes. We have done this because we feel that in a society where diversity is playing a greater and greater role, this material deserves special attention. Furthermore, we have replaced some of these boxes and added new ones with what we consider to be information of particularly high quality.

- We have integrated and referred to the biopsychosocial model even more in this edition. Because the data relating to developmental psychology are so extensive, we believe that this model helps you to better organize developmental data, and it also serves as an effective memory aid. We are likewise aware that it is impossible to understand development from a single perspective. That is, biological development has both psychological and social consequences; social development has both biological and psychological effects; and psychological development has both biological and social influences.

- In the chapter on general theories, we have added a section on the new concept of contextualism, which provides a framework for understanding.

- The new gender and identity section provides expanded coverage of the debates and relevant research applying Erikson's model of identity development to girls and women.

- Included is an increased presentation of recent research on parenting styles across cultures.

- A new section on sexual well-being in adolescence emphasizes the normative and healthy, as well as the problematic, aspects of sexuality in adolescence.

- A new section on sex education has been added.

- We offer updated research on factors that contribute to academic success and improvement among inner-city adolescents.

- Updated research on the working teen suggests the kinds of work experiences that can be beneficial.

- Given the importance of education in a modern society, we have significantly updated the section on schools and development.

- Because adolescent psychology is such a dynamic field, we have of course updated many of the descriptive data in this text.

Owner's Manual: A Guide to Teaching-Learning Features

You will enjoy and learn from this book to the extent that its topics, organization, and clarity make its contents meaningful to you. Helping you master the book's contents in an uncomplicated manner has been our most important pedagogical goal. To accomplish this task, we have built a number of features into each chapter:

ANTISOCIAL BEHAVIOR

As early as the early 1960s, Leonard Bernstein's *West Side Story* signaled society's concern that glib sociological excuses would only worsen the handling of behavior disorders. Public opinion and public institutions are turning back to a punishment model. The pendulum has swung away from the idealization of Rousseau's "noble savage," who would be good only if society did not overcontrol and frustrate him, toward today's bleak cynicism of the essential untreatability of the *Lord of the Flies*.

John Meeks and Allen Cahill, 1988

430

Chapter Outline
Outlines at the beginning of each chapter show what material will be covered.

List of Objectives
A carefully formulated list of objectives is presented after each chapter introduction to help guide your reading. After reading each chapter, you can use the objectives as a way to test your knowledge and to review the material.

Social relations gain increasing importance during the adolescent years. Friendships take on new meaning and the peer group becomes a more important source of influence. Nevertheless, parents and other adults continue to be influential in the lives of adolescents. The value and influence of peer and parental relationships during adolescence has often been debated. Recent research has helped to resolve some of these debates. The importance of parent and peer relationships and the ways in which peer relationships change during the adolescent years are the focus of this chapter.

After reading this chapter, you should be able to:

- Show how friends become increasingly important as the maturing adolescent begins to move beyond the immediate family.
- Explain why parents remain an important source of influence and support to adolescents.
- Describe the differences in friendship patterns that are characteristic of adolescent girls and adolescent boys.
- List positive influences that the peer group can have on an adolescent's growth.
- Discuss the structure of large peer groups and the subgroups within them.
- List the elements that make up a subculture, and relate these to the subcultures of adolescents in the United States.
- Explain the psychogenic, culture transmission, and behavioristic models of the origins of subcultures.
- Describe the common elements of teenage subcultures.
- Evaluate interactions in the classroom.
- Discuss these issues from an applied, a sociocultural, and your own point of view.

Friendships

THE ROLE OF PARENTS IN PEER RELATIONS

Children depend much more on their parents than on their friends for their emotional needs. In adolescence, however, sexual interests and the need to become an individual distinct from the family strengthens the influence of peers. The young person is now more likely to confide feelings and problems to close friends rather than to parents.

Progressively more time is spent in the company of friends than at home. Many parents find the increasing absences of their child from the home difficult, particularly when dealing with their first child. This shift highlights the importance of the parents' role in monitoring adolescent behavior (Flannery & others, 1999). Appropriate parental monitoring and guidance around peer issues can help to protect teens from some of the risks of increased peer influence (Bogenschneider & others, 1998; Flannery & others, 1999). Young adolescents can be quite discerning about whether their parents agree with each other on how to handle an adolescent (Johnson & others, 1991; Scaramella & others, 1999; Steinberg, 1987). Having an adolescent in the house requires that

A Sociocultural View

Research on cultural diversity has been integrated throughout the second edition of *Adolescent Development*. Highlights of this material are featured in the Sociocultural View boxes.

Key Terms

Because key terms are essential to your ability to understand material, they are highlighted and explained in the context of each chapter, and are briefly defined in the text's margins. The key terms are also listed and page referenced at the end of each chapter, and are defined again in a glossary at the end of the book.

An Applied View

These boxed features give examples to show how the material you are learning can be applied to real life situations, such as in a classroom or medical facility.

Marginal Questions

Questions that encourage you to think about pertinent issues and that promote discussion are placed in the margins of the text.

What's Your View?

Controversial issues related to developmental psychology are presented in these boxes for you to read and think over.

---

46 Part I ❖ Introduction to the Concept of Adolescence

young adults on their wedding nights, they are expected to perform with sexual competence. As many older married people will tell you, this seldom happened.

- *Responsibility.* In our society, children and youth are allowed (even encouraged) to be somewhat irresponsible. However, as soon as they leave home, get a job, or get married, they are expected to be totally responsible for their behavior.
- *Dominance.* Young people in our society are expected to switch suddenly from a totally submissive role as children living in the homes of their parents to complete independence and self-reliance. If they get married and have a baby right away (and most did in earlier times), they are expected to immediately assume total dominance over the life of their own child.

A SOCIOCULTURAL VIEW

Do the Samoans Have the Right Answers?

These abrupt changes, occurring as they do without much preparation during the adolescent period, put a great strain on the life of the individual. Can you think of better ways of introducing teens to sex (e.g., childbirth classes, sex education classes)?

Margaret Mead's studies of youths on the island of Samoa (1927/1949) convinced her that the Samoan culture's introduction of sex in the early teens is more natural. In that society, adolescent boys were expected to sneak into the huts of the girls they admired and have sex with them. After a trial period, if the girl found him as attractive as he found her, she agreed to marry him.

As a preparation for adult life in our society, Mead (1970) went so far as to suggest that we allow our youth to engage in trial marriages, in which teenagers are allowed to live and sleep together (in the home of one of their parents) and to practice sexual intercourse, provided birth control is used. Her argument was that because so many teenagers are engaging in sex anyway and because so many of them are becoming pregnant and/or contracting a disease from not using protection, they would be better off if they had the acknowledgement and advice of adults.

Albert Bandura's Theory [biopsychoSOCIAL]

observational learning Influence of modeling on personality development as stressed in Bandura's social learning theory.

Albert Bandura, one of the chief architects of social learning theory, has stressed the potent influence of modeling on personality development. He calls this **observational learning**. In a famous article on social learning theory, Bandura and Walters (1959) cite evidence to show that learning occurs through observing

---

400 Part IV ❖ Teens Who Have Serious Problems

AN APPLIED VIEW

Answers to "What Do You Know About Drugs?"

1. (d) All of the above. The use of drugs is as old as the history of humankind. The United States has witnessed special drug abuse problems in different periods. During the Civil War opium was used medically, and since its addictive properties were not clearly understood, many wounded soldiers became addicted. Following the Civil War, the practice of opium smoking became popular on the West Coast and spread to many urban areas. Throughout the century, there were periodic "drug scares" created by the use of cocaine at the turn of the century, heroin in the 1920s, marijuana in the 1930s, and heroin again in the 1950s. The 1960s saw a social explosion of drug use of all kinds, from LSD to heroin and marijuana.

2. (b) Through their friends. With the exception of alcohol, which is usually first used at home, most drug users are introduced to drugs by friends.

3. (b) Alcohol. Estimates are that about nine million Americans are alcoholics.

4. (b) Marijuana. In the past, marijuana was legally classified as a narcotic but it isn't now. Marijuana's effects are similar to stimulants, sedatives, or hallucinogens, and its actual effects depend on dose, frequency of use, set (personality and expectation of the user), setting (environment), and other factors. Morphine and heroin are legally and pharmacologically classified as narcotics. Methadone is a synthetic narcotic.

5. (c) Mescaline. All are stimulants except mescaline, which is a hallucinogen with effects similar to LSD.

6. (c) Mescaline. Physical dependence on mescaline (the drug derived from the mescal and peyote cactus) and many other hallucinogens has not been verified.

7. (d) MPA. MPA is not an acronym for any known drug. MDA, LSD, and STP are hallucinogens with similar effects. MDA (Mellow Drug of America) and STP (Serenity, Tranquility, Peace) are street drugs.

8. (d) All of the above. In particular, using nonsterile equipment is a serious hazard often overlooked by the drug user.

9. (b) To relieve discomfort. When people stop taking a drug that they are physically dependent on, they develop physical withdrawal symptoms (such as muscle spasms, vomiting, sweating, insomnia, and so forth). Taking the drug relieves the discomfort of withdrawal symptoms.

10. (d) Psilocybin. Psilocybin is a hallucinogen that has no accepted medical use. All the other drugs have at various times been used to treat narcotic addiction. When heroin was introduced in 1898, some people thought it had possibilities for treatment of "morphinism." Methadone, cyclazocine, and nalexone are used currently to block the high produced by heroin.

11. (e) All of the above. All these methods have been used successfully to treat drug abusers, and many have been used in combination.

possession Refers to illegally obtaining drugs from someone not sanctioned to distribute them.

Possession, dealing, and **trafficking** in drug substances are also distinguished legally. Illegal possession refers to obtaining drugs from someone not legally sanctioned to distribute them. Dealing is the sale of drugs on a small scale, usually carried out by a friend of the purchaser. Trafficking involves the sale of much larger amounts of drugs. Each of these violations of the law carries a different penalty.

Chapter 14 ❖ Initiation into Adulthood **479**

WHAT'S YOUR VIEW?

The Influence of Media and Advertisements on Teen Values

The leisure time created for identity development in this country has made teens a lucrative target audience for advertising. Not only do they have the time, but many possess the resources as well. In a school setting, they are a captive audience. The presence, and some might say the intrusion, of advertising and the media into the classroom has brought the issue to public attention.

Although advertising has long adorned vending machines, athletic scoreboards, and school yearbooks, the addition of television news programs such as *Channel One* has alarmed some social scientists. (*Channel One* is a news program oriented to teens, with ten minutes of news and two minutes of advertising. Subscribing schools receive equipment and agree to let

90 percent of their students watch the program daily.) Studying the effects of such a program on 10th-grade students in a Michigan school, Greenberg and Brand (1993) found that students viewing Channel One were more aware of the news but also that they reported more materialistic attitudes than nonviewers.

In his book, *Warning: Nonsense is Destroying America*, Vincent Ruggiero (1994) quotes an advertising vice president as admitting, "You've got to reach kids throughout the day—in school, as they're shopping in the mall . . . or at the movies. You've got to become part of the fabric of their lives" (p. 122).

In your opinion, what role should advertising and media play in education?

in the radical behaviors common in adolescence (i.e., heavy metal music, jargon language code, unusual fashions and hairdos, and so on). These cravings for extremes produce an inherent need for social survival skills, which teenagers cannot provide for themselves. Thus, there is a need for initiation. According to Ventura, initiation rites satisfy the craving while providing the necessary survival skills for life in this world.

Although obviously not appropriate for our society, ancient tribal rituals such as those of Yudia and Mateya appear to satisfy the craving for extremes. This satisfaction is accomplished through the extreme nature of the ceremony (involving danger and pain) while the survival skills are given, at least psychologically, along with the new status of adult. Having proven themselves worthy by passing the rite, the youth are now considered to be among the mature.

By contrast, Ventura says, American culture "denies the craving [and] can't possibly meet the need" (p. 47). As a result, adolescence is prolonged and the eventual transition to adulthood is almost impossible to pinpoint. Ventura claims that adolescents are forced to

❏ generate forms—music, fashions, behaviors—that prolong the initiatory moment, i.e., that cherish and elongate adolescence—as though hoping to be somehow initiated by chance somewhere along the way (p. 47).

Do you accept the idea that gang warfare is really a form of self-initiation into manhood?

Thus, the tendency toward self-initiating behavior arises. In the most frustrated of adolescents, this self-initiation takes on criminal characteristics. One need only glance at the headlines of any urban newspaper (e.g., the Los Angeles riots) for examples of this kind of criminal behavior.

Chapter Highlights

Each chapter conclusion is followed by a series of summary statements. These highlights are grouped according to chapter topic and help you review the chapter quickly and thoroughly.

What Do You Think?

Following the Key Terms, you will find a series of questions challenging you to demonstrate your knowledge of the chapter's content.

Suggested Readings

An annotated list of four or five books or journal articles to supplement the chapter reading can be found at the end of each chapter.

Student Study Guide

To help you fully learn the material, the second edition of this text includes a built-in student study guide. (Students need not buy a separate study guide—it is part of the book!) The student study guide is composed of:

- Chapter opening learning objectives
- End of chapter multiple-choice and fill-in-the-blank sample tests

---

32 Part I ❖ Introduction to the Concept of Adolescence

CHAPTER HIGHLIGHTS

What Is Adolescence Like Today?
- There are fewer adolescents in the 1990s than there used to be, and they make up a smaller proportion of the total population.
- Today's teens face many serious problems, but most manage to avoid difficulties and instead make valuable contributions to our society.
- Today's adolescents have more spending money than those of previous generations.
- Thinking back on your own adolescence can give you a deeper understanding of today's teenagers.
- It was once commonly believed that the teen years were a carefree time of fun and exploration; some now believe that adolescence is the most difficult of all life stages.
- Most experts, however, state that the majority of adolescents are happy and productive members of their families and communities.

What Was Adolescence Like in the Past?
- In ancient times, some philosophers believed that youths were frivolous and irresponsible, while others emphasized their growing intellectual skills and self-sufficiency.

- From the Middle Ages until the start of the twentieth century, strict discipline was believed necessary to force young people to take on adult responsibilities as early as possible.
- Two early twentieth-century concepts changed our view of adolescence: compulsory education and juvenile justice.
- Theories of adolescence as a separate life stage (and studying it with careful observation) also came into existence in the early twentieth century.

When Does Adolescence Begin and End?
- There is no general agreement about what marks the beginning and ending of adolescence.
- Physical, intellectual, social, and emotional factors all enter into the process of defining adolescence.
- In contrast to our own, some primitive societies have rites of passage that clearly mark the entry into adulthood.

What Are the Best Methods of Studying Adolescence?
- Methods of data collection include descriptive studies, manipulative experiments, and naturalistic experiments.
- There are four major time-variable research designs, each with its advantages and disadvantages.

KEY TERMS

age cohort 28	juvenile justice 17	one-time, one-group studies 28
case studies 25	longitudinal studies 28	retrospective accounts 18
compulsory education 17	manipulative experiments 26	self-report studies 25
correlational studies 26	menarche 23	sequential studies 30
cross-sectional studies 29	naturalistic experiments 27	sociocultural diversity 6
descriptive studies 25	observational studies 25	stereotype 3
		treatment 26

WHAT DO YOU THINK?

1. In your opinion, is adolescence today harder to live through than other periods of life? Easier? About the same?
2. How does adolescence today compare to your parents' teen years? Your great-great-grandparents' teen years?
3. What is the single most important sign that adolescence has started? Does gender make a difference? Race? Nationality?
4. Why has Western history treated teenagers so negatively?
5. Of all the research techniques described in this chapter, which is the best one? Why?
6. What are some things you could do to help adolescents better understand the causes of their own behavior?

Chapter 1 ❖ What Is Adolescence? **33**

SUGGESTED READINGS

Davis, J. (1991). *Checking out the moon*. New York: Orchard. This book, about a teen who is dealing with her parents' divorce, manages to include several other adolescent crises: leaving home, confronting cultural diversity, helping someone deal with being raped. This is an insightful book by an author who understands.

Golding, W. (1962). *Lord of the flies*. New York: Coward McCann. This tale of a group of teenage boys whose plane crashes on a Pacific island, killing the

adults, is *must* reading. You watch the clash of their values as they wend their way to the shocking ending.

Murry, V. M. (1990). *Black adolescence*. Boston: Hall. An annotated bibliography of the major current issues affecting African American teens today.

Wright, R. (1945). *Black boy*. New York: Harper & Row. The haunting autobiography of the novelist's adolescence in the Deep South. Its insights are entirely relevant to today's world.

GUIDED REVIEW

1. Which part of the adolescent population in the United States is NOT growing?
   a. White
   b. Native American
   c. Hispanic
   d. Asian/Pacific Islanders
2. Babies born to teen mothers are at heightened risk for
   a. low IQ.
   b. low birth weight.
   c. lower resistance to disease.
   d. heart disease.
3. Which of the following philosophers did NOT believe that adolescence is a positive stage of human development?
   a. Rousseau
   b. Plato
   c. Sophocles
   d. Aristotle
4. Parenting strategies that emphasize the development of decision making and self-determination most closely resemble the views of
   a. Hesiod.
   b. Plato.
   c. Rousseau.
   d. Aristotle.
5. The beginning and ending of adolescence is primarily determined by the individual's
   a. socioeconomic level.
   b. culture.
   c. educational level.
   d. physical development.
6. In a well-designed manipulative experiment, which of the following is true of both the experimental group and control group?
   a. They are both given the same treatment.
   b. They differ only after treatment.
   c. They are assigned by random sampling.
   d. They are each made up of at least 10 participants.
7. Infants, randomly divided into two groups, are exposed either to a colorful picture of a human face or to an equally colorful picture of random design. Researchers compare the amount of time that infants in each group look at the stimulus. In this experiment, which of the following is the treatment?
   a. the age of the infants
   b. the picture
   c. the length of time a picture is observed
   d. the experimental group
8. You are conducting a study in which you are solely an observer and do not administer a treatment to the participant. Which type of data collection method are you using?
   a. longitudinal study
   b. descriptive study
   c. manipulative experiment
   d. naturalistic experiment
9. Which of the following data collection methods is NOT an example of a descriptive study?
   a. case study
   b. manipulative study
   c. observational study
   d. self-report study

Note to Instructors: Supplementary Materials

An updated version of the Instructor's Resource Manual has been prepared by Deborah Margolis. Each chapter of the Instructor's Manual includes a chapter overview, learning objectives, lecture and classroom/student activities suggestions, and questions for review and discussion. The Instructor's Manual also includes multiple-choice test items for each chapter.

Acknowledgements

This book was produced through the cooperation of many publishing professionals. Joann Piet, acquisitions editor, provided overall guidance and Jon Hughes, publisher, facilitated our work in numerous ways. Our thanks to these and the rest of the excellent staff at Alliance Press.

INTRODUCTION TO THE CONCEPT OF ADOLESCENCE

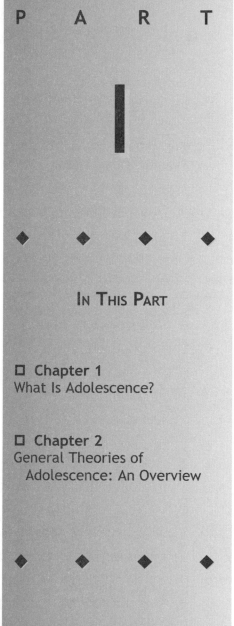

WHAT IS ADOLESCENCE?

The creative process of adolescent identity construction is directly analogous to . . . the process by which novelists and script writers create the fictional characters of novels and screenplays.

John Caughey, 1988

"Who are you?" said the caterpillar. Alice replied rather shyly: "I . . . I hardly know, sir, just at present—at least I knew who I was when I got up this morning, but I must have changed several times since then."

Lewis Carroll, *Alice in Wonderland*

stereotype
Generalizations about the characteristics—either positive or negative—of a group that are supposedly shared by its members.

What *is* adolescence like today? Some writers have suggested that like Alice, adolescents experience life as a constant swirl of painful adjustments. Is adolescence a time of topsy-turvy change, marked by abrupt emotional crises, or is this only a **stereotype**? This question has caused much debate among scientists who study this fascinating period of life.

Other important questions about the current nature of adolescence include: What part of our population is adolescent, and how is that changing? What changes are occurring in the racial and economic makeup of our adolescent population? What are some problems adolescents face today? What are some contributions adolescents make to our society? What are the economic values of these contributions? In this chapter, we give brief answers to these questions as a means of forming a useful definition of adolescence.

A second method that should help us understand this period better is to compare adolescence now with how it used to be, as seen by observers in earlier eras and by the reflections of those who are now adults. A third major issue focuses on when (at what age or with what event) this stage of life begins, and when it ends. A fourth important question is, "What are the best ways of answering questions about adolescence today?"

In this chapter we consider each of these questions. You should be forewarned, however, that defining this fascinating period of life is a complex business. There are many different facets of this developmental period, as individuals become more and more distinct from each other. The process of changing from a child to an adult is not easy to define. We will need this entire book to produce a truly adequate definition, but in this chapter we can at least get off to a good start. After reading this chapter, you should be able to:

- List several key factors that depict adolescence in American society.
- Identify ways in which your own adolescence was different from or similar to that of today's adolescents.
- Describe how adolescence was viewed in ancient times, during the Middle Ages, during the Age of Enlightenment, and in the present century.
- Explain the value of studying adolescence in two stages, early (11 to 14 years old) and late (15 to 18 years old), rather than as one stage.
- State what you believe to be the best definitions of the beginning and end of adolescence.
- Specify the importance of using scientific methods in the study of adolescent development, and improve your ability to analyze the tables and charts that are essential to this study.
- Define the three major data-collection methods and the four types of time-variable designs, and list the advantages and disadvantages of each.
- Discuss these issues, in this and every chapter, from an applied, a sociocultural, and your own point of view.

What Is Adolescence Like Today?

This section gives us a brief overview of adolescence as it now exists in our culture. This overview will only scrape the surface, however. As Petersen and associates (1993) have stated:

> ❏ There has been a literal explosion of research on adolescence in the United States as evidenced by the emergence of the Society for Research on Adolescence, several new journals focused on adolescence, and a dramatic increase in the number of researchers working in this field. Whereas much of the earlier work tended to be descriptive and more qualitative in nature, recent work has been enhanced by use of rigorous methodologies. For example, whereas writers about adolescence have speculated for some time about the influence of hormones on behavior, research in the past decade has systematically measured various aspects of hormone production and action and has examined these indicators relative to behavior. Similarly, studies of family interaction have been aided by the dramatic technological improvements in video equipment (p. 155).

As you read the descriptions of aspects of teen life presented in this chapter, and the data that support them, you can begin to put the pieces together into your own definition of what "adolescence" means.

WHAT PART OF OUR POPULATION IS ADOLESCENT?

How will the major population changes affect us in the next few decades?

One way to get a picture of adolescents is to examine their place in the population of the United States today. First of all, there are fewer of them than there used to be, and they compose a smaller part of the total population (see Figure 1.1 and Table 1.1). What other conclusions can you draw from these data?

Figure 1.1

Teenage population as a percent of total U.S. population.

Source: Data from the U.S. Bureau of the Census, 1990.

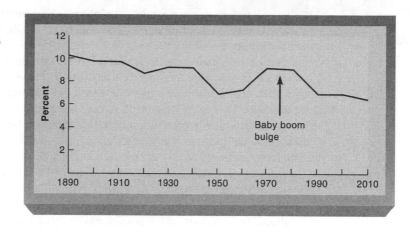

Table 1.1	A Numerical Picture of the American Adolescent Population (in millions)			
Year and Race	**10–14 Years**	**% of Total Population**	**15–19 Years**	**% of Total Population**
Total Population				
1980	18.2	8%	21.1	9%
1992	18.1	7%	17.9	7%
White				
1980	15.1	6%	17.6	8%
1992	14.4	6%	13.6	5%
African American				
1980	2.7	1%	3.0	1%
1992	2.8	1%	2.6	1%
Hispanic Origin				
1980	1.5	.6%	1.6	.7%
1992	2.7	1%	2.1	.8%
Asian, Pacific Islanders				
1980	.3	.1%	.3	.1%
1992	.7	.2%	.6	.2%
Native American				
1980	.1	.04%	.2	.08%
1992	.2	.07%	.2	.07%

Source: Statistical Abstracts of the United States, 1994.

We have included Table 1.1, with all its detailed information, because it says so much in a small amount of space. You will have to look at it closely, however, in order to understand the overview it is meant to portray. Can you answer these important questions by looking at Table 1.1: Is the adolescent population of the United States going up or down? Is it increasing or decreasing as a percentage of the whole population (which in 1980 was about 226 million and in 1992, about 255 million). How is this picture different for whites and nonwhites? Which is the fastest growing nonwhite population? Although the table won't yield the answer to this question, perhaps it is the most important one: How will these changes affect teens?

CRITICAL CHANGES IN THE SOCIOCULTURAL DIVERSITY OF THE ADOLESCENT POPULATION

sociocultural diversity
Many cultures coexist within a country such as the United States or Canada. The term "sociocultural" emphasizes social rather than biological or psychological differences.

Not only are there changes in the number and percentage of teens, but also in their racial and economic status. These changes will have sweeping ramifications for those who work with teens. In the early years of this country, it was hoped that the United States would become a large melting pot, which would gradually blend people of many races and cultures into one superior American culture. Now it seems more likely that we will need to build our strength through cooperation among several different cultures, all living alongside each other. This latter concept has become known as **sociocultural diversity**. Lipsitz (1991) summarizes these changes:

> ❏ Approximately 20 percent of America's children live in poverty. That figure is rising steadily toward 25 percent. Children are poorer than any other age group, and they are worse off than they were two decades ago. African-American and Hispanic children are two to three times more likely to be living in poverty than are White children. . . . Median family income, although rising, remains below 1970 levels. Family incomes of African-American and Hispanic children, however, have continued to decline. The median family income of White children is twice that of African-American children and more than 1f times that of Hispanic children.
>
> Between 1980 and 1988 the Hispanic population grew close to five times faster (34 percent) than the non-Hispanic population (7 percent). Differential numbers are riveting: Census estimates indicate that from 1985 to 2000 there will be 2.4 million more Hispanic children, compared with 1.7 million more African-American children, and 66,000 more non-Hispanic White children. If current population growth is sustained, Hispanics will become this country's largest minority by 2020 (p. 22).

Soon the term *minority population* will no longer refer only to nonwhites. By the year 2020, in many states, especially in the southwest, whites will cease to be in the majority (Petersen & others, 1993). Another major change is in the number of American families with children. The percentage dwindled from 45 percent in 1970 to 38 percent in 1980 to 36 percent in 1990.

A SOCIOCULTURAL VIEW

The Role of Ethnic and Cultural Differences

As more research on adolescents is done, our understanding of how different adolescents are from each other also grows. Younger adolescents are different from older adolescents, males are different in many ways from females, and there are differences due to ethnic and cultural background as well. It is increasingly important that we learn about these differences as our society becomes more diversified. Did you know, for example, that in 1998, whites will no longer outnumber nonwhites in the United States?

In the past, many people have believed that being nonwhite means being inferior. Among the stereotypes are the assumptions that nonwhites are more likely to be unmarried parents, to abuse drugs, to commit criminal acts, and to suffer mental illnesses. To what extent are these beliefs true, and to what extent are they prejudices? Where they are true, to what extent are they due to race and to what extent are they due simply to poverty?

Are the alleged strengths of nonwhites true? Do nonwhites have strong family ties, treat their elderly with compassion, have high religious participation, and support each other in adversity? Are these positive stereotypes supported by research?

How do African American, Latino, Asian, and Native American youths compare to each other? Are they similar in some ways and different in others? Over the course of this book, we will attempt to answer these questions, in part through the regular sections of the book, and in part through boxes like this one. You should ask yourself, as you read various research reports, "Does this research reflect most adolescents, or is it relevant only to the group under study?"

A way to further develop our picture of the modern teen is to ask the question, "What are some of the major problems American teens face today?" Before turning to that question, though, perhaps you would enjoy taking a moment to reflect on what you were like in early adolescence.

AN APPLIED VIEW

What Were You Like?

Following are some questions about your teenage personality that you might enjoy answering. Pretend you are in the eighth grade. Let your mind drift back in time, and imagine yourself sitting in your favorite classroom. Look around the room and see who is sitting there. Try to answer these questions as you would have then.

A. Are the following statements true or false?
1. Most of the other kids in the class are stronger than I am. T F
2. I am about as intelligent as anybody in this classroom. T F
3. I am certainly not one of the teacher's favorite students. T F
4. Most people would say I am above average in athletic ability. T F
5. I am probably one of the more attractive students in this class. T F
6. I am one of the shortest kids here. T F
7. I would say that I am more mature than most of my friends. T F
8. I am more popular than most of my classmates. T F
9. I am very moody and I seem to get upset easily. T F
10. I am unhappy with several of my physical traits. T F

continued

AN APPLIED VIEW

B. In the following multiple choice statements, choose the one that best fits you.

1. My knowledge of sex is
   a. practically nonexistent.
   b. much less than the other kids'.
   c. about the same as the other kids'.
   d. much greater than the other kids'.

2. At parties and dances, I am known as
   a. a real wallflower.
   b. shy, except with my close friends.
   c. outgoing.
   d. the life of the party.

3. I would describe my relationship with my parents as
   a. very loving; we really understand each other.
   b. friendly, but we sometimes have unpleasant fights.
   c. stormy about half the time.
   d. we are usually so angry at each other that we hardly talk at all.

4. My interest in my schoolwork is
   a. high.
   b. moderate.
   c. low.
   d. nonexistent.

5. Which of the following is true regarding my thoughts about death and physical illness?
   a. I am plagued with such thoughts.
   b. I frequently have such thoughts and they bother me from time to time.
   c. I don't think about them very much.
   d. I never think about them.

C. Fill in the blanks in the following sentences:
   1. The thing I would change most about my life is

   _____

   _____ .

2. My best friend is _____

   _____ .

3. My deepest secret is _____

   _____

   _____ .

4. My fondest memory is _____

   _____

   _____ .

5. The thing of which I am most ashamed is

   _____

   _____ .

D. Rank the following ten characteristics in the order in which they are true of you. Put a 1 in front of the characteristic that best describes you; put a 2 in front of your second most typical trait, and so on.

___Enthusiastic student ___Handsome/pretty
___Sports enthusiast ___Kind and generous
___Boy/girl crazy ___A loyal friend
___Introspective ___Well-mannered
___Humorous ___Reliable

There are several things that can be done with this list to enhance your understanding and empathy for adolescents. You and your friends or fellow students might want to compare answers, with each person reading their answers to each question, one at a time. You may wish to analyze the answers for clues about the definition of adolescence that we are seeking. On the other hand, you might share your answers with a small group of your most trusted friends, or even with your parents and siblings.

SOME PROBLEMS ADOLESCENTS FACE TODAY

❏ "These are the best years of your life! You'd better enjoy them now, because before you know it, you'll be weighed down with adult responsibilities!"

Although sex education does exist in some of our schools, the teenage pregnancy epidemic and the surge of sexually transmitted diseases seem to indicate that something more or different is needed.

Table 1.2	One Day in the Lives of American Children

Every Day in America:

- 8,441 teens become sexually active.
- 2,756 teens become pregnant.
- 1,340 babies are born to teen mothers.
- 2,754 babies are born out of wedlock.
- 638 babies are born to mothers receiving late or no prenatal care.
- 2,699 babies are born into poverty.
- 95 babies die before their first birthday.
- 2 children younger than 5 are murdered.
- 248 children are arrested for violent crimes.
- 176 children are arrested for drug abuse.
- 424 children are arrested for alcohol abuse or drunk driving.
- 4,500 children bring guns to school.
- 79 children are put into adult jails.
- 2,250 students ages 16 to 24 drop out of school.

Source: Children's Defense Fund, 1991, 1995.

Can you remember your parents saying this, or something like it? At some time during their teen years, most adults were probably advised not to "waste their youth." It used to be a common belief that adolescence was a carefree period, a stage of life when people would "sow their wild oats" before settling down to the demands of adult maturity.

Table 1.3	Births to never married 15-19 year old women as compared to total 15-19 year old women (in thousands)		
Year	Total	Number of Mothers	Mothers as % of Total
1990	8,049	451	5.6
1995	8,701	661	7.6

Source: U.S. Bureau of Census, 1997.

Some observers, however, believe that the teen years have become the *worst* time of life. Is adolescence an unusually problematic period of life? Are the changes that accompany it more abrupt and disruptive than those of earlier and later stages? Those who think so point to statistics such as those in Table 1.2 to support their case.

The increase in "children having children," that is, in unwed teenagers giving birth to children whom they are rarely prepared to care for, is especially dismaying. Table 1.3 shows the numbers and percentages of these births indicating an increase from 1990 to 1995.

Table 1.4	Percentage of Children Reporting Specific Problems in Previous Month				
Problem Area	9–10 Years (N = 111)	11 Years (N = 154)	12 Years (N = 169)	13 Years (N = 163)	14 Years (N = 82)
School	21.1	9.1	16.6	12.9	9.8
Siblings	18.6	12.3	10.1	10.4	13.4
Parent/child conflict	17.0	18.2	14.8	18.4	23.2
Friends	13.7	14.3	14.2	13.5	8.5
Parent/parent conflict	1.8	5.2	5.3	2.5	1.2
Household	8.5	1.3	2.4	1.8	1.2
Family health	5.3	3.2	3.6	3.7	1.2
Pets	5.3	1.9	1.8	0.6	1.2
Extracurricular activities	1.8	2.6	7.1	6.7	3.7
Boyfriend/girlfriend	0.9	1.3	3.0	3.7	7.3
Free time	1.8	3.2	1.2	2.5	1.2

From A. Spirito et al., "Common Problems & Coping Strategies Reported in Childhood and Early Adolescence" in *Journal of Youth and Adolescence,* 20 (5):531, 1991. Copyright 1991 © Plenum Publishing Corporation, New York, NY. Reprinted by permission of the publisher and the authors.

What do teenagers themselves believe their chief problems are? Spirito and associates (1991) asked 676 9- to 14-year-olds to identify the problems they had experienced in the last month. Table 1.4 gives a report of their answers. It is interesting to note the changes in the frequency of the various problem types with age. Only two (parent/child conflict and boyfriend/girlfriend problems) increase with age. Problems with siblings decline and then increase, parent/parent conflicts and extracurricular activity problems go up, then decline, and the rest decline with age.

Finally, it should be noted that when adolescents *do* have problems, the ill effects percolate throughout society and continue for years after the original problems occur. This is demonstrated clearly in the box titled "Costs of Preventable Adolescent Problems."

AN APPLIED VIEW

Costs of Preventable Adolescent Problems

School Dropout

- Each year's class of dropouts will, over their lifetime, cost the nation about $260 billion in lost earnings and forgone taxes.
- In a lifetime, a male high school dropout will earn $260,000 less than a high school graduate and contribute $78,000 less in taxes. A female who does not finish high school will earn $200,000 less and contribute $60,000 less in taxes.
- Unemployment rates for high school dropouts are more than twice those of high school graduates. Between 1973 and 1986, young people who did not finish high school suffered a 42 percent drop in annual earnings in constant 1986 dollars.
- Each added year of secondary education reduces the probability of public welfare dependency in adulthood by 35 percent.

Teenage Pregnancy

- The United States spent more than $19 billion in 1987 in payments for income maintenance, health care, and nutrition to support families begun by teenagers.
- Babies born to teen mothers are at heightened risk of low birth weight. Initial hospital care for low-birth-weight infants averages $20,000. Total lifetime medical costs for low-birth-weight infants averages $400,000.
- Of teens who give birth, 46 percent will go on welfare within four years; of unmarried teens who give birth, 73 percent will be on welfare within four years.

Alcohol and Drug Abuse

- Alcohol and drug abuse in the United States cost more than $136 billion in 1980 in reduced productivity, treatment, crime, and related costs.

Some Contributions Adolescents Make to Our Society

Obviously, adolescence can be a very difficult time for some people. Nevertheless our best evidence shows that this is definitely not the case for the majority. As Petersen and associates (1993) conclude:

> ❑ In the United States, so much attention has been given to problems in adolescence that there is little awareness of positive development and health at this age. This biased approach to adolescence has created a skewed focus on both research and programs. One could ask whether the intense focus on problems plays a role in producing them (p. 156).

Most adolescents are living healthy productive lives and are a wonderful asset to our country (Lerner, 1996). You may have noticed that this book devotes 3 of its 14 chapters to teen problems. We give these problems so much space not because one-fifth of American teens are in trouble, but because we want you to have a clearer picture of the ones who are. If you work with adolescents, for example, you are more likely to deal with those with problems. Nevertheless, we agree with Petersen and her colleagues that the great majority of teens live quite normal lives. Table 1.5 gives us a picture of these individuals who give valuable service to our society.

Can you think of other ways in which adolescents are beneficial members of our society?

Table 1.5	**Contributions of Today's Adolescents**

Today (and every other day this year), teenagers are engaged in many kinds of activities that enrich their own lives and those of the people around them. Here are some examples:

Join service-oriented clubs (e.g., Scouts, 4-H, Future Farmers of America)

Become members of Junior Achievement

Compete in athletic events

Become hospital volunteers or aides

Join Students Against Drunk Driving

Teach other teens in a peer tutor program

Counsel troubled teens on a hotline (telephone numbers that serve those at risk for substance, sexual, or physical abuse and/or those who contemplate suicide)

Serve food in shelters for the homeless

Volunteer at day care centers or health care facilities

Deliver newspapers or work at fast-food restaurants, supermarkets, farms, or at other seasonal or part-time jobs

WHAT'S YOUR VIEW?

What Do These Figures and Tables Really Mean?

As you might expect, throughout this book you will find tables and figures that reveal important data about adolescents. We have found that students who ask themselves questions about these tables and figures usually learn the science of adolescent development much better than those who don't. For example, what questions might you ask yourself about Table 1.1? Some examples would be, "In which age group (10–14 or 15–19) is there the most change? Why did this happen? What are the effects of this change?" You might wonder why, in looking at Figure 1.1, the baby boom started when it did, or what impact such a severe change in the number of births had on our society. We hope you will get in the habit of asking questions about everything you read in this book. You will understand it better, and we think you'll enjoy the topic more.

SPENDING HABITS OF TYPICAL TEENS

Another way to look at today's adolescents is to examine the things they value. Later in this book we will take a closer look at values, but here we can study their spending habits as one indication of what is important to them. Today's youth have quite a bit more spending money than those of previous generations. In Table 1.6, we see how males and females spend their money.

Table 1.6	Where Their Money Goes		
Males	**Spending per Week**	**Females**	**Spending per Week**
Food/snacks	$10.10	Clothing	$10.65
Clothing	$6.19	Food/snacks	$6.50
Entertainment	$4.35	Entertainment	$3.45
Records/tapes	$1.55	Cosmetics	$3.35
Grooming	$1.10	Records/tapes	$1.80

Source: Rand Youth Poll

Each of the tables and figures we have presented to you so far offers a snapshot of current adolescent life. Like a group of photos arranged on a table, they begin to portray the truth about today's adolescents. It is hardly a coherent picture, however. Perhaps if we were to look at views of adolescents from previous eras, we would get a better perspective.

What Was Adolescence Like in the Past?

Can you guess in what century the following statement was made?

> ❏ I see no hope for the future of our people if they are dependent on the frivolous youth of today, for certainly all youth are reckless beyond words. When I was a boy, we were taught to be discreet and respectful of elders, but the present youth are exceedingly wise and impatient of restraint.

This rather cranky statement was written by the Greek philosopher Hesiod in the eighth century B.C. It appears that teenagers were no more popular with civilization's earliest writers than they are with many people today. In what follows, we take you on a short excursion of the ways adolescents were viewed from ancient times to the present.

Before starting, there is one thing you need to understand. The idea that there are stages in life is a relatively new one, as you will see. Thus even when the ancients spoke of adolescence, they appear to have thought of it only as the time of puberty. Mental, personality, and spiritual traits of teenagers were not acknowledged. And adults do not seem to have liked teens much, either.

ANCIENT TIMES

Sophocles, another Greek philosopher and renowned teacher of the young, was no fan of adolescents. He wrote this in the fifth century B.C.:

> ❏ Our youth now love luxury. They have bad manners, contempt for authority; they show disrespect for their elders and love chatter in place of exercise. They no longer rise when others enter the room. They contradict their parents, chatter before company, gobble up their food and tyrannize their teachers.

Socrates' best-known student, Plato, took a more positive stand. Plato's view of the lifespan held that childhood is the time of life when the spirit (meaning life values) develops, and so children should study sports and music. In the teen years, reasoning ability starts to mature, and so studies should switch to science and mathematics. In understanding this, he was definitely the exception to the rule.

Why do you think there were so many negative attitudes toward adolescents in earlier times?

Plato's student, Aristotle, believed that during the teen years we develop our ability to choose, to become self-determining. This passage is not an easy one, however, and Aristotle felt it caused youth to be impatient and unstable, as is reflected in this quote from him:

> ❐ They are changeable and fickle in their desires, which are violent while they last, but quickly over . . . They have exalted notions, because they have not yet been humbled by life or learnt its necessary limitation; moreover their hopeful disposition makes them think themselves equal to great things—and that means exalted notions. . . . They love too much and hate too much, and the same with everything else. They think they know everything; and are always quite sure about it; this, in fact, is why they overdo everything (*Rhetoric*, Book 2, Chapter 12 [circa 300 B.C.]).

THE MIDDLE AGES

During the Middle Ages (roughly A.D. 400 to 1400), the concept of human development became extremely narrow. Klein (1990) describes just how narrow:

> ❐ People were either too young to be considered as adults, or else they were considered as adults. It is unlikely that children were treated by their parents in the loving, nurturing way that we now associate with childhood, because parents feared developing emotional attachments to young beings who might not live to see age ten or beyond. . . . Adulthood then, included those stages in one's life that today would be called childhood, adolescence and adulthood. Old age per se was essentially unknown in the Middle Ages, since people typically died by the age of 40 (p. 448).

In this era, children came to be regarded as "miniature adults." Philippe Aries, in his excellent book *Centuries of Childhood* (1962) summarizes the views of medieval writers:

> ❐ In medieval society the *idea* of childhood did not exist; this is not to suggest that children were neglected, forsaken or despised. The *idea* of childhood is not to be confused with affection for children; it corresponds to an awareness of the particular nature of childhood, that particular nature which distinguishes the child from the adult, even the young adult. In medieval society this awareness was lacking. That is why, as soon as the child could live without the constant solicitude of his mother, his nanny or his cradle-rocker, he belonged to adult society (p. 128).

Children rarely appear in paintings from those times, but when they do, they are always dressed in a younger version of their parents' clothes. It was generally agreed that the way to help them become mature adults was through strict, harsh discipline, so that they could overcome the natural evils of their childish personalities. This harsh view appears to be an accurate description of how youth was looked at in the period from the Romans to the Renaissance.

THE AGE OF ENLIGHTENMENT

The Age of Enlightenment lasted from the 1600s to the early 1900s. The first century of this period saw no major change from the previous ones. For example, Hesiod's and Socrates' observations were echoed by William Shakespeare:

> ❏ I would there were no age between ten and three-and-twenty, or that youth would sleep out the rest; for there is nothing in the between but getting wenches with child, wronging the ancientry [*elderly*], and stealing, and fighting.

This position lasted until the 1700s when Jean Jacques Rousseau argued forcibly through his book *Emile* that children and youth need to be free of adult rules so they can experience the world naturally. He believed most children to be like Native Americans, whom he referred to as "noble savages." He believed that both groups are instinctively good, that children grow into kind and insightful adults, just like the Native Americans, if they are not corrupted by civilization.

In early America, this view did not gain much support. Life at that time was not easy, and everyone was expected to work hard, including children. Most youths worked on farms, but as the population grew, more and more went into apprenticeships in the cities.

In the nineteenth century, however, a dual pattern began to emerge. By the 1840s, a middle class in America was clearly emerging from the large lower class. For the lower-class teenager, life consisted almost entirely of long hours of work. At that time, boys and girls made up almost 40 percent of the factory workers in the New England states. They went to work very early in the morning, and often worked 12- or even 14-hour days, taking only short breaks for breakfast and lunch. Many lived in boarding houses, and after their evening meal they went straight to bed to get rest from the totally exhausting day. Jordan (1987) describes the situation:

> ❏ The hazards to children [*working in factories*] were many. The most obvious was the long hours which children worked. The second was that the Industrial Revolution had moved the child worker away from the family circle into the world of adult factory life. In that regard, the child worker conformed to the regimen prescribed for adults, including, at its extreme, night work. The conventions of the period saw little wrong with twelve-hour days for workers of all ages.

Although the middle-class youths did not have to undergo this rigorous regime, their adolescence was also quite brief. It should be remembered that the average person did not live nearly as long as we do today, and so there was some urgency to get married and start bearing children.

In the latter half of the 1800s, the children of the poor continued in the old apprenticeship mold, but middle-class youth began to stay in school longer. Then, beginning in the early 1900s, two important changes occurred. First, the technical demands of the Industrial Revolution called for more extensive education. Second, the reform movement by the muckrakers (people who tried to introduce stricter labor and compulsory education laws) made it possible for poorer children to stay in school.

THE TWENTIETH CENTURY

compulsory education
Laws that require children to be in school between the ages of 6 and 16.

Adolescence as we know it today may be said to have started with the beginning of compulsory education. **Compulsory education** refers to laws that require children to be in school between the ages of 6 and 16. It was instituted in part to ensure that children would be freed from the terrible conditions of the sweatshops. However, another reason for this extended education was the business owners' desire to make sure that poor children had the skills that they needed in the industrialized world. Compulsory education was also designed to keep teens out of the workforce; because industrialization temporarily reduced the need for workers, keeping teens in school prevented them from competing with adult workers. As children were required to stay in school longer, the control over their lives by adults who were not their parents also grew.

juvenile justice
Includes special hearings, the confidentiality of records, and a separate jailing and punishment of youngsters.

Another aspect of life in the new century was the increasingly humane treatment of criminals. This was especially true of our handling of juvenile delinquents, which led to the concept of **juvenile justice**. Until the twentieth century, teenagers who had committed crimes were treated like adults. In 1899 the state of Illinois passed the first Juvenile Court Act, which stated that those 17 and under must be treated differently from adults. This act provided for special hearings, the confidentiality of records, and a separate jailing and punishment of youngsters.

The age of careful research also began at the turn of the century. At this time, those who were interested in understanding youth ceased *speculating* about adolescents and actually began to make careful *observations* of them. Here is an early example of such an observation, made by educator Irving King (1914):

❑ The girls are clearly beginning to look like young ladies, while the boys with whom they have thus far played on scarcely equal terms now seem hopelessly stranded in childhood. This year or more of manifest physical superiority of the girl, with its attendant development of womanly attitudes and interests, accounts in part for the tendency of many boys in the early teens to be averse to the society of girls. They accuse them of being soft and foolish, and they suspect the girls' whisperings and titterings of being laden with unfavorable comments regarding themselves (p. 12).

As a stage in human development, adolescence became better established during the Great Depression of the 1930s. During this period, teenagers had to leave school to take on much needed jobs to support their families. Many of the same groups who were concerned about child labor (physicians, educators, social workers, and psychologists) now grew concerned about the well-being of adolescents.

It was not until the 1950s, however, that adolescence really came into its own. Klein (1990) summarizes this trend:

> ❏ Industrialization occurring during the late 1800s created the need for a stage of adolescence; the Depression created the legitimized opportunity for adolescence to become differentiated from childhood and adulthood; and the mass media influence/blitz of the 1950s crystallized this life stage by giving it a reality all its own. For the first time in history, teenagers then had their own music—which their younger siblings were too young and their parents too skeptical and/or repulsed (by the purported "dance and sex" connotations of the very term "rock and roll") to enjoy—their own dances to accompany this music, their own movies, and even their own fads (p. 452).

The first half of this century saw the creation of numerous theories about the nature of adolescence. The major ones will be described in the next chapter.

To conclude this brief excursion into the historical point of view, let us say that skepticism about the nature of adolescence in the Western world has not died in the second half of this century. As noted adolescent sociologist Edgar Friedenberg remarked in his book *Coming of Age in America:*

> ❏ A great many young people are in very serious trouble throughout the technically developed world, and especially the Western world. Their trouble, moreover, follows certain familiar common patterns; they get into much the same kind of difficulty in very different societies (1967, p. 12).

These historical views certainly offer a gloomy picture of youth. However, each suffers from the same critical flaw: it reflects *subjective opinion*, not scientific measurement, which offers a more positive picture of youth.

This is not to say that subjective experience has no place in our considerations. For example, another reasonable way to look at adolescence is through the **retrospective accounts** of those who have lived through it. They are called retrospective because they involve a *backward* look at the period by a wide variety of individuals. What follows are some vignettes from the experiences of adults who were asked to "tell us of a typical incident from your adolescence." As you read them, see if they help you clarify the meaning of adolescence.

retrospective accounts
Reports made by individuals looking *backward* at previous periods in their lives.

Do Today's Adults Remember Their Teen Years as Difficult?

Betty, age 26: I remember getting caught by my mother with one of my girl friends in the closet. We were 11. Well, we were exploring each other's bodies—juvenile adolescent masturbating, I guess you'd call it. My mother just freaked out. She screamed, "What are you doing?" It was embarrassing.

We denied everything, and her whole reaction to the thing finally was to give me a pamphlet put out by the Tampax Company on "Your Growing Years." She wouldn't talk about it, though. She didn't punish me—I don't think she even told my father about it. I guess this must be just our secret.

Juanita, age 58: My life was dull when I was a teenager. The fifties were dull in general. I ran away from home—that I will admit to. I was 19 when my daughter was born in 1956.

I remember that in high school I made myself an outfit in sewing class: a pair of shorts and a middy top. That's when I was thin; I looked jazzy. I went to the beach with my girlfriend Maria. Jimmy, the guy I was going with at the time, called my parents' house and found out where I was. I was sitting on the beach smoking a cigarette when Jimmy came up from behind me and pulled the cigarette right out of my mouth. He said, "Don't you ever, ever! Don't you ever, ever act this way!" I should have married him. He tried to do a lot for me. I should have married him, but I played him along and I did him a lot of dirt.

Can you think of other types of hypocrisy common in secondary schools?

Tyrone, age 24: At the end of junior high school, I was picked to be bussed from my home in the black community to go to an all-white suburban school. Many of the other kids treated me well, but quite a few were bigoted. Some yelled names at me, and got especially offended when I would get a right answer in class. They seemed to feel that it might be all right if I came to their school, but that I should keep my mouth shut.

I made it to the varsity of the school's basketball team, and played pretty well. *That* was accepted by just about everyone. In my senior year, I was on the first string. We had a great team and we made it to the state semifinals, which was exhilarating. Suddenly everybody was my buddy. I must admit that it was very exciting, but I couldn't help thinking that there were more than a few hypocrites among my "new friends."

Glenn, age 41: In my hometown, there were two distinct sections—the upper middle-class WASPs and the lower middle-class immigrants from Eastern Europe. At our high school, the two groups rarely mixed. I was a WASP and regularly dated a girl who was, too.

One Friday in my junior year, a bunch of us were riding in a bus to a nearby university to see a play. I was riding with my girlfriend, and across the aisle sat Elaine, a gorgeous Polish girl whom I had often admired from afar. To my surprise, she kept looking at me in a flirting way.

AN APPLIED VIEW

Gathering Personal Stories

All of us have at least one personal story that comes to mind when we remember our adolescent years. These stories give us a clearer idea of what the teen years meant to our relatives and friends. Ask your parents, grandparents, aunts and uncles what they remember best when they think of their youthful years. Some of these stories will undoubtedly prove interesting and revealing to you. They might even be of interest to your friends or classmates.

We got to the play half an hour early. While we were waiting, Elaine came up to me and asked if I would like to see a special place she knew of. It turned out to be a lovely garden. We stayed there talking for some time and were late for the play. The teacher yelled at us, but at least we got to sit together.

Throwing caution to the winds, I rode home on the bus with her. She turned out to be much less shy than I had thought. We had a pretty heavy necking session, and as we neared home, a powerful orgasm swept over me. I was amazed and confused, because I didn't think that could happen unless you touched yourself. When I got off the bus, I was ecstatic. This ended abruptly when my girlfriend came over to me. She glared at me, and though I couldn't be sure, I thought she could tell what had happened. She started to speak, but tears filled her eyes, and she quickly turned and left.

I didn't sleep that night. I soon learned the real cost of the ride. My friends were furious at me, and obviously Elaine's friends had gotten to her, because she never spoke to me again. My girlfriend did speak to me, but only to tell me what a rat I was and to get out of her life. Adolescence is a part of my life which I have no desire to relive!

Eddie, age 21: I was 14 when I first did a crime. Me and a friend were going to bust into this house, because he saw where the lady hid the backdoor key, but then he chickened out. I went in at two in the morning, and I was hardly even scared. I went into the bedroom and I remember I was excited to hear the people breathing. I grabbed all the stuff on the dresser and knocked over the lamp. The man jumped out of bed and grabbed me—he was strong as hell. I thought I had had it, but then I bit him hard on the arm and he let go. I beat it fast. I still held on to the jewelry, but when I took it to a guy the next day, he said it was all cheap junk. I still don't know if it was or not. I been in some trouble since then, but never again for busting into houses!

Carlos, age 37: The incident I remember most vividly happened when I was in the eleventh grade. I guess every school has its arch radicals, and ours were Dan and Dolores, who went together. They were constantly making statements that infuriated everyone. We were together one night at a friend's house, and they were lecturing us about United States foreign policy. Everyone was making fun of their ideas, most particularly my girlfriend Sherene and me. Dolores had said something, I don't remember what now, and I made a sarcastic remark that made everyone laugh uproariously. She and Dan were very angry and stormed out of the gathering.

The next morning I was awakened at seven by a phone call from Dolores' mother screaming at me. "What have you people done to them?" Her voice was shrill with hysteria, and it took me several minutes to realize what she was saying. Dan and Dolores were dead. They had been found in Dan's car, parked in Dolores' garage. The garage door was closed and the running engine had asphyxiated them. It slowly became clear that Dolores' mother felt that I had somehow caused her daughter's death.

The next day, the official story was put out by her parents: The couple had been talking in the car, left the motor running to keep warm (it was winter), and they had fallen asleep and died there. I remember staring at Dolores' calm face as she lay in the casket, dressed in a lovely pink gown. I didn't know then and I don't know now whether her death was the result of a suicide pact or an accident, but the guilt that I felt that day is with me still!

These vignettes present some of the major themes of adolescence: sex, guilt, loyalty, parental conflict, friendship, confused values. We can look at the emotional intensity with which the events were experienced and say that adolescence is *sometimes* a period of great upheaval. Because these vignettes are retrospective, however, it is likely that the events described are somewhat exaggerated. Such memories may reflect a rather biased picture of what youth was really like.

So what is the answer to our question? We can tell you that modern scientists disagree with the negative stereotype of adolescence. Today, it is safe to say, adolescence is viewed by the experts as being difficult at times, but probably no harder to live through than any other period of life.

WHAT'S YOUR VIEW?

Adolescence: A Time of Turbulence?

After so many years of seeing adolescence as tumultuous, there has been quite a turnaround among adolescent theorists in the last few decades. Is it because teens in Western societies have changed? Do they differ that much from teens in less industrialized societies? Maybe so.

In their study of Moroccan adolescence, Davis and Davis (1989) asked adults in that country what word Moroccans use for the teen years. They call it the "age of *taysh*," which means a time of recklessness and rash, frivolous behavior. They certainly view adolescence as turbulent.

It could be that American adolescents are less tumultuous these days because our treatment of them better suits their needs. After all, it was not so many years ago that 13-year-olds routinely left school and went to work on farms and in factories, putting in 10- and 12-hour days, six days a week. They got married earlier and immediately started having families (and in other countries, many still do today). It may be that this prevented them from fulfilling their needs, and that staying in school for longer periods is better for them. Perhaps we are seeing less turmoil in their lives because, on the average, their lives are happier.

On the other hand, many theorists now believe that the traditional view of a turbulent adolescence was tainted because the more disturbed youth are always much more *visible*. Even in this book, which strives to present a balanced account of adolescence, you will read a lot about exceptional teens: those who are doing exceptionally well and those who have exceptional problems. This is because we assume most of our readers need to know more about these teens in order to deal with them successfully. Theorists hold that the average teen has not had a more tumultuous life than the average second grader or the average retiree, and that the more positive perception has been due to the tremendous increase in scientific research during recent years. What do you think?

Although we hear a great deal about the tumultuous existence of the adolescent, research indicates that it is not very different from other periods of life. Most teenagers manage to become contributing members of society.

Recent studies have revealed that part of the problem may be because adolescence has always been perceived as *one* stage. We now see that it must be viewed as at least two. In this book we divide studies into those of early adolescence (11 through 14 years old) and those of later adolescence (15 through 18 years old). This way, we can take a more specific look at what really goes on with these two quite different stages of life.

Research sponsored by Carnegie Corporation (1990), a leading source of scientific information on this subject, led to the following conclusion:

❏ Most American adolescents come through the critical years from ages ten to twenty relatively unscathed. With good schools, supportive families, and caring community institutions, they grow to adulthood meeting the requirements of the workplace, the commitments to family and friends, and the responsibilities of citizenship. Even under less-than-optimal conditions for healthy growth, many youngsters manage to become contributing members of society (p. 1).

In this book, we try hard to reflect this disparity, even as we summarize the various traits of youth. Now let us turn to the third question we need to look at, "When does adolescence begin and end?"

When Does Adolescence Begin and End?

Which of these adolescent experiences would you say best marks the start of adolescent development?
• When they begin to menstruate (girls); when they have their first ejaculation (boys).
• When their level of adult hormones rises sharply in the bloodstream.

- When they first think about dating.
- When their pubic hair begins to grow.
- When they are 10 years old (girls); when they are 11d (boys).
- When they begin developing an interest in the opposite sex.
- When they develop breasts (girls); when their genitals enlarge (boys).
- When they pass the initiation rites set up by society (among Christians, confirmation, for example, or in the Jewish faith, bar mitzvah and bas mitzvah).
- When they become unexpectedly moody.
- When they turn 13.
- When they form exclusive social cliques.
- When they think about being independent of their parents.
- When they worry about the way their bodies look.
- When they enter seventh grade.
- When they can determine the rightness of an action, independent of their own selfish needs.
- When they are influenced more by their friends' opinions than by what their parents think.
- When they begin to wonder who they really are.

Although there is at least a grain of truth in each of these statements, they don't help us much in defining the starting point of adolescence. For example, although most people would agree that menstruation is an important event in the lives of women, it really isn't a good criterion for the start of adolescence. First menstruation (called **menarche** [pronounced "men ar′ka"]) can occur at any time from 8 to 16 years of age. We would not call the menstruating 8-year-old an adolescent, but everyone would agree that the non-menstruating 16-year-old is.

menarche
The onset of menstruation.

Probably the most reliable indication is that point in time when there is a sharp increase in the production of the four hormones that most affect sexuality: progesterone and estrogen in females, testosterone and androgen in males. But determining this change would require taking blood samples on a regular basis, starting when youths are 9 years old. Not a very practical approach, is it?

It is also not so easy to determine the endpoint of adolescence. Instead of looking at biological tests, now you need to look at culture. Whether in the inner city in Detroit, in a small village in Africa, or in a suburban high school in Santa Paula, adolescence ends when youths are given access to and responsibility for adult activities. In one place or culture this could be age 15, in another, age 25. To define the dimensions of adolescence, then, you need to know what activities are required of an adult in that culture. When do you think adulthood is achieved? Perhaps you might try making up your own list.

Identifying the ages or events at which adolescence begins and ends is not a simple matter. To get a definitive answer, we need to study the question more closely from the standpoints of biology, psychology, sociology, and other sciences. We will do so in other chapters in this book. For now, let's say that adolescence begins for most of us at about 11 or 12 years old, and ends at about 19 years old. Important exceptions to this chronological definition will be discussed later in the book.

WHAT'S YOUR VIEW?

Rites of Passage into Adulthood

In his book, *The Men from the Boys* (1988), Ray Raphael makes the following statement:

❏ Traditional cultures throughout the world have often devised ways of dramatizing and ritualizing the passage into manhood and of transforming that passage into a community event. Through the use of structured initiation rites, these societies have been able to help and guide the youths through their period of developmental crisis. . . . Often, the trials a youth must endure are extreme.

Throughout their journey, the elders belabored them with firebrands, sticks tipped with obsidian and nettles. . . . They were beaten, starved, deprived of sleep, partially suffocated, and almost roasted. Water was forbidden, and if thirsty, they had to chew sugar cane. . . . All the time . . . the guardians gave them instructions about kinship, responsibilities, and duties to their seniors. At length, after some months, . . . the guardians taught their charges how to incise the penis in order to eliminate the contamination resulting from association with the other sex.

A series of great feasts then took place, and the initiates emerged decorated.

This is how the Busama youth in the highlands of New Guinea achieve adult respect. And what about in our culture? Here is one man's reaction today:

❏ I wish I had it that easy. Run through the fire, step on the coals—then it's over and done with. You're a man, everyone knows you're a man, and that's the end of it. For me it keeps on going on and on. The uncertainty of it—at any moment you could be out on the streets. It's all tied up with money. I've got to keep on fighting for money and respect. The fire never stops; I keep running through it every day (pp. 58–59).

Are we lacking a public acknowledgement of adulthood? Do we need a ritual? How do we recognize a person worthy of adult respect? What does one need to master to achieve adulthood? What do you think?

In the next section, we explore our fourth issue, the best methods of studying adolescence. We give you a brief summary of the techniques that have been used in this new scientific approach to the study of adolescence, together with actual examples of each technique.

What Are the Best Methods of Studying Adolescence?

Today we use many approaches to understanding human behavior. Each has its strengths and weaknesses; none is completely reliable. Most developmental psychologists employ one of three data-collection methods: descriptive studies, manipulative experiments, and naturalistic experiments. In the first type, information is gathered on subjects without manipulating them in any way. In the second two, an experiment takes place before the information is gathered.

Psychologists also use one of four time-variable designs: one-time, one-group studies; longitudinal studies; cross-sectional studies; and a combination of the last two that is called sequential studies. Each of these types of studies varies according to the effect of time on the results. Let us look more closely now at each of these aspects of research.

DATA-COLLECTION TECHNIQUES

Descriptive Studies

descriptive studies
Information gathered on subjects without manipulating them in any way.

self-report studies
Asking people their opinions about themselves or other people by use of interviews or questionnaires.

observational studies
Describing people simply by counting the number and types of their behaviors.

case studies
In-depth looks at individuals.

Descriptive studies are quite common. Most are numerically descriptive: how many 12-year-olds think abortion is wrong versus how many 17-year-olds think so; how much money the average 18-year-old woman spends per week; how many pregnant teenage girls were or were not using birth control; how positively or negatively a recent immigrant youth views his parents or school. Some descriptive studies (called **self-report studies**) ask people their opinions about themselves or about other people, using interviews or questionnaires. Other studies (called **observational studies**) describe people simply by counting the number and types of their behaviors. A third type, **case studies**, presents data on an individual or individuals in great detail in order to make generalizations about a particular age group.

An example of the case study approach is Mack and Hickler's *Vivienne: The Life and Suicide of an Adolescent Girl* (1982). After Vivienne's death, the researchers obtained the family's permission to read her diary, poems, and letters. They also interviewed her relatives, friends, and teachers in order to shed light on her thinking as she came closer and closer to committing this tragic act. Although their findings may explain the suicide only of this one person, their hope was to discover the variables that cause such a decision. Many of the best theories about adolescent life have been based on detailed case studies of small numbers of teens.

The case approach, in which one person is studied in great depth, is one of several methods used to better understand adolescent development.

correlational studies
Analyzing the relationship between two variables.

Many descriptive studies are done as **correlational studies**. These look at the relationship between two variables. Variables are traits such as height and intelligence that vary among individuals. If we should find, for example, that the correlation between these two variables (height and intelligence) is high, then we would know that the taller a person is, the higher the person's intelligence is. In reality, the correlation is low; that is, knowing a person's height tells us nothing about the person's intelligence. Suppose we wanted to know the correlation between height and weight in preadolescent boys and girls. If we found that the correlation is higher for girls, we would expect that height and weight are more closely related than they are in boys.

Descriptive studies have the advantage of generating a great deal of data. Because the sequence of events is not under the observer's control, however, causes and effects cannot be determined. For example, just because we know that mothers who give their babies a lot of attention have happier children does not mean that attention causes happiness. It could be that happy babies cause their mothers to want to give them more attention. It could be that the mother has a gene that makes her more likely to give attention and, when inherited by her child, makes it more likely to have a pleasant disposition. Knowing that attention and happiness are highly correlated does not tell us for certain that the one causes the other. It might, but we just cannot tell from descriptive research.

Manipulative Experiments

manipulative experiments
Experiments that keep all variables constant except one, which is carefully manipulated.

treatment
Action taken with an experimental group, but not with a control group, in order to measure its effects. Examples are instruction, medication, and therapy.

Well-designed **manipulative experiments** can answer our questions about causation. In the quest for the causes of behavior, psychologists have designed many experiments. In these experiments, they attempt to keep all variables (all the factors that can affect a particular outcome) constant except one, which they carefully manipulate. It is called a **treatment**. Differences in the results of the experiment can be attributed to the variable that was manipulated in the treatment. The experimental and control subjects must respond to some measuring instrument (e.g., a test, a questionnaire, a measure of heart rate) selected by the investigator in order to determine the effect of the treatment. Figure 1.2 illustrates this procedure.

E is the experimental group and C is the control group, which receives no special treatment. X stands for the treatment and b and a refer to measurements done before and after the experiment. Except for the treatment, there must be no differences between the two groups, either before or during the experiment. Otherwise, the results remain questionable. The most effective way to achieve this equality is by assigning subjects to the experimental or the control group on a strictly random basis. For instance, the subjects' names could be written on pieces of paper, placed in a bowl and shaken up, and then drawn two at a time, with the names faced down and arbitrarily placed, one in the experimental group and one in the control group. A better technique, if available, is to number each participant and use a computer's random number generator to determine placement of each subject.

Figure 1.2
The classic experiment.

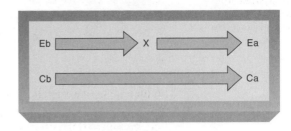

Even with all this care, it is difficult to ensure that the experimental and control groups are exactly the same except for the treatment that the experimental group will receive. For example, the control group may be in a room on the noisier side of the building. Therefore, experiments involving people should be referred to as "quasi-experiments," and the results should be suspect until the experiment has produced similar results several times.

An example of a manipulative experiment is a study by Dacey, Amara, and Seavey (1993) in which two similar groups of eighth graders were randomly selected from all eighth graders in two inner-city middle schools. One group was given the treatment, a series of 14 lessons in self-control, while the other group studied the traditional curriculum. At the end of this experiment, both groups were observed for any decrease in the number of pregnancies, the amount of drug use, or the number of dropouts. Since the experimental group did significantly better in terms of these criteria, we can assume that the 14 lessons in self-control were effective.

Though they often can lead us to discover the causes of human behavior, there are problems with manipulative experiments. How do you know that you will get the same results if you do the experiment a second time? Was the treatment similar to normal conditions? Do subjects see themselves as special because you picked them and thus react atypically? For these reasons, researchers often rely on naturalistic experiments.

Can you think of any other factors that could distort the results of a manipulative experiment?

Naturalistic Experiments

naturalistic experiments
Experiments where the researcher acts solely as an observer and does as little as possible to disturb the environment.

In **naturalistic experiments**, the researcher acts solely as an observer and does as little as possible to disturb the environment. "Nature" performs the experiment, and the researcher acts as a recorder of the results. (*Note*: Don't confuse these with descriptive studies that are done in a natural setting, such as a park; they are not experiments.) An example of a naturalistic experiment is the study of the effects of the northeast blizzard of 1978 (Nuttall & Nuttall, 1980). These researchers compared reactions of adolescents and their parents whose homes were destroyed to those who lived in the same area, but whose homes luckily suffered only minor damage.

Naturalistic experiments often increase our chances of discovering causes and effects in real-life settings. The main problems with this technique are that it requires great patience and objectivity, and that it is usually impossible to meet the strict requirements of a true scientific experiment.

TIME-VARIABLE DESIGNS

One-Time, One-Group Studies

one-time, one-group studies
Studies that are carried out only once on one group of subjects.

As the name implies, **one-time, one-group** studies are studies carried out only once on one group of subjects. For example, suppose we want to find out how a group of 16-year-old African American boys feels about premarital sex right now. We would present that group of teens with a questionnaire on the subject of premarital sex. This would be a one-time, one-group study. This type of study has many uses, but also has several serious weaknesses. Our example study cannot tell us about the effects of culture or gender, because it only examines boys of one race. It cannot tell us the effects of **age cohort** (age cohorts are groups of people born at about the same time), because it does not compare two or more age cohorts. Finally, it cannot tell us about changes among the boys as they grow older, because it only studies them at one point in time.

age cohort
A group of people born about the same time.

Longitudinal Studies

longitudinal studies
Observations of the same individuals at two or more times in their lives.

The **longitudinal study**, which observes the same individuals at two or more times in their lives, can contribute important information on such topics as the long-term effects of learning on behavior, the stability of habits and intelligence, and the factors involved in short- and long-term memory.

Although most childhood behavior disappears by adulthood, there has long been a suspicion that some adult traits develop steadily from childhood and remain for life. In his search for such stable characteristics, Benjamin Bloom notes in his classic work *Stability and Change in Human Characteristics* (1964) that the development of some human characteristics appears visible and obvious, while that of others remains shrouded in obscurity. The following are three growth studies that more than 300 persons have participated in for more than 30 years.

1. *The Berkeley Growth Study*, begun in 1928, was designed to study the mental, motor, and physical development of a sample of full-term healthy babies.
2. *The Guidance Study* took youngsters born in 1928 and 1929 and began to study them at 21 months of age. The aim was to study physical, mental, and personality development in a normal group.
3. *The Oakland Growth Study* of 200 fifth- and sixth-graders was designed to study many interrelations between developmental changes and behavior. The investigators tried to discover whether developmental changes affect an adolescent's potential.

One often-quoted longitudinal growth study was conducted by the Fels Research Institute (Kagan & Moss, 1962). The subjects were 45 girls and 44 boys, all white, whose personality development was traced from birth through early adulthood. The investigators conducted extensive interviews with both the children and their parents. Among the particular techniques used were:

- *Personality tests* given at regular intervals. The child was asked to react to a picture (the Thematic Apperception Test) or to a design such as a Rorschach inkblot. Trained persons analyzed responses for clues revealing personality, including motives, attitudes, and problems.
- *Observation of the mother* in the home with the child present, and also annual interviews with the mother.
- *Measurement of the intelligence* of both the mother and the father, using the Otis IQ test.
- *Regular observation of the child's behavior* in the home, in school, and at day camp. The child was also interviewed by workers.

Kagan and Moss summarize the obvious advantage of the longitudinal method when they note that it permits the discovery of lasting habits and of the periods in which those habits appear. A second advantage is the possibility of tracing those adult behaviors that have changed since early childhood.

Longitudinal research also presents many problems. It is expensive and often hard to maintain because of changes in availability of researchers and subjects. Changes in the environment can also distort the results. For example, if you began in 1960 to study changes in political attitudes of youth from 10 to 20 years of age, you would probably have concluded that adolescents become more radical as they grow older. Your findings would have been influenced by the war in Vietnam. The results of the same study done between 1970 and 1980 would probably not show this radicalizing trend.

Cross-Sectional Studies

cross-sectional studies
Comparing groups of individuals of several ages at the same time in order to investigate the effects of aging.

Cross-sectional study is a method that compares groups of individuals of various ages at the same time in order to investigate the effects of aging. For example, if you want to know how creative thinking changes during adolescence, you could administer creativity tests to groups of 10-, 12-, 14-, 16-, and 18-year-olds, and check the differences of the average scores of the five groups. Jaquish and Ripple (1980) did just this, but their subjects ranged in age from 10 to 84!

As you might expect, this method has problems as well. Although the effects of cultural change can be minimized by careful selection, it is possible that the differences you find may be due to differences in age cohort rather than maturation. Each cohort has had different experiences throughout its history, and this factor can affect the results, perhaps to a greater extent than do age differences. Figure 1.3 compares these two approaches.

Sequential (Longitudinal/Cross-Sectional) Studies

When a cross-sectional study is done several times with the same groups of individuals (such as administering creativity tests to the same five groups of youths, but at three different points in their lives), the problems we've mentioned before can be

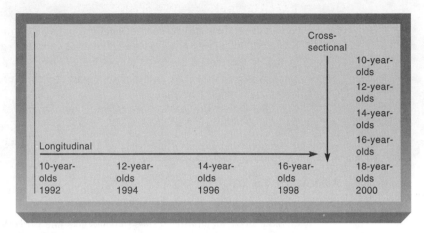

Figure 1.3
Comparison of the longitudinal and cross-sectional approaches.

sequential studies
A cross-sectional study done at different times with the same groups of individuals.

alleviated. Figure 1.4 illustrates such a **sequential study**. Although this type of research is complicated and expensive, it may be the only type that can answer important questions in the complex and fast-changing times in which we live.

Figure 1.5 shows how each data-collection method may be combined with each time-variable design. This section has mentioned a number of studies that combine these two aspects of research. Can you place these studies into the appropriate cell in Figure 1.5?

Creativity test			
Test I March 4, 1996	Test II March 4, 1998	Test III March 4, 2000	
Group A (12 years old)	Group A (14 years old)	Group A (16 years old)	**Mean score: Group A**
Group B (14 years old)	Group B (16 years old)	Group B (18 years old)	**Mean score: Group B**
Group C (16 years old)	Group C (18 years old)	Group C (20 years old)	**Mean score: Group C**
Mean score: 1996	Mean score: 1998	Mean score: 2000	

Figure 1.4
An example of a sequential study.

Time-variable designs	Data-collection techniques		
	Descriptive	Manipulated	Naturalistic
One-time, one-group			
Longitudinal			
Cross-sectional			
Sequential			

Figure 1.5
Combinations of data-collection techniques and time-variable designs.

C O N C L U S I O N S

Answering the question of what adolescence is like today is not simple. There are many important differences among the members of the age group from 11 through 18 years old (as we define the period). Historical evidence suggests that from ancient times parents have feared that their adolescent children were more unruly and less reasonable than they were in their youth. In line with this traditional view, some researchers suggest that it is normal for adolescents to be in a state of turmoil much of the time. Many others, however, find the majority of teenagers to be well-balanced, reasonably happy, and pleasant to work with.

The successful completion of the developmental tasks of adolescence is critical to the young person's future life. Supportive parents, good schools, and a caring community provide the best insurance of healthy growth and a competent entry into young adulthood. However, many teenagers who lack some or all of these advantages nevertheless are able to accomplish a successful journey through adolescence.

Some adolescents do have serious problems that may affect both their personal lives and society as a whole for many years. Teenage pregnancy and childbearing is one example; it impacts the girl and boy involved, their child, both their families, and the common need for state financial support in some cases.

It has only been within the present century that coherent theories of adolescent development have been formulated. Scientific methods must be used to verify or refute all aspects of such theories. Only evidence gathered in the real world and carefully analyzed can tell us whether these theories are correct.

The chapters that follow take a much closer look at the major theories, the studies of specific aspects of adolescent life, and the advice of experts in the field. We explain how this information can help those who wish to work with teenagers, as well as how it pertains to the several cultures that exist in our society. We believe that once you have a solid understanding of this material, you will be in an excellent position to understand and interact with adolescents successfully.

C H A P T E R   H I G H L I G H T S

What Is Adolescence Like Today?

- There are fewer adolescents in the 1990s than there used to be, and they make up a smaller proportion of the total population.
- Today's teens face many serious problems, but most manage to avoid difficulties and instead make valuable contributions to our society.
- Today's adolescents have more spending money than those of previous generations.
- Thinking back on your own adolescence can give you a deeper understanding of today's teenagers.
- It was once commonly believed that the teen years were a carefree time of fun and exploration; some now believe that adolescence is the most difficult of all life stages.
- Most experts, however, state that the majority of adolescents are happy and productive members of their families and communities.

What Was Adolescence Like in the Past?

- In ancient times, some philosophers believed that youths were frivolous and irresponsible, while others emphasized their growing intellectual skills and self-sufficiency.

- From the Middle Ages until the start of the twentieth century, strict discipline was believed necessary to force young people to take on adult responsibilities as early as possible.
- Two early twentieth-century concepts changed our view of adolescence: compulsory education and juvenile justice.
- Theories of adolescence as a separate life stage (and studying it with careful observation) also came into existence in the early twentieth century.

When Does Adolescence Begin and End?

- There is no general agreement about what marks the beginning and ending of adolescence.
- Physical, intellectual, social, and emotional factors all enter into the process of defining adolescence.
- In contrast to our own, some primitive societies have rites of passage that clearly mark the entry into adulthood.

What Are the Best Methods of Studying Adolescence?

- Methods of data collection include descriptive studies, manipulative experiments, and naturalistic experiments.
- There are four major time-variable research designs, each with its advantages and disadvantages.

K E Y   T E R M S

age cohort 28
case studies 25
compulsory education 17
correlational studies 26
cross-sectional studies 29
descriptive studies 25

juvenile justice 17
longitudinal studies 28
manipulative experiments 26
menarche 23
naturalistic experiments 27
observational studies 25

one-time, one-group studies 28
retrospective accounts 18
self-report studies 25
sequential studies 30
sociocultural diversity 6
stereotype 3
treatment 26

W H A T   D O   Y O U   T H I N K ?

1. In your opinion, is adolescence today harder to live through than other periods of life? Easier? About the same?
2. How does adolescence today compare to your parents' teen years? Your great-great-grandparents' teen years?
3. What is the single most important sign that adolescence has started? Does gender make a difference? Race? Nationality?

4. Why has Western history treated teenagers so negatively?
5. Of all the research techniques described in this chapter, which is the best one? Why?
6. What are some things you could do to help adolescents better understand the causes of their own behavior?

S U G G E S T E D   R E A D I N G S

Davis, J. (1991). *Checking out the moon*. New York: Orchard. This book, about a teen who is dealing with her parents' divorce, manages to include several other adolescent crises: leaving home, confronting cultural diversity, helping someone deal with being raped. This is an insightful book by an author who understands.

Golding, W. (1962). *Lord of the flies*. New York: Coward McCann. This tale of a group of teenage boys whose plane crashes on a Pacific island, killing the adults, is *must* reading. You watch the clash of their values as they wend their way to the shocking ending.

Murry, V. M. (1990). *Black adolescence*. Boston: Hall. An annotated bibliography of the major current issues affecting African American teens today.

Wright, R. (1945). *Black boy*. New York: Harper & Row. The haunting autobiography of the novelist's adolescence in the Deep South. Its insights are entirely relevant to today's world.

G U I D E D   R E V I E W

1. Which part of the adolescent population in the United States is NOT growing?
   a. White
   b. Native American
   c. Hispanic
   d. Asian/Pacific Islanders

2. Babies born to teen mothers are at heightened risk for
   a. low IQ.
   b. low birth weight.
   c. lower resistance to disease.
   d. heart disease.

3. Which of the following philosophers did NOT believe that adolescence is a positive stage of human development?
   a. Rousseau
   b. Plato
   c. Sophocles
   d. Aristotle

4. Parenting strategies that emphasize the development of decision making and self-determination most closely resemble the views of
   a. Hesiod.
   b. Plato.
   c. Rousseau.
   d. Aristotle.

5. The beginning and ending of adolescence is primarily determined by the individual's
   a. socioeconomic level.
   b. culture.
   c. educational level.
   d. physical development.

6. In a well-designed manipulative experiment, which of the following is true of both the experimental group and control group?
   a. They are both given the same treatment.
   b. They differ only after treatment.
   c. They are assigned by random sampling.
   d. They are each made up of at least 10 participants.

7. Infants, randomly divided into two groups, are exposed either to a colorful picture of a human face or to an equally colorful picture of random design. Researchers compare the amount of time that infants in each group look at the stimulus. In this experiment, which of the following is the treatment?
   a. the age of the infants
   b. the picture
   c. the length of time a picture is observed
   d. the experimental group

8. You are conducting a study in which you are solely an observer and do not administer a treatment to the participant. Which type of data collection method are you using?
   a. longitudinal study
   b. descriptive study
   c. manipulative experiment
   d. naturalistic experiment

9. Which of the following data collection methods is NOT an example of a descriptive study?
   a. case study
   b. manipulative study
   c. observational study
   d. self-report study

10. In order to investigate the changes in peer relation-ships during adolescence, a researcher administers questionnaires to groups of 10-, 15-, and 18-year-olds, checking for any differences between the average scores of the three groups. Which time-variable design is the researcher using?
    a. cross-sectional study
    b. sequential study
    c. longitudinal study
    d. one-time, one-group study

11. Because the ethnic and economic status of teens is changing, we will need to build our strength through cooperation among several different cultures, all living alongside each other. This concept has become known as _____ (sociocultural diversity/cultural identity).

12. The percentage of American adolescents that are _____ (nonwhite/white) is on the rise.

13. During the middle ages, children came to be known as _____ ("noble savages"/"miniature adults").

14. According to _____ (Aristotle/Socrates), adolescence is the time we develop our ability to make choices and become self-determining.

15. The modern day concept of adolescence started with the beginning of _____ (juvenile justice/compulsory education).

16. Researchers can better understand the meaning of adolescence by studying backward looks at the period by a wide variety of individuals. These are called _____ (historical case studies/retrospective accounts).

17. An individual between 15 and 18 years old is going through _____ (early/late) adolescence.

18. _____ (Descriptive/Correlational) studies do not enable the researcher to determine causes and/or effects.

19. Because one-time, one-group studies only study subjects of one age, they cannot tell us the effects of _____ (age cohort/current information).

20. In an experiment, the control group receives _____ (the treatment/no special treatment).

Answers

1. A, 2. B, 3. C, 4. D, 5. B, 6. C, 7. B, 8. D, 9. B, 10. A, 11. sociocultural diversity, 12. nonwhite, 13. "miniature adults", 14. Aristotle, 15. compulsory education, 16. retrospective accounts, 17. late, 18. Descriptive, 19. age cohort, 20. no special treatment

GENERAL THEORIES OF ADOLESCENCE: AN OVERVIEW

True, my theory is no longer accepted, but it was good enough to get us to the next one!

Donald Hebb (whose theory of intelligence was popular in the 1940s)

CHAPTER 2

CHAPTER OUTLINE

The purpose of Chapter 2 is to continue your introduction to the science of adolescent psychology, in this case through describing the ideas of several well-known theorists. We explain how theories assist us to a clearer understanding of adolescent psychology. Brief summaries of the theories of eleven major theorists are presented, with the position of each on the **biopsychosocial** causes of adolescent behavior.

biopsychosocial
That idea that development proceeds by the interaction of biological, psychological, and social forces.

After reading this chapter, you should be able to:
- Describe the purposes of theory-making and the relationship of theory to three other aspects of science.
- Explain biopsychosocial causes.
- List G. S. Hall's four stages of development.
- Itemize Freud's stages of development, together with his concepts of the functions and constructs of the human psyche. Also, explain how Freud's theory is augmented by his daughter Anna's ideas as well as by the more recent views of Peter Blos.
- Show how Ruth Benedict's and Margaret Mead's research disputed the biological explanation of adolescent behavior.
- Describe the contributions of Albert Bandura, Robert Havighurst, and Abraham Maslow to adolescent psychology.
- Detail the special importance Erik Erikson's psychosocial theory has for adolescence.
- Explain Richard Lerner's contextualism theory.
- Discuss these issues from an applied, a sociocultural, and your own point of view.

What Theories Do

For many people, the word *theory* means someone's guess about why something happens. For example, Bob might say, "It's my theory that Joe quit the team because he thinks we don't like him." Used in a textbook, *theory* often makes readers think of complicated arguments between experts—"Highbean's theory disagrees with Numbskull's, in that . . ."

In this book, we use the word differently. We believe theories are essential in psychology and serve several vital functions. Good theories perform the following functions:
- They are helpful tools for organizing a huge body of information. The published studies on adolescence number in the thousands. The results of these findings would be incomprehensible unless they were organized in some meaningful way. A theory is a shorthand description of this complexity. It forms a framework of "pegs" on which we can hang similar kinds of research findings.
- They help focus our search for new understandings. They offer guideposts in our quest for the truth about the complicated human body and mind.
- They not only describe findings, but also explain how these findings may be interpreted and judged. Theories offer building blocks that can help us understand which facts are important and which conclusions to accept.

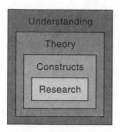

Figure 2.1
The four aspects of social science, and how they relate to each other.

construct
An idea about some aspect of the human being.

Can you suggest some examples of constructs, psychological or otherwise?

- They draw attention to major disagreements among scholars and scientists. By clarifying these differences, theories offer testable ideas that can be confirmed or refuted by research. The same facts can be interpreted differently. This can be confusing, but it is important that you learn to recognize these differences and to draw your own conclusions.

Theories do not stand alone—they are related to other aspects of science. At the basic level of all the sciences is *empirical research.* For social science, of which adolescent psychology is a part, empirical research means studying real people under carefully arranged conditions. From this research, scientists form constructs. A **construct** is an idea about some particular aspect of the human being. The ego and intelligence are two examples. Neither the ego nor intelligence are actual things—we cannot open up a brain and point to them—rather, they are concepts that we have created to explain functions of our personality (the ego) and of our minds (intelligence). Actually, personality and mind are constructs, too. It is *useful* to have these constructs, but we cannot truly observe them.

Constructs are used to build theories. A theory is a system of ideas (as we said earlier) that attempts to explain research findings by showing how the constructs are related. The ultimate goal of this process is to produce greater *understanding* of how and why we humans think, feel, and behave as we do. Figure 2.1 illustrates how these four factors of social science interact.

The Biopsychosocial Causes of Change

All theorists agree that humans change over time, and that there are biological, psychological, and social causes of this development. Theorists differ, however, on the *emphasis* they put on each of these three factors. Depending on how they think about human nature, theorists view change as being caused by one of the following:

- *Biological causes.* The genes are the major factor in our development from one discrete stage of life to the next. Change tends to be rather abrupt.
- *Psychological causes.* Changes in the personality and/or intellect are behind our development through life. Change depends on the way these internal states mature.
- *Social causes.* Development depends greatly on what is happening in our environment. Because so many things happen in the course of our lives, change tends to be a gradual, ever present process.
- *Some combination* of two of these factors or all three factors.

We have designated each of the theories that follow as belonging to one of these four "camps" by capitalizing some of the letters in the word *biopsychosocial.* As we said in Chapter 1, adolescent theorists agree that all these factors—biology, psychology, and social environment—play a role in human development. By capitalizing the part of *biopsychosocial* emphasized by the theorist, we show that each theory acknowledges the whole person, but gives more importance to one part than another.

In the sections that follow, we describe theories that have been considered the most comprehensive in the field of adolescent psychology. Several other highly regarded theories are not included because they pertain mainly to specific aspects of the mind. Examples are Jean Piaget (cognition); Lawrence Kohlberg and Carol Gilligan (morality); and Mary Belenky, William Perry, and Robert Selman (social cognition). These theories are considered in detail later in this book.

G. Stanley Hall's Theory [BIOpsychosocial]

G. Stanley Hall (1844–1924) is known as the father of adolescent psychology. Building on Charles Darwin's ideas about evolution, Hall constructed a psychological theory of teenage development. It was published in two volumes and entitled *Adolescence* (1904).

Hall posited four discrete periods of development, which he felt correspond to the four lengthy stages of development of our species: infancy/animal; childhood/anthropoid (humanlike apes); youth/half-barbarian; and adolescence/civilized.

- *From birth to 4 years.* In this stage children recapitulate the animal stage in which mental development is quite primitive. Sensory development and the development of sensorimotor skills are the most important aspects of this period.
- *From 4 to 8 years.* Hunting and fishing, using toy weapons, and exploring caves and other hiding places were common activities of childhood for boys at the turn of the century, but not for girls. Language and social interaction begin to develop rapidly, as they did during the nomadic period of the human race.
- *From 8 to 12 years.* This period corresponds to the more settled life of the agricultural world of several thousand years ago. This is the time when children are willing to practice and to discipline themselves; this is when routine training and drills are the most appropriate—especially for language and mathematics.
- *From 12 to 25 years.* **Storm and stress** (emotional upheavals) typify human history for the past 2,000 years. The same is true for youth. Adolescence is a new birth, for now the higher and more completely human traits are born.

storm and stress
Hall's description of emotional upheavals between the ages of 12 and 25 that also typify human history for the past 2,000 years.

Hall believed that each person's development passes through the same four stages the human species has. He thought that all development is determined by our genes. Development occurs in an unchangeable, universal pattern, and the effects of the environment are minimal. For example, Hall argued that some socially unacceptable behavior in children, such as fighting and stealing, is inevitable because it is rooted in our biology (genes). He urged parents to be lenient and permissive, assuring them that children must have this catharsis, and that when they reached the later developmental stages, these behaviors would simply drop out of existence.

He was a strong preacher against what he viewed to be teenage immorality, and was especially concerned that educators try to stamp out the plague of masturbation, which he considered to be running rampant among male youth.

Hall made a major point of the "contradicting tendencies" of adolescence— the experience of violent mood swings, which he felt tended to make youth more human. Some of the turbulent variations he noted are:

- Energy and enthusiasm vs. indifference and boredom
- Gaiety and laughter vs. gloom and melancholy
- Vanity and boastfulness vs. humiliation and bashfulness
- Sensitivity vs. callousness
- Tenderness vs. cruelty

Hall felt that the development of most human beings stopped short of this adolescent stage, in which appreciation of music and art are achieved. Most people seemed fixated at the third stage, in the dull routine of work. A social reformer, Hall believed that adolescence is the only period in which we have any hope of improving our species. He felt that placing teenagers in enriched environments would improve their genes, which their children would then inherit. Hence we could become a race of "superanthropoids."

Can you think of some reasons why Hall's ideas are so different from most adolescent psychologists today?

Most psychologists today argue that in this theory, we have an interesting but quite inaccurate picture of human social development. They believe Hall tried to force reality to fit an outmoded conception of evolutionary development. However, he was not alone. He shared the beliefs of turn-of-the-century social philosophers in Europe and America who assumed the evolutionary superiority of the civilization of the white man, or more specifically, the Anglo-Saxon male. His theory is considered wrong for several reasons. Although it may have described white American and European children of the early 1900s, it does not coincide with our knowledge of them today. Youths in his time may not have had much appreciation for civilized culture, but this was clearly due not to genetic imperfections but to such factors as having been forced to leave school to work on the farm. In addition, Hall's belief that changes can be passed on from one generation to the next through the genes is seen as scientifically false.

His theory most particularly does not present a true picture of adolescence. Because Hall looked only at American culture, and because most individuals in that culture did develop similarly, he mistakenly thought that genes were responsible for this similarity. Later studies of other cultures have shown wide differences in developmental patterns.

The central conception of adolescence as a period of storm and stress, rebellion, and sexual conflicts began with Hall; he saw struggle as adaptive. Such sweeping generalizations have not held up when it was discovered that most young people in other cultures do not seem to be in rebellion, especially in those cultures that adhere to familial, cultural, and religious value systems. This idea of adolescent storm and stress, although refuted by research (see the section on Mead and Benedict), lives on as a cultural stereotype.

WHAT'S YOUR VIEW?

Are the Teen Years an Unusually Stressful Time?

Although he was prejudiced, Hall is hardly the only adolescent psychologist who can be accused of bias in his thinking. In a fascinating study, Enright and others (1987) looked at 89 articles on adolescence published during two economic depressions and two world wars to see if these events had an influence on research. The results were striking:

❑ In times of economic depression, theories of adolescence emerge that portray teenagers as immature, psychologically unstable, and in need of prolonged participation in the educational

system. During wartime, the psychological competence of youth is emphasized and the duration of education is recommended to be more retracted [shorter] than in a depression (p. 541).

Is it likely that youths are viewed as immature during depressions to keep them from competing with adults for scarce jobs, and that their maturity is seen as greater during wartime because they are needed to perform such adult tasks as soldiering and factory work? If so, is this bias conscious or unconscious? What do you think?

Although Hall is to be admired for his efforts to bring objectivity to adolescent psychology through the use of empiricism, it has been suggested that he had several personal agendas of his own.

As you continue reading this chapter, see if you can spot any biases in the other theories.

WHAT'S YOUR VIEW?

An Early View of Adolescent Sexuality

G. Stanley Hall reported in his book *Life and Confessions of a Psychologist* (1923/1951) that sex was considered dirty when he was a child. He was taught that touching his genitals was a sin. He was particularly affected by a story his father told him about a boy who masturbated and had sex with women. According to the story, the boy caught a disease that ate away his nose and turned him into an idiot. For years after that, whenever he would get sexual feelings, Hall would touch his nose to make sure it was all right.

Throughout his adolescence and even into his college years, Hall was haunted by guilt feelings about his sexual responses. He would put bandages on himself to prevent an erection during his sleep. He even consulted a doctor about his "problem." He believed that these guilt feelings played a major role in the religious conversion that he experienced during his sophomore year. It was only in later years that he came to understand that his adolescence was completely ordinary. Do you suppose that these experiences account for his motivation to study and write about this stage of life?

Sigmund Freud's Theory [BIOPSYCHOsocial]

In more than one hundred years of psychological research, it is impossible to think of anyone who has played a larger role than Sigmund Freud (1856–1939). Even his most severe critics admit that his theory on the development of personality is a milestone in the social sciences. In fact, many people mistakenly think **psychoanalysis**, the name he gave to his theory, is the same as psychology.

Probably because of his experiences as a medical doctor, Freud doubted the reliability of people's testimony about themselves. He also distrusted behavior as a source of the truth. For him, the unconscious mind is the key to understanding the human being. It is here that the most important motives and values reside. Because many of the ideas in the unconscious are primitive, they are not acceptable to the conscious mind. For example, if an adolescent girl is furious with her mother, she may not be able to acknowledge it because she is not supposed to hate her mother. Only bad people do that.

psychoanalysis
Freud's explanation of the psychic development of humans. Also, his method of psychological therapy.

DEFENDING THE UNCONSCIOUS MIND

Freud believed that important information in the unconscious is kept from awareness by an array of **defense mechanisms** (Gay, 1988). These are unconscious attempts to prevent awareness of unpleasant or unacceptable ideas. Table 2.1 presents descriptions of some of the most common defense mechanisms.

defense mechanisms
Unconscious attempts to prevent awareness of unpleasant or unacceptable ideas.

Table 2.1	**Some Common Defense Mechanisms**
Repression	Unconsciously forgetting experiences that are painful to remember. Example: forgetting sexual abuse experiences.
Compensation	Attempting to make up for an unconsciously perceived inadequacy by excelling at something else. Example: learning to play the guitar if unable to make the basketball team.
Rationalization	Believing that a condition contrary to your desires is actually what you had wanted all along. Example: being glad the trip was cancelled because it would have been boring anyway.
Introjection	Adopting the standards and values of someone with whom you are afraid to disagree. Example: joining a gang.
Regression	Reverting to behaviors that were previously successful when current behavior is unsuccessful. Example: crying about getting a low grade in school with the subconscious hope that the teacher will change the grade.
Displacement	Expressing strong feelings about a certain person to someone less powerful. Example: yelling at a sister when angry at a teacher.
Sublimation	Making up for unfulfilled sexual drives by engaging in some

WHAT'S YOUR VIEW?

The Role of Defense Mechanisms

Defense mechanisms function to protect the conscious mind from the truth. They distort the realities that we find too painful to face. Because they mislead us, and because they require spending a lot of energy that could better be spent elsewhere, many psychologists feel we should try to eliminate defense mechanisms. Others argue that (especially for teens) some truths are just too painful to face, and therefore, at least in the short run, defense mechanisms are useful. They say that most adolescents go through difficult periods (e.g., breaking up with a first love) and that during these periods, defense mechanisms can provide a beneficial buffer for a vulnerable self-image. These theorists believe there will be time enough later to face reality. What do you think? Can you recall an instance in your adolescence when you used a defense mechanism to hide the truth?

CONSTRUCTS OF THE PERSONALITY

id
The simplest of Freud's mind structures; it operates only in the pursuit of pleasure.

ego
The central part of our personality, according to Freud. It is the (usually) rational part that does all the planning and keeps us in touch with reality.

superego
One of Freud's three structures of the psyche. It is comparable to the conscience.

Freud divided the mind into three constructs: the **id**, the **ego**, and the **superego**. These constructs appear at different stages of the young child's development:

The Id. This construct is the only one present at birth. It contains all our basic instincts, such as our need for food, drink, dry clothes, and love. It is the simplest of the structures, operating only in the pursuit of pleasure.

The Ego. The ego is the central part of our personality, the (usually) rational part that does all the planning. It keeps us in touch with reality. It begins to develop from the moment of birth. For instance, a baby boy learns to cry loudly if he wants his mother, and to not stop until she comes. This is the beginning of an ego.

We need the ego to live in the real world. The stronger the ego becomes, the more realistic and the more successful the person is likely to be.

The Superego. Throughout infancy, we gain a clear concept of what the world is like. Then, toward the end of the first year, our parents and others begin to teach us what they believe it should be like. They instruct us in right and wrong, and expect us to begin behaving according to their principles. This is the beginning of the superego and also the beginning of the never-ending battle between the desires of the id and the demands of the superego. The main job of the ego is to strive unceasingly for compromises between these two "bullies."

THE DEVELOPING PERSONALITY

For Freud, development means moving through five instinctive stages of life, each of which he assigns to a specific age range. Each is *discrete* from the others. Each stage has a major function, based on an erogenous zone: unless this pleasure center is stimulated appropriately (not too much, not too little), the

fixated
Stuck at a developmental stage and, therefore, unable to become a fully mature person.

person becomes **fixated** (stuck at that stage) and is unable to develop into a fully mature person. The five stages are:

- *The oral stage (birth to 1½ years old)*. The oral cavity (mouth, lips, tongue, gums) is the pleasure center. Its function is to obtain an appropriate amount of sucking, eating, biting, and talking.
- *The anal stage (1½ to 3 years old)*. The anus is the pleasure center. The function here is successful toilet training.
- *The phallic stage (3 to 5 years old)*. The glans of the penis and the clitoris are the pleasure centers in this stage and in the two remaining stages. That he named this stage the phallic stage reveals Freud's male bias. He deftly ignored half the population (Horney, 1967)! The major function of this stage is the healthy development of sexual interest. This is achieved through masturbation and an unconscious sexual desire for the parent of the opposite sex. Resolution of the conflicts caused by this desire (called the "Oedipal complex" in males and the "Electra complex" in females) is the goal. Freud believed that most males resolve this conflict at about 6 years old, but females do not resolve it until they enter adolescence.
- *The latency stage (5 to 12 years old)*. During this stage, sexual desire becomes latent (asleep). This is especially true for males, through the defense mechanism of introjection (see Table 2.1). They refuse to kiss or hug their mothers, and treat female age-mates with disdain. Because our society is more tolerant of the daughter's attraction to her father, the Electra complex is not resolved and girls' sexual feelings are less repressed during this stage.
- *The genital stage (12 years old and older)*. Adolescence brings about a surge of sexual hormones in both genders, which causes an unconscious recurrence of the phallic stage. Normally, however, youths have learned that desire for one's parent is taboo, and so they set about establishing relationships (bumblingly at first) with age-mates of the opposite sex.

Freud believed that if these five stages are not negotiated successfully, homosexuality or an aversion to sexuality itself results. (It should be noted that this concept is not popular among gays and lesbians, many of whom believe that their sexual orientation goes much deeper than this—see Chapter 9.) If fixation occurs at any stage, Freud believed, anxiety results, and defense mechanisms will be used to deal with it.

ANNA FREUD'S CONTRIBUTION

A psychoanalyst trained by her father, Anna Freud (1895–1982) believed that his definition of adolescence was too sketchy. She suggested that her father had overemphasized his discovery that sexuality begins not at puberty but in early infancy. Anna Freud spent the major part of her professional life trying to extend and modify psychoanalytic theory as applied to adolescence (1958, 1968, 1969a, 1969b).

Do these five stages make sense to you? Can you think of different ways to describe these years in terms of stages of development?

asceticism
A defense mechanism against the sexual, "sinful" drives of youth; often causes the teenager to become extremely religious.

intellectualization
A defense mechanism, discovered by Anna Freud, in which the adolescent defends against emotional feelings of all kinds by becoming extremely logical about life.

Anna Freud saw the major problem of adolescence as being the restoration of the delicate balance between the superego and the id, a balance that is established during latency and disrupted by puberty. The problems brought about by this internal conflict cause the adolescent to regress to earlier stages of development. A renewed Oedipal or Electra conflict brings about fears concerning sexuality that are entirely unconscious and often produce intense anxiety. Therefore, the unconscious defenses of the ego tend to multiply rapidly. The problem, of course, is that the use of these defense mechanisms causes new stresses within the individual and tends to further increase the level of anxiety.

Anna Freud discovered two primarily adolescent defense mechanisms: **asceticism**, in which, as a defense against the sexual, "sinful" drives of youth, the teenager frequently becomes extremely religious; and **intellectualization,** in which the adolescent defends against emotional feelings of all kinds by becoming extremely logical about life.

MORE RECENT PSYCHOANALYTIC VIEWS OF ADOLESCENCE

Researchers in the field of psychoanalysis have recently criticized the Freudian emphasis on the unconscious aspects of the id. They have argued that there is too much emphasis on sexuality and on the negative side of things. In these newer views, relationships with others, especially with the youth's parents, play a more important role (Kohut, 1984). This newer view holds that adolescents come to understand more objectively their parents' actual behavior and are upset that parents' actions do not fit with the ideals that the adolescents have been taught. For example, teens realize that their parents have often lied to them about Santa Claus, about how children are born, and about whether persons should cheat (on their taxes, for example). This forces the early adolescent into reexamination of ideals and values, which ultimately leads to reorganization of the self. Therefore, this view is often referred to as "self-psychology." Peers play a greater role in the formation of new values as the role of parents declines. This is not to say that parents are no longer important to teens. Healthy adolescents are those who make major adjustments in themselves without extinguishing the ties that bind them to their parents.

Peter Blos

Perhaps the best known of the newer psychoanalytic theorists is Peter Blos. He holds that changes in relationships with others, rather than physiological changes such as the development of the sexual system, play the most important role. He argues that Freud is right in saying that the latency period is relatively calm and that the changes in early adolescence do mark an increase in turbulence. However, these changes are not sexual so much as they are interpersonal. The early adolescent begins to form close relationships with friends of the same sex. In some cases, these early friendships do involve sexual experimentation, but they are mainly ways of defending the ego from the fearful changes in self-concepts that are going on at this time (more on the self-concept in Chapter 6).

As adolescents enter the middle period, they begin to relate more to opposite-sex friends. Many times these friends resemble the teen's parents physically or mentally. Because of the intensity of these new relationships, adolescents also tend to be "narcissistic," taking a strong interest in themselves, including their appearance and their thoughts. They become self-absorbed and very defensive about any criticisms. This is because their image to others is so intensely important to them in this new stage. As both Freuds suggested, defense mechanisms are likely to abound during this time. However, the positive effects of looking within oneself are also seen. Gradually, adolescents begin to reorganize their sense of self and come to a more mature resolution of their self-images. As later adolescence is reached, the typical person achieves more self-esteem and a clearer identity.

ego psychology
Psychoanalytic view that emphasizes the ego more and the id and superego less than Freud did.

Because the more recent psychoanalytic view emphasizes ego more and the id and superego less than the Freuds did, this view is also known as ego psychology. Perhaps the best-known proponent of **ego psychology** is Erik Erikson, whose theory will be considered later in this chapter.

Ruth Benedict's and Margaret Mead's Theory [biopsychoSOCIAL]

In the early decades of this century, while the theories of the psychoanalytic school were gaining considerable attention, anthropologists had only begun to study so-called primitive tribes in non-industrialized places like Africa and the Polynesian Islands. Not until the 1920s and 1930s did this research result in publication. When it did, however, it made a serious attack on the psychoanalytic position. Anthropologists were finding that behaviors on which psychoanalysts based their position simply did not exist in other cultures.

For example, 10-year-olds in some cultures are much more sexually active than in the United States. If human behavior differs from culture to culture, then clearly the idea that development is biologically determined cannot be right. These anthropologists, most notably Ruth Benedict (1887–1948) and Margaret Mead (1901–1978), argued that the behavior of the individual depends to a great extent on the environment in which he or she is raised.

The major difference between Western technological societies and undeveloped, isolated societies was in the flow of psychosocial development. The anthropologists saw development in industrialized societies as highly disconnected. In America, children are expected to act like children throughout an extended adolescence, and then, quite abruptly, they are expected to begin acting like adults. In the more primitive cultures, this process is much more gradual and less disruptive. Benedict (1950, 1954) specified three major areas in which cultural conditioning flows smoothly in primitive societies but is disconnected in Western societies:

- *Sexual role.* At least until recently, children in our society have been largely ignorant of the specifics of sexual behavior. They know little about sexual intercourse, childbirth, breast feeding, menstruation, and the like. Then, as young adults on their wedding nights, they are expected to perform with sexual competence. As many older married people will tell you, this seldom happened.

- *Responsibility.* In our society, children and youth are allowed (even encouraged) to be somewhat irresponsible. However, as soon as they leave home, get a job, or get married, they are expected to be totally responsible for their behavior.
- *Dominance.* Young people in our society are expected to switch suddenly from a totally submissive role as children living in the homes of their parents to complete independence and self-reliance. If they get married and have a baby right away (and most did in earlier times), they are expected to immediately assume total dominance over the life of their own child.

A SOCIOCULTURAL VIEW

Do the Samoans Have the Right Answers?

These abrupt changes, occurring as they do without much preparation during the adolescent period, put a great strain on the life of the individual. Can you think of better ways of introducing teens to sex (e.g., childbirth classes, sex education classes)?

Margaret Mead's studies of youths on the island of Samoa (1927/1949) convinced her that the Samoan culture's introduction of sex in the early teens is more natural. In that society, adolescent boys were expected to sneak into the huts of the girls they admired and have sex with them. After a trial period, if the girl found him as attractive as he found her, she agreed to marry him.

As a preparation for adult life in our society, Mead (1970) went so far as to suggest that we allow our youth to engage in trial marriages, in which teenagers are allowed to live and sleep together (in the home of one of their parents) and to practice sexual intercourse, provided birth control is used. Her argument was that because so many teenagers are engaging in sex anyway and because so many of them are becoming pregnant and/or contracting a disease from not using protection, they would be better off if they had the acknowledgement and advice of adults.

Albert Bandura's Theory
[biopsychoSOCIAL]

observational learning
Influence of modeling on personality development as stressed in Bandura's social learning theory.

Albert Bandura, one of the chief architects of social learning theory, has stressed the potent influence of modeling on personality development. He calls this **observational learning**. In a famous article on social learning theory, Bandura and Walters (1959) cite evidence to show that learning occurs through observing

What are some of the implications of Bandura's theory for people who are working with adolescents?

others, even when the observers do not imitate the model's responses at that time and get no reinforcement. Observational learning means that the information we get from observing other people, things, and events greatly influences the way we act. As Bandura and Walters note, *adolescents do not do what adults tell them to do but rather what they see adults do.* If Bandura's assumptions are correct, adults, through their actions, can be a powerful force in shaping the behavior of adolescents. The importance of models is seen in Bandura's interpretation of what happens as a result of observing others:

- Observation may lead to new responses, including socially appropriate behaviors.
- Observation may strengthen or weaken existing responses.
- Observation may cause the reappearance of responses that were apparently forgotten.
- Observation of undesirable behavior that is either rewarded or goes unpunished may result in undesirable behavior. The reverse is also true.

Bandura, Ross, and Ross (1963) studied what effects three presentations—live models, filmed human aggression, and filmed cartoon aggression—would have on preschool children's aggressive behavior. The live and filmed human adult models displayed aggression toward an inflated doll; in the filmed cartoon, a cartoon character displayed the same aggression. Later, all the children who observed the presentations were more aggressive than youngsters in a control group. Filmed models were as effective as live models in transmitting aggression. The research suggests that powerful, competent models are more readily imitated than models who lack these qualities.

In summary, Bandura holds that adolescent development does not happen in predictable stages but as a result of social stimuli from the environment. Because there are millions of these in a person's lifetime, he sees development as continuous, something that usually happens in small steps every day.

A SOCIOCULTURAL VIEW

Modeling Among Latino Youth

For most teens the family plays an important role, but for Latino adolescents this is especially so. Thus it is not surprising that they often choose a family member on whom to model their behavior and values. This quotation from a teenage girl is an example:

❏ As far as I can remember, my mother was strong and independent. She loved us so much that she protected us from the dangers of the barrio. She kept the family together as long as she could, and the traditions were a big part of her life. She is a very pretty woman with strong Mexican Indian features: high cheek-bones and a tired clear face. She is short, heavyset, and has a physically tired body. She always wore a little makeup and red lipstick. . . . "Mi madre" is the pride and joy of my life, and she is not only my mother but my closest friend (Kunjifu, 1985, p. 29).

Robert Havighurst's Theory [bioPSYCHOSOCIAL]

By the 1950s and 1960s, a new theory developed as a reaction to the earlier viewpoints. Robert Havighurst, a sociologist at the University of Chicago, became a major spokesperson. He suggested that each stage of life possesses specific **developmental tasks**, which lie midway between the needs of the individual and the goals of Western society. He defined these tasks as skills, knowledge, functions, and attitudes that are needed by an individual in order to succeed in life. As with Freudian theory, the inability to successfully negotiate any particular stage interferes with success at all subsequent stages.

developmental tasks
Skills, knowledge, and attitudes needed by an individual to succeed in life at each stage of life; these tasks lie midway between the needs of the individual and the goals of Western society.

For the adolescent stage, Havighurst (1951) describes nine developmental tasks:

• Accepting one's physique and accepting a masculine or feminine role.
• Forming new relations with age-mates of both sexes.
• Achieving emotional independence from parents and other adults.
• Achieving assurance of economic independence.
• Selecting and preparing for an occupation.
• Developing intellectual skills and concepts necessary for civic competence.
• Desiring and achieving socially responsible behavior.
• Preparing for marriage and family life.
• Building conscious values in harmony with an adequate scientific view of the world.

Although written more than 40 years ago, Havighurst's list holds up well today in Western cultures. Research has lent considerable support to Havighurst's theory, and educators and therapists have found his ideas useful, although the applicability of his ideas for other cultures has had little study.

Abraham Maslow's Theory [BIOPSYCHOSOCIAL]

Psychologist Abraham Maslow (1908–1970) spent most of his career examining the development of *needs*. Maslow (1968, 1971) argued that humans have basic instincts to which they should pay more attention. These instincts are not as strikingly evident in humans as they are in animals, but the message is there, and a psychologically mature person is able to listen.

Maslow believed that we have six such basic instincts. These instincts are manifested in the form of needs—some very strong and some much weaker. They appear in overlapping stages, with the basic needs present at birth and higher-order needs showing up as the person grows older. Furthermore, satisfaction of these needs is sequential; the basic needs must be met before later, more complex needs can be successfully fulfilled.

Table 2.2	Maslow's Hierarchy of Needs	
Type of Need	**Level of Need**	
1. Physiological needs. These are the needs we have for basic creature comforts: food, water, warmth, air, sex.	Deficiency	Low
2. Safety needs. We must feel we are free from threat to our lives. Safety needs include our desires for familiarity, regularity, and a secure home.	Deficiency	
3. Belongingness and love needs. We all want to feel that we belong somewhere and that at least one other person feels a sense of love and caring for us.	Deficiency	
4. Esteem needs. We need to feel that we are worthwhile and capable, and that the society we live in values our contribution to it.	Deficiency	
5. Self-actualization needs. We also need to be in touch with those resources that lie deep within us. These include imagination and creativity, our ability to experience great joy, and to make total use of our potential. Maslow suggests that we carry not only our past inside us, but also our future, in the sense that the direction of our growth lies within us and needs to be fulfilled.	Being	
6. Aesthetic needs. Very few people actually are aware of having an aesthetic need, which is the need to make an important contribution to humankind. It is the desire to have a deep understanding of the world around us and the purpose of life. This need exists on a high level, and only a few (e.g., Abraham Lincoln, Albert Einstein, Margaret Mead) experience it.	Being	High

From *The Farther Reaches of Human Nature* by Abraham H. Maslow. Copyright © 1971 by Bertha G. Maslow. Used by permission of Viking Penguin, a division of Penguin Books USA Inc.

hierarchy of needs
Sequential needs that Maslow believes overlap in stages. The basic needs are present at birth and higher-order needs show up as the person grows older.

deficiency needs
Those needs that *decrease* as they are attended to; they can be fully satisfied.

being needs
Needs that *increase* as they are attended to (also termed growth needs).

Maslow's description of the six needs is given in Table 2.2.

The insight of Maslow's **hierarchy of needs** is readily apparent. Clearly, if we are suffering from severe hunger pains, it is unlikely that we will be concerned with whether or not people respect us. In fact, we may be quite willing to steal, even if this brings us condemnation. Usually, when a person's needs in one particular level are unfulfilled, those needs must be preeminent. Until the lower-level needs are met, the person is not likely to be concerned with higher-level needs.

Maslow also made a distinction between **deficiency needs** and **being needs**. *Deficiency* needs are those that *decrease* as they are attended to; they can be fully satisfied. When some physical or psychological deficit occurs (thirst, loneliness) and there is an action that can be taken to eliminate it, the need is a deficiency need. *Being* needs, on the other hand, *increase* as they are attended to (thus they are also termed *growth* needs). An example of a being need is the appreciation of music; the more we come to like music, the more we desire the joys it can provide.

Maslow believed that earlier psychologists, especially Freud, overemphasized deficiency needs. They saw all motivation as an attempt to ease the discomfort of unfulfilled needs. The ultimate goal of human beings is a return to "nirvana," a needless state, they argued. Maslow felt that this theory explains only part of our behavior. It applies, he suggested, more to sick individuals and those whose low-level needs are unmet than to most people in civilized countries today. He argued that we should pay more attention to being needs, now that deficiency needs are normally well taken care of.

AN APPLIED VIEW

Reaching for Self-Actualization

Although the likelihood of an adolescent attaining Maslow's self-actualization is small, it is at this age that the quest should start. If we are to help young people strive to fulfill themselves, we must teach them that self-actualization can only be achieved by looking away from themselves and toward a commitment to a larger cause. Enabling teens to affiliate with volunteer groups such as those who assist nurses in hospitals is a good start. Organizing or managing a service organization is another way to provide teens with this experience. Becoming involved in these service activities can help young people begin to strive for self-actualization.

Most successfully creative people have met their deficiency needs. That is why they often seem not to care whether their environments are orderly.

Although the six needs develop in overlapping steps, the sequence contains notable exceptions. Some individuals become so involved with taking care of their deficiency needs that they are incapable of experiencing being needs. For example, some neurotic individuals come to be so threatened by their environment that they must constantly take care of their safety needs. Teens who become drug addicts, alcoholics, and compulsive eaters are examples of people fixated on deficiency needs. Other individuals seem so involved with being needs that they neglect their deficiency needs. Great artists and martyrs are examples.

Clearly, belongingness and esteem needs, the third and fourth levels, are of greatest relevance during adolescence. Yet teachers and parents often overlook these needs as they pressure a youth to be more self-actualizing. As Maslow's theory makes evident, teenagers cannot hope to begin fulfilling their potential until their lower-level, deficiency needs are reasonably well met.

Erik Erikson's Theory [bioPSYCHOSOCIAL]

Among other important books, Erik Erikson wrote *Childhood and Society* (1963). It is an amazingly perceptive and at times poetically beautiful description of human life. Erikson's view of human development derives from extensive study of people living in an impressive variety of cultures: Germans, East Indians, the Sioux of South Dakota, the Yuroks of California, and wealthy adolescents in the northeastern United States (1959, 1968). His ideas also stem from intensive studies of historical figures such as Martin Luther (1958) and Mahatma Gandhi (1969). He sees human development as the interaction between your genes and the environment in which you live.

According to Erikson, human life progresses through a series of eight stages. Each stage is marked by a crisis that needs to be resolved before the individual can move on. Erikson uses the term *crisis* in a medical sense: an acute period during illness, at the end of which the patient takes a turn for the worse or better. At each life stage, the individual is pressured—by internal needs and the external demands of society—to make a major change in a new direction.

The ages at which people go through each of the stages vary somewhat, but the sequence of the stages is fixed. The ages of the first five stages are exactly the same as in Freud's theory (Erikson is an ardent student of his theory). Like Maslow, Erikson believes that the stages overlap.

Each of the crises involves a conflict between two opposing characteristics. Erikson suggests that successful resolution of each crisis should favor the first of the two characteristics, although its opposite must also exist to some degree. Table 2.3 gives an overview of his psychosocial theory.

Table 2.3		Erikson's Theory of Psychosocial Development
Stage	**Age**	**Psychosocial Crisis**
1	Infancy (birth to 1d)	Trust vs. mistrust
2	Early childhood (1d to 3)	Autonomy vs. shame, doubt
3	Play age (3 to 5)	Initiative vs. guilt
4	School age (5 to 12)	Industry vs. inferiority
5	Adolescence (12 to 18)	Identity and repudiation vs. identity confusion
6	Young adult (18 to 25)	Intimacy and solidarity vs. isolation
7	Adulthood (25 to 65)	Generativity vs. stagnation
8	Maturity (65+)	Integrity vs. despair

Source: Data from E. Erikson, *Childhood and Society,* 2d ed., 1963. W. W. Norton & Company, Inc., New York, NY.

It is necessary to have experienced each crisis before proceeding to the next. Inadequate resolution of the crisis at any stage hinders development at all succeeding stages unless special help is received. When a person is unable to resolve a crisis at one of the stages, Erikson suggests that "a deep rage is aroused comparable to that of an animal driven into a corner" (1963, p. 68). This is not to say that anyone ever resolves a crisis completely. It is important to note that Erikson's description of the eight stages of life is a picture of the ideal, and that no one ever completes the stages perfectly. However, the better a person does at any one stage, the more that person progresses. Let us look at each stage more closely.

Basic Trust Versus Mistrust (Birth to 1½ Years Old)

In the first stage, which is by far the most important, a sense of basic trust should develop. For Erikson, trust has an unusually broad meaning. To the trusting infant, it is not so much that the world is a safe and happy place, but rather that it is an orderly, predictable place; it contains causes and effects that can be anticipated. For Erikson, then, trust flourishes with warmth and care, but it might well include knowledge that one will be punished for disobeying rules. (Actually, Erikson strongly opposes punishment of any kind for infants.)

If the infant is to grow into a person who is trusting and trustworthy, regularity must exist in the infant's early environment. The child needs variation, but this variation should occur in a regular order that the child can learn to anticipate. For example, the soft music of an FM radio provides regular changes in sound level and the movement of a colorful mobile hanging over a child's crib provides predictable visual variations.

Some children begin life with irregular and inadequate care. Anxiety and insecurity have a negative effect on family and other relationships so important to the development of trust. When a child's world is so unreliable, we can expect mistrust and hostility, which under certain circumstances can develop into anti-social, even criminal, behavior. Of course, not all such people become criminals.

It is also possible to gain basic trust in infancy and then lose it later. Sometimes people who have not suffered an injurious childhood can lose their basic sense of trust because of damaging experiences later in life.

Autonomy Versus Shame and Doubt (1½ to 3 Years Old)

When children are about 1½ years old, they should move into the second stage, characterized by the crisis of autonomy versus shame and doubt. This is when children begin to gain control over their bodies and usually when toilet training is begun.

Erikson agrees with other psychoanalysts that toilet training has far more important consequences in one's life than control of one's bowels. The sources of generosity and creativity lie in this experience. If children are encouraged to explore their bodies and environment, a level of self-confidence develops. If they are regularly reprimanded for their inability to control excretion, they come to doubt themselves. They become ashamed and afraid to test themselves.

Excretion regulation is not the only goal in this stage. Children of this age usually start learning to be self-governing in all their behaviors. Although some self-doubt is appropriate, general self-control should be fostered at this stage.

INITIATIVE VERSUS GUILT (3 TO 5 YEARS OLD)

The third crisis, initiative versus guilt, begins when children are about 3 years old. Building on the ability to control themselves, children now learn to have some influence over others in the family and to successfully manipulate their surroundings. They should not merely react, they should initiate.

If their parents and others make children feel incompetent, however, they develop a generalized feeling of guilt about themselves. In the autonomy stage they can be made to feel ashamed by others; in this stage, they learn to make themselves feel ashamed.

INDUSTRY VERSUS INFERIORITY (5 TO 12 YEARS OLD)

The fourth stage corresponds closely to the child's elementary school years. Now the task is to go beyond imitating ideal models and to learn the elementary technology of the culture. Children expand their horizons beyond the family and begin to explore the neighborhood.

Their play becomes more purposeful, and they seek knowledge in order to complete the tasks that they set for themselves. A sense of accomplishment in making and building should prevail. If it does not, children may develop a lasting sense of inferiority. Here we begin to see clearly the effects of inadequate resolution of earlier crises.

As Erikson puts it, the child may not be able to be industrious because "he may still want his mother more than he wants knowledge." Erikson suggests that the typical American elementary school, staffed almost entirely by women, can make it difficult for children (especially boys) to make the break from home and mother. Under these circumstances, children may learn to view their productivity merely as a way to please their teacher (the mother substitute) and not as something good for its own sake. Children may perform in order to be "good little workers" and "good little helpers" and fail to develop the satisfaction of pleasing themselves with their own industry.

IDENTITY AND REPUDIATION VERSUS IDENTITY CONFUSION (12 TO 18 YEARS OLD)

identity crisis
Erikson's term for the situation, usually in adolescence, that causes us to make major decisions about our identity.

The main task of the adolescent is to achieve a state of identity. Erikson, who originated the term **identity crisis**, uses the word in a special way. In addition to thinking of identity as the general picture one has of oneself, Erikson refers to it as a state toward which one strives. If one were in a state of identity, the various aspects of one's self-image would be in agreement with each other; they would be identical. Ideally, a person in the state of identity has no internal conflicts whatsoever.

Repudiation of choices is another essential aspect of reaching personal identity. In any choice of identity, the selection we make means that we have *repudiated* (turned down) all the other possibilities, at least for the present. When youths cannot achieve identity, when identity confusion ensues, it is usually because they are unable to make choices.

As Biff, the son in Arthur Miller's *Death of a Salesman*, says, "I just can't take hold, Mom, I can't take hold of some kind of life!" Biff sees himself as many different people; he acts one way in one situation and the opposite way in another—a hypocrite. Because he refuses to make choices and shies away from commitments, there is no cohesiveness in his personality. He is aware of this lack but is unable to do anything about it. (We will have much more to say about Erikson's views on this stage in Chapter 6.)

INTIMACY VERSUS ISOLATION (18 TO 25 YEARS OLD)

In the sixth stage, intimacy with others should develop. Erikson is speaking here of far more than sexual intimacy. He is talking about the essential ability to relate one's deepest hopes and fears to another person and to accept another person's need for intimacy in turn.

Each of us is entirely alone, in the sense that no one else can ever experience life exactly the way we do. We are imprisoned in our own bodies and can never be certain that our senses experience the same events in the same way as another person's senses. Only if we become intimate with another person are we able to understand and have confidence in ourselves. During this time of life, our identity may be fulfilled through the loving validation of the person with whom we have dared to be intimate.

GENERATIVITY VERSUS STAGNATION (25 TO 65 YEARS OLD)

Generativity means the ability to be useful to ourselves and to society. Like in the industry stage, the goal here is to be productive and creative. Productivity in the industry stage, however, is a means of obtaining recognition and material reward. In the generativity stage, productivity is aimed at generating something useful to others. Thus, the act of being productive is itself rewarding, regardless of whether recognition or reward results.

Furthermore, there is a sense of trying to make the world a better place for the young in general, and for one's own children in particular. During the middle of this stage many people become mentors to younger individuals, sharing their knowledge and philosophy of life. When people fail in generativity, they begin to stagnate; they become bored and self-indulgent, unable to contribute to society's welfare. Such adults often act as if they were their own child.

INTEGRITY VERSUS DESPAIR (65 YEARS OLD AND OLDER)

When people look back over their lives and feel they have made the wrong decisions or, more commonly, that they have too frequently failed to make any decision at all, they see their life as lacking integration. They feel despair at the impossibility of "having just one more chance to make things right." They often hide their terror of death by appearing contemptuous of humanity in general, and of people of their own religion or race in particular. They feel disgust for themselves.

To the extent that adults have been successful in resolving the first seven crises, they achieve a sense of personal integrity. Adults who have a sense of integrity accept their lives as having been well spent. They feel a kinship with people of other cultures and of previous and future generations. They have a sense of having helped to create a more dignified life for humankind. They have gained wisdom. Can you see now the importance of stage five—identity and repudiation versus identity confusion—to all the stages that follow?

Richard Lerner's Theory [BIOPSYCHOSOCIAL]

Finally in this chapter, we examine a more recent theory of adolescence. Known as **contextualism**, (Lerner & Walls, 1999; Lerner, 1993; 1995; in press a; in press b; Muuss, 1996) this theory is the product of the thinking of Richard Lerner (b. 1946), who holds the Bergstrom Chair in Applied Developmental Psychology at Tufts University. His position is more a criticism of the previous theories and research on adolescence than it is a complete theory in its own right. Nevertheless it is having a considerable impact on the field of human development in general and on adolescent psychology in particular.

Contextualism holds that all previous theories of adolescence are inadequate because they fail to emphasize the interaction that always occurs between adolescents and the environment in which they are growing. Some previous theories emphasize the internal state of the person (for example, Piaget and the two Freuds), whereas other theories have emphasized the environment (for instance, Benedict and Mead, Bandura, and of course B. F. Skinner's theory of behaviorism). Lerner would place Erikson's theory in the first category, because he feels that, although it speaks of interactions with the environment, it emphasizes internal processes in the person.

The goal of contextualist researchers is to improve our understanding of the interactions among the factors and sub-factors that influence the internal state of the child as well as the surrounding environment. The theory emphasizes the idea that these two general factors (the external and the internal) are not only constantly affecting each other, each is actually imbedded in the other. Thus the

contextualism
Richard Lerner's theory that to understand an adolescent, we must carefully describe the ever-present interaction between the person's internal state and the environment in which that person is operating. The internal state and the environment are embedded in each other.

child is always changing because of the influence of the environment, and the environment surrounding the child is always changing because of the actions taken by the child. The two are imbedded in each other and cannot be understood except in terms of the constant interactions between them.

Take, for example, the concept of parenting style and its effect on a child. There are many studies of the reactions children have to such parenting styles as authoritarianism (see Chapter 7). It is assumed that this parenting style *causes* certain reactions in a child. However, it may be that because of such variables as genetic inheritance and socioeconomic level, the child tends to be irritable. This irritability, in turn, may well affect how the parents view their role: they may find parenting to be an unpleasant and unrewarding task. As a result, they become more authoritarian, punishing the child for minor infractions, which could make the child even more irritable. Each component continually affects and re-affects the other.

Contextualism presents four major factors in development. These are:
- the physical setting (for example, the classroom or the home).
- the social environment (the teachers and peers that the child experiences in the classroom environment, for instance).
- the personal characteristics of the child (the child's physical appearance and manner of speaking would be examples).
- time (in the short term, the child's daily schedule, and in the long term, the effects of history on the physical and social setting).

Contextualism argues that any statements about developmental patterns are suspect because the context in which these patterns occur is always being altered.

Thus development cannot be seen in a simplistic way. Lerner argues that we must no longer subscribe to theories that claim to be universal and timeless. How a person moves through a particular age period will depend on each of the four factors cited. To be accurate, a description of development through an age period (life stage) must be decidedly more intricate than has been the case with the unidimensional theories of the past.

probabilistic
The term used by contextual theorists to describe their predictions about behavior. We can never say how people will behave in a certain situation, but only how they will probably behave.

Furthermore, the contextualist takes the position that statements about a child must be **probabilistic**. The more accurately we can define the four developmental factors, the more confidence we have in our predictions about how a person is likely to behave. Of course, we can never know everything about these factors. Probabilistic statements can only be made about the course that any particular person's life is *likely* to take. The goal is to make statements like those of meteorologists predicting the weather: "People like Jimmy are 80 percent more likely to quit smoking as a result of Program X."

To get a better understanding of the complexity that contextual analysis calls for, let us take a closer look at one of the factors, the social environment. More specifically, let us look at some examples of the family variables that Lerner and his colleagues suggest are crucial to a contextual theory:

- *Family climate.* What is the relationship between the parents? Are they loving and supportive of each other or are they constantly bickering? Is there much sibling rivalry?
- *Family structure.* Imagine what a difference it makes whether the family is an old-fashioned one typified by two parents, three or more children, a grandparent, and perhaps an uncle or aunt living in the same house, or whether it is a modern-style family, more commonly containing two children and only one parent in the home.
- *Socioeconomic factors.* Is the family lower, middle, or upper class in terms of its income and the education of the parents? Are both parents employed? Does the type of work the parents do allow them to be home at the same time every day, or are they on varying schedules? How does the child view the family's socioeconomic level?
- *Family chores.* Are members of the family satisfied with the assignment of their household jobs, or do some feel that they are overburdened?
- *Geographical location.* Obviously it makes a big difference whether the family is residing in an urban, rural, or suburban setting. The type of house will also affect the lives of those who live in it.

You can see that there are many possible components affecting the development of the family members at any one time. Many of these factors have been studied, but usually only one at a time and usually in only one direction.

bidirectional research
Research that takes into account the two-way nature of person-environment interactions. The two are always affecting each other.

Contextualism calls for **bidirectional** research. For example, imagine a study that looks at how a student's learning ability is affected by the type of classroom the child studies in. According to Lerner, not only are children in school affected by the way their classroom is arranged, but the classroom arrangement is also affected by the types of students and the teacher in it. Several children whose behaviors tend to be negative will affect how all the other children in the classroom learn.

goodness of fit
The term contextual theorists use to describe the quality of the interaction between people and their environments. If this interaction produces desirable results, there is "goodness of fit."

Another important contextual concept is **goodness of fit**. This aspect of the theory refers to the relationships among the traits of the child (or adolescent, or young adult, or senior citizen) and the biological, psychological, and social demands of the environment in which the person is found. Thus Lerner moves away from concern over what variables are to *blame* for development and toward the idea of how well matched they are. If the person's traits fit well with the psychosocial environment, the person experiences healthy development, but if this fit is negative, then mental and/or physical illness is likely to result. (Notice that this prediction is put in probabilistic terms.)

Another example: two adolescents are attracted to each other and believe they are in love. Then abruptly they break up. The adolescents may try to figure out who is at fault for their parting of ways. The contextualist theorist, on the other hand, would look at this relationship and see that, over time, the couple found that the traits each possesses did not fit well together. No one is to blame for the breakup; the relationship simply lacked goodness of fit.

Figure 2.2

Lerner's dynamic interactional model of adolescent development.

From R. Lerner's "Dynamic Interactional Model of Adolescent Development," from *America's Youth in Crisis.* Copyright © 1995 by Sage Publications, Inc. Reprinted by permission of Sage Publications, Inc.

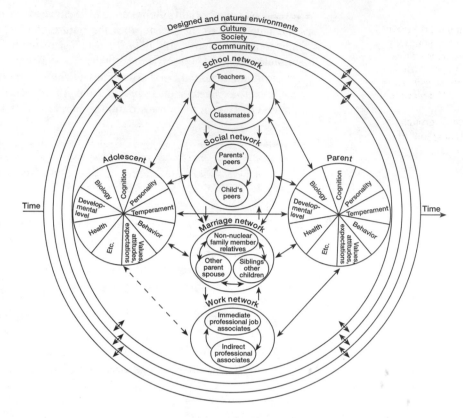

Lerner has designed a model of the interactions of the various factors affecting adolescent development. The latest version of his model is presented in Figure 2.2. In the circle on the left are the various factors that influence the life of the adolescent. On the right, we see that the same factors affect the behavior of the parent. These two persons interact with each other through the mediation of four major networks (centered in the figure). These are:

- the school network, which has interactions between teachers and classmates as the primary focus.
- the social network, which is influenced by the peers of the adolescent and of the parents.
- the marriage network, which powerfully affects the actions of all members of the family.
- the work network, which focuses on associates of the parents as well as of the adolescent, and the indirect associates of both.

All these factors are surrounded by the larger environment of community, society, culture, and designed and natural environments. Running through the whole (from left to right in this diagram) is the concept of time. Notice that with the exception of time, all the arrows are double-headed. This is meant to emphasize the bidirectional, interacting nature of every relationship in the theory.

developmental diversity
The concept espoused by contextual theorists that we ought to pay particular attention to the rich and beneficial differences in the ways people of various ethnic groups develop throughout life.

Several other important implications of contextualism need to be mentioned briefly. One of these has to do with the nature of **developmental diversity**. Contextualism highlights the idea of individual differences and pays particular attention to the rich and beneficial diversity that exists among ethnic groups. In a world in which the number of nonwhite persons is rapidly ascending, this is a most relevant distinction. In his book, *Theories of Adolescence* (1996), Muuss states the relevance of this implication well:

❐ As children possess different genetic dispositions, grow up in diverse settings, contribute uniquely to their own development, these contextual factors do shape children differently, thus turning the focus towards diversity . . . Each individual child is unique and becomes increasingly more unique as development progresses through the course of life (p. 351).

This emphasis on the importance of diversity in understanding human development is particularly crucial at this point in American history. There is considerable evidence that adolescent psychology has paid disproportionate attention to the developmental patterns of white middle-class children and has neglected minorities (Miller & Lerner, 1994; Muuss, 1996). Of great concern is the finding that African Americans, in particular, are being studied less and less. Compared to whites, African American adolescents experience more of the risk factors that adolescent psychologists most commonly study, such as delinquency, dropout rates, and drug abuse. Therefore, an understanding of the unique contexts with which African Americans must deal is vital to combating these problems. Lerner's theory forces us to take a more careful look at these factors.

Another implication of his theory has to do with the use we make of the advice of "experts." Just as parents, for example, cannot rely completely on the experiences that other parents pass on to them because the circumstances in each family are different, they also must be cautious in adopting what they read in newspapers, magazines, and even serious books on the subject of parenting. "There are no simple solutions" is the message from Lerner and his colleagues.

This last ramification of contextual theory points to its biggest weakness, according to critics (Muuss, 1996). They argue that if we are to believe the contextualists, human development is so complicated that it will be many years before we can offer meaningful recommendations to teachers, social workers, and parents. Those who find fault with Lerner find his theory to be too demanding. They believe that he has been too critical of theorists who have gone before him, in particular, Erik Erikson. After all, critics argue, Erikson does pay considerable attention to the social as well as the psychological variables that affect development.

The goal of adolescent psychology is to produce generalizations that are ultimately correct and useful to those working in the field. Nevertheless, Lerner's assertions should make us more aware of the need for a complex, more interactive general theory of adolescent psychology.

In Table 2.4, you will find a list of the theorists considered in this chapter, together with an indication of which aspect of biopsychosocial causes they emphasize.

Table 2.4	Developmental Theorists and Their Positions on the Major Source(s) of Change
Theorist	**Source(s) of Change**
G. Stanley Hall	BIOpsychosocial
Sigmund and Anna Freud	BIOPSYCHOsocial
Peter Blos	BIOPSYCHOsocial
Ruth Benedict	biopsychoSOCIAL
Margaret Mead	biopsychoSOCIAL
Albert Bandura	biopsychoSOCIAL
Robert Havighurst	bioPSYCHOSOCIAL
Abraham Maslow	BIOPSYCHOSOCIAL
Erik Erikson	bioPSYCHOSOCIAL
Richard Lerner	BIOPSYCHOSOCIAL

WHAT'S YOUR VIEW?

Cultural Biases of the Theorists

Each of the theories you have read in this chapter was designed by scientists who have been influenced by the culture of their time and place: Freud lived primarily in Vienna, which was dominated by a male patriarchy; Karen Horney, who questioned Freud's stance on many issues, lived in the rebellious times of the Weimar Republic in Berlin; typical of the American culture in the 1960s is Blos's belief that boys' development is directed toward control and dominance of the physical world, but that girls' development is directed toward boys and hence relationships. Arising from the same culture and academic history, Erikson, too, believes "anatomy is destiny," that one's gender determines one's social role. What do you think? Are all theories tainted to some extent by cultural bias? For example, do you feel there are many essential differences in adolescent development for boys and girls, or is this view the result of the dominant social view?

C O N C L U S I O N S

This chapter has covered a great deal of important ground. It has introduced you to the framework on which, in many ways, the rest of the book depends. Although there are many theories we might have chosen to include in this chapter (some of which we will discuss later), the ones we have presented here have played or are playing major roles in the development of adolescent psychology.

There is a lot to remember, but we will be coming back again and again to these seminal ideas, and this will help you gain a firm understanding of them. In these first two chapters, you have received an overview of the fascinating study of adolescent development. Now we begin to take a much closer look at each of the major aspects of teen life with the powerful spotlights of theory and research. In the next section (Chapters 3–6), we review psychological factors. In the following sections (Chapters 7–14), we analyze social factors. Throughout these chapters, biological issues are also considered.

C H A P T E R   H I G H L I G H T S

What Theories Do

- Good theories help to organize information, to focus our search for understanding, to describe and explain interpretations and conclusions, and to recognize major disagreements among scholars and scientists.
- A theory attempts to explain empirical research by showing how constructs are related.

The Biopsychosocial Causes of Change

- While all theorists believe that there are biological, psychological, and social causes of human development, they differ on the emphasis they put on each of the three factors.
- The biological camp believes that genes are the major factor in our development.
- The psychological view believes that changes in personality and/or intellect guide our development.
- The social view is that the environment plays the major role in human development.
- There are theorists who believe that it is a combination of two or three of these factors that guide human development.

G. Stanley Hall's Theory [BIOpsychosocial]

- There are four periods of development that all human beings pass through: birth to 4 years; 4 to 8 years; 8 to 12 years; and 12 to 25 years.
- Hall believed that development, being determined by our genes, occurs in an unchangeable, universal pattern, with the environment having little effect.
- Hall's interpretation of adolescent development was greatly influenced by his observation that it is a period of storm and stress.

Sigmund Freud's Theory [BIOPSYCHOsocial]

- The unconscious mind is the key to understanding human beings.
- Important information in the unconscious mind is kept hidden through an array of defense mechanisms.
- The mind is divided into three constructs: the id, the ego, and the superego, each of which appears at different stages of a child's development.
- Personality development is divided into five instinctive stages of life, each stage serving a major function: oral, anal, phallic, latency, and genital.
- Failure to pass through a stage of development results in fixation, which halts a person from becoming fully mature.
- Anna Freud believed that the delicate balance between the superego and the id being disrupted by puberty causes the adolescent to regress to earlier stages of development.
- Peter Blos has made a number of contributions that have modernized psychoanalytic theory.

Ruth Benedict's and Margaret Mead's Theory [biopsychoSOCIAL]

- In their studies of primitive cultures, Benedict and Mead found that human behavior differs from culture to culture, therefore proving that development cannot be determined solely by biology.
- Three areas of development that are different among cultures are sexual roles, responsibility, and dominance.

Albert Bandura's Theory [biopsychoSOCIAL]

- Learning occurs through observing and modeling the actions of others.

Robert Havighurst's Theory [bioPSYCHOSOCIAL]

- Each stage in development has specific developmental tasks, skills, knowledge, functions, and attitudes that are needed by a person in order to succeed in life.

Abraham Maslow's Theory [BIOPSYCHOSOCIAL]

- A human being has needs that appear in overlapping stages, beginning with deficiency needs at birth and progressing to being needs as the person develops.
- A hierarchy of needs develops—basic needs (deficiency needs) must be fulfilled before concern can be given to higher-level needs (being needs).

Erik Erikson's Theory [bioPSYCHOSOCIAL]

- Human life progresses through eight psychosocial stages, each of which is marked by a crisis and its resolution.
- Although the ages at which a person goes through each stage vary, the sequence of stages is fixed. Stages may overlap, however.
- A human being must experience each crisis before proceeding to the next stage. Inadequate resolution of the crisis at any stage hinders development.

Richard Lerner's Theory [BIOPSYCHOSOCIAL]

- The contextualism theory claims that all previous theories of adolescence are inadequate because they fail to emphasize the interaction occurring between adolescents and their environment.
- The internal state of the adolescent and the external environment do not just affect each other, the two are imbedded in each other.

K E Y   T E R M S

asceticism 44
being needs 49
bidirectional research 57
biopsychosocial 36
construct 37
contextualism 55
defense mechanisms 41
deficiency needs 49

developmental diversity 59
developmental tasks 48
ego 42
ego psychology 45
fixated 43
goodness of fit 57
hierarchy of needs 49
id 42

identity crisis 53
intellectualization 44
observational learning 46
probabilistic 56
psychoanalysis 41
repudiation 54
storm and stress 38
superego 42

W H A T   D O   Y O U   T H I N K ?

1. What is your reaction to the statement, "The truth or falseness of a theory is not important. All we care about is its usefulness"?
2. Some people say that in his concept of human development Freud emphasizes sexuality too much. What do you think?
3. What similarities can you detect among the social theories of Bandura and Havighurst?
4. Are any one of Maslow's types of needs most relevant during adolescence?

5. Is it possible for a person to be deeply intimate with another person and still be in a state of identity confusion?
6. Which theory is most useful to the person working with adolescents: the psychological, the social, or the psychosocial theories?
7. Which theory best fits your experience as an adolescent?
8. Do you notice any pattern in Table 2.4? Can you offer an explanation for it?

S U G G E S T E D   R E A D I N G S

Clark, R. W. (1980). *Freud*: The man and the cause. New York: Random House. This is one of the most judicious and even-handed books written about the father of psychoanalysis.

Erikson, E. (1958). *Young man Luther*. New York: W. W. Norton. Martin Luther was the main force behind the Protestant Reformation. In Erikson's penetrating analysis of the causes of Luther's actions, we have a wonderfully clear example of his ideas about adolescence in general, and negative identity in particular.

Skinner, B. F. (1948). *Walden two*. New York: Macmillan. Many students find it hard to see how Skinner's behaviorism would function in everyday life. In this novel, we see how a community based on his principles would operate. In fact, for a while at least, several such communities really existed. This is a good way to understand his theory.

Tyler, A. (1986). *The accidental tourist*. New York: Knopf. This is the story of a man who tries desperately to avoid the bumps of life. It offers an excellent opportunity to evaluate each of the theories presented in this chapter by analyzing the main character's life from the standpoint of each theory.

G U I D E D   R E V I E W

1. By drawing attention to major disagreements among scholars and scientists, theories provide
   a. testable ideas that can be confirmed or refuted by research.
   b. helpful tools for organizing a huge body of information.
   c. a focus in our search for new understandings.
   d. an explanation of how findings may be interpreted.

2. G. Stanley Hall believed that an individual's development was determined by his or her genetic makeup. Therefore, Hall's theory of development stresses the importance of _____ causes.
   a. psychological
   b. developmental
   c. biological
   d. social

3. After losing the election for class president, Jeff decided to join the service club. Jeff was exhibiting
   a. sublimation.
   b. displacement.
   c. rationalization.
   d. compensation.

4. When the professor returned Sarah's exam with a failing grade, Sarah was so angry that she came back to her dorm room and yelled at her roommate. Sarah displayed
   a. compensation.
   b. rationalization.
   c. projection.
   d. displacement.

5. According to Freud, boys refuse to show much affection toward their mothers and scorn their female peers when they are going through which stage of development?
   a. latency
   b. phallic
   c. anal
   d. oral

6. According to Freud, which of the following keeps us in touch with reality?
   a. id
   b. ego
   c. superego
   d. psychic censor

7. Who believed that the major problem of adolescence was restoring the delicate balance between the superego and the id?
   a. Sigmund Freud
   b. Anna Freud
   c. Peter Blos
   d. Albert Bandura

8. Bandura's social learning theory stresses the influence of _____ on personality development.
   a. modeling
   b. the environment
   c. belongingness and love needs
   d. conflict

9. In Maslow's hierarchy of needs, what is the highest level need?
   a. aesthetic
   b. psychological
   c. esteem
   d. self-actualization

10. Beth likes to act and is always cast in the school plays. She also plays two sports and is involved in student government. As a result of all her activities, her school work is slipping. She knows that she must drop at least one activity but because she likes them all, she can't decide which one she should drop. Beth is in which of Erikson's psychosocial stages?
    a. autonomy versus shame and doubt
    b. industry versus inferiority
    c. initiative versus guilt
    d. identity and repudiation versus identity confusion

11. _____ (Constructs/Theories) are helpful tools for organizing a huge body of information.

12. At the basic level of all the sciences is _____ (empirical/descriptive) research.

13. A theorist who emphasizes _____ (psychological/biological) causes holds that the changes in personality and/or intellect are the major factors of human development.

14. Hall believed that adolescence was a time of _____ (positive modeling/storm and stress).

15. According to _____ (Erikson/Freud), the unconscious mind is the key to understanding the human being.

16. Unconsciously forgetting experiences that are painful to remember, such as forgetting sexual abuse experiences, is known as _____ (repression/regression).

17. Anthropologists such as Benedict and Mead found that behaviors on which psychoanalysts based their position simply did not exist in _____ (other cultures/women).

18. In Maslow's hierarchy of needs, _____ (belongingness and esteem needs/self-actualization needs) are of greatest relevance during adolescence.

19. According to Erikson, the ages at which people go through each developmental stage _____ (are fixed/vary somewhat), but the sequence of stages _____ (is fixed/varies somewhat).

20. Erikson believed that the main task of adolescence is to achieve a state of _____ (intimacy/identity).

Answers

DEVELOPMENTAL PATTERNS

PHYSICAL DEVELOPMENT

Gretchen, my friend, got her period. I'm so jealous, God. I hate myself for being so jealous, but I am. I wish You'd help me just a little. Nancy's sure she's going to get it soon, too. And if I'm last, I don't know what I'll do. Oh, please, God, I just want to be normal!

Judy Blume, *Are You There, God? It's Me, Margaret*, 1970

CHAPTER 3

CHAPTER OUTLINE

If you want to understand adolescence, you surely need to know quite a bit about puberty. In this chapter, you learn about some of the myths of puberty. We explain its biological basis, the sequence of events that make it up, the contrast of changes for males and females, and the effects of timing. Finally, we explore the concept of body image.

After reading this chapter, you should be able to:

- List three common myths of puberty and be able to refute them.
- Identify the important parts of the male and female reproductive systems and explain their functions.
- Describe the endocrine system and the hormones that influence pubertal development.
- Explain genetic and nutritional influences on pubertal development.
- List the normal sequence of events in puberty for males and females.
- Contrast male and female development in puberty.
- Describe the influence of timing on individual adolescents' emotional reactions to the physical changes of puberty.
- Specify how our culture overemphasizes the importance of physical attractiveness and how this overemphasis affects male and female adolescents.
- Relate your own experience of puberty to factors of preparation, timing, and body image as presented in the research.
- Discuss these issues from an applied, a sociocultural, and your own point of view.

Early Studies of Puberty

puberty
A relatively abrupt and qualitatively different set of physical changes that normally occur at the beginning of the teen years.

Since humans began writing about the experience of living, they have speculated about **puberty**. The scientific study of puberty, however, started only at the turn of this century when, because of child labor laws, teenagers were increasingly kept from working so they could stay in school. Teachers and psychologists became more interested in the way children developed—much more so than were the factory owners who had governed the lives of so many teenagers in the 19th century.

The adolescent theorists of the early 20th century (e.g., Boas, 1911; Bourne, 1913; Burnham, 1911; King, 1914) had far less data available than we do today, and their opinions about puberty were largely subjective. For example, King stated that for girls, puberty peaks at 12, and for boys, at 13. His evidence was that in Boston in 1913, there were fewer deaths per thousand among 12-year-old girls and 13-year-old boys than for any other female and male age group, respectively. Because relatively few died at these ages, King argued, this must be when their "vital force" was highest, and thus was an indication of the onset of puberty.

delayed puberty
When the stages of pubertal change do not begin until a significant time after the normal onset.

This hypothesis is wrong. For one thing, the age at which the death rate is lowest changes from year to year and from place to place. Biological factors such as the timing of puberty are now known to be much more stable than that. King also suggested that the major cause of **delayed puberty** was "excessive social interests, parties, clubs, etc. with their attendant interference with regular habits of rest and sleep" (p. 25). We now know that this opinion is also wrong.

The Common Myths of Puberty

How do most myths get started?

There are few terms that conjure up images of adolescence as much as the term *puberty*. Adults often remember puberty as a period of awkwardness in early adolescence; and photos of the early teen birthdays often confirm those remembrances! There is a lot of personal or common sense knowledge about puberty, but only recently have psychologists started to look at puberty as more than a "given" of adolescence. As with any body of knowledge that is originally built on personal impressions, the period of puberty has not always been represented in its truest form.

MYTH 1: PUBERTY STARTS AT ONE POINT IN TIME

In fact, puberty is a process that takes place over several years. The term *puberty* comes from the Latin word *pubescere*, which means "to grow hairy." Biologically, puberty is a series of gradual biological changes that lead to reproductive maturity (Rowe & Rodgers, 1989). That is, these changes make the adolescent able to reproduce.

Psychologists argue that puberty is not just a biological process, that it is more than just hormonal changes. The individual's social life is also involved (Kreipe & Strauss, 1989; Petersen, 1987). Growing mental abilities and evolving relations with family and friends interact with biological changes. For example, a hormonal change, which may cause temporary feelings of depression, can be made even worse by social change such as an argument about going to a party (Petersen). Puberty, therefore, is viewed as a *biopsychosocial* concept. *Bio* refers to biological change. *Psycho* refers to psychological adjustment. *Social* refers to the influence of family, peers, and community (Wilson & Keye, 1989).

MYTH 2: PUBERTY STRIKES WITHOUT WARNING

If you listen to parents, it may sound as if the process of puberty starts without warning. "Whatever happened to our wonderful little child?" they often lament. In fact, all three biopsychosocial factors are in a constant state of growth that is even more rapid during adolescence than in the first two years of life (Vaughan & Litt, 1990).

The seeds of this rapid pubescent change are present in the developing fetus (Calderone, 1985). Film of a male fetus at 29 weeks of age (taken with an ultrasound scanner, a device that detects images through the mother's abdomen) demonstrates that even as the fetus is developing, his reproductive system is functioning. It functions as a series of reflexes that mimic the adults' reproductive system—complete with penile erection—occurring in cycles during sleep (Calderone, 1985). The reproductive system is fully present in the infant, but **hormones** suppress it until early puberty.

hormones
Chemical agents produced by the endocrine system that trigger physical change such as puberty.

endocrine system
The ductless glands such as the pituitary, thyroid, and sex glands that secrete directly into the bloodstream.

MYTH 3: PUBERTY IS THE RESULT OF RAGING HORMONES

To untangle the myth that young adolescents become highly emotional because of increased hormone flow, let's first consider what hormones are. Hormones are chemicals released by the **endocrine system**. They act as messengers that trigger physical change. There is some truth in this myth: hormones *do* play a large role in the biological changes of puberty. However, enough is known about the hormones to be able to pinpoint which ones are responsible for which changes. The notion that the hormones produce all changes, from mood swings to acne, is an overstatement of their power. Hormones do not work in isolation. Other forces, such as social interaction, are also involved (Smith, 1989). The adolescent is always interpreting biological changes mentally, so it is hard to determine what part of puberty is strictly biological (Kreipe & others, 1989; Lerner & Foch, 1987).

Your Reproductive System

Much more is known today about many aspects of puberty. We know more about the organs of our reproductive system and how these organs function together. And we are learning how to present this knowledge to adolescents effectively.

How well do you know your own reproductive system? Take the test in the Applied View box, then read the following sections on the female and the male sexual systems.

AN APPLIED VIEW

How Well Do You Know Your Own Reproductive System?

Most of us think we understand the workings of sex and reproduction well enough, yet when asked to define the various parts of our sexual system, we don't do very well. How high would you rate your knowledge?

If you would like to learn how much you really know (and this knowledge is important, if only because the adolescents you deal with may ask you questions about it), take this test. Put an M on the blank line after each item that is part of the male sexual system, an F after each that is part of the female sexual system, or M/F if it is both. The correct answers may be found by examining Figures 3.1 and 3.2 and reading the accompanying text.

M, F, or M/F		M, F, or M/F		M, F, or M/F	
Bartholin's glands	_____	Hymen	_____	Scrotum	_____
Cervix	_____	Labia majora	_____	Testes	_____
Clitoris	_____	Labia minora	_____	Ureter	_____
Cowper's glands	_____	Mons pubis		Urethra	_____
Epididymis	_____	(mons veneris)	_____	Uterus	_____
Fallopian tubes	_____	Ova	_____	Vas deferens	_____
Fimbriae	_____	Ovary	_____	Vulva	_____
Foreskin	_____	Pituitary gland	_____		
Glans penis	_____	Prostate gland	_____		

The highest possible score on this test is 24. This test has been given to groups of sophomores and graduate students. The sophomores, whose mean age is 18, averaged 19 on the test. The graduate students, whose mean age is 27, averaged 13! Why do you suppose the older students did worse? Do you think that adolescents would benefit from a better knowledge of their own and the opposite gender's sexual systems?

THE FEMALE SEXUAL SYSTEM

The parts of the female sexual system are defined here and are illustrated in Figure 3.1.

- *Bartholin's glands:* a pair of glands located on either side of the vagina. These glands provide some of the fluid that acts as a lubricant during intercourse.
- *Cervix:* the opening to the uterus, located at the inner end of the vagina.
- *Clitoris:* comparable to the male penis. Both organs are extremely similar in the first few months of life, becoming differentiated only as sexual determination takes place. The clitoris is the source of maximum sexual stimulation and becomes erect through sexual excitement. It is above the vaginal opening, between the labia minora.
- *Fallopian tubes:* tubes through which the ova travel from the ovary to the uterus. A fertilized egg that becomes lodged in the Fallopian tubes (called a Fallopian pregnancy) cannot develop normally and, if not surgically removed, will cause the tube to rupture.
- *Fimbriae:* hairlike structures located at the opening of the oviduct that help move the ovum down the Fallopian tube to the uterus.
- *Hymen:* a flap of tissue that usually covers most of the vaginal canal in virgins.
- *Labia majora:* the two larger outer lips of the vaginal opening.
- *Labia minora:* the two smaller inner lips of the vaginal opening.
- *Mons pubis or mons veneris:* the outer area just above the vagina that becomes larger during adolescence and on which the first pubic hair appears.
- *Ova:* the female reproductive cells stored in the ovaries. These eggs are fertilized by the male sperm. Girls are born with more than a million follicles, each of which holds an ovum. At puberty, only 10,000 remain, but they are more than sufficient for a woman's reproductive life. Since usually only one egg ripens each month from her mid-teens to her late forties, a woman releases less than 500 ova during her lifetime.

Figure 3.1

The female sexual system.

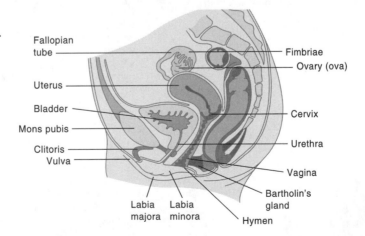

- *Ovaries:* glands that release one ovum each month. They also produce the hormones estrogen and progesterone, which play an important part in the menstrual cycle and pregnancy.
- *Pituitary gland:* the master gland located in the lower part of the brain. It controls sexual maturation and monthly menstruation.
- *Ureter:* a tube connecting the kidneys with the bladder.
- *Urethra:* a canal leading from the bladder to the external opening through which urine is excreted.
- *Uterus:* the hollow organ (also called the womb) in which the fertilized egg must implant itself for a viable pregnancy to occur. The egg attaches itself to the lining of the uterus from which the unborn baby draws nourishment as it matures during the nine months prior to birth.
- *Vulva:* the external genital organs of the female.

THE MALE SEXUAL SYSTEM

The parts of the male sexual system are defined here and are illustrated in Figure 3.2.

- *Cowper's glands:* located next to the prostate glands. Their job is to secrete a fluid that changes the chemical balance in the urethra from an acidic to an alkaline base. This fluid proceeds up through the urethra in the penis where it is ejaculated during sexual excitement just preceding the sperm-laden semen. About a quarter of the time, sperm also may be found in this solution, sometimes called preseminal fluid. Therefore, even if the male withdraws his penis before he ejaculates, it is possible for him to deposit some sperm in the vagina, which may cause pregnancy.
- *Epididymis:* a small organ attached to each testis. It is a storage place for newly produced sperm.
- *Foreskin:* a flap of loose skin that surrounds the glans penis at birth, often removed by surgery called circumcision.
- *Glans penis:* the tip or head of the penis.

Figure 3.2
The male sexual system.

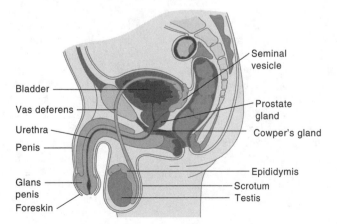

- *Pituitary gland:* the master gland controlling sexual characteristics. In the male it controls the production of sperm, sexual excitement, and the release of testosterone (and thus the appearance of secondary sexual characteristics such as hair growth and voice change).
- *Prostate glands:* glands that produce a milky alkaline substance known as semen. In the prostate the sperm are mixed with the semen to give sperm greater mobility.
- *Scrotum:* the sac of skin located just below the penis, in which the testes and epididymis are located.
- *Testes:* the two oval sex glands suspended in the scrotum that produce sperm. Sperm are the gene cells that fertilize the ova. They are equipped with a tail-like structure, which enables them to move with a swimming motion. After being ejaculated from the penis into the vagina, they attempt to swim through the cervix into the uterus and then into the Fallopian tubes where fertilization takes place. If one penetrates an egg, conception occurs. Although the testes regularly produce millions of sperm, the odds against any particular sperm penetrating an egg are enormous. The testes also produce testosterone, the male hormone that affects other aspects of sexual development.
- *Ureter:* a tube connecting the kidneys with the bladder.
- *Urethra:* a canal that connects the bladder with the opening of the penis. It is also the path taken by the preseminal fluid and sperm during ejaculation.
- *Vas deferens:* a pair of tubes that lead from the epididymis up to the prostate. They carry the sperm when the male is sexually aroused and about to ejaculate.

AN APPLIED VIEW

Talking About the Hot Topics of Puberty

When we talk to adolescents about the myths of puberty (or about menstruation, or wet dreams, or any other aspect of physical development), most of us want to be helpful, which leads to a tendency to lecture. This is a mistake.

Because puberty is so great and rapid a change, because it seems so mysterious, and especially because it is so linked to sexuality, the average teenager is embarrassed by the subject. Most have fearful feelings about it. They frequently wish they could just forget about puberty. Thus they have a hard time hearing us when we insist on explaining things to them.

On the other hand, the profound changes that their bodies are undergoing make it difficult to neglect the topic. They often feel a need to express their confusion and fear. Those of us who work with them can do no greater service than to encourage them to express those concerns. We should avoid the temptation to calm their fears with our explanations, which often makes them feel we are not listening.

It is much better, at least at the beginning of such a discussion, to respond to their remarks empathetically. "That must make you feel weird." "Yes, I remember feeling just like that." "It's not easy to deal with that, is it?" When you respond this way, sooner or later you will be asked for your opinion. Now your help will be welcomed and, therefore, more helpful.

The Biological Basis of Puberty

THE ENDOCRINE SYSTEM AND THE HORMONAL BASIS OF PUBERTY

pituitary gland
The main endocrine gland that secretes gonadotropins into the bloodstream, which, upon reaching the brain, stimulate the production of more hormones in other glands.

gonadotropins
The hormones secreted by the pituitary gland.

The endocrine system is regulated by endocrine glands. The main endocrine gland is the **pituitary gland**. The pituitary gland is attached to the base of the brain. The process of hormone secretion from the pituitary gland is influenced by other hormones secreted by the part of the brain called the hypothalamus (Sroufe & Cooper, 1988). The pituitary gland, known as the master gland, secretes hormones into the bloodstream that travel to the brain. These pituitary hormones are known as **gonadotropins**. When the gonadotropins reach the brain, they stimulate the production of more hormones in other glands (Petersen, 1987). Hormones regulate the growth of the reproductive system. A hormone is also produced by the human body that, when combined with a by-product of the liver and kidneys, stimulates physical growth (Carlson, 1994).

Amazingly, the whole endocrine system is in place at the time of birth. Researchers speculate that right before birth the fetus tests the system, but then the system is suppressed (Petersen, 1987). During the first 7 to 9 years of life the body suppresses any release from the endocrine glands. Apparently it is programmed into the system that early childhood is not a time for reproductive maturity.

At around age 7, a gradual activation process occurs. At this point the body becomes ready for hormonal change. The system emits some hormones, resulting in changes in hormonal balance that occur during sleep. In both boys and girls, gonadotropins called LH (luteinizing hormone) and FSH (follicle-stimulating hormone) are dispatched in tiny bursts during sleep (Nottelmann & others, 1987a). Although you cannot see any difference in the child, this process is a preview of change to come.

testosterone
The hormone produced by the testes that triggers the physical changes of puberty in boys.

Soon the endocrine system becomes ready to function fully. In boys the increased level of LH eventually stimulates the testes to produce the hormone called **testoster**one. Testosterone, in turn, triggers the changes well associated with puberty for boys: genital growth, increased body hair, change of voice. The increase in FSH in boys stimulates the growth of testicles and the production of sperm.

In early adolescence (ages 10 to 12), girls' bodies store LH. FSH makes the ovaries release hormones that trigger changes associated with puberty for girls: breast development, increased body hair, and menstruation. This series of pubertal changes ultimately ushers in biological adulthood, but it happens gradually (Thornburg & Aras, 1986).

The endocrine system is often called a feedback system. This is because glands only release hormones when they receive feedback from the system that it is ready. In adulthood, the system is active and steadily functioning. Figure 3.3 depicts this process from fetus to adulthood.

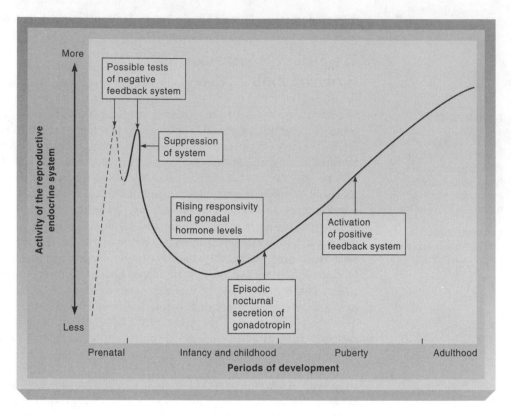

Figure 3.3

Schematic illustration of changes in the endocrine system from prenatal to adult development.

From A. C. Petersen, "The Nature of Biological-Psychosocial Interactions," in R. M. Lerner, *Biological-Psychosocial Interactions in Early Adolescence.* Copyright © 1987 Lawrence Erlbaum Associates, Inc., Hillsdale, NJ. Reprinted by permission of publisher and author.

Puberty is a time of tremendous physical growth and change. This growth includes changes in height, in skeletal and body composition/weight distribution, and in secondary sex characteristics (pubic hair, breast development, change of voice). Also included are the reproductive capacities: spermatogenesis, which is the male ability to produce sperm, and ovulation, the female ability to release ova (the eggs that can be fertilized by sperm). In addition to the hormonal changes noted here, these events are influenced by two other factors: genes and nutrition.

GENETIC INFLUENCES

There has long been controversy in psychology over where the sources of development lie (Dixon & Lerner, 1992). Essentially the debate centers on whether inborn (hereditary) traits or acquired (learned) processes are responsible for development. Francis Galton (1822–1911), Charles Darwin's first cousin, presented his concept of a sharp distinction between the role of nature versus the role of nurture in human development. Galton believed that whereas nature offers the potential for development, one's experience affects the realization of one's potential (Gottlieb, 1992). Today development is viewed by many psychologists as a biopsychosocial process, with biological, psychological, and social-cultural forces influencing development. Biological components include the role of our genes in development; psychological elements include all aspects of personality and cognitive development; and social elements include family, school, and peer influences (Dacey & Travers, 1996).

To understand the way heredity affects puberty, researchers often study identical or monozygotic twins—twins who are produced from the fertilization of one egg (*mono* and *zygote*). These children are compared to fraternal or dizygotic twins—twins that are produced from the fertilization of two (*di*) separate eggs. This helps establish the direct effect of genes.

Rowe and Rodgers (1989) reviewed many such studies. They concluded that most of the changes in puberty appear to be caused by genes. Genes have their greatest influence on the increased rate of growth. The amount of body hair (Tanner, 1990) and the timing of menarche (first menstruation) (Thornburg & Aras, 1986) are also affected by genes.

NUTRITIONAL INFLUENCES

The impact of nutrition on physical development during puberty has received much attention in the United States (Brooks-Gunn, 1988; Rees & Trahms, 1989; Thornburg & Aras, 1986). However, in a recent article Hartley (1999) suggested that even more attention should be paid to the specific nutritional needs during puberty and adolescence. The attention to the impact of nutrition is probably due to two factors that primarily reflect female puberty: the **secular trend** in Western countries and the increased attention to the impact of **eating disorders**. Let us examine each more closely.

The secular trend refers to the decreasing age of the onset of puberty, including a significant *drop* in the average age at which females in a particular country reach menarche. In Western countries, the average age of menarche has declined about three months per decade over the past 100 years. In the United States, 17 years was the average age in the late 18th and early 19th centuries (Vaughan & Litt, 1990). Today the average age of onset is 12.5 years. Most researchers feel that improved nutrition, sanitation, and health care are responsible for the trend, and that we are now at a period of leveling off (Brooks-Gunn & others, 1985). We think nutrition is involved because girls must typically achieve a certain proportion of body fat before they can menstruate (Frisch, 1988). Studies of female athletes and dancers have shown that a lack of fat can delay menarche or can stop menstruation after it has begun (Brooks-Gunn, 1987).

Studies of eating disorders have also shown that nutrition can impact menstruation (Thornburg & Aras, 1986). For example, studies of anorexia nervosa have added greatly to our knowledge of the impact of malnutrition on pubertal development. Anorexia nervosa is an eating disorder whereby a person (generally female) drastically reduces food intake.

secular trend
The decreasing age of the onset of puberty in females in Western countries, particularly the average age of menarche.

eating disorders
Disorders such as anorexia nervosa and bulimia nervosa characterized by drastically reduced food intake or episodic binge eating.

A SOCIOCULTURAL VIEW

Poverty, Not Ethnicity, Affects Secular Trend

It appears that environmental factors play a role in the secular trend. More clearly seen in Western societies, the secular trend has been documented in industrialized countries and in some Nigerian and Bangladeshi children from affluent families. Research on these children indicates a connection between better nutrition and health care and earlier physical maturation (Haq, 1984; Imobekhai, 1986). Boys and girls from industrialized societies enter adolescence earlier and stretch out their adolescence. In less developed countries, children reach puberty at a later age, often just before they reach adulthood.

Clearly the ultimate cause is poverty. Research indicates that for children raised in poverty, the secular trend for height falls behind their peers by one generation. Ethnicity, however, does not seem to play a role in the secular trend. The differences in physical growth for children living in and out of poverty confirm the importance of health, nutrition, and living environment in stimulating or detracting from physical development (Balk, 1995).

---

amenorrhea
The absence or suppression of menstruation, often seen in anorexics and bulimics.

Studies have shown that when an anorexic girl loses 15 percent of her body weight, she will generally stop menstruating (**amenorrhea**) (Brooks-Gunn & Petersen, 1984). At this time, there is no research that looks specifically at the impact of weight loss on male puberty. This may be because male sexual development does not depend directly on fat reserves as does female sexual development (see Figure 3.4). We will have much more to say about eating disorders in Chapter 11.

Whereas pubertal development in females is associated with increases in body fat, in males the pubertal weight spurt is due primarily to increased muscle (Vaughan & Litt, 1990). Female sexual development ultimately centers on the capacity for *gestation* (to be pregnant) and *lactation* (to be able to breast-feed a baby)—both are processes that require stores of body fat for fuel (Savin-Williams & Weisfeld, 1989). In a society that values slimness, the changes in the female reproductive system are especially disconcerting to early adolescent girls. Not surprisingly, therefore, eating disorders are most prominent in female populations—about 19 out of 20 cases (Thornburg & Aras, 1986). However, recent research suggests that eating disorders are more prevalent in males than previously believed (Moulton & Roach, 1998).

Several kinds of eating disorders seem to be on the rise these days. Can you think of any reasons why?

In reviewing research on the impact of nutrition on puberty in both males and females, Rees and Trahms (1989) note that eating too little is not the only problem. Obesity may cause the body to try to keep itself abnormally fat, instead of utilizing that fat for the growth spurt that is characteristic of adolescence.

The Sequence of Events in Puberty

Puberty is predictable in terms of the sequence of changes in the sex characteristics that occur in any healthy adolescent. What is unpredictable is the timing

Figure 3.4

Percentage of body weight represented by fat.

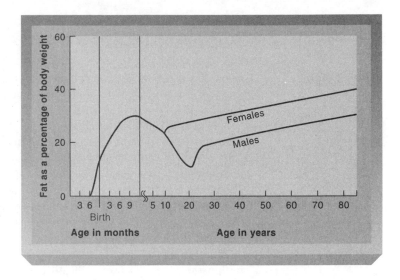

(Brooks-Gunn, 1988; Brooks-Gunn & Warren, 1988). Although the order in which change occurs (the sequence) is the same for most teens, it is important to note that an adolescent may experience rapid bodily change in one area and not in another (Vaughan & Litt, 1990). The sequences of bodily change presented in Tables 3.1 and 3.2 may vary somewhat from individual to individual. Mainly, however, these sequences of change are standard throughout the world.

A SOCIOCULTURAL VIEW

Eating Disorders and Body Image

Recent research indicates that a person's environment—where and how that person grows up—contributes to anorexia nervosa and bulimia (French & others, 1994). Female teenagers who felt negatively about their family cohesion, communication processes with their parents, and overall family satisfaction were more at risk for eating disorders than were teenage girls who rated their family relationships positively. Males who were considered at risk for eating disorders, however, did not rate their family characteristics any differently than did male teenagers who were not considered at risk (Leon & others, 1994).

Western society also sends out strong messages to teenagers, boys and girls, about body shape and slenderness (Toro & others, 1994) (see also the section on body image on p. 90). Many teenagers experience stress about their body sizes and shapes caused by pressure from advertising, verbal messages, and social situations. Teenagers, in particular, seem to be more susceptible to weight loss advertising pressure.

The development of eating disorders is due to several factors. It is important to keep in mind that a person's family environment, peers, and the media all contribute to the misperception of the ideal body. Psychologists today need to be aware of the influence of social pressures on teenagers' perceptions of body image and appearance. We should help adolescents resist societal pressure to conform to unrealistic standards of appearance. We should provide guidance on nutrition, on realistic body ideals, and on achievement of self-esteem, self-efficacy, interpersonal relations, and coping skills (Nagel & Jones, 1992).

The Contrast of Male and Female Pubertal Change

Look at a photo of any junior high school class. You can't help but notice that many of the girls are taller and more developed than the boys. This difference has existed for a very long time.

   The growth spurt for females occurs almost two years earlier than for males. Females generally start a growth spurt around age 10.5, and males generally start a growth spurt around age 12.5. For both sexes the growth spurt lasts approximately two years. The development of pubic hair generally happens about nine months earlier in females than in males. In addition, the development of breasts gives a more publicly noticeable sign of pubertal development in girls than does development of the male testes, which occurs only six months later than female breast development (Diamond & Diamond, 1986).

   In an effort to quantify the changes in sexual maturity both between and within the sexes, Tanner (1990) devised the *Sexual Maturity Rating Scale* (SMR), which has been used extensively in research concerning adolescent development (Dorn & others, 1990; Nottelmann & others, 1987b; Westney &

Table 3.1	**Sequence of Male Pubertal Changes**

- Testicular enlargement.
- Change in texture and color of skin on scrotum.
- Increased growth of penis (usually around 12.5 years).
- Growth of pubic hair.
- Growth spurt of height; enlargement of body through muscle tone (usually occurs l year after testicular enlargement).
- First ejaculation of seminal fluid (also known as a wet dream or nocturnal emission).
- Seminal fluid is usually infertile until further development of prostate gland and seminal vesicles.
- Growth of axillary hair (arms, legs, chest).
- Growth of facial hair (usually upper lip first, then lower lip, and then chin).
- Breaking of voice (due to the increased length of the vocal cords that follows the growth spurt of the larynx).
- Breast change. Yes, in males, too. There is an increase in the diameter of the areola (circle around the nipple), and with some boys there is temporary enlargement of the breasts. This is relatively unnoticeable and will go away with time. Male breasts that increase in size in a more pronounced manner indicate a condition called gynecomastia. We will discuss this later in the chapter.
- More penile growth, and enlargement of prostate gland and seminal vesicles resulting in more potent sperm.
- Hair growth from pubic area to navel.
- Start of recession of hairline on scalp.

From J. M. Tanner, "Sequence & Tempo in the Somatic Changes in Puberty" in M. Grumbach, *Control of the Onset of Puberty*, 1974. Reprinted by permission of M. M. Grumbach.

Table 3.2	Sequence of Female Pubertal Changes

- Growth spurt (height and weight increase; fat is added to hips).
- Growth of breasts (called breast buds in the earliest stage). This includes some enlargement of the areola.
- Growth of vagina and change of pH balance of vaginal mucus (Thornburg & Aras, 1986).
- Growth of pubic hair. This usually occurs around 6 months after breast buds; although Tanner (1990) estimates that one-third of all girls develop pubic hair before breast buds.
- Continued growth of breasts and pubic hair.
- Growth of the uterus.
- Enlargement of ovaries.
- Menarche. This usually occurs when the growth spurt for height is declining most rapidly.
- The advent of female fertility. This usually occurs 12 to 18 months after menarche. Adolescents cannot assume, however, that they are sterile at first menarche, as there is great individual variation.

From J. M. Tanner, "Sequence & Tempo in the Somatic Changes in Puberty" in M. Grumbach, *Control of the Onset of Puberty*, 1974. Reprinted by permission of M. M. Grumbach.

others, 1984). The scale presents sequences of development for genital development in males, breast development in females, and pubic hair development in both. It is often presented as a series of pictures or sketches that the adolescent, parent, or medical practitioner uses to describe the state of an adolescent's development in each area. The scale has been criticized for its implicit message that all development proceeds in a certain sequence (Brooks-Gunn & others, 1985); however, it is a useful tool for gathering highly personal information from adolescents.

Table 3.3 is a composite of the SMR stages (Petersen & Taylor, 1980). Figure 3.5 portrays sketches of Tanner's (1990) stages of adolescent penis and scrotum development. Figure 3.6 depicts sketches of Tanner's stages of adolescent breast development.

MENARCHE

menarche
The onset of menstruation.

Menarche (the onset of menstruation, pronounced "men-ark-ah") has been studied more than any other aspect of puberty (Dorn & others, 1999; Brooks-Gunn & others, 1985; Brooks-Gunn & Warren, 1989; Kitahara, 1983). The general findings are that menarche evokes a myriad of responses from individual girls. Most describe both positive and negative feelings (Petersen, 1983). As with other biological changes in adolescence, the individual's perception of menarche depends on her experience. When girls are well prepared for menarche, they are more likely to formulate positive impressions of the experience (Brooks-Gunn, 1987). Kim and Smith (1998) found that earlier menarche was associated with more family stress in late childhood and more conflict between mother and daughter in early childhood. Therefore, it is

AN APPLIED VIEW

Using the Sexual Maturity Rating Scale

Professionals who need to communicate with each other about an adolescent's development often must include specific data on the precise state of the youth's bodily development. Such a need may occur among doctors, educators, psychologists, or others in our field. For example, researchers often require very precise definitions of development when looking at the relationship between such variables as nutrition and body development. A nurse may require emergency medical advice over the phone and may need to describe the patients physical development in detail.

It is not sufficient to make such statements as "physically in early adolescence" or "has reached pre-adult development." Tanner's SMR scale allows us to be more accurate than we could be without it, because it gives us agreed-upon reference points for making clear descriptions. Looking at the scale, can you remember at what ages you reached the various stages? Were you early or late? Did it matter?

possible that reaction to menarche is related not only to this developmental milestone but also to early family functioning.

Rierdan and Koff (1985) looked at the relationship between the age at which menarche occurred and the person's feelings about it. They asked 87 college women to recall the date on which their menarche occurred. They then asked these women to estimate the percentage of peers that reached menarche before them, and to rate their own experience of menarche on a scale from positive to negative. They found that the women's sense of when menarche appeared in

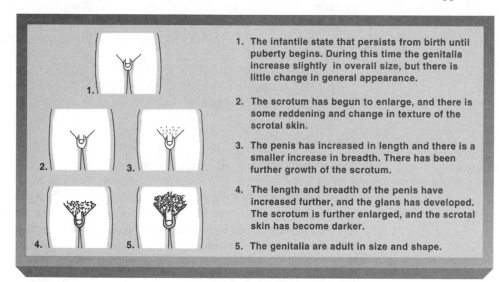

1. The infantile state that persists from birth until puberty begins. During this time the genitalia increase slightly in overall size, but there is little change in general appearance.

2. The scrotum has begun to enlarge, and there is some reddening and change in texture of the scrotal skin.

3. The penis has increased in length and there is a smaller increase in breadth. There has been further growth of the scrotum.

4. The length and breadth of the penis have increased further, and the glans has developed. The scrotum is further enlarged, and the scrotal skin has become darker.

5. The genitalia are adult in size and shape.

Figure 3.5
Sketches of Tanner's stages of male adolescent penis and scrotum development.

Table 3.3	Composite of the Sexual Maturity Rating Scale	
Stage	**Characteristic**	
	Genital Development in Males / **Pubic-Hair Development in Males and Females** / **Breast Development in Females**	

Stage	Genital Development in Males	Pubic-Hair Development in Males and Females	Breast Development in Females
1	Testes, scrotum, and penis are about the same size and shape as in early childhood.	The vellus over the pubes is not further developed than over the abdominal wall; in other words, there is no pubic hair.	There is elevation of the papilla only.
2	Scrotum and testes are slightly enlarged. The skin of the scrotum is reddened and changed in texture. There is little or no enlargement of the penis at this stage.	There is sparse growth of long, slightly pigmented, tawny hair, straight or slightly curled, chiefly at the base of the penis or along the labia.	Breast bud stage. There is elevation of the breast and the papilla as a small mound. Areolar diameter is enlarged over that of stage 1.
3	Penis is slightly enlarged, at first mainly in length. Testes and scrotum are further enlarged than in stage 2.	The hair is considerably darker, coarser, and more curled. It spreads sparsely over the function of the pubes.	Breast and areola are both enlarged and elevated more than in stage 2 but with no separation of their contours.
4	Penis is further enlarged, with growth in breadth and development of glans. Testes and scrotum are further enlarged than in stage 3; scrotum skin is darker than in earlier stages.	Hair is now adult in type, but the area covered is still considerably smaller than in the adult. There is no spread to the medial surface of the thighs.	The areola and papilla form a secondary mound projecting above the contour of the breast.
5	Genitalia are adult in size and shape.	The hair is adult in quantity and type with distribution of the horizontal (or classically "feminine") pattern. Spread is to the medial surface of the thighs but not up the linea alba or elsewhere above the base of the inverse triangle.	Mature stage. The papilla projects but the areola is recessed to the general contour of the breast.

comparison to their peers was critical to how they felt about the experience. Women who *felt* they were on time were most happy; and those who felt they were early recalled menarche most negatively. However, the researchers also found that 45 percent of the women who were objectively on time (by the date they gave for their menarche) perceived themselves as being either late or early. Recent research by Siegel & others (1999) supports this earlier finding about the importance of timing of puberty for adolescents. Based on research with a multiethnic sample, Siegel found that timing of puberty relative to peers was related to depression for both males and females. This finding suggests that

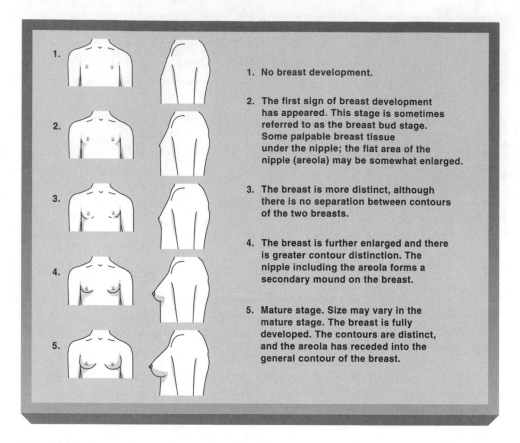

1. No breast development.

2. The first sign of breast development has appeared. This stage is sometimes referred to as the breast bud stage. Some palpable breast tissue under the nipple; the flat area of the nipple (areola) may be somewhat enlarged.

3. The breast is more distinct, although there is no separation between contours of the two breasts.

4. The breast is further enlarged and there is greater contour distinction. The nipple including the areola forms a secondary mound on the breast.

5. Mature stage. Size may vary in the mature stage. The breast is fully developed. The contours are distinct, and the areola has receded into the general contour of the breast.

Figure 3.6
Sketches of Tanner's stages of female adolescent breast development.

when adolescents feel that they are developing ahead or behind their peers they are at greater risk for depression.

MENSTRUATION

Although reactions to menarche have been extensively studied, only recently have studies been done on menstruation as it occurs during the adolescent years. In particular, two health problems associated with menstruation have been studied: **dysmenorrhea**, or menstrual cramps, and **premenstrual syndrome (PMS),** a series of behavioral, emotional, and physical symptoms that occur regularly around a week before menstruation (Fisher & others, 1989; Wilson & Keye, 1989). These two conditions had previously been considered problems that only adult women experience, but studies now show that adolescents experience both conditions in the same frequency and intensity as adults.

In the case of dysmenorrhea, Wilson and Keye (1989) found that 91 percent of their sample of 88 girls age 15 to 18 experience cramps and of that group, 64 percent of the girls rated the cramps as moderate to severe. Dysmenorrhea is

dysmenorrhea
Menstrual cramps.

premenstrual syndrome (PMS)
A series of behavioral, emotional, and physical symptoms that occur to many women around a week before menstruation.

AN APPLIED VIEW

What Was Your Reaction to Menarche?

Typical Responses of College Women

"I felt grateful. It meant I was normal. There was always that chance that I'd be the last one to get it, or not get it until I was 17."

"I was scared and embarrassed, mostly embarrassed. I felt like everyone who looked at me could tell I was wearing one of those awful Kotex pads!"

"I was terrified that my mother would tell my father and he'd tell my brothers. He would only tell them so they'd be more understanding, but I knew they'd only make fun of me."

"I just didn't want to think about it. It was the 'curse.'"

"All I could think of was what I would feel if someone saw me buying sanitary napkins at the drugstore."

"I laughed. I don't know why. I just thought it was funny."

"My mother and my school had prepared me for it, and although it was a surprise, it was kind of exciting to feel like I was a woman."

"When it happened, my mother called my older sister, my aunt, a female friend of hers, and me together in our den. She said, 'I have an announcement to make. We are all women here.' Everyone knew what she meant, and it was like a celebration. It made me feel really proud!"

"When I finally got it (I was 14), my mother told my father, who came to me and said, 'I hear you got your period, Judy. Congratulations.' Then he shook hands with me. I thought I would die with embarrassment!"

"My mother must have neglected to tell me the whole story in advance (I was only 11), because when I got it, I said, 'Well, I certainly am glad *that's* over with!'—thinking it only happened once!"

now known to be caused by the release of a hormone called prostaglandin by the uterus during menstruation. Many females can be treated for cramps by an antiprostaglandin. However, in their study Wilson and Keye found that girls viewed cramps as just part of the normal female experience that is untreatable. Wilson and Keye note that misinformation about dysmenorrhea and the lack of attention to adolescents' complaints about menstrual cramps may lead to a negative view of menstruation that is passed along generationally and in peer groups.

Premenstrual syndrome had also been widely assumed to be present only in adults. Fisher and others (1989) studied 207 high school females age 16 to 18 and found that three-fourths experience PMS as at minimum general discomfort. Like adults, the adolescents reported symptoms of backaches, headaches, water retention, fatigue, and feelings of depression or being blue during the week before menstruation. Some think that PMS may be the result of social expectancy or knowing that one is *supposed* to feel like that before menstruation. However, even if this is the case, Fisher and her colleagues contend that adolescent complaints of PMS should be taken as seriously as adult complaints. This is not to say that medication is always needed, but the possibility should not be discounted either. In a recent attempt to create an educational program to increase knowledge about PMS among adolescents, Chan and Chang (1999) found that 94 participants from four high schools in Hong Kong had significantly increased knowledge scores and reported a reduction in symptoms following their PMS educational program. This finding suggests that increased information and better understanding of what to expect, as well as adult willingness to talk about PMS, might mediate the effects of PMS.

MALE CONCERNS

There really isn't an event in male puberty that has been studied like female menstruation has. Ejaculation or the first wet dreams may be similar events in terms of reproduction, but they are not similar in terms of social recognition or acceptance (Brooks-Gunn & Petersen, 1984; Diamond & Diamond, 1986). This may be because ejaculation is often paired with masturbation, another subject that researchers are hesitant to study. This does not mean that boys have no concerns about puberty. One study of adolescents' perception of health problems showed that adolescents (male and female) have far more questions about their bodies and health than medical practitioners expected (Levenson & others, 1987).

gynecomastia
Inappropriate physical development marked by male breast growth.

varicocele
A condition occurring in male puberty in which one testicle is noticeably larger than the other.

Two conditions that sometimes occur during male puberty have received attention by the medical community: **gynecomastia** and **varicocele**. Gynecomastia is the enlargement of male breasts. When it happens temporarily in the sequence of other pubertal changes, it is called Type I (Silber, 1985). When the breast enlargement persists, it is Type II. Type II gynecomastia can be a symptom of marijuana use, liver disease, or testicular disease. It can also be simply the result of hormonal imbalance. In most cases it can be treated.

Varicocele—one testicle noticeably larger than the other—is another condition that can be of concern to pubescent boys. It occurs in about one-fourth of all adolescent boys (Silber, 1985) and often the difference is so slight that boys don't even detect it. If it is severe, it can usually be treated with surgery.

The Effects of Timing on Puberty

A label such as "late bloomer" is often used to describe adolescents whose physical maturation lags behind that of the majority. Such labels are so much a part of how we talk about development that it is easy to classify adolescents as early, average, or late maturers. In fact, if we only look at physical signs, classification is simple. However, this would fail to take into account what is probably the most important factor: what adolescents themselves think about the timing of their puberty (Williams & Dunlop, 1999; Blyth & others, 1985; Brack & others, 1988; Brooks-Gunn, 1988; Gargiulo & others, 1987). Indeed, normalcy is in the eye of the beholder!

Adolescents compare themselves to their own circle of peers (Finkelstein & others, 1999; Williams & Dunlop, 1999; Siegel & others, 1998; Gargiulo & others, 1987) and form judgments about themselves based on that information. For example, a boy who is the smallest on the football team may perceive himself as being a late maturer, when another boy the same size whose friends are not football players might see himself as average. A study of female ballet dancers found that the girls did not see themselves as late developers (although their bodies would be classified as such on the Tanner scales). This was because their reference group was other ballet students (Gargiulo & others).

AN APPLIED VIEW

What Was Your Reaction to Wet Dreams?

Typical Responses of College Men

"I don't remember."

"I remember thinking, 'Anything that feels this good must be a sin.' Sure enough, a friend told me he asked a priest, and was told that if you want it to happen, that's a mortal sin. I really hated that guy for telling me!"

"I was really embarrassed. I couldn't think of a way to keep my mother from seeing it on my sheets."

"I thought it was strange, but my older brother said it was normal and after that it was no big deal."

"I enjoyed it, although I was sort of curious about why it was happening."

"I thought they were fun! I didn't feel any anxiety, but I wondered what my mother must have thought."

"They made me very anxious, and I just wished they would stop!"

> Did you feel positively, negatively, or neutral about the timing of your own puberty?

How the adolescent appears to peers, family, and community will affect how she or he will be treated by those people (Zakin & others, 1984). The adolescent who looks older will usually be treated as being older, even by individuals who know the adolescent's actual age. Since physical maturity does not equal cognitive maturity (Orr & others, 1988), being treated as an adult can be disruptive for both males and females.

There are some differences in the way that timing of pubescent change affects boys and girls. Studies show that for "early" boys, pubescent traits such as being more muscular, taller, and having a deeper voice are perceived (by adolescents) to be positive (Brack & others, 1988). However, a recent study of delinquent behaviors in boys suggests that both early and late maturers reported a wider range of delinquent behaviors than their more typically developing peers Williams & Dunlop, 1999).

For "early" girls, whose pubertal changes include larger breasts and hips and increased body fat, these changes are often perceived negatively (Brooks-Gunn, 1988; Rierdan & others, 1988; Zakin & others, 1984). Early-maturing girls often have poor body images because they compare themselves to adult ideals (Brooks-Gunn, 1988; Duncan & others, 1985). Sometimes early-maturing girls engage in behaviors such as dating (Kim & Smith, 1998) and smoking that are more characteristic of late adolescents. In doing so, they often believe they are joining a reference group that will help them feel more normal. Some research has found that both male and female early-maturing adolescents are more likely to be interested in the opposite sex (Kim & Smith, 1998; Smith & others, 1985). It is difficult to tell if these adolescents seek out such experiences or if they simply attract those experiences to them because of their appearance.

EIGHT DIFFERENT 14-YEAR-OLDS

> **normal range of development**
> The stages of pubertal change occurring at times that are within the normal range.

In this section, eight adolescents, four males and four females, are compared to illustrate the differences that often occur among children even though they are all in the **normal range of development**. Each adolescent is 14 years old. The first female and male are early maturers, the second are average maturers, and the third are late maturers. They all fall within the typical range of all adolescents. The fourth female and male represent average adolescents of one hundred years ago.

The Early-Maturing Female: Ami

At 5 feet 5 inches and 130 pounds, Ami is considerably bigger than her agemates. Her growth accelerated when she was 8 years old, and by the time she was 10, her maximum growth spurt crested. She is still growing taller but at a slower rate. Her motor development (coordination and strength) had its greatest rate of increase two years ago. She is stronger than her agemates, but her strength and coordination have reached their maximum.

She started menstruating three years ago, at age 11, and her breasts are already in the secondary (adult) stage. Her pubic and underarm hair are also at an adult stage.

Ami is confused about the way her body looks. She feels conspicuous and vulnerable because she stands out in a crowd of her friends. Her greater interest in boys, and their response, often causes conflicts with other girls. They envy the interest the boys show in her more mature figure. She often has negative feelings about herself because she is "different." Other girls tend to avoid her now because her early maturation makes her seem older than they are. In later adolescence, she may experience some difficulties; she may find herself in situations (such as with drugs, sex, or drinking) for which she is not yet prepared. She may also find that her dancing partners are not as tall as she is.

In summary, we can say that although Ami is experiencing difficulties with her early maturity, she will begin to feel better about herself as she approaches age 16.

The Average-Maturing Female: Beth

Although Beth is also 14, she is different in almost every way from Ami. She represents the typical adolescent today in the sense of being average in her measurements and physical change. It is clear that there is no "average" adolescent from the standpoint of personality and behavior.

Beth is 5 feet 3 inches tall and weighs 120 pounds. She reached her maximum growth spurt two years ago and is also starting to slow down. She is presently at the peak of her motor development.

Her breasts are at the primary breast stage; she is beginning to need a bra, or thinks she does. She started menstruating two years ago. She has adult pubic hair, and her underarm hair is beginning to appear. She feels reasonably happy about her body, and most, but not all, of her relationships with her peers are reasonably satisfying. Although she does have some occasional emotional problems, they are not related to her physical development as much as are Ami's.

The Late-Maturing Female: Cathy

Cathy is at the lower end of the normal range of physical development for a 14-year-old girl. She is only 4 feet 8 inches tall, weighs 100 pounds, and is just beginning her growth spurt. She is not too happy about this, because she feels that the other girls have advantages in relationships with boys.

Cathy's breasts are at the bud stage; her nipples and encircling areolae are beginning to protrude, but she is otherwise flat-chested. She has just begun menstruation. Pubic hair growth has started, but as yet no hair has appeared under her arms.

Other girls tend to feel sorry for her, but they also look down on her. She is more dependent and childlike than the others. She feels a growing dislike for her body, and she is becoming more and more introverted and self-rejecting because of it. At this stage her immaturity is not a great disability; at least she is more mature than some of the boys her age. As she reaches later adolescence, her underdeveloped figure may be a more serious source of unhappiness for her if she accepts conventional standards of sexual desirability.

The Average Adolescent Female of One Hundred
Years Ago: Dorothy

Although records of adolescent physical development of one hundred years ago are inadequate, we can be certain about some of the data. Dorothy, who was typical for her time, was physically much like Cathy is now. At 4 feet 7 inches, she was 1 inch shorter, and at 85 pounds, she weighed 15 pounds less than Cathy.

At age 14, Dorothy would still have had four years to go before her peak of motor development, and she would not have started to menstruate for another year. In all other physical ways she looked a great deal like Cathy. The major difference between the two girls is that while Cathy is unhappy about her body's appearance, Dorothy, who was typical, felt reasonably good about hers.

The Early-Maturing Male: Al

Al finds that at 5 feet 8 inches tall, he towers over his 14-year-old friends. He reached his maximum growth spurt approximately two years ago and weighs 150 pounds. He is now about two years before the peak of his motor development. His coordination and strength are rapidly increasing but, contrary to the popular myth, he is not growing clumsier.

As adolescents reach their peak of motor development, they usually handle their bodies better, although adults expect them to have numerous accidents. It is true that when one's arms grow an inch longer in less than a year, one's eye-hand coordination suffers somewhat. However, the idea of the gangling, inept adolescent is more myth than fact.

Al's sexual development is also well ahead of that of his age-mates. He already has adult pubic hair, and hair has started to grow on his chest and underarms. He began having nocturnal emissions almost two years ago, and since then the size of his genitals has increased almost 100 percent.

Because our society tends to judge male maturity on the basis of physique and stature, Al's larger size has advantages for him. He is pleased with his looks, although once in a while it bothers him when someone treats him as though he were 17 or 18 years old. Nevertheless, he uses the advantages of his early maturity whenever possible.

His friends tend to look up to him and consider him a leader. Because size and coordination often lead to athletic superiority, and because success in school sports has long meant popularity, he has the most positive self-concept of all the adolescents described here, including the females. He has a good psychological adjustment, although he is sometimes vain, and is the most confident and responsible of this group. He engages in more social activities than the others, which also occasionally gets him into trouble, because he is not psychologically ready for some of the social activities in which he is permitted to participate.

The Average-Maturing Male: Bob

Interestingly, Bob is exactly the same height (5 feet 3 inches) as his average female counterpart, Beth. At 130 pounds, he outweighs her by 10 pounds. He is currently in the midst of his maximum growth spurt and is four years away from reaching the peak of his coordination and strength.

His sexual development began about a year ago with the start of nocturnal emissions, and he is just now starting to grow pubic hair. As yet he has no hair on his chest or under his arms. However, his genitals have reached 80 percent of their adult size.

Bob gets along well with his age-mates. He is reasonably happy with the way his body has developed so far, although there are some activities that he wishes he could excel in. Most of the attributes that he aspires to are already possessed by Al, whom he envies. This causes few problems, because Bob still has every reason to hope his body will develop into his ideal physical image.

The Late-Maturing Male: Chuck

Chuck is also similar in stature to his counterpart, Cathy. They are both 4 feet 8 inches tall, although at 90 pounds, Chuck is 10 pounds lighter than Cathy. He is a year and a half away from his maximum growth spurt and must wait six years before his motor development will peak.

Chuck's sexual development is also lagging behind those of the other two boys. His genitals are 50 percent larger than they were two years ago, but as yet he has no pubic, chest, or underarm hair. He has not yet experienced nocturnal emissions, although these are about to begin.

Of the adolescents described here, Chuck is the least happy with his body. His voice has not yet changed, and he is much smaller and not as strong and coordinated as the other boys. They tend to treat him as a scapegoat and often ridicule him. He chooses to interact with boys who are younger than himself and is attracted to activities in which mental rather than physical prowess is important, such as chess and band. He avoids girls, almost all of whom are more physically mature than he. This lack of heterosexual experience may later affect his self-concept.

Chuck lacks confidence in himself and tends to be dependent on others. He was of almost average size in grammar school and now feels he has lost prestige. He frequently does things to gain the attention of others, but these actions seldom bring him the acclaim he craves. Probably as a result, he is more irritable and restless than the others and engages in more types of compensating behaviors.

The Average Adolescent Male of One Hundred Years Ago: Dan

At 4 feet 7 inches, Dan was shorter than Chuck by 1 inch, and weighed the same, 90 pounds. He trailed Chuck in sexual development by two years and, at age 14, his genitals had increased only 20 percent in size.

However, the major difference between Chuck and Dan lies in self-satisfaction. While Chuck is extremely unhappy about the way he is developing, Dan was as happy as Bob is now because he was quite average for that time. Although we cannot know what Dan's relationship with peers was like or how he viewed himself, we can guess that these were quite similar to Bob's.

The preceding descriptions illustrate the great variability in adolescent growth and in adolescent responses to growth. It should be kept in mind, however, that self-image and peer relationships are not entirely determined by physique. Reaching puberty "on time" does not guarantee a charmed life. Adolescents who have clarified their values and set their own standards are not likely to be overly affected by the timing of pubertal changes or the peer approval or disapproval brought about by that timing.

Figure 3.7 details the age ranges considered normal for development. Adolescents who experience these changes earlier or later may have no medical problem, but it is probably a good idea to consult a doctor. If a glandular imbalance exists, the doctor can usually remedy the problem with little difficulty.

Body Image

body image
How people believe they look to others.

Body image (how people believe they look to others) is an important part of a person's self-concept and self-esteem (see Chapter 6). This is especially true during adolescence. Adolescents of both sexes are concerned about the appearance of their bodies (Botta, 1999; Nowak, 1998; Dacey, 1986; Elkind, 1984; Winship, 1991). Some investigators have found that adolescents' judgments of their physical appearance is the *most* important factor in their self-esteem (Harter, 1990a; Simmons & Blyth, 1987). Susan Harter quotes one of the girls in her study: "What's really important to me is how I look. If I like the way I look, then I really like the kind of person I am" (Harter, p. 367). Studies done in Turkey, England, and Finland show results similar to those obtained with American teenagers (Çok, 1990; Davies & Furnham, 1986; Wright, 1989).

At this time of rapid body change, young people are most apt to be dissatisfied with their appearance (Attie & Brooks-Gunn, 1989; Brooks-Gunn & Reiter, 1990; Harter, 1990a; Koff & others, 1990; Wright, 1989). Brooks-Gunn and Warren (1988) found that breast development has a positive influence on the

Figure 3.7
Age ranges of
normal adolescent
boy development.

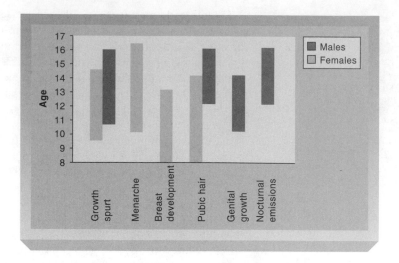

AN APPLIED VIEW

Dealing with Early or Late Development

Adolescents' perceptions of their bodies during puberty have been demonstrated to be affected, sometimes quite negatively, by their own and others' assessments. Peterson (1983) mentions an increased risk of sexual abuse for the early-maturing female. Significantly early or late development has been linked to depression and eating disorders in both boys and girls (Siegel & others, 1998; Rierdan & others, 1988), so it is important for

practitioners working with adolescents to be aware of these reactions to adolescent physical change (or lack of it). If you know a teen who is significantly early or late in body development, be on the lookout for psychological problems. If you find evidence of such problems, make arrangements for the youth to have access to appropriate professional attention.

body image of girls just entering puberty, but suggest that the normal increase in body fat that occurs at this time has a negative effect. Older adolescents are more likely to accept and approve of their bodies, but unhappiness with their physical appearance is still very common among girls in the later teen years (Attie & Brooks-Gunn, 1989; Heilbrun & Friedberg, 1990; Klemchuk & others, 1990).

Many studies indicate that during adolescence, looks become more critical for girls than for boys (Allgood-Merten & others, 1990; Simmons & Blyth, 1987; Steiner-Adair, 1990). Girls are far more likely to believe themselves to be too fat, when in fact their weight is normal for their height. Girls are also far more likely to be dieting to lose weight. In one study, which included 1,373 high school girls and boys of diverse racial, geographical, and economic backgrounds, girls were four times more likely than boys to be trying to lose weight (Rosen & Gross, 1987).

According to your own observations, was body image more important when you were an adolescent or is it more important today?

Though girls may be more likely to be on diets than boys, recent studies (Nowak, 1998; Pope & others, 1999) indicate that both boys and girls are sensitive to issues of body image and societal expectations. Nowak (1998) found that boys trying to lose weight tended to reduce their intake of sweets and increased their intake of healthy snacks. Girls generally reduced their overall intake of food, skipping meals and avoiding core foods. Overall Nowak found that both the boys and girls involved in the study were dissatisfied with their bodies and considered themselves overweight.

In spite of feminism's criticisms in the 1960s and 1970s that women are treated as sex objects, girls today still consider their physical attractiveness their most important asset (Brooks-Gunn & Reiter, 1990; Koff & others, 1990; Pierce, 1990; Sidel, 1990). Because girls place so much value on appearance, it is not surprising that almost every study finds that adolescent girls at every stage are more dissatisfied with their bodies than are boys (e.g., Freedman, 1986; Pliner & others, 1990; Rosen & Gross, 1987; Steiner-Adair, 1990; Wright, 1989). As a result, their self-esteem suffers. Adolescent girls have lower self-esteem than boys and generally feel less attractive (Allgood-Merten & others, 1990; Pliner & others, 1990).

Erikson's (1968) discussion of the formation of identity and self-concept in adolescence (see Chapter 6) suggests that many females view their bodies primarily as a means of attracting others, while males focus on competence and strength (Lerner & others, 1976). Gilligan (1982) would argue with Erikson's viewpoint. According to her, boys are socialized to be more competitive, girls are socialized to be concerned with relationships. Gilligan claims that Erikson's theory is based on the male model and, therefore, does not take into account the developmental needs of females. Gender differences in socialization, then, contribute to different views regarding body image. It is not the case that adolescent females simply view their bodies as a means of attracting others. In all aspects of their lives, adolescent females frame their identity on their relationships with other people in their lives.

Other recent studies, such as that by Koff and others (1990), confirm that boys in early adolescence are increasingly pleased with their bodies as they grow, while the opposite is true for young girls. It is not known whether poor body image is one of the causes or one of the results of teenage depression, but it is almost always present in depressed teens (Cyranowski & others, 2000; Allgood-Merten & others, 1990; McCarthy, 1990; Noles & others, 1985; Rierdan & others, 1988).

THE CULTURAL PREFERENCE FOR THINNESS

Teenagers in some activities (such as ballet) and jobs (such as modeling) where extreme thinness is a requirement are particularly at risk, both for inadequate nutrition and for eating disorders (Benson & others, 1989; Braisted & others, 1985; Carruth & Goldberg, 1990; Lundholm & Littrell, 1986; Rucinski, 1989). Stein and Reichert (1990) suggest that worrying about weight and dieting have become so common among adolescents that such concern is considered "normal."

Although their bodies may be within the normal range in every way, some adolescents perceive body image negatively. This is most likely to happen to those who have early or late physical development.

The current cultural ideal of the female body in Western countries demands extreme slenderness. There is much evidence to suggest that there is an increase in incidence of eating disorders with our societal focus on low fat, low cholesterol diets (Hartley, 1999). The belief that "fat is bad, thin is beautiful" is learned long before adolescence. Maloney and others (1989) report that children as young as five believe it is important to avoid being fat. They found that 45 percent of the 318 third to sixth graders in their study wanted to lose weight, and 37 percent had tried to do so. As Striegel-Moore and others (1986) put it, "from childhood, girls are encouraged to be concerned with their appearance, and may even worry about their weight" (p. 250).

Body images promoted by the entertainment and fashion industries emphasize the importance of physical appearance and provide standards that are unrealistic for most teenagers. The media promotes the equation of thinness with perfection (Hartley, 1999). Hours of watching the popular MTV (Music Television) channel, which stresses physical features over personal attributes and qualities, can leave little doubt in the adolescent's mind about what is attractive in both young men and women (Kaplan, 1990). Magazines aimed at an audience of young girls continue to focus on the significance of physical attractiveness in both advertising and editorial content, as a recent study by Pierce (1990) empirically confirms.

Only a minority of young girls are directly involved in the subculture of beauty pageants, but many more are affected by watching them on television. Beauty pageants remain profitable for the sponsors, the participants, and the media. An estimated 55 million Americans, 75 percent of whom are females, tune in each year to the Miss America Pageant. In addition to this top pageant, the Miss Universe, Miss USA, Miss Teen USA, and all the state and local pageants that serve as preliminary eliminations required for participation at the national level, add up to a very large number of young women in bathing suits walking up and down runways and being judged almost solely on how they look. For example, in 1987 some 80,000 contestants tried to qualify for the Miss America pageant in local and state contests. Although a talent of some kind is a requirement for participation,

> ❑ No matter how well they juggle or sing or play "Malagueña" on the accordion, we all know that it is their measurements and pretty faces and how well they turn that are really being judged. And the message is not lost on the young women who are watching and trying to figure out who they are, what they want to be, and how they will get there (Sidel, 1990, p. 27).

THE ATHLETIC BODY

At the age of 15, gymnast Christy Henrich narrowly missed making the United States Olympic team. Following the trials, a judge told Christy that she should lose a few pounds—at the time, Christy was 4 feet 10 inches and weighed 90 pounds! Comments such as these led to Christy's having anorexia and bulimia. At one point Christy was down to 47 pounds. In July 1994, as her parents and coach were helping her recover, Christy died of multiple organ failure at the age of 22. Her coach claims that gymnastics and its emphasis on slimness are not

A SOCIOCULTURAL VIEW

Dyeing to Win

In the drive to be chosen "most beautiful," it may be necessary to transform not only one's body but also one's ethnic image. In the Miss California pageants during the late 1980s, Marlise Ricardos, a young Latino woman, tried three times to win the title, but was unsuccessful. It was only after she changed her hair color from brunette to blond and wore blue contact lenses when she competed that she was selected to represent California in the 1988 Miss America contest in Atlantic City. As journalist Anne Taylor Fleming observed, "She's the ultimate self-made competitor, a chemically 'sun-streaked' miss who rid herself of both pounds and ethnic identity to please pageant judges" (Sidel, 1990, p. 23).

On the other hand, more members of minority groups *have* been winning pageants in recent years as evidenced by the multiracial background of the 1995 Miss USA, Chelsi Smith. Are these contests more fair, or are they as racist as they used to be?

At the same time, Heather Whitestone, Miss America 1995, is deaf. She communicates verbally and through sign language. She is the first Miss America to have a significant physical disability. Because of her role, Ms. Whitestone traveled around the country and, through this experience and exposure, made many people more aware of, and sensitive to, people living with physical challenges. Does this example indicate that American society is becoming more open to the reality that people who are physically challenged deserve opportunities to achieve? Or is Ms. Whitestone likely to be just a rare exception?

responsible for Christy's illness and death. He claims her intense drive to win contributed to her eating disorder. The gymnastic training system encourages young female athletes to keep their weight down, and many gymnasts live on diets of fruits, laxatives, and painkillers. In 1976 the average age of a U.S. gymnast was 18, with an average weight of 106 pounds. In 1992, the average age of a U.S. gymnast was 16, with an average weight of 83 pounds—a drop of 23 pounds! Cathy Rigby, a former gymnast who won a gold medal at the 1968 Olympics, fought a 12-year battle with anorexia and bulimia. Now recovering from her illness, Cathy states that gymnasts are trained to be slim through fear, guilt, and intimidation (Ryan, 1995). This trend appears to continue. In a recent study of college athletes, Picard (1999) found that females involved in sports that stress a lean figure, place restrictions on weight, and compete at higher levels demonstrated more symptoms of disordered eating and were at higher risk for developing eating disorders.

Adolescents engaged in athletic activities that require a particular body shape have been studied by developmental psychologists. Such athletes include runners, Nordic skiers, ice skaters, cheerleaders, gymnasts and dancers (Garner & others, 1998; Benson & others, 1989; Braisted & others, 1985). The demands are quite clear (especially in classical ballet): Participants must maintain a very low body weight to conform to the standards, and must also devote many hours to practice to maintain those standards. Female dancers are more likely than non-dancers to be late maturers, and many of them fail to have normal menstrual periods. Whenever the requirements call for thinness, many youths rely on poor nutrition and other unhealthy practices such as laxatives and vomiting (see Chapter 11) to achieve a lower than normal weight (Carruth & Goldberg, 1990; Lundholm & Littrell, 1986; Rucinski, 1989). Such athletes are at elevated risk

AN APPLIED VIEW

Look Before You Leap

In each of the Applied View boxes in this chapter, we have recommended a course of action that you might take with teenagers who have some physical problem. We believe it is very important that you be sensitive to physical as well as other needs. However, speaking to adolescents about their bodies can be tricky.

During the vulnerable teen years, adolescents are especially sensitive about the topic of physical development. Such discussions are further complicated by gender roles. Obviously, most adolescent girls find it harder to talk about menstruation with a man than with a woman, and most adolescent boys find it harder to talk about nocturnal emissions with a woman than with a man (although they find it hard to talk about wet dreams with anybody!). The possibility that you might be misunderstood in these areas is even greater if you are not the same gender as the teen that you wish to counsel. For example, despite your most sincere intentions, a teen might view your behavior as sexual harassment.

Therefore, we advocate that you "look before you leap" in this aspect of dealing with teens. That is, we think you should be especially cautious in the way you handle such discussions. It is usually a good idea to get the advice of someone who is more experienced than you before tackling this delicate problem area.

for eating disorders. Garner and others (1998) recommend that adults working with these athletes should be cognizant of warning signs and that these warning signs should never be ignored. The impact of the ideal body image though often focused on females has not been limited to adolescent girls. Within the past few years a number of college wrestlers have died as a result of techniques that were commonly used by competitive wrestlers to try to drop weight categories (which result in self-starvation and dehydration). These tragic deaths became the impetus for NCAA rule changes in wrestling at the college level. Although Dale and Landers (1999) who studied weight related behaviors in wrestlers found that concerns with weight seemed tied entirely to the demands of wrestling and did not generalize into eating disorders beyond the season. They did find that the in-season disordered eating behaviors were similar to those of individuals suffering from bulimia.

> What other problems, psychological as well as physical, can teenagers have when trying to maintain an athletic body?

In an interesting study addressing male body image, Pope and others (1999) suggest that by looking at children's action figures we can see major changes in societal ideals of male body image over time. These researchers found that over the last 30 years, popular male action figures have grown much more muscular. Presently, if considered proportionally to the real human body, these toys exceed the size of the largest body builders.

Has all this information about physical development made you think about your own body image? Perhaps you would like to have a more exact picture of how you feel about your body. If you are counseling teenagers, perhaps you would like a more precise means of measuring their body images. The two Applied View boxes titled "How I Feel About My Body" provide a means for comparing aspects of your body with those of others, and for recording how you feel about each of these comparisons. After you complete this exercise, compare the placement of your marks in each of the two columns. Do you feel good about the comparison? Are you being realistic?

AN APPLIED VIEW

How I Feel About My Body (Female)

My age _____   My height _____   My weight _____   My waist size_____

	Compared to Others			How I Feel About It		
	Larger than average	Average	Smaller than average	Good	Bad	Neither
1. My height	_____	_____	_____	_____	_____	_____
2. My weight	_____	_____	_____	_____	_____	_____
3. My bust	_____	_____	_____	_____	_____	_____
4. My hips	_____	_____	_____	_____	_____	_____
5. My legs	_____	_____	_____	_____	_____	_____
6. My nose	_____	_____	_____	_____	_____	_____

	Compared to Others			How I Feel About It		
	Better than average	Average	Less than average	Good	Bad	Neither
7. The quality of my hair	_____	_____	_____	_____	_____	_____
8. The beauty of my eyes	_____	_____	_____	_____	_____	_____
9. My face in general	_____	_____	_____	_____	_____	_____
10. My figure in general	_____	_____	_____	_____	_____	_____
11. My clothes	_____	_____	_____	_____	_____	_____
12. My cleanliness	_____	_____	_____	_____	_____	_____
13. My posture	_____	_____	_____	_____	_____	_____
14. My sports ability	_____	_____	_____	_____	_____	_____
15. How coordinated I am	_____	_____	_____	_____	_____	_____
16. The quality of my skin	_____	_____	_____	_____	_____	_____
17. My strength	_____	_____	_____	_____	_____	_____
18. My teeth	_____	_____	_____	_____	_____	_____
19. My overall looks	_____	_____	_____	_____	_____	_____

AN APPLIED VIEW

How I Feel About My Body (Male)

My age _____   My height _____   My weight _____   My waist size_____

	Compared to Others			How I Feel About It		
	Larger than average	Average	Smaller than average	Good	Bad	Neither
1. My height	_____	_____	_____	_____	_____	_____
2. My weight	_____	_____	_____	_____	_____	_____
3. My nose	_____	_____	_____	_____	_____	_____
4. My penis	_____	_____	_____	_____	_____	_____
5. The amount of hair on my face	_____	_____	_____	_____	_____	_____
6. The amount of hair on my body	_____	_____	_____	_____	_____	_____

	Compared to Others			How I Feel About It		
	Better than average	Average	Less than average	Good	Bad	Neither
7. My teeth	_____	_____	_____	_____	_____	_____
8. The quality of my hair	_____	_____	_____	_____	_____	_____
9. My eyes	_____	_____	_____	_____	_____	_____
10. My face in general	_____	_____	_____	_____	_____	_____
11. My strength	_____	_____	_____	_____	_____	_____
12. My physique (build)	_____	_____	_____	_____	_____	_____
13. My voice	_____	_____	_____	_____	_____	_____
14. My chest	_____	_____	_____	_____	_____	_____
15. My clothes	_____	_____	_____	_____	_____	_____
16. My cleanliness	_____	_____	_____	_____	_____	_____
17. My sports ability	_____	_____	_____	_____	_____	_____
18. How coordinated I am	_____	_____	_____	_____	_____	_____
19. The quality of my skin	_____	_____	_____	_____	_____	_____
20. My overall looks	_____	_____	_____	_____	_____	_____

C O N C L U S I O N S

We wish that all children could go through the period of puberty and reach the other side with a normal, healthy body and body image. We wish they could negotiate adolescence so successfully that they become energetic, self-confident adults.

Unfortunately, we know that this is not always the case. Some youths suffer from a negative self-concept because, although they differ from the norm only slightly in their body development, they perceive this as catastrophic. Others really do deviate significantly from the norm because of some physiological problem. Of particular concern for females is our society's obsession with thinness. Taken together, the various aspects of puberty can cause the adolescent quite a bit of chagrin.

The only solution is to give adolescents some perspective on these problems. The good news is that just when they need it, most adolescents develop improved mental abilities that enable them to view themselves more realistically. In the next chapter, we explore this exciting new development.

C H A P T E R   H I G H L I G H T S

Early Studies of Puberty

Theories of adolescence in the early 20th century were largely based on personal bias because little empirical data existed.

The Common Myths of Puberty

- Puberty takes place over several years and has biological, social, mental, and emotional aspects.
- Growth is continuous from infancy, and the changes of puberty are more gradual than many believe.
- It is difficult to determine which aspects of puberty are strictly biological.

Your Reproductive System

- Those who work with adolescents need complete knowledge of the reproductive systems of both sexes.
- The structure and functions of female and male systems are explained and illustrated.

The Biological Basis of Puberty

- The pituitary hormones, or gonadotropins, regulate the growth of the reproductive systems.
- The physical aspects of puberty include changes in height, bone density, height/weight distribution, and the development of secondary sex characteristics and reproductive capacity.
- Studies of identical and fraternal twins show that many pubertal changes are directly affected by genes.
- Good nutrition has produced the secular trend, which refers to, among other things, a steady drop in the average age of menarche over the last 100 years. This trend now appears to be leveling off.
- Poor nutrition, sometimes caused by eating disorders, can delay or suppress menstruation.

The Sequence of Events in Puberty

- The order of physical changes in puberty is largely predictable, but the timing and duration of these changes are not.

The Contrast of Male and Female Pubertal Change

- The pubertal growth spurt occurs almost two years earlier in girls than in boys.
- No event in male development is directly comparable to the female menarche.
- Genital development is somewhat earlier in girls than in boys.
- Tanner's Sexual Maturity Rating Scale, often presented as a series of pictures, helps to identify the adolescent's development in each area.

The Effects of Timing on Puberty

- The normal range in pubertal development is very broad, and includes early, average, and late maturers.
- The adolescent's own perception of being normal has more influence on self-esteem than does objective normality.
- Maturity of appearance affects whether adolescents are treated appropriately for their age.
- Early maturing is a positive experience for boys, but may be negative for girls.
- Late maturing is difficult for both boys and girls.

Body Image

- Self-concept and self-esteem are strongly affected by how we believe others see us, especially in adolescence.
- Adolescents are often highly critical of their own and others' physical appearance.

- Girls are far more likely than boys to suffer from a poor body image, particularly in the area of weight.
- Boys are most concerned with appearing strong and muscular.
- Participation in athletics and dance may lead to excessive concern about being the proper body type and may carry dangers of unhealthy nutritional practices or drug use.

K E Y   T E R M S

amenorrhea 76
body image 89
delayed puberty 67
dysmenorrhea 82
eating disorders 75
endocrine system 68

gonadotropins 73
gynecomastia 84
hormones 68
menarche 79
normal range of development 85
pituitary gland 73

premenstrual syndrome (PMS) 82
puberty 67
secular trend 75
testosterone 73
varicocele 84

W H A T   D O   Y O U   T H I N K ?

1. Should children be taught about their body functions in school? Should this teaching include sexuality? If so, at what grade should it start?

2. What was the beginning of puberty for you? Why do you think so?

3. Can you think of any myths about puberty other than the ones described in this chapter?

4. The sequence of events in puberty is rather fixed. Why do you think this is so?

5. What, if anything, should be done about the cultural preference for thinness?

6. What body image concerns did you have as a teen? Ask your parents and grandparents about theirs. Do differences between your attitudes and their attitudes indicate that social-cultural norms of the times affect how adolescents feel about their bodies?

S U G G E S T E D   R E A D I N G S

Blume, J. (1970). *Are you there, God? It's me, Margaret*. New York: Bradbury. Although written for teens, this book has a wealth of insights into pubertal change, at least for females. Blume really understands.

Brooks-Gunn, J., Petersen, A., & Eichorn, D. (Eds.). (1984). Time of maturation and psychosocial function in adolescence. *Journal of Youth and Adolescence*, 14 (3 & 4). A much more detailed explanation of the physical side of puberty than has been offered in this chapter.

McCoy, K., & Wibbelsman, C. (1987). *The teenage body book*. Los Angeles: The Body Press. An excellent reference book for teenagers and those who work with them!

G U I D E D   R E V I E W

1. All of the following are part of both the male and female reproductive systems, EXCEPT the
   a. urethra.
   b. vas deferens.
   c. ureter.
   d. pituitary gland.

2. The opening of the uterus is called the
   a. clitoris.
   b. cervix.
   c. hymen.
   d. vulva.

3. What part of the male sexual system produces sperm and testosterone?
   a. testes
   b. scrotum
   c. prostate glands
   d. Cowper's glands

4. Hormones are chemicals that act as messengers to trigger physical change. Hormones are released by the
   a. pituitary gland.
   b. vas deferens.
   c. endocrine system.
   d. prostate glands.

5. Which of the following statements regarding puberty is TRUE?
   a. Puberty strikes without warning.
   b. Puberty is the result of raging hormones.
   c. Puberty starts at one point in time.
   d. None of the answers is correct.

6. What gland is also known as the master gland?
   a. pituitary gland
   b. Cowper's gland
   c. Bartholin's gland
   d. prostate gland

7. Adolescents who are more dependent and childlike than others, who feel a growing dislike for their bodies, and who become more introverted and self-rejecting because of this dislike, are likely to be
   a. early-maturing males.
   b. early-maturing females.
   c. average-maturing males.
   d. late-maturing females.

8. Which of the following is true of the secular trend?
   a. It refers to adolescents entering puberty at a later stage.
   b. It will continue at its current pace.
   c. It is the result of improvements in nutrition, medicine, and health care.
   d. It is relatively old.

9. An anorexic girl will generally stop menstruating once she loses what percentage of her body weight?
   a. 5 percent
   b. 15 percent
   c. 25 percent
   d. 40 percent

10. Which of the following is NOT a myth of puberty?
    a. Puberty occurs without warning.
    b. Puberty is the result of high hormone levels.
    c. Puberty involves more than just hormonal changes.
    d. Puberty starts at one point in time.

11. The clitoris is comparable to the male _____ (penis/testes).

12. In boys the increased level of LH eventually stimulates the testes to produce the hormone called _____ (estrogen/testosterone).

13. The _____ (vulva/clitoris) is the external genital organ of the female.

14. Whereas pubertal development in females is associated with increases in body _____ (muscle/fat), in males the pubertal weight spurt is due primarily to increased body _____ (muscle/fat).

15. The onset of menstruation is called _____ (menarche/amenorrhea).

16. With regard to the effects of timing on puberty, the _____ (early-maturing male/late-maturing female) has the most positive self-concept.

17. The advent of female fertility usually occurs _____ (at the same time as/12 to 18 months after) first menstruation.

18. In order to study the way heredity affects puberty, researchers often conduct _____ (twin studies/case studies).

19. The Sexual Maturity Rating Scale has been criticized for its implicit message that all development proceeds _____ (at a specific age/in a certain sequence).

20. There are two conditions that sometimes occur during male puberty that have received attention by the medical community: _____ _____ (dysmenorrhea and premenstrual syndrome/gynecomastia and varicocele).

Answers

1. B, 2. B, 3. A, 4. C, 5. D, 6. A, 7. D, 8. C, 9. B, 10. C, 11. penis, 12. testosterone, 13. vulva, 14. fat, muscle, 15. menarche, 16. early-maturing male, 17. 12 to 18 months after, 18. twin studies, 19. in a certain sequence, 20. gynecomastia and varicocele

COGNITIVE DEVELOPMENT

I was about twelve when I discovered that you could create a whole new world just in your head! I don't know why I hadn't thought about it before, but the idea excited me terrifically. I started lying in bed on Saturday mornings till 11 or 12 o'clock, making up "my secret world." I went to fabulous places. I met friends who really liked me and treated me great. And of course I fell in love with this guy like you wouldn't believe!

Susan Kline, an eighth-grade student

CHAPTER OUTLINE

Adolescence is a complex process of growth and change. Because biological and social changes are so much the focus of attention, changes in the young adolescent's ability to think often go unnoticed. Yet it is during early and middle adolescence that thinking ability reaches its highest level—the level of abstract thought. To understand how abstract thought develops, we have to know more about the intellect itself.

In this chapter we investigate the stages of cognitive development. We also describe research on social cognition, information processing, and critical and creative thinking.

After reading this chapter, you should be able to:

* Describe Piaget's four main stages of cognitive development.
* Explain the cognitive development that takes place in early and late phases of the formal operational stage.
* List Flavell's seven aspects of the transition from childhood thinking to adolescent and adult thinking.
* Describe the major elements of social cognition as defined in this chapter by egocentric thinking, the imaginary audience, and the personal fable.
* Explain the elements of information processing, including perception, storage, manipulation, and retrieval of information.
* Identify the differences between critical thinking and creative thinking, and discuss the importance of each.
* Show why the use of metaphor is important in adolescence, especially for creative thinking.
* Discuss these issues from an applied, a sociocultural, and your own point of view.

Stages of Cognition

THE FIRST THREE STAGES—PIAGET

cognitive structures
Specific quantitative and qualitative mental abilities that develop in stages as a child's intelligence develops.

Just as Freud was concerned with the structures of the personality, the Swiss biologist/psychologist Jean Piaget (1896–1980) sought to understand **cognitive structures** (1948/1966, 1953; Flavell, 1977, 1982).

Piaget is the foremost contributor to the study of intellectual development. Beginning his scholarly career at 11 (!), Piaget published numerous papers on birds, shellfish, and other topics of natural history. As a result, the diligent Swiss was offered the curator's position at the Geneva Natural History Museum. He was only 15 at the time and turned it down to finish high school. He received his Ph.D. in biology at the age of 21, and wrote more than fifty books prior to his death.

Piaget began his research on cognitive development in 1920 when he took a position in the Binet Laboratory in Paris. He was given the task of standardizing a French version of an English language test of reasoning ability, which enabled him to observe how children responded to the questions. He discovered that there were similarities in the wrong answers given by each age group. For example, 5-year-olds give a wrong answer for one reason, whereas older age groups give the same wrong answer for other reasons. This discovery led Piaget to the idea that children of different age groups have different thinking patterns. Prior to 1920, little research had been done on the nature of intelligence. Most scientists viewed children as miniature adults who used adult thinking methods, but used them poorly. Scientists felt that as information is poured into the child, mental maturity slowly develops. Piaget discovered that specific *abilities* must be acquired before the child's intellect can fully mature; information alone is not enough. Gradually, the child learns to manipulate information *mentally*. Furthermore, he observed that this mental ability develops in stages, each one preparing the way for the development of the next.

Piaget's background in biology formed the basis for his view of the role of intellect. If you have ever seen an episode of *The Undersea World of Jacques Cousteau*, you have probably been amazed by the way ocean dwellers biologically adapt themselves to their particular environments. The puffer fish swells to twice its size when threatened by an enemy. The anglerfish uses its dorsal fin as bait to lure smaller fish to its mouth. Nature is filled with examples of superb adaptations of animals to their environments. But nature has not fitted humans to any specific environment. It has equipped us to adapt to most environments through the gift of intellect. Like the swelling of the puffer fish, human intellect seemed to Piaget to be another of nature's examples of biological adaptation.

The Sensorimotor Stage (From Birth to About 2 Years Old)

> The Swiss psychologist Jean Piaget used his scientific training to observe the cognitive development of his children. From these observations came a very powerful explanation of how the human mind develops.

sensorimotor stage
Piaget's first stage, during which mental operations are not yet possible.

mental structures
The blueprints in our minds that equip us to affect our environment. They are the tools of adaptation.

schemata
Patterns of behavior that infants use to interact with the environment.

In the **sensorimotor stage**, intellect matures through the growth of increasingly effective **mental structures**. These structures can best be defined as the blueprints that equip us to affect our environment. They are the tools of adaptation. At birth, all babies simply reflect the environment. When a specific event (a stimulus) occurs, infants react automatically to it. Sucking and crying are examples of these reflexes.

At birth, the child possesses several **schemata**. Schemata are patterns of behavior that infants use to interact with the environment. Figure 4.1 depicts this relationship. Soon after birth, infants begin to develop new schemata for looking, for grasping, for placing objects in their mouths.

Figure 4.1
Piaget's concept of schemata. Mental structures held in the mind direct and control our behaviors.

In addition to schemata, the infant has at birth two basic tendencies or drives that affect intellectual functioning throughout his or her lifetime. They are **organization** and **adaptation**, and they govern the way we use our schemata to adjust to the demands of the environment.

organization
An innate tendency that causes us to combine our schemata more efficiently.

adaptation
An innate human tendency to adapt to the environment.

Can you think of a third basic tendency, beyond organization and adaptation, that infants have at birth?

- *Organization.* Our innate tendency to organize causes us to combine our schemata more efficiently. The schemata of the infant are continuously reorganized to produce a coordinated system of higher-order structures. When new schemata are acquired, they are integrated into existing schemata. Consider children learning to throw a ball. They may understand the various parts of a good throw, but they will not throw the ball well until they have integrated these parts into a smooth and efficient movement.
- *Adaptation.* The second tendency in all human beings is to adapt to the environment. Adaptation consists of two complementary processes: **assimilation** and **accommodation**. We assimilate when we perceive the environment in a way that fits our existing schemata. That is, we make reality fit our minds. We accommodate when we modify our schemata to meet the demands of the environment. That is, we make our minds fit reality.

assimilation
Perceiving the environment in a way that fits our existing schemata. That is, we make reality fit what already exists in our minds.

accommodation
Modifying our existing schemata because we cannot make our perception of the environment fit. What we are seeing is so new and different that we must change in order to adapt to it; that is, we learn.

We try to assimilate as much as possible because it is easier than accommodation. It is easier to see situations, events, and objects as something we understand and can work with. When asked to describe an unfamiliar object, we say: "It looks like an orange" or "It's hard like a rock." To make these comparisons, we perceive the object in a way that fits what we know—we look for similarities. In the same way,

(You are able to read these words, because you altered your perception of them mentally to suit the structures in your mind.)

You may have physically assimilated this idea by turning this book upside down. When we mentally alter what we see, we *recognize* or rethink it and thus assimilate it.

Does __ . . . __ . ._. . 88 * * * 8 __' " ? To answer this question written in code, you have to learn the key. In doing so, you are adding to your existing schemata. This is accommodation. All mental activity uses assimilation and accommodation.

It is necessary to balance these processes. A person who is incapable of assimilation, such as a mentally retarded person, does not have the capacity to take advantage of previous experiences. A person who is incapable of accommodation is unresponsive to his or her environment. The rigid schizophrenic is an example.

One final point: the processes of organization, adaptation, assimilation, and accommodation are not restricted to the sensorimotor stage. They apply equally well to all human existence. They undergo a definite refinement, however, during the adolescent years.

The Preoperational Stage
(From 2 to About 7 Years Old)

preoperational stage
Piaget's second stage, during which the ability to represent objects symbolically in the mind begins.

operations
Mental events that take the place of actual behavior.

As infants grow older and begin to encounter more elements of their world, schemata are combined and rearranged into more efficient structures. During the **preoperational stage**, higher forms of psychological structures are beginning to develop. These structures are called **operations**.

According to Piaget, adult thinking is composed of numerous operations that enable the individual to *manipulate the environment*. Operations are mental, internalized actions, similar to programs in a computer. Programs enable the computer to manipulate the data fed into it in various ways. Mental operations do the same. In fact, mental operations are able to take things apart and reassemble them without actually touching them. It is important to note that operations include the ability to *reverse* what has been done, either mentally or in actuality. This is a vital capacity of the human mind. We can think not only about how things are done, but also how they might be undone.

The onset of operations may be seen when the child imitates the behaviors of others. Imitation involves the ability to remember what someone else has done earlier, and copy that behavior. On the other hand, true operations such as *inductive reasoning* (deriving a general rule from particular actions) and *deductive reasoning* (deciding on a particular action on the basis of a general rule) have not yet developed. Instead, the preoperational child uses transductive reasoning. In this, children go from particular to particular, without trying to attain any generalizations. For example, Amy may believe that radiators burn you because she was burned by one, and that stove tops are safe because she has not yet burned herself on one.

The Concrete Operational Stage
(From 7 to About 11 Years Old)

concrete operational stage
Piaget's third stage, during which children become concerned with why things happen. The intuitive thinking style of the preoperational stage is replaced by elementary logic.

In the **concrete operational stage**, children now become concerned with *why* things happen. The intuitive thinking style of the preoperational stage is replaced by elementary logic, as children begin to develop operations that enable them to form more complex mental actions on concrete elements of their world. For example, now children can determine the total number of objects in two groups (five pencils and three pencils) by adding them together in their heads. At the beginning of the concrete operational stage, however, most children can only do this by imagining pictures of the two groups and counting the total.

Logical thinking requires an understanding of the physical properties of the world. Knowing the correct answer to the question, "Which is heavier, a pound of feathers or a pound of lead?" depends on the child's understanding of the concept of density in relation to measurement. Preoperational children cannot understand that the weight is the same, regardless of the density of the objects. The operations necessary for this understanding are developed in the concrete operational stage.

Early in this stage, children come to understand class inclusion relationships. This means they realize that tulips are a kind of flower, and that roses are in that class, too. Until the end of this stage, however, they still make mistakes about class inclusions. For example, if you ask, "If all the tulips in the world were to die, would there be any more flowers?" many children will say, "No."

When children acquire these operations, they become able to logically solve problems by the use of elementary deductive reasoning (deciding on a particular action on the basis of a general rule). However, their thinking style still needs refinement. Ask the 9-year-old, "How would things be different if we had no thumbs?" and he is likely to respond, "But we *do* have thumbs!" The concrete operational child does not consider possibilities that are not real. The tools of thought are assembled but still need the refinement that takes place in the formal operations stage.

THE FORMAL OPERATIONS STAGE—PIAGET

formal operations stage
Piaget's fourth stage, during which children are able to perform abstract operations entirely in their minds.

During adolescence, thinking style takes flight. It is called the **formal operations stage** because adolescents begin to be able to think about the *form* an argument may take, and not just its content. In his original thinking, Piaget (1948/1966) felt that this stage takes place from around 11 to about 14 years old. Later Piaget (1972) came to believe that it only begins in these years, and he hypothesized a second stage, from 15 to about 19 years old, during which further refinements take place.

early formal operations stage
The first stage of formal operations (from 11 to about 14), in which abstract thought, logic, metacognition, and hypothetical reasoning occur.

The Early Formal Operations Stage

The **early formal operations stage** begins in adolescents who are from 11 to about 14 years old. Formal operations, at the early and at the later stages, consist of four major aspects of human thinking: **abstract thinking; logic; metacognition; and hypothetical reasoning**. Let us first discuss these four as they occur in early adolescence.

abstract thinking
Thinking that allows reality to be represented by symbols that can be manipulated mentally.

Abstract Thought Formal operations expand thought to the abstract. Think for a moment about the meaning of the word *abstract*. It's hard to define, isn't it? Here are two simple examples of it. When a child is young, she learns to describe a number of objects as red. Only slowly does she come to *abstract* the concept of redness from these specific instances. Now think of the abstract concept of democracy. If you dig deep enough into its definition, you will inevitably come to such concrete concepts as placing a marked ballot in a voting box, and telling your local representatives how you want them to vote. *Democracy* is an abstraction of all these actions.

logic
Thinking that is more orderly and systematic.

metacognition
"Thinking about thinking," or being able to analyze one's own thoughts.

In learning to deal with abstractions, we are actually allowing reality to be represented by symbols that can be manipulated mentally. This is similar to the way that data is represented by electromagnetic code that can be programmed in the computer. Early adolescents begin to understand the complexities of symbol systems such as music and math. They realize that words can have double meanings. There is a definite lack of sophistication, however, and they often misunderstand subtler meanings.

hypothetical reasoning
Forming conclusions based on hypothetical possibilities.

AN APPLIED VIEW

Defining Democracy

In order for adolescents to improve their ability to perform formal operations, they need to be aware of what abstract thinking is. Furthermore, they need to have plenty of practice in doing it. You can help them with both of these tasks by asking them to explain the meaning of such abstract terms as *citizenship, truth, beauty, honor, creativity,* and *adolescence.*

Suppose we use the term *democracy.* Ask a group of students to suggest examples of democracy; try to get fifteen or twenty examples. Have them group these ideas in three piles: those that are clearly democratic, those

that are mostly democratic, and those that are less democratic. Now ask the students to list the criteria that distinguish between the first and last pile. In this way, through induction, they will better understand democracy.

As you will see, the definitions of each of the items that the students generate ultimately depend on concrete actions. By helping your students realize what they are doing when they define these terms and how this kind of thinking differs from concrete operations, you will be giving them opportunities to grow in their formal thinking skills.

Logic Thinking in the formal stage becomes more orderly and systematic. Most eight-year-olds would be unable to answer the question, "If Jane's hair is darker than Susan's, and Susan's hair is darker than Mary's whose hair is darkest?" Although eight-year-olds are capable of ranking the characters by darkness of hair, they are unable to manipulate facts concerning imaginary people. They can deal with the concrete concept of an individual person, but not with the idea of its form.

Metacognition This term means "thinking about thinking." Children reaching early adolescence are able for the first time to analyze their own thoughts. They realize that sometimes they do and say things for unconscious reasons, and can decipher their own motives. They can retrace the train of thought they took in trying to solve a problem (in fact, this is what makes them better able to be logical). They can spot thinking errors, and restart the problem-solving process.

Hypothetical Reasoning Most early adolescents become capable of forming conclusions based on hypothetical possibilities. Answering the question, "What would things be like if it rained up?" involves mentally picturing rain rising from the ground. This mental picture is contrasted with reality and various conclusions produced.

At this stage, it becomes possible to think about problems even in the absence of real data. Resolutions of problems that cannot be true can nevertheless be evaluated to see if they offer any clues. While children can sometimes get the right answer to a question by trial and error, early adolescents realize that it is much more efficient to proceed according to a more careful strategy devised beforehand.

The Later Formal Operations Stage

later formal operations stage
The second stage of formal operations (age 15 to 19), which includes the development of propositional logic, individual thinking patterns, and scientific reasoning, and the ability to comprehend systems of symbols.

The **later formal operations stage** begins in adolescents from 15 to about 19 years old. This second phase differs from the first phase not only in quantity (problem solving is done more quickly and efficiently) but also qualitatively (new skills are mastered). Evidence for this stage of Piaget's theory comes from research by Higgins-Trenk and Gaite (1971) and Arlin (1975), although some questions have been raised by Fakouri (1976), and Cropper and associates (1977). In this section, we look at the same four aspects of formal operations as they apply to older adolescents.

Abstract Thought We said before that formal operations means dealing with symbols. In later adolescence, it becomes possible to deal with systems of symbols. Like notes are used to stand for sounds, musical notation can stand for the complex ways those sounds are to be made (soft or loud, sharp or mellow, with or without vibrations). Another comparison is between long division (mastered in early adolescence) and trigonometry (mastered by many in late adolescence). It is at this stage that many youths become capable of understanding political cartoons and religious symbolism. They also gain ability not only in problem solving but in problem finding (Arlin, 1975).

Logic Early adolescents begin to recognize the rules of logic. It is not until later adolescence, however, that the subtler forms of this discipline may be mastered. This higher level set of rules is known as *propositional logic*.

Do you ever do any metacognition? What would it be like to do so?

Metacognition In the previous stage, most teens can look back at their thinking and spot errors in it. In this stage they get better at analyzing their thought processes as they work through a problem. They begin to notice trends or patterns in their thinking and learn to compensate for them.

AN APPLIED VIEW

Cognitive Therapy and Drug Treatment

Cognitive therapists use metacognition all the time. For example, a cognitive therapist might ask one member of a drug rehabilitation group to explain how she thinks she got into the problems that she is dealing with. If she is addicted to cocaine, she is asked to explain why she thinks she started using the drug in the first place. Then the rest of the group analyze her thinking. They point out aspects of her thinking that are logical and those that are irrational. They try to look for trends or patterns in her thinking and to seek alternatives that will help her deal more successfully with her drug habit.

Hypothetical Reasoning Here is where we see the most striking change from the early formal operations stage. Now many adolescents are able to think like a scientist (one at the beginning level, at any rate). They are able to establish a plan for solving a problem. They are likely to investigate more than one source of data, and can think of multiple possible causes. They are able to conduct their study with little or no prejudice toward the outcome (of course this is not to say they *always* do so). They are able to apply the rules of logic. And finally, they are better at acting on solutions.

The Role of Gender The question might be asked, "Are males or females more likely to attain the formal operations stage?" There have been numerous studies of the relationship between IQ and gender, but little has been done from the standpoint of formal operations. Clearly more research is needed before we can make any sound conclusions on this question.

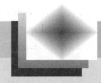

A SOCIOCULTURAL VIEW

Did Piaget Have a Bias?

Barbara Rogoff (1990) argues that Piaget's theory depends excessively on scientific-type problems. A society, or an individual for that matter, might not value scientific reasoning. Additionally, the way in which one learns might affect cognitive development. Tomasello and associates (1993) state that cultural differences in learning manifest themselves in three ways throughout one's cognitive development: through imitative learning, instructed learning, and collaborative learning. Imitative learning relies on one's perspective of what is being learned; instructed learning relies on alternating and coordinating the perspective between instructor and learner (a concept referred to as intersubjectivity); and collaborative learning relies on mutual reflection between instructor and learner with an integrated perspective.

Goodnow and Collins (1990) realized the need for a different definition of intelligence after using Piagetian tasks with Chinese and African American children. While administering three Piagetian tasks to both educated and illiterate African adults in the Ivory Coast, Tapé (1990) found two styles of thinking: an experimental-analytic style (like formal reasoning) and an experiential style (pragmatic and action-oriented logic). The first type asks the question *how* and the latter type asks the question *why*. He found that the illiterate adults used the latter type of logic to solve the Piagetian tasks.

Throughout the world, the sequence of Piaget's stages has been found to be the same, though the timing varies. Piaget's theory of formal operations, however, has only partially been confirmed both in single cultural and multicultural studies. Cross-cultural studies have shown that schooling up to the secondary level is necessary but not sufficient to succeed at formal operational tasks (Shea, 1985). This is undoubtedly because the tasks relate to physics, chemistry, or mathematics as they are taught in school. Even Piaget himself (1972) acknowledged the contradictory evidence and qualified his position by stating that all adults have the capacity for formal operational reasoning, but this capacity needs favorable circumstances to be manifested.

CRITICISMS OF PIAGET

Piaget's ideas, as widely accepted as they have been, have not escaped criticism and suggested revisions (Case, 1987; Flavell, 1982; Gelman & Baillargeon, 1983; Kuhn, 1984; Mandler, 1983). For example, some psychologists dispute his belief that cognitive development proceeds through a series of distinct stages, each of which contains important changes in the way that children think. These psychologists believe that cognitive development is a gradual process (This will be explained in the next section), that it is not all-or-nothing; that is, children are not all preoperational or all concrete operational in their thinking (Rest, 1983).

In one of the most famous studies challenging Piaget, Gelman (1979) trained 5-year-olds to examine rows of equal-length sticks. Later, when asked to pick rows of equal-length sticks from rows of unequal, they were successful. This seems to show that preschoolers are more competent than Piaget thought. These youngsters may have been ready for more complex tasks and just needed the push that Gelman gave them.

In spite of these and similar criticisms, Piaget's ideas remain remarkably popular and his work is a landmark among studies of cognitive development. Gelman and Baillargeon (1983) sum up the general feeling among professionals: "It will be hard, very hard, to do as well as Piaget" (p. 220).

VARIABLES IN COGNITIVE DEVELOPMENT—FLAVELL

Psychologist John Flavell (1977), on the basis of his own studies and his review of the literature, has suggested that there are seven aspects of the transition from childhood to adolescent and adult thinking. The following sections describe these seven aspects from the standpoint of an imaginary adolescent.

We would like you to picture the following scenario. Your local television station has decided to sponsor a new program called *Young Women Today*. It has already hired a 29-year-old woman to host the show, a person who has had considerable experience in the talk show format.

In addition, the station wants to hire a 17-year-old high school student as an assistant hostess. This has been announced at the local high school. Because the employers are sure that many female students will apply for this job, they have designed a simple application form. With it, they hope to identify the better candidates whom they will then interview. The application form asks that the applicants describe two characteristics that would make them an especially good choice for the job. Here are Flavell's seven aspects of the childhood-adolescent cognitive transition, analyzed through this scenario.

real versus the possible
The ability of the adolescent to imagine possible and even impossible situations.

Real Versus the Possible

For Piaget (1948/1966), probably the most important cognitive change from childhood to adolescence has to do with the growing ability of the adolescent to imagine possible and even impossible situations (**real versus the possible**).

What are some of the implications for teenagers of the distinction between real versus possible thinking?

Elementary school children tend to approach problems by examining the data firsthand and attempting to make guesses about the solution on the basis of the first piece of information they happen to look at.

This is no longer the case for 17-year-old Ellen, who would very much like to get the television show job. As it happens, she is the daughter of an unemployed actor. She wonders whether she should mention this. She thinks that it may help her, both because her father has been in show business and because her family could use the money. On the other hand, she suspects that the television people might think that if her father has been unsuccessful in show business, she may be, too. Whereas a 9-year-old child might put this fact down on the application, hoping that it might make some difference, Ellen tries to think of all the possible ramifications of this piece of information before deciding whether to use it. Ellen is obviously a formal operational thinker.

AN APPLIED VIEW

What Kind of Elephant Is That?

The capacity to envision "the wider world of possibility" often depends on our ability to visualize, that is, how well we are able to picture something vividly in our minds. For example, middle class and wealthy children have been found to be better at visualizing than are poor children, probably because of the greater variety of stimuli in more privileged homes. Think of the disadvantages that poor adolescents have when they are unable to imagine themselves wearing a cap and gown at graduation, or sitting behind a desk with a good-paying white-collar job.

You can help children become better at visualizing through a great number of techniques. For example, ask the teens you are working with to visualize their names in bold black letters. When they have the picture clearly in their minds, ask them if they can spell their name backwards by looking at their mental picture. Another technique is to ask them to visualize an animal—an elephant, for example. Ask them if they can tell if it's an Asian elephant (with ears the same size as its head) or an African elephant (with ears twice as big as its head). Ask them if they can count the wrinkles on the elephant's trunk. Exercises like these, which promote the ability to see mental images, also help teens to see the differences between the real and the possible. You may want to take a course in visualization to learn other techniques. Such courses are often given in art departments of high schools or colleges.

---

empirico-inductive method A method of problem solving used by young children, in which they look at available facts and try to induce some generalization from them.

hypothetico-deductive method A method of problem solving used by adolescents, in which they hypothesize about the situation and deduce from it what the facts *should be* if the hypothesis were true.

Empirico-Inductive Versus Hypothetico-Deductive Methods

A further major difference between concrete and formal operational thinkers is in their *method* of problem solving. Younger children use an **empirico-inductive method**; that is, they are likely to look at available facts and try to induce some generalization from them. Adolescents, on the other hand, are likely to use a **hypothetico-deductive method** to hypothesize about the situation, and then deduce from it what the facts *should be* if the hypothesis were true. Adolescents tend to look at what might be, in two senses: they attempt to discover several possibilities in a situation before starting to investigate it empirically. Then they try to imagine the possible outcomes of each.

Whereas the younger child might be satisfied with simply putting down two possible characteristics on the application, Ellen is likely to think of what the ramifications are for each of a number of possible characteristics. For example, she might hypothesize that if the station has a reputation for being concerned with poor people, they would be sympathetic with the fact that her father is out of work. On the other hand, if the station has a reputation for fierce competition, it might look negatively on this information, thinking that perhaps her whole family are losers.

Intrapropositional Versus Interpropositional Thinking

intrapropositional thinking
The ability to think of a number of possible outcomes that could result from a single choice.

interpropositional thinking
The ability to think of the ramifications of *combinations* of propositions.

Elementary school-age children, especially older ones, may well be able to analyze the relationships among several aspects of a single choice. This would be **intrapropositional thinking**. It is only in later adolescence, however, that they are able to think of the ramifications of *combinations* of propositions. This is **interpropositional thinking**.

For example, Ellen may decide that her father's unemployment will hurt her, but she may balance this by stating that she herself has maintained a paper route successfully for the past four years. Interpropositional thinking can become infinitely more complex than intrapropositional thinking. Logicians (students of logical thinking) have discovered many aspects of the interrelationships that can occur between sets of propositions.

An important aspect of this complexity is the ability to think logically about statements that may have no relationship to reality whatsoever. As Flavell (1977) states, "Formal operational thinkers understand that logical arguments have a disembodied, passionless life of their own" (p. 106). Most persons find that thinking about abstract concepts—for example, the laws of trigonometry—is more difficult than thinking about how to build a table, but the laws of logic are the same for both. We become aware of this during adolescence and adulthood.

Combinations and Permutations

During adolescence, the thinker becomes able to realize that systematically generating combinations (A with B, A with C, B with C, etc.) can aid in thoroughly examining the possible solutions to a problem.

Ellen will probably sit down and draw up a list of all her good characteristics before even starting to decide which of these would be best to put on her application. To improve her creativity, she might write all these characteristics on separate sheets of paper, put them in a hat, and draw them out in pairs to see if she comes across a pair that she thinks will be unusually appealing. Without such a technique, she might never have thought of that particular pair.

Inversion and Compensation

Suppose you have before you two containers with an equal amount of water in each, hanging evenly from each side of a balance scale. Obviously, if a cup of water is added to the container on the right, it will sink to a level lower than the container on the left. If you were asked to make the containers even again, you

Because she has reached late adolescence, Ellen approaches the job interview with more complex thought patterns than she would have a few years earlier.

would probably recognize that it could be done in one of two ways: withdraw a cup of water from the container on the right (**inversion**), or add a cup of water to the container on the left (**compensation**).

Concrete operational thinkers usually will not recognize that there is more than one possible way to solve the problem. Having solved it one way, they no longer pay attention to the task. Because formal operational thinkers are able to imagine both inversion and compensation as being useful in solving a problem, they have a better chance of coming up with other fruitful solutions.

Ellen would be using compensation if she mentioned her newspaper job to balance her father's unemployment. She would be using inversion if she simply left out the fact that her father was unemployed.

Information-Processing Strategies

Older thinkers are not only more likely to have a large array of problem-solving strategies, but are also more likely to attempt to devise a plan to use this array. Such a plan is an **information-processing strategy**.

Ellen's problem is to discover what major criteria the television people have in mind in selecting their talk show host. A good strategy might be for her to visit and interview several individuals in the community who know those criteria.

Consolidation and Solidification

The changes from childhood to adulthood mentioned so far have been qualitative. There are also two quantitative aspects of this transition. The mental gains that are being made are slowly **consolidated**. Not only are the improved problem-solving techniques learned, but they are employed in a wider variety of

inversion
The cognitive ability to recognize an inequality of quantity and then subtract from the greater amount to create an equality.

compensation
The cognitive ability to recognize an inequality of quantity and then add to the lesser amount to create an equality.

information-processing strategy
The process of understanding information and acting on that understanding.

consolidation
A quantitative cognitive change from childhood to adulthood, whereby improved problem-solving techniques are employed in a wider variety of situations and with greater skill.

Table 4.1	Overview of Flavell's Variables in Intellectual Development
Childhood: Concrete Operational Thinking	**Adolescent: Formal Operational Thinking**
1. The real	1. The possible
2. Empirico-inductive method	2. Hypothetico-deductive method
3. Intrapropositional thinking	3. Interpropositional thinking
4. Unable to see combinations and permutations	4. Able to see combinations and permutations
5. Can use inversion or compensation	5. Can use inversion and/or compensation
6. Poor information-processing strategies (weak plans)	6. Good information-processing strategies (clear plans)
7. Mental gains not consolidated and solidified	7. Mental gains consolidated and solidified

From J. Flavell, "Cognitive Development" in *Child Development*, 53, 1–10, 1982. Copyright © 1982 The Society for Research in Child Development, Inc., Chicago, IL.

solidification
Cognitive growth in which the thinker is more certain and confident in the use of the newly gained mental skills and is more likely to use them in new situations.

situations, and with greater skill. These gains are also **solidified**; that is, the thinker is more certain and confident in the use of the newly gained mental skills and is more likely to use them in new situations.

If Ellen were 12 years old instead of 17, she may have tried some of the tactics discussed here, but she might very well have given up quickly and just put down any two traits that seemed acceptable to her. It is Ellen's greater experience with these thinking styles that gives her the motivation to persevere at the task. Table 4.1 offers an overview of Flavell's variables.

Although Piaget and those like Flavell who have followed in his footsteps have greatly expanded our understanding of the human mind, they have left many fascinating questions unanswered. For further insights into the adolescent mind, we turn now to several other aspects of cognitive functioning, the first of which will be social cognition.

ZONE OF PROXIMAL DEVELOPMENT-VYGOTSKY

One of the most astute psychologists ever to study cognition was the Russian Lev Vygotsky (1896–1934). A contemporary of Piaget's, his studies of Russian children and adolescents led him to posit a much more important role for social action than did Piaget. His basic hypothesis was that we humans gain most of our cognitive growth from our interactions with other people. This growth comes when we try to achieve tasks that are beyond our ability. When someone comes to our assistance and shows us what to do, our intellectual abilities develop.

zone of proximal development
In Vygotsky's theory, the area at the upper edge of our present abilities in any task in which cognitive growth takes place.

Vygotsky (1933; 1962; 1978) suggested that for any type of task, each person has a **zone of proximal development**. At the lower limit of that zone, we are capable of working independently. At the upper end of the zone, we need help. For example, a 14-year-old may possess excellent math skills, but would need advice in solving complex algebraic equations. Vygotsky was one of the first to realize that we can learn a great deal about how we think by watching how people solve problems in pairs. In this situation, especially when the problem is above the lower end of each problem solver's zone of proximal development, the two people are most likely to talk aloud about their thinking. This proves to be a much better way of studying thinking than merely asking subjects to describe their thinking as they perform tasks.

Social Cognition

"This is terrible!" thought Juan, as he stared at his nose in the mirror. "A zit, right on the tip, blazing like a neon sign. No way I can go to school for the next few days. Everyone would laugh themselves silly. Damn, this means missing the game too!"

Tracy's boyfriend Todd is more than 20 minutes late. Tracy is certain he has forgotten about her. "I wish I could ask Mom what to do, but of course I can't," she moaned. "There isn't anyone to talk to—they all think I am stupid for having a boyfriend like Todd. If he doesn't come soon I'll have to break up with him for good!"

social cognition
The ability to think critically about interpersonal issues, which develops through age and experience. Used to make sense of other people and to decide how to interact with them.

You may smile as you read the comments of Juan and Tracy, remembering some similar incident from your own adolescence. It is quite normal for adolescents to begin to think about how they relate to others. This is known as **social cognition**, discussed briefly here and in Chapter 5. One important part of this cognitive aspect of the adolescent's life is called **egocentric thinking**. Adolescent egocentrism, a term coined by Elkind (1978) refers to adolescents tendency to exaggerate the importance, uniqueness, and severity of social and emotional experiences. This trait, which begins in early adolescence and becomes more prominent during the middle teen years (Elkind & Bowen, 1979), is composed, according to Elkind, of two specific factors: imaginary audience (Montgomery & others, 1996; Rycek & others, 1998; Vartanian, 1997) and personal fable (Siegel & Shaughnessy, 1995; Vartanian, 1997).

egocentric thinking
Adolescent's style of thought in which they think more about themselves, and watch themselves as though from above.

THE IMAGINARY AUDIENCE

Because he is now in the formal operations stage, Juan is able to imagine what others might be thinking of him. Because he is so interested in himself, this imagination often goes to extremes. He can literally think that everyone around him is noticing everything he does and every aspect of his appearance in fine detail. He is "on stage" most of the time, in front of an **imaginary audience** (Montgomery & others, 1996; Rycek & others, 1998; Vartanian, 1997).

imaginary audience
One aspect of adolescent egocentric thinking in which adolescents think everyone is looking at them. The adolescent feels "on stage" most of the time, in front of an imaginary audience.

THE PERSONAL FABLE

personal fable
An aspect of adolescent egocentric thinking in which adolescents make up stories about themselves. Most of these fables have two aspects: uniqueness and invulnerability.

Can you remember having a personal fable? Did you have more than one?

perspective taking
The ability to see the environment from someone else's point of view.

Many adolescents seem to make up stories about themselves, which become their own **personal fables** (Siegel & Shaughnessy, 1995; Vartanian, 1997). Most of these fables have two aspects. Adolescents have a sense that they are unique and therefore are extremely different from people around them. This means that they cannot believe that anyone else could understand how they're feeling. This is especially true if the other person is older or younger or of a different gender or race.

Thus Tracy must suffer with her thoughts of Todd's insensitivity all by herself because she cannot believe that her mother or friends would understand her feelings, much less be willing to help her. She truly believes that she's alone in her misery, and is certain that her mother could never understand.

The other aspect of the personal fable is a belief that the adolescent is all-powerful and cannot be hurt or destroyed by anything. Obviously this is not true, but adolescents' dreams of success can lead them to believe that this unrealistic situation is true.

Teenagers often have problems working with others because of their egocentric thinking. The concept of adolescent egocentrism was first introduced by psychologist David Elkind (1967, 1976, 1978, 1985; Elkind & Bowen, 1979). Adolescent psychologist Robert Selman (1976, 1980) has added a great deal to Elkind's theory with his concept of **perspective taking**. This ability to take the perspective of other people, which begins to flourish during the formal operations stage, interacts in important ways with adolescent egocentrism. Perspective taking is an important part of cognitive development, but because it is an even more important part of moral development, we will tell you more about it in the next chapter. In fact, social cognition itself relates to several other facets of adolescent life, and so will be considered in several of the remaining chapters of this book.

We turn now to a different way of looking at adolescent cognition. It is known as *information processing*.

Information Processing

Information processing is a topic frequently studied by cognitive psychologists. "The information processing perspective centers on how individuals differ in the ways they store, manipulate, and retrieve information. As with the Piagetian perspective, information processing assumes that individuals take active roles in their development. Unlike Piaget's theory, however, information processing theory does not offer universal stages of development. It concerns itself instead with the *processes* involved in intellectual performance" (Alexander, 1985). There are many implications for the study of information processing. If we take a moment to think about most of the daily activities that we take for granted, we can begin to see the importance of this field of study. Take, for example, driving a car or finding our way home at the end of the day. We do these and many other activities without a tremendous amount of conscious thought or energy, and yet we succeed as a result of information processing.

AN APPLIED VIEW

Video Games and Cognitive Development

The emergence of video games as a popular pastime among children and adolescents in the late 1970s created a firestorm of controversy. Parents and educators were warned that there may be detrimental consequences to prolonged exposure. Surgeon General C. Everett Koop cautioned that video games could contribute to the development of psychosis (Funk, 1993). Amidst all this controversy is little empirical evidence to support the claims of the negative impact of video games (Dill & Dill, 1999; Subrahmanyam & Greenfield, 1994). In fact, a recent study by Blumberg (1999) suggests that video games can actually have a positive impact on learning, attention, motivation and performance. For example, there has been study on divided attention and video games. Divided attention is necessary for many vocations such as air traffic controller, as well as for everyday tasks such as driving a car or playing certain sports. In a recent study involving both expert and novice video game players, strategies of divided attention were shown to improve in both categories of subjects (Greenfield & others, 1994).

Video games are a type of symbolic play representing movement within actual space. The effect of video games on spatial skills has recently been studied (Okagaki & Greenfield, 1994; Subrahmanyam & Greenfield, 1994). Action video games were shown to improve spatial relations skills more than an educational word game did. Although males of all ages have better

spatial abilities (Greenfield, 1994), researchers found no difference in the rate of improvement in spatial skills between males and females playing six hours of the game Tetris after they had not played any video game during the previous year (Okagaki & Greenfield). This result was obtained in spite of higher pretest scores of spatial skill for the male subjects. In general, males have more experience in video game playing than do females (Greenfield). If spatial skills are acquired and not genetically determined, video games could be an excellent method to develop these skills and could contribute to gender balance in this domain of cognitive development (Subrahmanyam & Greenfield).

It does appear that video games have been viewed too negatively; they do serve to socialize children to interact with artificial intelligence. A greater cause for concern may be that most video games contain violent themes. Wiegman and Van Schie (1998) found that children, especially boys who preferred aggressive video games were more aggressive and less prosocial in their own play. Boys tend to enjoy that type of aggressive competition whereas girls do not. Since video games may contribute to cognitive development and prepare individuals for skills related to computer technology, more games with active female characters and with nonviolent themes should be designed to ensure equal preparation for survival in this age of technology (Greenfield, 1994).

information processing
The study of how children (and adults) perceive, comprehend, and retain information.

automatic processing
Information-processing procedures that occur in situations that are consistent and provide a lot of opportunity for practice.

controlled processing
An information-processing procedure in which new information lacking consistency in rules and sequence is manipulated, therefore making specific attention to each step a necessity because there is no established pattern.

Information processing is another approach to the study of the human mind. Information processing refers to the ways through which information is received, processed, and understood. It is made up of the procedures through which memory is created and learning takes place, processes that most of us take for granted.

Two important aspects are automatic and controlled processing (Ackerman & others, 1986). **Automatic processing** occurs in situations that are consistent and provide a lot of opportunity for practice, such as driving a car. Such procedures are characteristically quick and effortless. They also, generally, become easier and faster with practice.

Controlled processing involves the manipulation of new information. In addition, controlled processing implies the need for conscious attention to a task. Generally speaking, controlled processing is used when there is a lack of consistency in rules and sequence, therefore making specific attention to each step a necessity because there is no established pattern. Controlled processing takes place when the information to be processed does not provide the opportunity for practice that leads to speed and ease. Working out an unfamiliar formula in math class is an example.

Frey and Rosch (1984) looked at the effect of *receptivity* to information presented; that is, whether the person feels good, bad, or neutral about it. They were interested in the type of information that would and would not be well received. They found that people are more open to new information if it either supports a decision that they have made or if the decision that they have made can be changed.

AN APPLIED VIEW

How We Can Help Teens Be More Receptive to New Information

How can we help people to be more receptive to new information? We need to show them how information could be relevant to their lives, even to the point of making them change their minds about something. This presentation should usually be in a small group setting. For instance, suppose a teenager has been engaging in unsafe sex. Then he is presented with information that demonstrates that unless he refrains from sex altogether or uses protection, he has a good chance of becoming a father and may even contract AIDS.

This information alone is not likely to change his mind. It is necessary for him to discuss the beliefs that he has about his sexual practices, and thus reveal these stereotypes to the others in the group. In doing so, he begins to see how faulty his thinking is.

You will find that most of the time, when irrational beliefs are discussed in a group, the members of the group become more open to change because they hear what others have to say. When teens are more open to change, then the new information suddenly becomes more relevant to their opinions, and they are able to incorporate it into their value system.

Learning is the process by which our nervous system, and consequently our behavior, is changed by our experiences. The experiences we have change the structure of the nervous system, altering neural circuits that contribute to the way we perceive the world. Perceptual learning refers to the ability to learn to recognize stimuli that have been seen before. The major function of this form of learning is the ability to identify and categorize objects and situations. Perceptual learning occurs primarily by changes in the sensory association cortex in the brain. For example, learning to recognize objects by sight occurs in the visual association cortex; learning to recognize certain sounds occurs in the auditory association cortex (Carlson, 1994).

FACTORS THAT INFLUENCE DEVELOPMENT OF INFORMATION PROCESSING

Information-processing capabilities seem to develop and improve with age, but not in stages similar to other cognitive functions (Knight & others, 1985). Adolescents have a greater ability to make use of memory strategies and other information-processing techniques than do younger children. Also, most older children have learned more information than have younger children. This makes them better able to understand metaphors and to detect contradictions between two sets of facts (Keil, 1984).

All this in turn probably explains to a great degree the developmental differences found in social values between younger and older teens. Apparently the older adolescent's ability to deal with more complex social and ethical issues (e.g., what is moral sexual behavior?) is attributable in part to the increasingly complex level of information processing.

Another issue that has been investigated is the speed with which a person handles information. (This is a much more important issue for Western cultures than elsewhere. In most non-Western cultures, slow, calm, deliberate thinking is more valued.) Speed is frequently studied using simple tasks that require little or no complex thought so that the speed of the task and the nature of the task do not become confused. There are two basic points of view with respect to speed of information processing. One suggests that it is dependent on the type of material being considered, for example, math versus music. The other notion is that it is related to the mental ability of the person and not dependent on the specific content of the task. Levine and others (1987) found that speed, like complexity of information processing, increases with age. Adolescents in general perform cognitive tasks more quickly than younger children, and older adolescents are faster than younger adolescents. (Unfortunately, this process does not continue into old age.)

A number of studies have been conducted to investigate the relationship between speed of information processing and intellectual ability. Reaction times on simple cognitive tasks were compared for gifted subjects and their average counterparts (Cohn & others, 1985; Jensen & others, 1989). Data from these studies demonstrate that the higher the level of intellectual functioning, the greater the speed of processing.

Critical Thinking

critical thinking
The ability to think logically, to apply this logical thinking to the assessment of situations, and to make good judgments and decisions.

Critical thinking is making judgments after assessing a situation. The integration of several skills helps an individual make judgments. Pierce and associates (1988) found the factors that play a part in these judgments include the ability to:

- Make inferences from observations.
- Recognize assumptions.
- Think deductively.
- Make logical interpretations.
- Evaluate arguments by recognizing the difference between a weak and a strong position.

In short, critical thinking involves the ability to think logically, to apply this logical thinking to the assessment of situations, and to make good judgments and decisions. As Moore and Parker (1986) put it, it is "the correct evaluation of claims and arguments" (p. 2). In a recent study, McCarthy and Tucker (1999) found that teaching logic to a group of adolescents improved their critical thinking ability. This finding could have practical applications for school curricula focused on developing critical thinking skills.

Guilford's distinction between convergent and divergent thinking (1975) is relevant here, because critical thinking is made up of these two abilities.

convergent thinking
In problem solving, when we *converge* or close in on the one correct answer.

Convergent thinking is used when we want to solve a problem by *converging* or closing in on the correct answer. For example, if we were to ask you to answer the question, "How much is 286 times 469," you probably could not produce it immediately. However, if you used a pencil and paper or a calculator, you would almost certainly arrive at the same answer as most others trying to solve the problem. There is only one correct answer.

divergent thinking
The type of thinking used when the problem to be solved has many possible answers; especially important in creative thinking.

Divergent thinking is just the opposite. This is the type of thinking used when the problem to be solved has many possible answers, for example, "What are all the things that would be different if it were to rain up instead of down?" Other divergent questions might be, "What would happen if we had no thumbs?" and "What should we do to prevent ice buildup from snapping telephone lines?" Divergent thinking can be right or wrong, too, but there is considerably greater leeway for personal opinion than with convergent thinking. Not all divergent thinking is creative, but it is more likely to produce a creative concept. To be a good critical thinker, it is not enough to analyze statements accurately. Often you will need to think divergently to understand the possibilities and the implications of those statements, too.

Because early adolescents are entering the formal operations stage of intellectual development, they become vastly more capable of critical thinking than younger children. As they move through the teen years, adolescents grow in their ability to make effective decisions. This involves five types of newly acquired abilities (Moore & Parker, 1986):

Phase 1. Recognizing and defining the problem.

Phase 2. Gathering information.

Phase 3. Forming tentative conclusions.

Phase 4. Testing tentative conclusions.

Phase 5. Evaluation and decision making.

As you may know, there is an ongoing controversy among educators as to whether we should spend more time teaching students basic information or teaching thinking skills that can be applied more generically. Critical thinking is one of these general abilities; creative thinking and moral reasoning are two others. As we discussed in the first part of this chapter, most young teens have reached the developmental stage at which the capacity for abstract and critical thinking exists. But can youths be taught to do it better? A number of studies, summarized in Idol (1987) and in Pierce and associates (1988), indicate that this is definitely possible.

In the next section, we will make a distinction between critical and creative thinking, but it is important that we not make too great a distinction, especially at this age level. Paul (1987) describes this concern well:

> ❑ Just as it is misleading to talk of developing a student's capacity to think critically without facing the problem of cultivating the student's rational passions—the necessary driving force behind the rational use of all critical thinking skills—so too it is misleading to talk of developing a student's ability to think critically as something separate from the student's ability to think creatively. . . . The imagination and its creative powers are continually called forth (p. 143).

Creative Thinking

Is it basically a good thing for a person to be a creative thinker? Are there some ways in which people could get in trouble through creative thinking?

> ❑ This is the story about a very curious cat named Kat. One day Kat was wandering in the woods where he came upon a big house made of fish. Without thinking, he ate much of that house. The next morning when he woke up he had grown considerably larger. Even as he walked down the street he was getting bigger. Finally he got bigger than any building ever made. He walked up to the Empire State Building in New York City and accidentally crushed it. The people had to think of a way to stop him, so they made this great iron box which made the cat curious. He finally got inside it, but it was too heavy to get him out of again. There he lived for the rest of his life. But he was still curious until his death, which was 6,820,000 years later. They buried him in the state of Rhode Island, and I mean the whole state.

Ralph Titus, a seventh-grade student

The restless imagination, the daring exaggeration, the disdain for triteness that this story demonstrates—all are signs that its young author has great creative potential. With the considerable knowledge we now have about how to foster creativity, this boy could develop his talents to his own and society's great benefit.

Creative thinking appears to have many elements—divergent thinking, fluency, flexibility, originality, and remote associations. One element that seems to be of special importance in adolescence is the use of metaphor.

THE USE OF METAPHOR

A metaphor is a word or phrase that comes to stand for another word or phrase, either by comparison or analogy (Morris, 1971). An example would be calling someone a blockhead. The word *block* implies that the person is so stupid that his or her head must be made of a block of wood. Common sense suggests a relationship between metaphor use and creativity. Using a metaphor in speech involves calling attention to a similarity between two seemingly dissimilar things. This suggests a process similar to divergent thinking (Jaquish & others, 1984; Kogan, 1973, 1983; Wallach & Kogan, 1965).

AN APPLIED VIEW

Practice Makes Perfect

Isn't it a shame that the wonderful ability to make metaphors declines during the school years? Although we have suggested why this happens, we also know that practice makes perfect. Therefore, giving students a chance to practice making metaphors is one of the best ways to contribute to their ability to think creatively.

Jim Betts, a seventh grade social studies teacher in New Orleans, runs a monthly contest in his class. He asks students to put up examples of metaphors on the bulletin board. Some examples have been, "About as reliable as a politician's promises," and "Wiser than an Abraham Lincoln." At the end of the month, he has the class vote for the best metaphor of the month. A small prize is offered, but he finds that the approval of the students is more important than any prize he can offer.

Kogan (1983) believes that the use of metaphor can explain the difference between ordinary divergent thinking and high-quality divergent thinking. A creative person must be able to think of many different things from many different angles. In addition, she or he must also be able to compare them in uniquely different ways.

Although metaphors are typically first used by older children and adolescents, researchers are now looking at the symbolic play of very young children to see how the ability to make metaphors is learned (see Kogan, 1983, for a good review). The early imaginative play of children is now being viewed as a precursor both of metaphor use and of creativity.

One of the best sources on the study of the role of metaphor has been the work done by Howard Gardner and his associates at Harvard University. His seminal book *Art, Mind, and Brain: A Cognitive Approach to Creativity* (1982) offers many insights into the process.

Gardner has based his research on the theories of Jean Piaget, Noam Chomsky, and Claude Levi-Strauss. He states that "these thinkers share a belief that the mind operates according to specifiable rules—often unconscious ones—and that these can be ferreted out and made explicit by the systematic examination of human language, action, and problem-solving" (1982, p. 4).

Gardner's main efforts have focused on the relationship between the art of normal and brain-damaged children and adolescents, and their understanding of metaphor. He describes talking to a group of youngsters at a Seder (the meal many Jewish people eat to commemorate the flight of the Hebrews from Egypt). He told them how, after a plague, Pharaoh's "heart was turned to stone." Each child interpreted the metaphor differently, but only the older ones could understand the link between an object (a stone) and a personality trait (the inability to feel one's emotions). Younger children were more apt to apply magical interpretations (for example, God or a witch) to explain a heart turned into stone. Gardner believes that the development of metaphoric language in students is as sequential as the stages proposed by Piaget.

Gardner and his wife (Gardner & Winner, 1982) examined such metaphors as a bald man having a "barefoot head" and an elephant being seen as a "gas mask." They found clear changes in the level of sophistication as children grow older. Interestingly, there appear to be two opposing features:

- When you ask students to *explain* figures of speech, they get steadily better at it as they get older. There is a definite increase in this ability as the child reaches maturity.
- However, very young children seem to be best at *creating* their own metaphors. Furthermore, their own metaphors tend to be one of two types:

❏ Children who make their metaphors on visual resemblances may approach experience largely in terms of the physical qualities of objects. On the other hand, children who base their metaphors on action sequences may view the world in terms of the way events unfold over time. We believe that the difference may continue into adulthood, underlying diverse styles in the creation and appreciation of artistic forms (p. 164).

These researchers believe that the spontaneous production of metaphors declines somewhat during the school years. This is probably because the child, having mastered a basic vocabulary, has less need to "stretch the resources of language to express new meanings" (p. 165). In addition, there is greater pressure from teachers and parents to get the right answers, so children take fewer risks in their language. Gardner and Winner point to the *Shakespeare Parallel Text Series*, which offers a translation of Shakespeare's plays into everyday English ("Stand and unfold yourself" becomes "Stand still and tell me who you are"), as a step in the wrong direction. "If, as we have shown, students of this age have the potential to deal with complex metaphors, there is no necessity to rewrite Shakespeare" (p. 167).

When asked their diagnosis of 11-year-old Tommy, many psychology students label him emotionally disturbed, and recommend that he be taken from his parents and institutionalized for his own good. Actually, this is an old but true case report, and the boy described in it eventually became a famous inventor—Thomas Alva Edison!

As you can see from these details about Edison's childhood, it is often hard to spot the children who have the most creative potential unless you know what you are looking for. In the next section we discuss this.

OBSTACLES AND AIDS TO CREATIVITY

We may agree that creativity is a valuable trait and should be fostered, but how? A number of theorists have offered excellent suggestions (e.g., Adams, 1986; Treffinger & others, 1983), but educator Ralph Hallman's suggestions (1967) on the obstacles and aids to creativity are still classic. According to him, several persistent obstacles to creativity are:

AN APPLIED VIEW

Guidelines for Creative Problem Solving

Would you like to become a more creative problem solver or to help adolescents to think more creatively? Here are some suggestions that should help (Dacey, 1986):

- Avoid the filtering out process that blocks problems from awareness. Become more sensitive to problems by looking for them. For example, examine your relationships with those with whom you live. Any problems there?
- Never accept the first solution you think of. Generate a number of possible solutions, then select the best from among them.
- Beware of your own defensiveness concerning a problem. When you feel threatened by a problem, you are less likely to think of creative solutions to it. For instance, is defensiveness causing you to filter out problems?
- Get feedback on your solutions from others who are less personally involved.
- Try to think of what solutions someone else might think of for your problem.
- Mentally test out opposites to your solutions. When a group of engineers tried to think of ways to dispose of smashed auto glass, someone suggested trying to find uses for it instead. Fiberglass was the result!
- Give your ideas a chance to incubate. Successful problem solvers report that they frequently put a

problem away for a while, and later on the solution comes to them. It is clear that they have been thinking about the problem on a subconscious level, which is often superior to a conscious, logical approach.

- Diagram your thinking. Sometimes ideas seem to fork, like the branches on a tree, with one idea producing two more, each of which produces two more, and so on. Diagramming will let you follow each possible branch to its completion.
- Be self-confident. Many ideas die because the person who conceived them thought they might be silly. Studies show that females have been especially vulnerable here.
- Think about the general aspects of a problem before getting to its specifics.
- Restate the problem several different ways.
- Become an idea jotter. A notebook of ideas can prove surprisingly useful.
- Divide a problem, then solve its various parts. For example, problems with a roommate may involve false assumptions, miscommunication, and unwillingness to talk these things over.
- Really good ideas frequently require some personal risk on the part of the problem solver. In this we are like the turtle, which can never move forward unless it sticks its neck out.

- *Pressures to conform.* The pressure on us to follow standardized routines and inflexible rules is probably the major inhibitor. Authoritarian parents, teachers, and managers who demand order are responsible for the demise of a great deal of creative talent.
- *Ridicule of unusual ideas.* This destroys our feelings of worth and makes us defensive and compulsive.
- *An excessive quest for success and the rewards it brings.* An over-concern with material success is often the result of trying to meet the standards and demands of others in order to obtain the rewards they have to give. In the long run, this distorts our view of reality and robs us of the strength of character to be creative (Amabile & others, 1986).

- *Intolerance of a playful attitude.* Innovation calls for "playing around" with ideas, a willingness to fantasize and pretend, and a healthy disrespect for accepted concepts. Often creative persons are seen as childlike and silly and their activity as wasteful, but these are only appearances. As Hallman remarks, "Creativity is profound fun."

In addition to recommending that we avoid these obstacles, Hallman urges that we promote the following aids, in ourselves and others.

- *Engage in self-initiated learning.* Most people who are in charge of others (managers, teachers, parents) find it hard to encourage others to initiate and direct their own learning. After all, this is certainly not the way most people were taught. We fear that if our subordinates (employees, students, children) are given greater freedom to explore reality on their own, they will learn wrong things or will not learn the right things in the proper sequence. We must put less emphasis on learning "the right facts" and more on learning how to learn. Even if our subordinates do temporarily mis-learn a few things, in the long run the practice in experimentation and imagination will benefit them (and us) greatly.
- *Become deeply knowledgeable about your subject.* Only when we make ourselves fully familiar with a particular situation can we detach ourselves enough to get an original view of it.
- *Defer judgment.* It is important to make wild guesses, to juggle improbable relationships, to take intellectual risks, to take a chance on appearing ridiculous. Refrain from making judgments too early.
- *Be flexible.* We need to shift our point of view; dream up new ideas for things; imagine as many solutions to a particular problem as possible.
- *Be self-evaluative.* At the time we come up with a creative idea, we are always a minority of one. History is replete with examples of ideas that were rejected for years before people began to realize their worth. Therefore, in order to be a creative person, we must know our own mind and be independent of the judgment of others. In order to become a good judge of our own thinking, we need to practice making many judgments.
- *Ask yourself lots of open-ended questions.* One extensive study showed that 90 percent of the time, the average teacher asks questions to which there can be only one right answer, which the teacher already knows. Questions that pique curiosity and allow many possible right answers were asked only 10 percent of the time. We should realize that we were probably taught that way and take steps to rectify the tendency.
- *Learn to cope with frustration and failure.* Thomas Edison tried more than 2,000 combinations of metal before he found just the right kind for the electric element in his first light bulb.

CREATIVITY, GIFTEDNESS, AND THE IQ

As Feldman (1979) has pointed out, there have been many studies of "giftedness," but only a few studies of exceptionally creative, highly productive

Child prodigies are distinguished by the passion with which they pursue their interests.

adolescents. He believes that this unfortunate situation is attributed mainly to "the foremost figure in the study of the gifted," Lewis M. Terman. Terman (1925) was well known for his research on 1,000 California children whose IQs in the early 1920s were 135 or higher. Terman believed these children to be the "geniuses" of the future, a label he kept for them as he studied their development over the decades. His was a powerful investigation, one that was followed by scholars and popular writers alike.

Precisely because of the notoriety of this research, Feldman argues, we have come to accept a *numerical* definition of genius (an IQ above 135), and a somewhat low one at that. Feldman notes that the *Encyclopedia Britannica* now differentiates between two basic definitions of genius: the numerical one fostered by Terman; and the concept first described by Sir Francis Galton (1879), of "creative ability of an exceptionally high order as demonstrated by actual achievement" (p. 148).

Feldman says that genius, as defined by IQ, really only refers to **precociousness**—doing what others are able to do, but at a younger age. **Prodigiousness** (as in child prodigy), on the other hand, refers to a young person who is *qualitatively* higher in ability than other children. This is a different concept from that of simply being able to do things sooner. Further, prodigiousness calls for a highly unusual matching of high talent and an environment that is ready for and open to creativity. If such youthful prodigies as Mozart in music or Fischer in chess had been born two thousand years earlier, they may well have grown up ordinarily. In fact, if Einstein had been born *fifty* years earlier, he might have done nothing special—particularly since he did not even speak well until he was five!

So if prodigies are more than just quick learners, what is it that truly distinguishes them? On the basis of his intensive study of three prodigies, Feldman urges that

precociousness
The ability to do what others are able to do, but at a younger age.

prodigiousness
The ability to do what is qualitatively better than the rest of us are able to do.

❑ Perhaps the most striking quality in the children in our study as well as other cases is the passion with which excellence is pursued. Commitment and tenacity and joy in achievement are perhaps the best signs that a coincidence has occurred among child, field, and moment in evolutionary time. No event is more likely to predict that a truly remarkable, creative contribution will eventually occur (1979, p. 351).

In summary, critical and creative thinking are similar in that they both employ convergent and divergent production. The main difference between them is that whereas critical thinking aims at the correct assessment of *existing* ideas, creativity is more aimed at the invention and discovery of *new* ideas. While each requires a certain amount of intelligence, creativity also depends on such traits as metaphorical thinking, an independent personality, and, as Feldman points out, a joyfulness in the process.

C O N C L U S I O N S

As you have seen, cognitive development is a complex matter, one that we understood very little about prior to this century. The intellect clearly develops in stages. Contrary to earlier beliefs, thinking in childhood, adolescence, and adulthood are qualitatively different from each other. Furthermore, there are a number of other aspects of cognitive development: social cognition, information processing, egocentric thinking, critical thinking, and creative thinking.

We humans often use our thinking abilities to make decisions about how we should treat others and how we feel we should be treated by them. The study of how thinking affects social relations is known as morality and we will investigate this trait, together with the closely associated trait of spirituality, in the next chapter.

C H A P T E R   H I G H L I G H T S

Stages of Cognition

- Piaget focused on the development of the cognitive structures of the intellect during childhood and adolescence.
- The infant and child pass through Piaget's first three stages: sensorimotor, preoperational, and concrete operational.
- Piaget's highest stage of cognitive development, that of formal operations, begins to develop in early adolescence.
- Flavell suggests that there are seven aspects of transition from childhood to adolescent and adult thinking: the real versus the possible, empirico-inductive versus hypothetico-deductive, intrapropositional versus interpropositional, combinations and permutations, inversion and compensation, information-processing strategies, and consolidation and solidification.

Social Cognition

- Adolescents focus much attention on themselves. This is referred to as egocentric thinking.

- Adolescents tend to believe that everybody is looking at them. This phenomenon is called the imaginary audience.
- Many adolescents also hold beliefs about their own uniqueness and invulnerability. This is known as the personal fable.

Information Processing

- Theories of information processing are influential at present.
- These theories focus on individual differences in the storage, manipulation, and retrieval of information, not on stages of cognitive development.
- Two important aspects of information processing are automatic processing and controlled processing.

Critical Thinking

- Critical thinking skills include the ability to make inferences from observations, to recognize assumptions, to think deductively, to make logical interpretations, and to evaluate weak and strong positions in an argument.

- Critical thinking combines both convergent thinking, in which there is only one correct answer, and divergent thinking, in which there are many possible answers to a problem.
- Effective decision making, a formal operational process, is a part of critical thinking.

Creative Thinking

- Creative thinking includes divergent thinking, fluency, flexibility, originality, and remote associations.
- In adolescence, the understanding and use of metaphor appears to be an important aspect of creative thinking.
- Conventional schooling often has a dampening effect on students' willingness to risk creative, metaphorical thinking.
- Terman's longitudinal study of 1,000 children with IQs of 135 and above began in the 1920s and continues to influence beliefs about giftedness.
- Criticism of genius, as defined by IQ, states that IQ indicates only precociousness but cannot account for prodigiousness.

K E Y   T E R M S

W H A T   D O   Y O U   T H I N K ?

1. When you read about the progress the mind makes during adolescence, can you remember these changes happening to you? Describe some of your memories of those times.

2. Which of Flavell's seven aspects of intellectual development do you think is the most important during adolescence? Why?

3. Which of the two types of information processing, controlled or automatic, do you use most? Why do you think that is so?

4. What are some examples of egocentric thinking, perhaps drawn from your own experience?

5. Studies show that most people believe they are below average in creativity, which cannot be true, of course. By definition, half of all people are above average. Where do you fit on this trait compared to your acquaintances? Why?

6. Do you believe you can free up your creative abilities? How can you start? Why don't you?

SUGGESTED READINGS

Clavell, J. (1981). *The children's story*. New York: Delacorte Press. Illustrates the way children and youth tend to accept things without question, whereas adults are more likely to fear change and be suspicious of any deviation from the norm.

Erikson, E. (1969). *Gandhi's truth*. New York: Norton. This is one of the best examples of psycho-history, which is the biography of a person as seen from the two disciplines of psychology and history. Mahatma Gandhi was one of the most brilliant figures to ever live. His quest to free India from British domination makes for good reading. The stories about the forces that influenced his youthful thinking are particularly instructive.

May, R. (1975). *The courage to create*. New York: Norton. A brilliant analysis of why most people don't create, and how we might get the courage to do so.

Potok, C. (1967). *The chosen*. New York: Fawcett. This is the story of a boy whose father is a rabbi in a strict Hassidic sect of the Jewish religion. It chronicles the struggle he has over his desire to be a good student and still be "normal."

GUIDED REVIEW

1. Billie learns that it is okay to lift her stuffed dog by the tail, but not the family's real pet dog. In Piagetian terms, Billie has
   a. organized.
   b. assimilated.
   c. accommodated.
   d. differentiated.

2. Stephen's grandfather, "Pop-pop", always wears a tweed hat. Upon seeing a man in a store who is wearing a tweed hat, Stephen calls out "Pop-pop" but soon discovers that this man is not his grandfather. This is an example of
   a. learning.
   b. assimilation.
   c. accommodation.
   d. equilibration.

3. Which of the following statements regarding adaptation is TRUE?
   a. It is easier to assimilate than it is to accommodate.
   b. It is easier to accommodate than it is to assimilate.
   c. It is easier to form schemata than it is to form mental structures.
   d. It is easier to form mental structures than it is to form schemata.

4. What cognitive stage extends from birth to about 2 years old?
   a. formal operations stage
   b. concrete operational stage
   c. sensorimotor stage
   d. preoperational stage

5. Four aspects of thinking—abstract, logic, metacognition, and hypothetical reasoning—occur during the
   a. sensorimotor stage.
   b. formal operations stage.
   c. preoperational stage.
   d. concrete operational stage.

6. In what way do later formal operations differ from early formal operations?
   a. quantitatively only
   b. qualitatively only
   c. both quantitatively and qualitatively
   d. neither quantitatively or qualitatively

7. Answering the question, "What would the world be like if our noses were on the top of our heads?" best illustrates the concept of
   a. logic.
   b. abstract thought.
   c. metacognition.
   d. hypothetical reasoning.

8. The real is to _____ as the possible is to _____.
   a. preoperational thinking; concrete operational thinking
   b. concrete operational thinking; preoperational thinking
   c. concrete operational thinking; formal operational thinking
   d. formal operational thinking; concrete operational thinking

9. Which of the following is an example of adolescent egocentric thinking?
   a. worrying that everyone is staring at the pimple on your face
   b. arguing with your parents about curfew
   c. gossiping about the boy who won the school election
   d. not wanting to walk your younger brother to school

10. Gardner's research on the use of metaphor is based on the theory of
    a. Jean Piaget.
    b. Noam Chomsky.
    c. Claude Levi-Strauss.
    d. all of the above

11. Sucking and crying are examples of _____ (operations/reflexes).

12. Piaget believed that adolescents are in the _____ (preoperational/formal operational) stage of development.

13. According to Piaget, an individual becomes concerned with why things happen during the _____ (concrete operational stage/late formal operational stage).

14. Twelve-year-old Meghan demonstrates her ability of _____ (metacognition/abstract thought) when she realizes that words can have double meanings.

15. _____ (Flavell/Elkind) suggested that there are seven aspects of the transition from childhood to adolescent and adult thinking.

16. The concept of an imaginary audience is an example of adolescent _____ (egocentric thinking/hypothetical reasoning).

17. _____ (Convergent/Divergent) thinking is used when we want to solve a problem by closing in on the correct answer.

18. An individual who is able to do things at a younger age than most people is said to be _____ (prodigious/precocious).

19. Pressures to conform, an excessive quest for success, and intolerance of a playful attitude are obstacles to _____ (creativity/social cognition).

20. _____ (Critical thinking/Creative thinking) involves making judgments after assessing a situation.

Answers

CHAPTER 5

MORALITY AND SPIRITUALITY

All morality consists in a system of rules, and the essence of all morality is . . . in the respect which the individual acquires for these rules.

(Piaget, 1932/1965, p. 13)

Although children are capable of behaving morally, most of them are unable to think about what it truly means to be moral until they reach adolescence. There is an essential connection between the way we think and the way we think about morality. Cognitive changes of adolescence often encourage spiritual growth as well. We may say that morality has two parts: how we think about our behavior toward others (moral judgment), and whether we are motivated to behave as we believe we ought. We open our study of morality by looking at the first part of this definition.

After reading this chapter, you should be able to:

- Explain Piaget's concepts of the practice of and awareness of rules as they develop through childhood and adolescence.
- List Kohlberg's three stages and six levels of moral development and explain his theory of how moral development occurs.
- Describe Gilligan's criticisms of Kohlberg's theory, and list her three stages of moral development.
- Define the concepts of moral components, social cognition, and perspective taking.
- Relate Perry's theory of intellectual/ethical development to the theory of Belenky and her associates.
- List Fowler's seven stages of the development of spiritual faith.
- Discuss these issues from an applied, a sociocultural, and your own point of view.

The Scientific Study of Moral Judgment— Piaget

Best known for his theory of cognitive development (see Chapter 4), Piaget has also received widespread acclaim for his ideas on the moral judgment of the child. Before he began his scientific observations (1932/1965, 1948/1966), morality was seen primarily as a philosophical problem. Piaget defined morality as understanding rules and following them by choice. He studied morality by systematically observing children play the game of marbles (as well as by asking them questions about a number of stories).

The game, as you probably know, calls for each player to place a marble in a two-foot-wide circle drawn in the dirt. Then, by turns, each player rolls a marble from behind a line into the circle. Any marbles knocked out of the area belong to the shooter (see Figure 5.1). When there are no longer any marbles in the circle, each player must put in another marble. This social situation provides a good chance to watch the development of morality, since children of most ages and in many different countries play the game and are able to talk about their understanding of its rules.

Piaget suggested that we can look at the development of moral judgment in two ways: the way children *practice* the rules and the degree of their *awareness* of those rules. This distinction is made because children are often able to follow rules without understanding the reasons for them.

Figure 5.1
The game of marbles.

THE PRACTICE OF RULES

The four stages of rule practice suggested by Piaget are as follows:

- *Stage 1: The individual stage (up to 3 years).* In this period, children begin to grasp the mechanics of the game; they realize that there is regularity in the world and that this regularity may be seen in games. Once the ritual of the game is mastered, young children often want to go on to other games.

  However, youngsters at this stage have no sense of "oughtness" about the rules. They imitate older children and have no awareness that the rules are necessary to play the game. The game itself is not a social activity and is often played alone. Enjoyment is derived only by imitation of the practice of the game.

- *Stage 2: The egocentric stage (4 to 7 years).* Now we can see the beginning of responsibility to follow the rules, although children rarely object if there is more than one winner. The major change at this stage is the awareness that it is important to play the game with other children. Players are beginning to enjoy the social interaction, but their own desires are still uppermost. At this stage, Piaget says, "The very nature of the relation between child and adult places the child apart, so that his thought is isolated, and while he believes himself to be sharing the point of view of the world at large, he is really still shut up in his point of view" (1932/1965, p. 36).

- *Stage 3: The cooperation stage (8 to 11 years).* By now children care about the definition of the rules, which are fixed and common for all players. Manual dexterity in the play of the game is secondary to winning. Even if luck plays a part, the child cares more about winning than about playing well. Cooperation is essential at this stage, but it is still largely a matter of necessity. The main thing is to rigidly follow the rules and to ensure that no one cheats.

- *Stage 4: The codification of the rules (12 years and up).* The change that takes place at this stage is of great importance to teachers and others who work with adolescents because the mental functioning of adolescents takes a sharp new direction: they are beginning to reason hypothetically and abstractly. They can respond, for example, if asked, "What would happen if you were made the judge who determines what will happen to anyone who breaks the rules of your school?" They are able to imagine a large variety of possibilities in any situation.

At this stage, teens become deeply interested in the *reasons behind* the rules and can more readily imagine different and perhaps better rules for the game. They become more consistent in their practice of the game because of their new interest in the **codification** (establishing the details) of rules in general. Adolescents take great pleasure in anticipating all possible cases in a game and in developing rules to cover any of those possibilities.

codification
The tendency of adolescents to establish detailed rules.

AN APPLIED VIEW

Just a Little Game of Cards

I recall that when I was in junior high school, I often got together with some friends to play the game of blackjack, or twenty-one. This card game is quite simple but has a lot of variations. My friends and I began discussing the rules and writing down our agreements as to what would happen if any of the many contingencies were to occur. For three afternoons in a row we talked on and on, and the discussions were so enjoyable that we never did play the game.

We were clearly at the codification stage. Teachers and others working with adolescents should be aware of this strong need of teenagers to codify rules. If teens have a role in making the rules, they are *far more likely to obey them*! Some examples would be asking the class to suggest topics for a writing assignment, to redesign the classroom setup, and to identify consequences for misbehavior.

At any of these stages except the codification stage, children may not be consciously aware of the rules they are following. Piaget suggests that there are three stages in the development of the consciousness of rules.

THE AWARENESS OF RULES

- *Stage 1: Individualism (up to 5 years).* As children learn to play a game, they usually know that some things are allowed and some are forbidden. They also have the sense that these structures apply all the time. But this is the extent of their awareness of regulations. They are primarily interested in doing whatever they want to do and if that fits well with their playmates' interests, fine; if not, that's fine, too.

- *Stage 2: Heteronomy (6 to 9 years).* Piaget asked the children he studied three questions: "Can rules be changed?" "Have the rules always been the same?" and "How did the rules begin?" He discovered that children in this age group have an exaggerated respect for the sanctity of rules. Most are firmly convinced that rules in general have been handed down from authority figures like fathers, or sometimes from a politician or religious leader. Children are aware that rules can change, but believe that it is solely in the power of one of these authorities to change them.

Why do you suppose Piaget chose the word heteronomy to describe stage 2?

This stage parallels the development of the child's superego (described in Chapter 2). Children now identify strongly with those in authority; some see rules as sacred and absolutely unchangeable. Their sense of self is closely tied with their sense of the adults in their family and society. Just as mystics often cannot differentiate between their ideas and those that they believe to be God's, children at this stage cannot distinguish between their own (often mistaken) interpretations of the rule, and rules that actually have been imposed from above.

- *Stage 3: Autonomy (10 to 12 years).* In this stage the dictates of adults and older children are left behind, and the rules become the tools of the player. Players may suggest rules at any time, and if these changes are accepted by all the players, then the game may be altered accordingly.

Adolescents at this age realize that the generations before them have also made changes to rules. They can understand, for example, that the game of marbles was probably invented hundreds of years earlier by children playing with rounded pebbles, and that the rules of the game must have been continuously modified over time.

A SOCIOCULTURAL VIEW

Does Democracy Develop Naturally in All Adolescents?

Piaget believed that the tendency toward codification of the rules is a natural outgrowth of each child's maturation. As a result, he sees biological evidence that as adolescence begins, there is a natural tendency to believe in *democracy rather than authoritarianism*, regardless of the culture in which one is raised.

He came to this view while watching children of several nationalities (although all of his subjects were Europeans). Early adolescents have an instinctive sense of equality among all participants in games. Obviously some ideas are considered more reasonable than others, and an individual counts on the group to recognize these

differences. He or she also expects the players to prohibit unfair innovations because these would make the game less a matter of skill than it should be. There is a strong sense that each person, whether a good player or not, is entitled to one vote and that the group has every right to make changes in the governance of the game they are playing. Thus, because codification develops naturally, Piaget reasoned, every youth has an innately positive attitude toward the principles of democracy!

A positive attitude toward democracy is genetic? What an exciting idea! Does it fit with your observations of young teens in the Western hemisphere?

For Piaget, the development of moral judgment is closely linked to the development of intellectual ability. Thus, good character is largely a matter of good thinking. Taking up where his research ends is the brilliant work of the late Harvard psychologist, Lawrence Kohlberg (1927–1987).

Levels and Stages of Moral Reasoning—Kohlberg

Why do you suppose Kohlberg's first studies were all of male adolescents? Do you think this biases his research?

Kohlberg's first studies were of male adolescent moral judgment (he referred to it as moral reasoning). He saw adolescence as a critical time in the development of moral reasoning because of the cognitive changes that Piaget discovered. Kohlberg studied moral reasoning by asking adolescents to solve hypothetical moral dilemmas. The most famous dilemma that Kohlberg devised is the Heinz dilemma (see the Applied View box titled "The Moral Dilemma").

AN APPLIED VIEW

The Moral Dilemma

To discover the structures of moral reasoning and the stages of moral development, Kohlberg (1970) employed a technique called the *moral dilemma*, in which a conflict leads subjects to justify the morality of their choices. In one of the best known dilemmas, a husband needs a miracle drug to save his dying wife. The druggist is selling the remedy at an outrageous price, which the woman's husband cannot afford. He collects about half the money and asks the druggist to sell it to him more cheaply or to allow him to pay the rest later. The druggist refuses. What should the man do: steal the drug or permit his wife to die rather than break the law? By posing these conflicts, Kohlberg forces us to project our own views.

One caution on the use of moral dilemmas to advance a person's reasoning level: Kohlberg suggested, on the

basis of his research, that people can only advance from one level to the next. That is, it is impossible for a person to move from level 2 to level 4 directly—it is necessary to go through level 3 first. Before you try to advance an adolescent's moral reasoning, you must find out at what stage she or he is operating.

This can cause problems for some instructors who may be anxious to bring students up to their own level of reasoning. If Kohlberg was right, it will be futile to try to advance an adolescent over intermediate levels. Kohlberg suggests a method for bringing students to a higher level: discuss moral dilemmas with students and by using probing questions and by pointing out errors in thinking, guide them to an understanding of the next level.

preconventional level
In Kohlberg's theory of moral reasoning, people at the preconventional level are concerned with avoiding punishment and gaining satisfaction.

conventional level
In Kohlberg's theory of moral reasoning, the level at which the person wants to fulfill society's expectations and be fair to all.

Kohlberg classified adolescents' responses to the dilemmas on the basis of the reasons they gave for their solutions. Two adolescents could give the same solutions and be rated at different stages if their explanations revealed different thought processes. Kohlberg suggested that there are three basic levels of moral reasoning: the preconventional, conventional, and postconventional levels.

LEVELS OF MORAL REASONING

At the **preconventional level**, the person is concerned with avoiding punishment and gaining satisfaction. At the **conventional level** (the one into which most adults fall), the person wants to fulfill society's expectations and to be fair

postconventional level
In Kohlberg's theory of moral reasoning, people at the postconventional level are concerned with moral principles that they have thought carefully about and chosen as their own.

to all. At the **postconventional** level, the person is concerned with moral principles that have been carefully thought about and chosen. Kohlberg divided each level into two more specific stages; thus there are six stages altogether. Table 5.1 lists the levels and stages, with an example of a moral dilemma and typical solutions at each stage.

Most young children and most delinquents are at stages 1 and 2. Most adults are at stages 3 and 4. Kohlberg estimated that 20 to 25 percent of American adults are at the postconventional stages (5 and 6), with only 5 to 10 percent ever reaching stage 6. Since his studies have turned up very few stage-6 adolescents, Kohlberg grouped stages 5 and 6 together (Lickona, 1977). Figure 5.2 summarizes these findings.

Table 5.1	Kohlberg's Stages of Moral Judgment

The dilemma: Al, age 14, sees his brother Jimmy, age 10, steal money from their mother's purse. Should Al tell Mom what Jimmy did?

Preconventional Morality

Stage 1: Obedience. Child is self-centered and has strict pleasure-pain orientation. *Al:* "I wouldn't tell Mom—Jimmy would only get even with me later. It's better not to get involved."

Stage 2: Instrumental. Trade-offs and deals are made, but only if the child sees something in it for her- or himself. Need satisfaction is still uppermost, but an awareness of the value of reciprocity has begun. *Al:* "It's better if I don't tell. I do bad things sometimes, and I wouldn't want Jimmy squealing on me."

Conventional Morality

Stage 3: Good boy, nice girl morality. Child is eager for approval of others and wants to maintain good relations. *Al:* "It's better to tell on him. Otherwise, Mom might think I was in on it."

Stage 4: Authority and social order. Child now seeks approval of society in general, but has rigid ideas as to what rules are; child has a law and order mentality. *Al:* "I have no choice but to tell. Stealing just isn't right."

Postconventional Morality

Stage 5: Social contract. Person makes contracts and tries hard to keep them; attempts to keep from violating the will or rights of others; believes in the common good. *Al:* "I'll try to persuade Jimmy to put the money back. If he won't, I'll tell. I hate to do it, but that money belongs to Mom, and he shouldn't have taken it."

Stage 6: Universal principles. Person shows obedience to social rules, except where they can be shown to contradict universal justice. The principles of pacifism, conscientious objection, and civil disobedience fall into this category. *Al:* "The most important thing is that Jimmy comes to see he's being unfair to Mom. Telling on him won't help out. I'm going to try to show him why he's wrong, then I'll help him earn money to pay Mom back without her knowing."

Excerpt from *Essays on Moral Development: The Psychology of Moral Development, Volume II* by Lawrence Kohlberg. Copyright © 1984 by Lawrence Kohlberg. Reprinted by permission of HarperCollins Publishers.

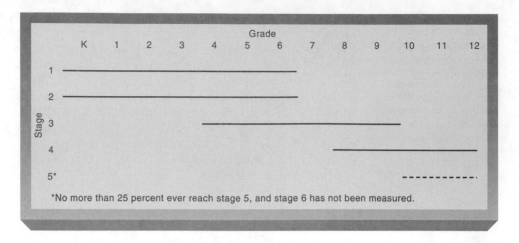

Figure 5.2

The average grade in school when each of Kohlberg's stages prevails.

Note: This figure is summarized from the findings of Lickona (1977).

At the lower stages, people act to *avoid punishment by others;* at the higher ones, they act to *avoid self-condemnation* (the punishment they impose on themselves, like guilt). Kohlberg proposed that what people believe guides how they will act. The reasons for conduct differ at the various levels. Higher-stage subjects might say that they don't cheat because to do so would not fit with how they want to be. Lower-stage subjects might say they don't cheat because they're afraid of being punished. Krebs (1967) found that whereas 70 percent of preconventional subjects cheated on a test, 55 percent of conventional subjects cheated, and only 15 percent of postconventional subjects cheated.

One interesting experiment, the Milgram electric shock obedience test (Turiel, 1974), was used to investigate the willingness of persons at the six stages to inflict pain on others. Prior to the experiment, the morality stage of each subject was assessed. The subjects were told to give shocks of increasing severity to a learner in order to get the person to learn a task quickly. The learner was actually an actor who put on a convincing performance. Most postconventional subjects refused to participate, quit when the victim expressed pain, or said they wanted to quit but felt obligated to fulfill their agreement. The rest of the subjects at the other moral stages willingly continued the experiment.

Kohlberg found that the stage of a person's moral reasoning remains fairly constant, regardless of the content of the dilemma. That is, those who reason at stage 2 tend to do so no matter what the nature of the problem is. Kohlberg's scoring showed that approximately half of an individual's statements about a moral dilemma fell into one stage and the rest generally fell into the two adjacent stages. The same stages are also found in other cultures; the sequence of development is the same, although the speed of development is faster and is more likely to proceed to higher stages in some cultures than in others (Kohlberg, 1984).

In terms of morality, people tend to move from avoiding punishment by others to avoiding self-condemnation.

Kohlberg's stages are invariably sequential; that is, persons cannot get to a higher stage without moving sequentially through the lower stages. There are no specific ages for any of his stages. Like Piaget, Kohlberg believed that any of the stages can be present in adults, even level 1 (for example, a psychopathic killer). He also believes that persons can become fixated at any of the levels. For example, persons at level 2 who are badly mistreated may find it impossible to move on to level 3 because they are so cynical about fellow human beings.

How Moral Development Occurs

What motivates change from a lower to a higher stage? Kohlberg suggested three possibilities:

- *The psychoanalytic explanation.* Freud and his followers (see Chapter 2) propose that the young child develops a set of beliefs about morality based on the ideals of the adults around him. This set of beliefs is internalized and is called the superego. If the child's superego is strong, the child will gradually rise to the highest levels of morality. However, this theory has proved difficult to validate through research.
- *The religious explanation.* Moral behavior is a response to one's conscience—a combination of learned values and the innate ability to discern right from wrong. When the child's learning is compatible with his or her conscience, the child will tend to be "good." If the teachings have not been "proper" (in line with religious values), the child will have a tendency to be "bad." Like the superego theory, research evidence is insufficient to support this position.

Kohlberg, like Piaget, saw conscience largely as a matter of decision-making ability. This is different from the Freudian view. Kohlberg believed that such factors as intelligence, self-esteem, and the ability to delay gratification (also

Which of these three explanations do you find most compelling?

principle of justice
In Kohlberg's moral theory, refers to our inherited potential to recognize when we are being fair or unfair with each other.

called willpower) are likely to play a major role in a person's stage of morality. Philosophical beliefs, like religion and personality characteristics, Kohlberg argued, tend to have only a minor effect on morality.

• *The genetic factor explanation.* The premise is that each person brings a genetic predisposition to her or his social behavior. Kohlberg called this the **principle of justice**. It refers to our inherited potential to recognize when we are being fair or unfair with each other—a concept similar to the religious position. The justice principle states that all human beings are equal in value and that fairness in human interrelations is essential. It is a basic part of our nature and exists universally. Kohlberg felt that justice is the natural result of living in any community. This position is similar to Piaget's theory that democracy is the natural state of human beings.

CRITIQUE OF KOHLBERG'S THEORY

Kohlberg's theory has been criticized on a number of counts. The most obvious criticism arises from the empirical evidence that very few adults reach stages 5 and 6; on a worldwide scale only 1 to 2 percent reach levels 5 and 6. Those mixing stages 4 and 5 account for 6 percent (Snarey, 1985). Children are typically at levels 2 and 3. Adults typically are at levels 3 and 4. Hence, the main difference between children and adults seems to be that adults no longer talk about personal likes and dislikes and begin talking about social institutions and systems.

In addition, Kohlberg's premise that formal operational thinking is necessary for genuine moral understanding has not been confirmed conclusively (Haan & others, 1982). As we mentioned before, not only do few adults reach formal operational thinking, but context may determine whether formal operational thinking takes place. One's moral stage may depend on what one is thinking about (Miller & Bersoff, 1989).

A SOCIOCULTURAL VIEW

The Effects of Culture on Moral Reasoning

Moral reasoning and moral development appear to be influenced by the culture in which an individual is raised. Reid (1990) has proposed that a person's view of the world affects moral reasoning. Reid stated that there are cultural differences in the application of principles, in justifications for behaviors, and in ethical types.

Other researchers (Bersoff & Miller, 1993; Miller & others, 1990) have found that Indian culture maintains a broader and more stringent view of social responsibilities than does American culture. When an individual under emotional duress acts immorally, Americans are less forgiving than are Indians.

Huebner and Garrod (1991) argue that Western theories of moral reasoning do not adequately explain moral development in other cultures. In Buddhist culture, for example, Buddhist philosophy influences moral development—a philosophy quite different from Western tradition. An examination of morality, they argue, is dependent on the world in which one is to be moral. Can you think of other ways that culture may affect moral reasoning?

Kohlberg's theory is alleged to be biased by class and schooling. It has been found that urban, middle-class groups throughout the world score higher on Kohlberg's tests of moral development (Snarey, 1985). It could be surmised that schooling increases self-reflection and, hence, moral thinking. But the more negative conclusion is that because Kohlberg ties his theory to Piaget's stages of cognitive development, less-educated, lower-class, and rural people are at lower cognitive levels as well as lower levels of moral development (Liebert, 1984).

Some critics have concluded that Kohlberg's methodology is at fault. They see his moral dilemmas as verbal justifications of moral ideas. Thus, his methodology is dependent on argumentation skills. This in turn may explain the class and schooling bias; obviously, schooling encourages development of verbal skills. But the question remains—does Kohlberg's methodology reveal how someone acts or how someone says he or she would act?

Another important question about morality is, "Is there a difference between males and females in stages of morality?" Yes, according to research reviewed by Kohlberg (1970). Whereas the average male reaches stage 4 (law and order), the typical female develops no higher than stage 3 (good girl). This difference has been the subject of hot debate. The reasons for this discrepancy in levels of morality of men and women was first investigated by Kohlberg's Harvard colleague Carol Gilligan (1977).

Moral Development and Gender—Gilligan

Gilligan believes that Kohlberg's theory penalizes women for their greater sensitivity to what others think. Gilligan summarizes the situation: "Herein lies that paradox, for the very traits that have traditionally defined the 'goodness' of women, their care and sensitivity to the needs of others, are those that mark them as deficient in moral development" (1977, p. 484). Thus she argues for the importance of the **principle of caring**.

principle of caring
In Gilligan's moral theory, the traits of care and sensitivity to the needs of others based on relationships, which so often typifies women's moral judgments.

Kohlberg (1970) argued that the difference is the result of men having greater practice in moral problem solving. Gilligan (1983) feels that this is not true. She believes the problem lies with male bias inherent in Kohlberg's theory itself:

❑ As long as the categories by which development is assessed are derived within a male perspective from male research data, divergence from the masculine standard can be seen only as a failure of development (p. 114).

This bias results in an emphasis on what is fair and in a lack of attention to other aspects of interpersonal relations.

There are ways in which Kohlberg's and Gilligan's theories of morality are similar to each other and also ways in which they are different from each other. In which ways do you think they are most different?

Gilligan (1983) argued that in addition to male bias, Kohlberg's scoring system is suspect because it is based only on responses to hypothetical examples. In her groundbreaking research, she studied moral development by examining the reasoning of women contemplating an abortion, clearly a serious and difficult moral decision.

Gilligan's lengthy interviews lead her to conclude there are three distinct levels of female moral development, with a specific period of transition between each. She found the following sequence of development in women's moral growth.

Levels of Female Moral Reasoning

I. Individual survival. At this level, reasoning is limited strictly by concern for herself. The woman contemplating an abortion justifies it by saying, "I just don't want a baby, that's all. It would not be good for me now." Or she says, "I just want this baby," even though she has no means to support and nurture it.

IA. Transition from selfishness to responsibility.

II. Self-sacrifice and social conformity. As a result of the first transition, which often occurs during adolescence, the woman moves from selfishness to self-sacrifice. A sense of responsibility and concern for others now dominates her thinking. Even if she wants to have the child, a woman might abort it because "it would not be a good life for the baby, and its father wants me to get rid of it." Another example might be the woman who does not want the child but has it anyway "for the child's sake and/or for the father's sake." Gilligan feels that female lives have traditionally been governed by males. Females, therefore, feel powerless. Because of this, females at this stage justify their position by exalting the life of sacrifice.

IIA. From goodness to truth.

III. Nonviolence. In the second transition, which for many never takes place, women come to recognize their powerlessness as being more a matter of attitude than of necessity. Such a woman learns to "verify her capacity for independent judgment and the legitimacy of her own point of view" (p. 502). Her moral decision now includes her own needs as well as those of others. Now the criterion is to be nonviolent, to cause as little hurt to self and others as possible. "It is true my boyfriend wants me to have an abortion and I can see why he feels that way, but I think it would cause me too much pain, so I'm not going to do it." Such decisions are more difficult than those made at the first two levels because they are more complex.

Table 5.2 compares Gilligan's and Kohlberg's theories.

Since her original work, Gilligan has gone on to further explore real-life moral reasoning. Her recent work has answered many theoretical questions. Among them is the question, "Are there differences between male and female moral reasoning, and if so, why?"

The Origin of Gender Differences in Morality

The care orientation that Gilligan describes places an emphasis on the *interdependence of relationships* (see Chapter 6). Kohlberg's justice orientation emphasizes rights and fairness. Gilligan finds that women and girls often use the

Table 5.2	Comparison of Gilligan's Morality of Care and Responsibility and Kohlberg's Morality of Justice	
	Morality of care and responsibility	**Morality of justice**
Primary Moral Imperative	Nonviolence/care	Justice
Components of Morality	Relationships Responsibility for self and others Care Harmony Compassion Selfishness/self-sacrifice	Sanctity of individual Rights of self and others Fairness Reciprocity Respect Rules/legalities
Nature of Moral Dilemmas	Threats to harmony and relationships	Conflicting rights
Determinants of Moral Obligation	Relationships	Principles
Cognitive Processes for Resolving Dilemmas	Inductive thinking	Formal/logical-deductive thinking
View of Self as Moral Agent	Connected, attached	Separate, individual
Role of Affect	Motivates care, compassion	Not a component
Philosophical Orientation	Phenomenological (contextual relativism)	Rational (universal principle of justice)
Stages	I. Individual Survival IA. From Selfishness to Responsibility* II. Self-Sacrifice and Social Conformity IIA. From Goodness to Truth* III. Morality of Nonviolence	I. Punishment and Obedience II. Instrumental Exchange III. Interpersonal Conformity IV. Social System and Conscience Maintenance V. Prior Rights and Social Contract VI. Universal Ethical Principles

*Marks a transition stage.

From M. Brabeck, "Critical Thinking Skills & Reflective Judgement" in *Journal of Applied Developmental Psychology*, 4:23–24. Reprinted with permission of Ablex Publishing Corporation, Norwood, NJ.

ethic of care in making decisions about their lives (Gilligan & others, 1988; Gilligan & others, 1990). This position has been further supported in recent research by Skoe (1998) who suggests the importance of care based morality for healthy human development. Gilligan feels that the ethic of care stems from the *attachment* of children to their mothers (Gilligan & Wiggins, 1987). Through attachment, children learn about the give-and-take nature of relationships. Girls learn to influence their mothers through connection, laying the foundation for the ethic of care.

Boys learn an ethic of justice through their early relationships with their mothers. Boys learn to influence their mothers by asserting themselves—striking out on their own. Gilligan says that boys become more aware of the *inequality* that is a part of the mother/child relationship: mothers are powerful and children are dependent (Gilligan & Wiggins, 1987). Both girls and boys experience attachment and inequality in their early relationships with their mothers, but it is attachment that is most important to girls and inequality that is most important to boys. This results in the female preference for care and connection and the male preference for justice and fairness, according to Gilligan.

CRITIQUE OF GILLIGAN'S THEORY

Gilligan's initial work was conducted in response to Kohlberg's theory of moral development (Gilligan, 1982). She felt that Kohlberg's dilemmas were not sensitive to interpersonal elements of moral reasoning (care). This, she argued, resulted in females being rated at Kohlberg's third stage of moral reasoning, as opposed to the fourth stage for males. One researcher (Walker, 1984; 1989) reviewed 108 studies of moral reasoning and found that gender differences in Kohlberg's stages did not exist. Instead, Walker found that level of education accounted for most of the differences. Therefore, Kohlberg's assumptions of a gender difference and Gilligan's assumption of bias against women appear to be inaccurate (Boldizar & others, 1989; Gibbs & others, 1984).

Nevertheless, strong support for Gilligan's position comes from recent studies (Skoe, 1998; Skoe & Gooden 1993). These researchers designed an interview schedule called the *Ethic of Care Interview* (ECI). As expected, females score significantly higher than males on it. Scores on this instrument were correlated with content analyses of real-life moral dilemmas of 23 boys and 23 girls, all 11 to 12 years old. More girls than boys described personal life dilemmas. In validation of Gilligan's theory, the researchers found that "girls tended to be concerned about hurting others and maintaining friendships, whereas boys tended to be concerned about leisure activities, such as sports, and avoiding trouble" (p. 154). Bookman (1999) suggests that many criticisms of Gilligan's work are over-simplifications and misrepresentions. Bookman believes that Gilligan's work should help us understand gender development as

A SOCIOCULTURAL VIEW

Attachment Among Zawiya Youths

Davis and Davis (1989) studied adolescents in Morocco and found that both male and female Zawiya youths valued reciprocity and mutual respect over abstractions of ethics. This finding supports Gilligan's theory. The youths' responses to moral dilemmas revealed their culture's ethics in which reciprocal obligations to friends and family are central. This evidence would seem to support an ethic of connectedness and caring rather than one of autonomy and independence.

AN APPLIED VIEW

Morality and Teenaged Pet Ownership

Researchers asked young adolescents to write an essay in response to the question, "In case of fire, what one possession would you save and why?" (Gage & Christensen, 1990). One-quarter of the 1,265 responses were about saving a pet. The researchers used Kohlberg's coding scheme to rate the essays. They found that most essays were in stage 2, hedonism. Below are two essays by 13-year-olds that Gage and Christensen received. Do you agree that they should be rated in stage 2?

❑ In case of fire, the one possession I would save is my dog. My dog is always by my side when I am sick or nobody else is around, and she always protects me when nobody else is home. She is fun to play ball and tug-of-war with and take walks and play tag with. She is always in a happy mood. Another and the most important reason I would want to save her is because she is my best friend. She is easy to talk to and never gives away my secrets. She is a great partner and very helpful. She would also

need my help getting out of the house, so we would need each other in a way. She would probably do the same thing for me if she were me and I were she.

❑ If there were a fire in my house, the first thing I would save besides my family, would be my parakeet, Humphrey. I got him for Christmas and have spent a lot of time training him and building up a friendship that I hope will last forever. Secondly, I would save him because I am not through training him. I want to see him grow up and learn the ultimate trick—talking. Finally, I would save him because he is the only thing of real value that I own. He is sitting on this pen making faces at me, trying to get me to play with him—maybe he finds this paper hard to write.

Into what other stages might you put these two responses? How would Carol Gilligan rate them? When you were 13, did you have a pet? Do either of these essays reflect your own feelings about your pet when you were that age?

a complex process that must be understood within a cultural and historical framework. Interestingly, Miller and Bersoff (1999) who also focus on the importance of cultural factors, recognize the contributions of Gilligan but believe that the model does not sufficiently acknowledge the importance of cultural factors.

Carol Gilligan's work has also been criticized (Brabeck, 1989; Philibert, 1987) as to how well her findings can be generalized. Since her research consists of interviews with small numbers of women, it is unclear if it represents all women (Puka, 1989). Other researchers note that Gilligan's theory of justice and care isn't really new: Aristotle in the third century b.c. recognized both ethics (Waithe, 1989). Gilligan's use of justice and care, as ethics preferred by men and women respectively, also isn't new: Lacan wrote about this years earlier (Brabeck & Weisgerber, 1988). And even though Gilligan says that both justice and care are equally important (Gilligan & Attanucci, 1988), some researchers have concerns that care is presented as being preferable to justice (Waithe, 1989). (Yet, Kohlberg's hierarchy of stages assumes that an abstract principle of justice is preferable.) Others think that we are led to believe that justice and care are mutually exclusive (you can have only one or the other) (Hare-Mustin, 1987; Higgins, 1989; Puka, 1989). These researchers say that both ethics should be equally valued in our society (Brabeck, 1989).

Moral Components—Rest

Another researcher who has built on Kohlberg's foundation is James Rest (1983). Rest and his colleagues (1999) have recently proposed some new ideas about moral development. They suggest that morality can be divided into two categories: macromorality and micromorality. Macromorality involves formal societal structures like institutional rules, while micromorality involves everday face to face relations. These researchers believe that Kohlberg's framework is most useful for understanding macromorality. Rest (1983) asks the important question, "By what thinking process does a person solve a moral dilemma?" Rest has found that the path to a solution is usually composed of four **moral components**:

moral components
In Rest's moral theory, the four steps that are used in the process of solving moral dilemmas.

- *Component I: Interpreting the situation.* The first step is to figure out how different plans of action would affect the people involved. This is social cognition (which we will discuss in detail later in this chapter). The ability to think critically about interpersonal issues develops through age and experience. Rest also feels that emotional arousal influences how we might interpret a situation. A person who is insensitive to the needs of others may have difficulty with this component.
- *Component II: Figuring out the ideal moral course of action.* The second step is to determine what ought to be done. This involves balancing personal wishes and **social norms** (what the society we live in says is right). A person who is unable to think about all the complexities of a situation may have difficulty with this component.

social norms
The behavior that society says is appropriate.

- *Component III: Deciding what to actually do.* The third step is to select a path of action that is close to the ideal but is realistic to all circumstances. Rest believes this involves decision-making ability and moral motivation. Decision

"Figuring out the ideal moral course of action" is the second of Rest's four moral components.

making is usually talked about as a bargaining process that happens in a person's mind: weighing all the variables of a situation as costs and benefits.

Rest thinks that the decision-making model fails to include the emotional component of moral situations, and so it must be considered along with motivation. Motivation to act in ways that an individual thinks are moral develops throughout the lifespan. This is difficult to pinpoint. It may come from a desire to help others or from empathy for others. People who find their moral values to be in conflict with other values (like wanting to fit in with a peer group that has different values) may have difficulty with this component.

- *Component IV: Implementing the plan.* The final step is to actually do what has been decided. As Rest points out, good intentions can be quite different from good deeds. To carry out a plan of moral behavior, Rest notes that an individual needs perseverance and self-regulation. A person who carefully decides on a course of action but loses sight of the goal when implementing the plan may have difficulty with this component.

Rest developed a test to capture the first three components of moral behavior. It has often been used with adolescents (Lapsley & others, 1984; Lonky & others, 1988). As did Kohlberg, Rest seeks to understand moral reasoning. His test, called the *Defining Issues Test* (DIT), uses six moral dilemmas, including Kohlberg's story about Heinz. The person first reads a dilemma, then reads 12

AN APPLIED VIEW

Suggestions for Promoting Moral Growth in the Classroom

1. Focus on establishing the classroom as a community where the participants will live and learn together in an atmosphere of respect and security.
2. Provide opportunities for the class to have a voice in establishing the rules of the classroom.
3. Choose punishments that relate to the offense, stressing where possible the effect of the student's action on the group.
4. Make distinctions between criticisms of academic work and criticisms of behavior, and rules for the good order of the school and rules affecting justice and human relations.
5. Provide opportunities for peer group work.
6. In stories and discussion of everyday experience, help the students to consider the feelings of other real or fictional persons.
7. Role play experiences from daily life, events that lead to disappointments, tensions, fights, joys,

in order to provide opportunities for the pupils to see the event from perspectives other than their own.

8. Discuss with the class what they consider fair and unfair classroom procedures and relationships.
9. Frequently take time to listen to each student's responses to questions of moral judgment, and stimulate discussions that will provoke higher-stage reasoning, using literature, film, and life experiences.
10. Avoid making judgments about moral development on the basis of behavior. People at six different stages might perform the same action, but for different reasons.

Reprinted from *Moral Development: A Guide to Piaget and Kohlberg*, by Ronald Duska and Mariellen Whelan. Copyright © 1975 by The Missionary Society of St. Paul the Apostle in the State of New York. Used by permission of Paulist Press, Mahway, NJ.

statements about issues that are involved in solving the dilemma. The person rates the importance of each statement in solving the dilemma. An example of an issue statement is, "Whether a community's laws are going to be upheld" (Rest, 1983, p. 584). The person goes through this process with all six dilemmas. An evaluation of all responses determine an individual's stage of moral development.

Rest's DIT assumes that moral thinking is based on principles of how people should cooperate in social situations. The higher the score on the DIT, the better the person is at understanding rights and duties in a social setting. Rest and his associates (1999) have recently updated the DIT to create the DIT2. Rest believes that the DIT2 is a more powerful instrument than the DIT. The primary changes in the revised test are shorter more clearly written test items, and a new method of analysis.

As we mentioned previously, social cognition is a major contributor to moral behavior. As the Native American saying states, "One must walk a mile in another man's moccasins in order to know why he does what he does." That is, we cannot recognize a person's motivations unless we have had similar experiences. Let us take a closer look at this concept.

Social Cognition

social cognition
The ability to think critically about interpersonal issues, which develops through age and experience. Used to make sense of other people and to decide how to interact with them.

Social cognition is the means by which people make sense of other people and decide how to interact with them (Fiske & Taylor, 1984). Given adolescents' interest in themselves and in peers, it is not surprising that social cognition plays a critical role in adolescent thinking. As mentioned in Chapter 4, it is during adolescence that the ability to think in the abstract emerges. This ability combines with an increase in social experience and the result is a major change in the way adolescents can understand themselves and others. In Morocco researchers found that adolescent social cognition levels correlated with educational levels; in other words, youths with advanced education did better on the average than those with a limited education (Davis & Davis, 1989).

The adolescent is also able to entertain conflicting views. For example, if a young boy has a fight with a friend, he'll often declare, "I'm not your friend anymore." And that's exactly what the child means: he does not have the cognitive capacity to understand that he can both be angry at his friend and still be friends with him. The young child cannot hold two such conflicting views in his head at once. In adolescence, the ability to analyze his own thinking will allow him to make that leap. Adolescents, because of their interest in maintaining social connections, can misuse this capacity. This can lead to exaggerated conclusions. How many times have you heard a teenager say something like, "I thought he was a really nice guy, but now I can see that he is a total jerk!"

THE ROLE OF SCHEMAS

schema
Mental representations of our roles, of other people and their roles, and of the situations in which they might interact.

Fiske and Taylor (1984) have stated that research on social cognition emphasizes two areas. The first is the study of how we store and sort social information in our minds. We develop **schemas**, which are mental representations of our roles, of other people and their roles, and of the situations in which they might interact. These schemas function like files into which we sort and classify new information. Examples would be "people whom I'd better be nice to" or "situation in which I ought to cooperate/share."

When we interact with the world, we are constantly receiving new information. We take that information and sort it into our cognitive files. Like any filing system, we build new folders when we come across information that won't fit into the folders we already have. This process happens unconsciously and it helps us store information efficiently. Schemas also guide what we expect to see. When we come across something that in some ways seems familiar, we look into our files to see how it fits with what we know. For example, an adolescent girl has a schema about what adults are like based mostly on her parents (this is assimilation; see Chapter 4). When she interacts with a new adult about the same age as her parents, that schema will shape her initial impressions of the new adult. The schema shapes what the adolescent expects, which in turn shapes how she behaves. But how do schemas change as we have new experiences?

SOCIAL COGNITION AND THE PROCESS OF CHANGE

The second area of social cognition is the study of the process of change (known as accommodation; see Chapter 4). This studies how people use schemas and what makes schemas change or stay the same. Let's again consider the same adolescent. If she is bused to a suburban school and meets a lot of adults who are very different from her parents, will her expectations (schemas) of adults change? How does cognitive functioning change in response to experiences with others? These are the issues that researchers studying social cognition try to sort out.

Traditional theories of cognitive development describe how people's interactions with the world alter their way of thinking about things. Piaget's theory, for instance, emphasizes an individual's hands-on experiences with *things*, not people. It is only when Piaget talks about moral development and older children that he focuses on social experiences and the importance of other people in development.

Who in your opinion is more right, Elkind or Lapsley? What is your reasoning?

Similarly, Elkind's classic work (1967) focuses on how an adolescent's cognitive development shapes social development. As you recall from Chapter 4, Elkind described the advent of formal operations as leading to two adolescent thinking traits. Teens come to think that everyone is thinking about and watching them (imaginary audience) and they also are filled with a feeling of invincibility—that they can rise above anything (personal fable).

perspective taking
A child's personal theory about what other people are like, which affects the way she or he relates to other children. This view of what other people are like changes over time.

Lapsley (1993, 1989) thinks that Elkind's theory overlooks the fact that social experiences are important throughout life, not just starting in adolescence. He argues that to understand adolescent social cognition, we need to think about how it develops in childhood as well. Lapsley's ,position is further supported by Vartanian's (1997) recent research looking at the ego development process of adolescents. Robert Selman's theory of **perspective taking** focuses on the importance of social experience.

Perspective Taking—Selman

Robert Selman (1980) believes that the way a child relates to other children depends on that child's personal theory about what other people are like. Furthermore, the child's view of what other people are like changes over time. The child's *perspective* on being a person develops with age. The peer review section in Chapter 8 includes a summary of Selman's five levels of perspective taking at each of five age ranges, together with the implications for friendship of each. Note that a major theme that flows through the levels is the growing realization that other people have different points of view, and that these must be taken into consideration when dealing with them.

Selman's work, together with what we know about formal operations, helps explain how the adolescent's imaginary audience comes to be. When an adolescent is at Selman's level 3, there is a high degree of self-consciousness that pushes the adolescent to use her or his new cognitive capabilities to create the imaginary audience. Lapsley and Murphy (1985) propose that the imaginary audience becomes a way for the adolescent to *anticipate the reactions of others*.

This is different from Elkind's definition of the imaginary audience, which says that adolescents attribute their own perspective to others. At the same time, teens weave together their ability to think hypothetically (formal operations) and their ability to analyze relations with others. As a result, the imaginary audience seems to emerge.

How does Selman's work explain the personal fable? Again Lapsley and Murphy note that the self-reflective awareness of level 3 helps adolescents understand themselves. This leads to an appreciation of their own mental powers that can result in a feeling of great power. This sense of power leads to what Elkind has called the personal fable. However, when adolescents enter Selman's level 4, they realize that there are limits to their ability to control their own thoughts. This pushes them to abandon the personal fable.

In one study, Lapsley (1989) measured adolescents' ability to monitor their own behavior, their use of imaginary audience, and the use of their personal fables. He found that high self-monitors had higher scores on imaginary audience and on personal fables. This supports his theory that imaginary audience and personal fables are actually tools to sharpen adolescent social cognitive skills—not retreats from that development.

Another aspect of adolescent social cognition is the emergence of relative thought. The awareness of their own ability to come to their own understandings about the world sparks adolescents to see knowledge as *manufactured by*

Teenagers deal with the uneasiness of relative thought by simultaneously emphasizing their own individuality and becoming ardent members of some club or cult.

the individual. Therefore they see knowledge as person-relative (Lapsley, 1989). This realization is "typically accompanied by a sense of uneasiness that is hard to shake off " (Chandler, 1975, p. 172). In his classic work, Michael Chandler notes that there are two main ways that adolescents deal with this uneasiness: they emphasize their own individuality and at the same time, many also become ardent members of some club or cult. Both defenses help them deny the loneliness they feel. By celebrating individuality, the adolescent is saying, "We're all different and I think it's great . . . it's not at all scary." By becoming a committed group member, the adolescent is saying, "Even though we may all seem different, really we're all very much alike. Look how much I'm like my friends." As they develop, most are able to reconcile these two extreme reactions, and realize that some knowledge may be relative and some may not.

This concept of relativity—what is true depends on how you view it—plays a growing role in adolescent thinking, and especially how they think about morality. Two research efforts, one led by William Perry and the other by Mary Belenky and her associates, have made excellent contributions to our understanding of this thinking.

Intellectual/Ethical Development—Perry

Perry (1968a, 1968b; 1981) studied the intellectual/ethical development of several hundred Harvard College male students aged 17 to 22. The results of these studies led Perry to suggest a sequence of intellectual and ethical stages that typically occur during the transition from late adolescence to early adulthood. This sequence consists of nine positions on the nature of truth, which progress from belief in absolute authority to the recognition that one must make commitments and be responsible for one's own beliefs. A recent study by Zhang

(1999) comparing Chinese and American students indicates that Perry's stages may not be universal. Zhang found that the intellectual/ethical development of the Chinese university students did not follow the pattern outlined by Perry, whereas the intellectual/ethical development of the American university students did. This finding suggests that intellectual/ethical development might vary as a function of culture and educational system.

PERRY'S STAGES AND POSITIONS

dualism
The first of Perry's three intellectual and ethical stages in which "things are either absolutely right or absolutely wrong."

relativism
The second of Perry's three intellectual and ethical stages in which "anything can be right or wrong depending on the situation; all views are equally acceptable."

commitment
The third of Perry's intellectual and ethical stages in which "because of the available evidence and my own understanding of my values, I have come to new beliefs."

Perry's nine positions are divided into three broader stages: **dualism, relativism,** and **commitment**.

Dualism **("Things are either absolutely right or absolutely wrong.")**
- *Position 1:* The world is viewed in such polar terms as right versus wrong, we versus they, and good versus bad. If an answer is right, it is absolutely right. We get right answers by going to authorities who have absolute knowledge.
- *Position 2:* The person recognizes that some uncertainty exists, but ascribes it to poorly qualified authorities. Sometimes individuals can learn the truth for themselves.
- *Position 3:* Diversity and uncertainty are now acceptable but considered temporary because the authorities do not know what the answers are yet. The person becomes puzzled as to what the standards should be in such cases.

Relativism **("Anything can be right or wrong depending on the situation; all views are equally acceptable.")**
- *Position 4a:* The person realizes that uncertainty and disagreement are often extensive, and recognizes that this is legitimate—"anyone has a right to an opinion." It is possible for two authorities to disagree with each other without either of them being wrong.
- *Position 4b:* Sometimes the authorities (such as college professors) are not talking about right answers. Rather, they want students to think for themselves, supporting their opinions with data.
- *Position 5:* The person recognizes that all knowledge and values (including even those of an authority) exist in some specific context. It is therefore relative to the context. The person also recognizes that simple right and wrong are rare, and even then, they exist in a specific context.
- *Position 6:* The person realizes that because we live in a relativistic world, it is necessary to make some sort of personal commitment to an idea or concept as opposed to looking for an authority to follow.

Commitment **("Because of the available evidence and my own understanding of my values, I have come to new beliefs.")**
- *Position 7:* Persons begin to choose the commitments that they will make in specific areas. This is where social cognition is particularly relevant. The growth in cognitive ability that usually comes with adolescence permits a much deeper awareness of all the ways adolescents might relate to fellow human beings.

- *Position 8:* Having begun to make commitments, the person experiences the implications of those commitments and explores the various issues of responsibility involved.
- *Position 9:* The person's identity is confirmed through the various commitments made. There is a recognition of the necessity for balancing commitments. Perry (1981) describes this ninth position:

> ❏ This is how life will be. I will be whole-hearted while tentative, fight for my values yet respect others, believe my deepest values are right yet be ready to learn. I see that I shall be retracing this whole journey over and over—but, I hope, more wisely (p. 276).

Some students move through these stages and the nine positions within them in a smooth and regular fashion; others, however, are delayed or deflected in one of three ways:

1. **Temporizing**: Some people remain in one position for a year or more, exploring its implications, but hesitating to make any further progress.
2. **Escape**: Some people use opportunities for detachment, especially those offered in positions 4 and 5, to refuse responsibility for making any commitments. Since everyone's opinion is "equally right," the person believes that no commitments need be made, and thus escapes from the dilemma.
3. **Retreat**: Sometimes, confused by the confrontation and uncertainties of the middle positions, people retreat to earlier positions.

If young adults can avoid these traps, they should be able to achieve the commitments that are the hallmark of the mature person. Although some have criticized Perry's theory (see Brabeck, 1984; Kitchener & King, 1981), there is about it a certain common sense. At least, his ideas seem to fit well with our combined 34 years of experience teaching college students. Now let us turn to the research that was spurred by Perry's work.

"Women's Ways of Knowing"

In a collaborative study, four female psychologists (Belenky & others, 1986) set out to answer the question, "Do women's ways of knowing develop differently from men's? If so, how do females come to learn and value what they know?" Their study was rooted in Perry's work and also the work of Carol Gilligan.

Belenky and her associates conducted a series of lengthy and intense interviews with 135 women of diverse socioeconomic backgrounds. They found five different categories of ways in which women know and view the world. While some of the women interviewed clearly demonstrated a progression from one category to the next, the researchers contend that, unlike Perry and Gilligan, they are unable to discern a progression of clear-cut stages. Although the implications for ethical/intellectual development in adolescent females are not yet clear, it should be obvious that this research is going to have an impact on adolescent psychology, so we thought you should know the findings.

temporizing
An aspect of Perry's theory of ethical development in which some people remain in one position for a year or more, exploring its implications, but hesitating to make any further progress.

escape
Perry's term for refusing responsibility for making commitments. Since everyone's opinion is "equally right," the person believes that no commitments need to be made and thus escapes from the dilemma.

retreat
According to Perry's theory of ethical development, when someone retreats to an earlier ethical position.

FIVE FEMALE PERSPECTIVES ON KNOWING

The five perspectives are:

1. *Silence.* Females in this category literally described themselves as deaf and dumb. These women feel passive and dependent. Like players in an authority's game, they feel expected to know rules that don't exist. Their thinking is characterized by concepts of right and wrong, similar to the males in Perry's first stage. Questions about their upbringing revealed family lives filled with violence, abuse, and chaos. The researchers noted that "gaining a voice and developing an awareness of their own minds are the tasks that these women must accomplish if they are to cease being either a perpetrator or victim of family violence" (p. 38).

2. *Received knowledge.* These women see *words* as central to the knowing process. They learn by listening and assume truths are received from authorities. They are intolerant of ambiguities and paradoxes, always coming back to the notion that there are absolute truths.

   Received knowers seem similar to the males that Perry described in the first stage of dualism, but with a difference. Perry's subjects feel a great affiliation with the knowing authority. The women of this perspective are awed by the authorities but are far less affiliated with them. In contrast to the males of Perry's study, received knowers channel their energies and increased sense of self into the care of others.

3. *Subjective knowledge.* In their study, the researchers noted that women in the subjective knowledge category often had two experiences that pushed them toward this perspective: some crisis of male authority that sparked a distrust of outside sources of knowledge, and some experience that confirmed a trust in themselves. Subjectivists value their intuition or firsthand experience as their best source of knowledge and see themselves as "conduits through which truth emerges" (p. 69). The researchers note that subjectivists are similar to males in Perry's second stage (relativism) in that they embrace the notion of multiple truths.

4. *Procedural knowledge.* The women of this perspective have a distrust of both knowledge from authority and their own inner authority or intuition. The perspective of procedural knowledge is characterized by an interest in how you say something rather than what you say. They also have a heightened sense of control. This category is similar to Perry's relativist stage, where students learn analytic methods that authorities sanction, except that it emerges differently in women because they don't affiliate with authorities.

5. *Constructed knowledge.* The hallmark of constructed knowers is an integration of the subjective and procedural ways of knowing (perspectives 3 and 4). Women of this perspective note that "all knowledge is constructed and the knower is an intimate part of the known" (p. 137). They feel responsible for examining and questioning systems of constructing knowledge. Their thinking is characterized by a high tolerance of ambiguity and internal

contradiction. Indeed, the women whose ways of knowing were of this perspective often balance many commitments and relationships as well as ideas.

Perry's original research dealt only with male subjects, and the research of Belenky and her associates was only with females. Alishio and Maitland-Shilling (1984) have focused on gender difference in ethical/intellectual growth as it related to sexual identity and interpersonal relations. In their study, 31 females and 29 males were interviewed using Perry's method in the four content areas of occupational choice, interpersonal relationships, sexual identity, and religion and ego development.

Results of the study indicate that the two genders differ in the ways they form their personalities in late adolescence. Males focus on occupational issues while females are primarily concerned with interpersonal issues. The two genders also differ in their approach to interpersonal relationships. Men approach relationship development as a dimension of achievement and autonomy, while women center on issues of sexuality, trust, and intimacy. The process is more complementary for females than for males. The results from this study seem to support Gilligan's (1982) view that developmental theory is deficient in its explanation of personality and identity formation in women. In *Knowledge, Difference and Power: Essays Inspired by Women's Ways of Knowing* (Goldberger & others 1996) the original authors have edited a collection of essays that continue the discussion begun in the first book. The essays investigate the issues of race, class, gender and access and look at how these variables overlap to support and differ from the basic premise of *Women's Way of Knowing*.

WHAT'S YOUR VIEW?

Some Important Questions About the Nature of Morality

The work of Piaget, Kohlberg, Gilligan, Rest, Selman, Perry, and Belenky and her associates has greatly advanced our knowledge of intellectual and moral development in the late adolescent and early adult years. It has also produced much controversy. Many questions remain to be answered. For example, does socioeconomic level make any difference? What about cultural background? Is gender very important in some cultures and unimportant in others? Is morality really only a matter of cognitive growth (those who behave well do so because they think well)? What do you think?

The capacity to introspect, to examine one's thoughts and feelings and reason about them, which begins during adolescence, leads many young people to questions of a spiritual nature (Elkind, 1980). Just as with moral development, a person's spiritual growth is greatly affected by cognitive changes. In the next section, we examine the spiritual side of adolescent life.

Spirituality and the Development of Faith

❏   There is but one true philosophical problem, and this is suicide: Judging whether life is or is not worth living amounts to answering the fundamental question of philosophy.
Albert Camus, *The Myth of Sisyphus*, 1955

Spirituality relates to the meaning of life—whether life is worth living and why it is worth living. The attempt to better understand the reasons for living may include striving to know the intentions of a Supreme Being. It may include reading inspired books like the Bible or the Koran, or trying to discern the purposes and goals of some universal life force by studying such subjects as the history of biological changes in species. Spirituality includes all our efforts to gain insight into the underlying and overriding forces of life. For many, it is the only justification for morality (Dacey & Travers, 1991).

In a recent investigation, Baker (1989) asked teenagers on five continents what was important to them. He found *the search for a deeper understanding of the meaning of life* strong in many of the young people he interviewed. This was especially common in highly industrialized countries:

> ❐ Buffeted by dread and joy, urged by parent, teacher and priest to make the "right" choices and thus enjoy to the full a rare, once-only gift, inundated by potent images of many, often incompatible "lifestyles": the young may indeed be seen at the very deepest level as impelled to find a faith, a point of rest and defense, a touchstone by which they can accept or reject, love or hate, act or not act. They yearn to belong, to join with their peers in some high enterprise, and thus to release the extraordinary passion and energy youth commands—or had better command, lest it turn erratic and self-destructive (p. 13).

Berman (1990) found that some young people are immersed in the desire to acquire the trappings of wealth or live only for the sensation of the moment. Others, however, speak of the importance of spirituality in their lives. One girl who is committed to an Eastern religion called Siddha Yoga stated, "If I hadn't discovered this religion, I would be dead . . . I was suicidal for, like, years" (p. 197). A young man who had given up drugs and become a Christian spoke of the sharp division of lifestyles he observed around him:

> ❐ There's no middle-of-the-road around here, there's no sitting on the fence. You're either living for your money, surfing, sexual pleasure, how many girls or how many guys you can score, or you're living a moral life that has meaning and substance to it . . . there's lots of Christians and lots of people that are just the opposite and very little in-between (p. 204).

RELIGIOUS BELIEF AND PARTICIPATION

Religious belief and participation may be one dimension of spirituality, but as Parks (1986) points out, faith is more than just belief in religious doctrines. Faith is an activity that seeks and composes meaning from all of one's experience. As such, it is a vital activity of adolescents and young adults. Vitz (1977) observed that faith in the humanistic psychology and therapies of the 1960s and 1970s served for many as a substitute religion. Today, therapeutic practitioners themselves are realizing that there are some problems beyond the scope of mental health professionals. Many are themselves once more recognizing the value of religion (Butler, 1991). Furthermore, Aalsma and Lapsley (1999) suggest that religiosity has been consistently associated with positive mental health outcomes. Therefore they suggest that practitioners should explore resources of religious tradition as part of a therapeutic program.

One thing therapists cannot provide is the sense of community and commitment to something greater that many adolescents and adults find in church membership. Some teens gain a sense of community belongingness and support as well as spiritual values through participation in church. For those troubled youths who suffer from addiction or from living with an addicted family member, there is help from membership in the 12-step recovery groups such as Alcoholics Anonymous, Narcotics Anonymous, and Alateen.

For some, the inspiration gained from the example set by an admired person, present or historic, serves as a model for their own growth in spiritual values. Smith (1999) recently found that adolescent religious participation was found among other belonging variables (connection to others, close family and participation in extracurricular activities) to be a significant predictor of civic and political involvement in young adulthood. This is an important finding as we grapple with ways to help encourage the development of productive, prosocial members of society. Many of today's young people state that they have no heroes to inspire them, but others have a devotion to popular music stars that approaches religious proportions (Baker, 1989). Navone (1990) argues that whether or not we are conscious of it, we all do have heroes who shape our values, and it is very important that we learn to choose ones who are worthy.

THEORIES OF SPIRITUALITY

Early in this century, Freud dismissed the importance of the spiritual dimension of human life and taught that religions were no more than mass delusions (Wulff, 1991). It was in part because of a strong disagreement about the existence and value of the spiritual that Carl Jung, a favorite student and colleague of Freud's, broke permanently with him. Jung went on to study many religions and devised an explanation of spirituality that emphasized its development only in the second half of life.

- *Jung's theory of spirituality.* According to Jung, life has two major parts. In the first half of life, a person develops individuality through thinking, feeling, sensing, and intuiting. By the age of 35, we have become truly individual and are quite different from one another. The second half of life has a goal opposite from the first half. We now move away from separateness and toward a new wholeness. In the process, introspection marks the beginning of adult spirituality.
- *Frankl's theory of spirituality.* Viktor Frankl first began to realize the importance of his spiritual nature during World War II, when he was imprisoned in a concentration camp. He managed to stay alive in the camp for six years, even though he knew his entire family had been murdered by the Nazis. In 1978, he reaffirmed his belief in the human need for meaning beyond selfish interests and materialism (Frankl, 1978). Frankl believes that all humans are born with an "unconscious religiousness" as well as a basic conscience (Wulff, 1991).

Frankl's theory of spiritual development (1970) describes three interdependent stages. In each stage there is a predominant dimension.

1. The first stage is the *somatic* (physical) dimension. Its dimension is directed toward keeping the individual and the species alive, through eating, drinking, staying warm, etc. This dimension exists at birth and continues throughout life.
2. The *psychological* dimension begins to form at birth and by early adulthood is fully developed. This dimension includes instincts, drives, and needs for interacting with the environment.
3. The *noëtic* dimension has its roots in childhood but develops primarily in late adolescence. It is spiritual not only in the religious sense, but also in its search for the meaningfulness of life. Frankl sees the freedom to make choices as uniquely human and as the basis of responsibility. Reason and conscience are also based in the noëtic dimension.

Frankl believes that development in the somatic and psychological dimensions is a result of the sum of different influences that shape an individual. In contrast, the noëtic is *more than the sum of its parts*. It makes adults responsible for inventing and reinventing themselves despite any failure of parents or upbringing. Here in the noëtic dimension, a person can overcome early training and come to a higher level of spiritual thought and moral behavior (Dacey & Travers, 1991). Using Frankl's framework, Shantall (1999) spent extensive time with a group of Holocaust survivors and found that like Frankl, these survivors were able to find faith and meaning in suffering when suffering was seen as a challenge for good to overcome evil.

* *Fowler's theory of faith.* James Fowler, a theologian and professor of human development, builds on the work of Piaget, Erikson, Kohlberg, Perry, Gilligan, and Levinson. He believes that spirituality and faith can develop only within the scope of intellectual and emotional growth that the individual has attained. He has described faith as developing in six stages (Fowler, 1984) and has recently added a seventh, the stage of primal faith, at the very

AN APPLIED VIEW

Individualizing Instruction in Faith

Adolescence is a very important time for spiritual development. Many Protestant and Catholic teens are preparing to make their confirmation, and many Jewish boys and girls are getting ready for their bar or bas mitzvah. A growing number of the religious education programs that prepare these youths are using the ideas of James Fowler as they design their programs.

However, these programs may not be considering the spiritual developmental stage of the youths. If you are teaching in such a program, you should question your students along the lines of Fowler's stages. In this way,

you can ascertain the spiritual level they have reached. Then you will be better able to individualize your instruction, depending on the level of your students. Most likely you will have to decide if your students have reached stage 4. If they have not, avoid using symbolism in your lessons because the students will be incapable of understanding poetic-conventional faith. Thus, Fowler's stages fit well with Piaget's concept of formal operations, which are required for symbolic thinking to occur.

beginning of human life (Fowler, 1991a). Fowler (1996) further believes that faith helps to sustain us through personal and social change. A person must reach a minimum age before going from one stage to the next. Many people never reach Fowler's two highest levels. Following are Fowler's seven stages:

1. *Primal faith* begins even before birth and consists of trust in the parents and other very early relationships.
2. *Intuitive-projective faith* emerges in early childhood as the child acquires language. Very young children focus on the surface quality of religion and view religion as magic.
3. *Mythical-literal faith* develops at a minimal age of 5 or 6. The source of faith during this stage is verification of facts, but the facts come from authorities such as parents or teachers. During this stage, for example, the story of Adam and Eve is taken literally.
4. *Poetic-conventional faith* requires an age of 12 or 13. It is characterized by an awareness of symbolism and more than one way of knowing the truth. Fact and authority remain important, but young people at this stage are more selective about which authorities they consider valid.
5. *Individuating-reflective faith* can begin as early as 18 or 19 years of age. Here a synthesis of beliefs, and individual responsibility for those beliefs, appears. At this stage, personal experiences play a major part in one's faith.
6. *Conjunctive faith* (called *paradoxical-consolidation faith* in Fowler's earlier writings) is characterized by an integration of symbols, rituals, and beliefs. This stage can begin as early as 30 years of age. At this age one can understand that other people might have different but equally valid ways of approaching such complex questions as the supernatural. The individual at this stage considers humans as members of the same universal community and values the kinship of all.
7. *Universalizing faith,* Fowler's highest stage, requires an age of at least 40. According to Fowler, this stage, like Kohlberg's final stage of moral development, is rarely reached. Those who do attain this stage do more than recognize the mutuality of existence—they also act on it. At this stage, people are beyond needing approval from others for their actions (Dacey & Travers, 1991; Fowler, 1986; 1991a; 1991b).

In contrast to Jung's belief that spirituality could develop only in the second half of life, both Frankl and Fowler observe spiritual needs at all age levels. According to Fowler's theory, adolescents may have a poetic-conventional faith by as early as age 12 and may begin developing the synthesis and sense of personal responsibility for their own guiding beliefs well before they are 20.

The involvement of many young people in religious cults from the 1970s to the present (Hunter, 1998; Mosatche, 1983; Work, 1989) confirms that the search for deeper meaning in life is important to adolescents. Shelton (1983) states that although adolescents often find their relationship with the institutional church difficult, they "overwhelmingly express religious needs and find personal value in religious commitment" (p. 6). Hunter (1998) believes that adolescents are attracted to cults for a variety of reasons. For example, identity confusion, alienation from family, weak cultural and religious ties and feelings of power-lessness in a world that feels overwhelming and out of control.

Adolescence is a particularly sensitive period for either the development of religious faith or disengagement from the religion practiced by the family. The growing ability to think abstractly often prompts the young to reconsider beliefs that they had accepted easily at an earlier age. Social relationships with both peers and parents may also play a part. Ozorak (1989) questioned 390 adoles-cents about their religious beliefs and practices. The group included 176 boys and 214 girls who were Christian, Jewish, or non-religious. She found that those who were emotionally close to their families and belonged to a religion with a strong group identity were least likely to have made a religious change. Those who did not practice the same religion as either parent reported stronger connectedness to their peers.

Willits and Crider (1989) report on an 11-year follow-up of 331 students who were part of a group of 1,100 questioned about their religious participation and belief as high school sophomores in 1977. The question asked by this study was whether those who had stayed in the faith of their upbringing through adoles-cence retained this commitment in adulthood. Like those in Ozorak's group, these adolescents were likely to have close emotional ties to their parents. By the time of the second survey, when the participants were in their mid-twenties, the church attendance of either parent was no longer related to their own religiosity. However, the religious involvement of their spouses was now an important element in the choice of church commitment.

Both of these studies and others we have looked at in this chapter suggest that with or without membership in an organized religious body, young people do have a strong need for making meaning in their lives.

C O N C L U S I O N S

The research on moral development began in a restricted way: it sought only to explain moral judgment—how we *think about* morality. Piaget and Kohlberg relied on physical games like marbles and intellectual games like solving moral dilemmas. Subsequent psychologists have attempted to use more realistic approaches to the study of moral thought and have tried to understand actual moral behavior based on moral thinking.

The close relationship of morality to cognitive development is clear in the early research. William Perry's three stages of intellectual/ethical development and the five perspectives found in "Women's Ways of Knowing" make this relationship even clearer. We have also seen the important role cognition plays in the development of spirituality and faith.

You can see in this chapter how extensive the research on morality and spirituality has been, no less with teens than with other age groups. This makes it clear that these traits are considered a very important aspect of adolescent development. Although the studies indicate that most teens take this part of their lives seriously, you will find considerable evidence later in this book (especially in Chapters 9, 11, 12, and 13) that we adults could be helping them much more than we are. Anyone who wishes to help religious, educational, and other community leaders help teens solve their problems should pay close attention to the ongoing research on morality and spirituality.

Thus far in this book, we have examined physical, cognitive, moral, and spiritual growth. In the next chapter we look at one more trait that emphasizes internal development: the self and identity formation.

C H A P T E R   H I G H L I G H T S

The Scientific Study of Moral Judgment—Piaget
- Piaget studied the development of morality by observing Swiss children playing the game of marbles.
- He suggested that we can look at the development of morality in two ways: the way children actually practice the rules and the degree of their awareness of those rules.

Levels and Stages of Moral Reasoning—Kohlberg
- Kohlberg completed numerous studies of the development of moral reasoning, most of which involved teenage boys.
- He suggested that there are three levels of moral reasoning: preconventional, conventional, and postconventional. Each of his levels is divided into two stages.
- One of Kohlberg's major contributions is his conclusion that morality develops largely in the same way that cognition develops.

Moral Development and Gender—Gilligan
- Gilligan has suggested that female morality matures in a different way from that of male morality.
- Females, she believes, are more concerned with interpersonal relationships than they are with abstractions like Kohlberg's principle of justice.
- Thus she sees female morality developing in three phases: self-interest, self-sacrifice, and a concern for nonviolence.

Moral Components—Rest
- Rest has offered the view that moral decisions consist of four moral components.

Social Cognition
- The science of social cognition deals with the means by which we make sense of other people and with the ways we decide how to interact with them.

Perspective Taking—Selman
- Our interactions with others are influenced, as Selman points out, by the person's perspective-taking ability.

Intellectual/Ethical Development—Perry
- Perry also sees morality largely as the development of intellectual traits.
- According to Perry, these traits mature from dualism through relativity to commitment.

"Women's Ways of Knowing"
- The Belenky group found that female development involves five "ways of knowing."

Spirituality and the Development of Faith
- Several investigations have concluded that a religious and spiritual life are of great importance to most adolescents.
- Three theorists offer useful conceptions of how spirituality and faith mature. Carl Jung divides our spiritual life into two main components, whereas Viktor Frankl finds there are three.
- Closely related is the development of faith, which James Fowler separates into seven stages.

K E Y   T E R M S

codification 134
commitment 152
conventional level 136
dualism 152
escape 153
moral components 146

perspective taking 150
postconventional level 137
preconventional level 136
principle of caring 141
principle of justice 140
relativism 152

retreat 153
schema 149
social cognition 148
social norms 146
temporizing 153

W H A T   D O   Y O U   T H I N K ?

1. What is the most important aspect of moral development?
2. Which of Piaget's stages of moral development did you find to be most interesting? Why?
3. Which of Kohlberg's stages of moral development did you find to be most interesting? Why?
4. Which of Gilligan's stages of moral development characterizes most of your friends? Why do you think so?
5. In what ways is male morality different from female morality?
6. What would be some of the problems in trying to teach morality?
7. Do you believe that it is important to have religious beliefs? Spiritual beliefs? Why?

SUGGESTED READINGS

Baldwin, J. (1965). *Going to meet the man*. New York: Dial Press. This is Baldwin's first collection of stories, which are filled with moral dilemmas that involve racial relations. It is easy to read and hard to forget.

Bradbury, R. (1962). *Something wicked this way comes*. New York: Knopf. This book about two boys works as allegory, as fantasy, and as suspense story. The boys' growth in awareness as they learn of time, death, and good and evil is sensitively shown.

Dreisler, T. (1964). *An American tragedy*. New York: Signet Classics. This fast-paced tale of youthful love and lust presents a wide spectrum of moral dilemmas. It follows one moral decision to its horrible ending.

Frankl, V. (1970). *Man's search for meaning*. New York: Simon and Schuster. A powerful book dealing with the fundamental issues of the meaningfulness of life.

Haley, A. (1976). *Roots*. New York: Doubleday. This book allows the reader to really feel what it must have been like to be enslaved. It demonstrates the vital role that religion and spirituality can have, even when existence is almost impossible.

Power, F. C., Higgins, A., & Kohlberg, L. (1989). *Lawrence Kohlberg's approach to moral reasoning*. New York: Columbia University Press. A collection of articles looking retrospectively at models of moral education using Kohlberg's theory of moral development.

Shakespeare, W. *The merchant of Venice*. This classic play outlines the underlying subtleties of justice and the difference between concrete and abstract principles of right and wrong, good and evil.

GUIDED REVIEW

1. What is the correct sequence of Piaget's stages of rule practice?
   a. egocentric, individual, codification of the rules, cooperation
   b. individual, egocentric, cooperation, codification of the rules
   c. individual, codification of the rules, egocentric, cooperation
   d. codification of the rules, egocentric, cooperation, individual

2. Two youngsters are arguing about the rules of a game they are playing. The first one's position is that "rules are rules," and the second one's position is that "we can make up our own rules if we want to." According to Piaget, at which level of moral thinking is the first child?
   a. heteronomy
   b. individualism
   c. autonomy
   d. post-moral

3. How did Kohlberg study moral reasoning?
   a. by watching children play marbles
   b. by examining the reasoning of women contemplating abortion
   c. by asking adolescents to solve hypothetical moral dilemmas
   d. by following the development of students at ages 7, 12, and 19

4. Kohlberg suggests three levels of moral development. They are
   a. individualism, interpersonal, and community.
   b. obedience, social, and universal.
   c. preconventional, social, and universal.
   d. preconventional, conventional, and postconventional.

5. Even though James knew it was against the law to trespass, he sat in on the protest against testing on animals because of his conviction that such testing is wrong. James is operating at what stage of moral development?
   a. good boy morality
   b. authority and social order
   c. social contract
   d. universal principles

6. Gilligan argues for the importance of the principle of caring in response to Kohlberg's theory of moral development, which Gilligan suggests is biased toward
   a. North Americans.
   b. college students.
   c. upper class Americans.
   d. males.

7. Research on social cognition is concerned with
   a. development of identities.
   b. schemas and the process of change.
   c. recognition of moral dilemmas.
   d. understanding perspective taking.

8. Rest suggests that all of the following are moral components EXCEPT
   a. figuring out the ideal moral course of action.
   b. deciding what to actually do.
   c. interpreting the situation.
   d. evaluating the success of the plan.

9. Which of Fowler's stages of faith begins even before birth?
   a. primal faith
   b. intuitive-projection faith
   c. universalizing faith
   d. mythical-literal faith

10. Perry suggested a sequence of intellectual and ethical development that can be divided into
    a. 5 stages.
    b. 3 stages.
    c. 9 stages.
    d. 6 stages.

11. _____ (Kohlberg/Piaget) defined morality as understanding rules and following through one's own choice.

12. _____ (Individual stage/Egocentric stage) is the first stage of rule practice.

13. Adolescents are in the _____ (codification of the rules/ cooperation) stage of rule practice.

14. At Kohlberg's postconventional level of moral reasoning, the person is concerned with _____ (wanting to fulfill society's expectations and be fair to all/moral principles that have been thoughtfully and carefully chosen).

15. Kohlberg refers to the concept that each person brings a genetic predisposition to social behavior as the _____ (principle of justice/principle of caring).

16. Gilligan's care orientation emphasizes the _____ (independence/interdependence) of relationships.

17. _____ (Self-sacrifice and social conformity/Nonviolence) is the highest level of female moral reasoning, according to Gilligan.

18. Rest's Defining Issues Test assumes that moral thinking is based on principles of how people should _____ (reason/cooperate) in social situations.

19. Selman suggested that our interactions with others are influenced by the person's _____ (perspective-taking ability/ability to tolerate diversity).

20. _____ (Spirituality/Religious belief) is concerned with the meaning of life—whether life is worth living and why it is worth living.

Answers

1. B, 2. A, 3. C, 4. D, 5. D, 6. D, 7. C, 8. D, 9. A, 10. B, 11. Piaget, 12. Individual stage, 13. codification of the rules, 14. moral principles that have been thoughtfully and carefully chosen, 15. principle of justice, 16. interdependence, 17. Nonviolence, 18. cooperate, 19. perspective-taking ability, 20. Spirituality

THE SELF AND IDENTITY FORMATION

The Chinese . . . are reported to have a way of writing the word "crisis" by two characters, one of which signifies "danger," and the other "opportunity."

Louis Wirth

self-concept
Those beliefs, attitudes, and thoughts about the self that are descriptions about one's physical, social, and psychological qualities.

The self has been of major interest to psychologists, especially those interested in adolescents, for many decades. William James (1890/1950) presented a model of the self that remains useful in understanding the links between the development of **self-concept** and personal identity. According to James, self-understanding consists of a person's beliefs, attitudes, and thoughts about the self. Some of these thoughts and beliefs are descriptions about one's physical, social, and psychological qualities (e.g., what I look like, what I am good at, and how I feel). James refers to these as the "me." In this chapter, we discuss "me" characteristics as we talk about self-concept. Other thoughts and beliefs about the self are referred to by James as the "I." These are beliefs about how one changes over time yet remains the same individual, how one is different from others, and how one is able to act independently. This set of beliefs is important to the development of **personal identity**. The development of a clear, realistic, and integrated self-concept lays the basis for identity development. Adolescent self-concept and identity formation include a need to achieve a sense of uniqueness and distinctiveness while also remaining connected. The challenge is to construct a unique self that will retain the support of significant others and society.

personal identity
The set of beliefs about the self concerning how one changes over time yet remains the same individual, how one is different from others, and how one is able to act independently.

The adolescent period involves physical, cognitive, and social change. The body looks and feels different. The person thinks differently, judges right or wrong differently, and engages in different types of social relationships. Identity gives adolescents an integrated blueprint of what they will do with their bodies, their minds, and their relationships. Identity involves exploration (a loosening of parental control) and commitment (self-defined and integrated). Independence need not be the only end product; interdependence and connection have an important part in maturity for both men and women. James also recognized self-evaluation or **self-esteem** as an important dimension of the self. According to James, self-esteem is related to the evaluation of one's success in meeting set goals. Identity development also occurs within the larger societal context, where cultural identity, sexual identity, and the impact of gender roles are also central to the adolescent's self-definition and self-evaluation. In this chapter, we look at the development of self-concept, self-esteem, gender role, and personal, cultural, and sexual identity during adolescence.

self-esteem
Evaluating one's success in meeting set goals.

After reading this chapter, you should be able to:

- Explain how the cognitive changes that take place during adolescence affect the development of self-concept.
- Describe, according to psychoanalytic theory, how individuation contributes to the development of the self during adolescence.
- Describe, according to self-in-relation theory, how interpersonal relationships are important to the self during adolescence.
- Describe the development of self according to social learning theory.
- Identify two social factors associated with the development of self-esteem.
- Provide several explanations for the differences found in self-esteem levels of adolescent girls and boys.
- Explain the relationship between self-concept and social competence during adolescence.

- Describe, according to Erikson's model, how identity is achieved during adolescence.
- Describe several criticisms of Erikson's theory when applied to female identity development.
- List the stages involved in the development of cultural identity.
- Define three aspects of gender role.
- Demonstrate awareness of gender schema and how it influences adolescent behavior.
- Discuss these issues from an applied, a sociocultural, and your own point of view.

Many psychologists view the self as the foundation of mental health. According to Carl Rogers (1961), for example, people feel good about themselves when their view of themselves fits well with the feedback they get from those around them. Psychological difficulties develop when parents, teachers, siblings, and friends are unwilling to accept a person the way he or she is. Rogers went so far as to argue that "unconditional positive regard," which means total acceptance of the child as a person, is necessary to the development of a psychologically healthy self.

Early experiences with caregivers are critical to the development of the self, according to many theorists (Beck, 1976; Bowlby, 1988; Kohut, 1977; Mahler, Pine, & Bergman, 1975; Rogers, 1961). A lack of support and nurturance from parents and other caretakers usually results in an unhealthy self-concept. In a warm and caring home, with responsive and sensitive caretaking (Ainsworth & others, 1978), children are likely to develop positive feelings of self-worth and self-acceptance. Positive feelings of self-worth contribute to continued healthy psychological development and help reduce the negative effects of stress while growing up. The teen who confronts the multiple changes of adolescence with a healthy sense of self is better prepared to meet the challenges that accompany this period of rapid physical and psychological change. Although these theorists help us understand early influences on self-concept development, they do not help us understand how the self-concept changes during adolescence. For that understanding, we need to consider the many changes that take place during adolescence.

Let us begin with a look at the impact of cognitive changes on the self.

How Cognitive Changes in Early and Late Adolescence Affect Self-Concept

The adolescent's self-concept changes in many ways throughout the adolescent years. That is, how an individual describes "who I am" at age 12 will be very different from the self-description at age 19. The changes that take place in self-concept during those years are due in large part to the cognitive changes that are also taking place. Growth in abstract and hypothetical thinking, as described by

Piaget (see Chapter 4), affect the way in which adolescents describe themselves. Advances in social-cognitive processes, including social comparison, perspective taking, and self-awareness, also contribute to the changing self-concept. Cognitive abilities to compare oneself with others emerge in middle childhood and increase in adolescence as the teen encounters more complex and varied groups of individuals with whom to compare the self. According to Selman (1980), abstract thinking contributes to developments in perspective taking (or the ability to understand a situation from another person's point of view). Perspective taking enhances self-understanding, because the adolescent can step outside an immediate situation and look at herself or himself as others would.

WHAT'S YOUR VIEW?

Adolescent Self-Concept

Picture in your mind three of your closest friends, people with whom you have grown up (if your siblings are close friends, by all means include them). In your mind, imagine a little movie of each one as they were at age 8, at age 14, and as they are today.

At which of these times were they most happy? Can you see any pattern? Was there an age at which all liked themselves best? Try this exercise on yourself. Do you fit the pattern? Is adolescence a better or worse time for self-concept? What's your opinion?

Some psychologists believe that adolescents' self-concepts are frequently troubled (see Chapter 11). Rosenberg (1985), for example, argues that adolescents are more likely than younger or older persons to develop an unhealthy self-concept. Offer and associates (1981), on the other hand, provide evidence that most adolescents are happy and have positive self-concepts. Harter (1999; 1993) believes that the more advanced reasoning skills of the adolescent influence self-concept in both positive and negative ways. Advances in social comparison, for example, lead to the realization that one is not as good as one's peers in many activities. With social cognitive advances that permit critical self-evaluation according to the standards of others, the adolescent may suffer from an awareness that others hold negative views of the adolescent. Advanced social reasoning skills can also help adolescents behave in more socially acceptable ways, enhancing their evaluation by others (Harter, 1990; 1999). In a study of 198 adolescents, McLun and Merrell (1998) found self concept scores to be associated with parenting style (see Chapter 7). Authoritative parenting was associated with the highest self-concept scores. This finding suggests that supportive parenting contributes to the development of feelings of self adequacy in teens.

For some adolescents, at least, changes in the way they think about themselves contribute to occasional emotional problems such as depression and anxiety. Harter (1990;

The self-concept becomes more differentiated as adolescents participate in various social roles and view themselves differently in each role. Some adolescents wonder, "Which is the real me?"

1999) described four changes in self-concept during adolescence and the emotional risks of those changes.

• A person's self-concept becomes more *abstract* during adolescence. This increase in abstract thinking is what Piaget described as formal operational thought. Early adolescents are likely to answer the question of "Who am I?" with concrete descriptions of the self. They describe physical characteristics, such as being tall or short; simple feelings, such as being happy; or group membership, such as being a girl, a son, or a student. These characteristics can often be seen or touched and are easily tested. That is, by simply looking in the mirror you can check whether you are tall or short. Consistent with the emergent importance of peer relationships and cognitive capacities that allow for comparisons of self with others, early adolescents' self-descriptions also include qualities relevant to social interactions, such as social skills and social appeal (Damon & Hart, 1988).

Older adolescents are more likely to describe internal characteristics of the self, such as emotions, wishes, and motivations. Instead of simply looking in the mirror to see if they are tall or short, older adolescents look inward to explore complex feelings like embarrassment or depression. Late adolescent descriptions of self may include the individual's belief system, moral standards, personal philosophy, inner thought processes, and life plans (Damon & Hart, 1988).

These descriptions are more difficult to test than external ones. By looking in the mirror, you cannot know whether you are really kind or brave or considerate. For some adolescents, this is problematic: most adolescents are able to accurately describe their inner selves, but others develop unrealistic self-concepts.

- The self-concept becomes more *differentiated* as the person enters adolescence. Early adolescents describe themselves as having a greater number of social roles than younger children do. Harter (1990) found that most adolescents describe themselves in eight different roles. These include student, worker, athlete, close friend, and roles involving their general social relationships, romantic relationships, physical attractiveness, and conduct/citizenship.

In addition, adolescents begin to view themselves differently in each role. Who they are with their parents may be very different than who they are with a girlfriend or boyfriend or best friend. The problem this can present for some adolescents is determining which of these selves is the "real me." Harter (1990; 1999) suggests that those in middle adolescence (ages 14 to 15) are most troubled by conflicting views of the self. Cognitively, they have the ability to recognize that they are behaving differently with different people. They do not yet understand why, and this worries them. The more advanced cognitive skills of late adolescents help them realize that it is normal to behave differently with different people and help them integrate the diverse views of self into a more abstract conceptualization. Late adolescents might be able to understand and label their changing behavior with peers and families as part of the abstract concept of mood.

ideal self
The self that one would like to be. The development of the ideal self is made possible by hypothetical thinking.

- An *ideal* or *imagined self-concept* develops during adolescence. The **ideal self** is the self that one would like to be. The development of the ideal self is made possible by hypothetical thinking, which is another characteristic of formal operations (see Chapter 4).

Difficulty can arise for adolescents if a large difference exists between the actual self and the ideal self. When you view yourself as being very different from the way you would like to be, you are more likely to be unhappy, disappointed, and even depressed. Adolescents' self-concepts are likely to change frequently as their ideal and actual views of themselves change. The difference between the ideal and actual self is believed to be greatest in mid-adolescence, probably because the ideal self is often inflated in some teens. As adolescence ends, most of us start a career or a family, both of which have a way of reducing an overblown ego down to more realistic proportions.

Although most psychologists believe that a large difference between the real and ideal self is to be avoided, Markus and Nurius (1986) argue that the development of the ideal self can also be helpful. The ideal self can give you direction and motivation to work toward the person you would like to be.

introspection
Inward-looking, or thinking about what is going on in one's mind.

- The adolescent becomes more *introspective*, or inward-looking. In contrast with the unquestioning self-acceptance of childhood, the adolescent begins to ask, "Who am I, really?" Piaget's stage of formal operations is necessary for **introspection** (thinking about what is going on in one's own mind). Sometimes adolescents become very concerned with how they appear to others. They may become unable to tell the difference between their own

AN APPLIED VIEW

Who Are You?

If adolescents typically have a hard time knowing who they are, what can be done to help them? Well, one simple rule to follow is: most people get better with practice. Whether you are a teacher, a counselor, a parent, or just a friend, you can provide a teen with practice at getting to know herself or himself by asking, in any number of ways, "Who are you?"

Here are a few sample questions you might ask during a discussion or counseling session, or give as a written assignment:

- "What makes you happier than anything else?"
- "If you could be anyone other than yourself, who would you be? Why?"
- "Imagine yourself 10 years from now. Tell me what you are like."
- "If you'll tell me a secret about yourself, I'll tell you one about myself."

- "Imagine you are standing in a room near an open door, and out in the hall some acquaintances of yours are talking about you. They do not know you can hear them. One says, 'Ya know, I really like [your name], because she . . .' What do you think this person would probably say? Now someone says, 'Maybe you're right, but the one thing I can't stand about [your name] is . . .' What do you think this person would probably say?"

We especially like this last method, because it allows young people to speak of a really good and a really bad aspect of their self-concept without having to take credit or blame for it. This often frees people to speak more honestly, because they can believe it is not them who is speaking. Psychologists refer to this as a projective technique, because the self is being "projected onto someone else."

thoughts and what others are thinking about them. They may feel that "everyone is looking at me" (the imaginary audience). Adolescents sometimes also view themselves as extremely different from everyone else. They do not believe that anyone can understand them because their problems are so different, and they complain that their parents will never understand them (the personal fable).

Individuation, Interdependence, and the Development of the Self

individuation
The first separation process, which takes place in early childhood when young children realize that they are separate and different from their parents.

second individuation
According to Blos, this occurs in adolescence when the individual grows from a dependent child to an independent adult.

According to traditional psychoanalytic theory (Blos, 1979) (see also Chapter 2), the self develops through a series of separations. In the beginning, the newborn has no concept of being a separate self from the parents. Separation, however, is necessary to the development of the self. The first separation process, called **individuation**, takes place in early childhood. This is the time at which young children come to realize, through all the things that happen to them, that they are separate and different from their parents.

Blos (1979) believes there is also a **second individuation**, which takes place in adolescence. Through this second individuation, the individual grows from a dependent child to an independent adult. In time, adolescents develop clearer pictures of themselves as separate and different from their parents. To do this, adolescents must distance themselves emotionally from their parents. Parents must no longer be viewed as having all the answers. Adolescents should no longer feel that they have to do what their parents say. Adolescents will question and may reject their parents' ideas and values.

This can make the adolescent feel anxious, angry, and guilty at times. According to Blos, a stable and independent sense of self develops by late adolescence. This happens after the adolescent has developed firm boundaries and positive feelings about being separate from parents and others. At last, adolescents begin to be at peace with themselves.

Lapsley and Rice (1988) also view adolescence as a second individuation period. Using James's (1890/1950) distinction between the self as "I" and "me" (described in the introduction to this chapter), Lapsley and Rice describe early adolescence as important to the development of the "I." Younger adolescents are most concerned with their ability to act and think as independent persons, experiencing feelings of uniqueness and power. The imaginary audience and personal fable (Elkind, 1981), described in Chapter 4, are characteristic of this period. Feelings of uniqueness and unrealistic power diminish by late adolescence. According to Lapsley and Rice, the changes that occur in the adolescent's sense of self are related to psychodynamic processes and cognitive advances in self-understanding. From a psychodynamic perspective, the self-centeredness of the younger adolescent is a way of coping with feelings of anxiety that occur as the adolescent gains independence from parents. Adolescents need to feel powerful in order to feel secure without parental support. They develop a rich fantasy life of personal relationships to make up for the loss of parental closeness and to prepare for intimate relationships with peers. Cognitively, younger adolescents have a new awareness of their ability to control their own thoughts and feelings and are able to think about their own thinking, contributing to feelings of power and uniqueness. The older adolescent has gained a greater awareness of the perspective of others and is consequently less self-centered. As cognitive development proceeds, an awareness of the power and limitations of one's thinking processes increases.

According to Lapsley and Rice (1988), parents, siblings, and peers need to be emotionally supportive of adolescents during the process of individuation. Parents need to not only be sensitive and understanding, but also provide the adolescent with opportunities to develop independence. Such opportunities should provide a "gentle push" or offer enough challenge to promote independence, but not be so difficult that self-esteem will suffer.

Are you independent or interdependent?

The goal of individuation is the development of the self as an independent person. This view has been challenged in recent years. Some social scientists do not believe that this model fits the development of women and people living in other cultures. **Interdependence**, rather than independence, is important to the lives of many adults. People depend on one another for help and define themselves as members of a group, such as a family, a church, or a community. Connection with other people helps rather than interferes with the development of the self. Nobles (1976) points out that African tradition, unlike the traditions of Europe and America, places little importance on individuality. Cooperation,

interdependence
When people depend upon one another for help, and define themselves as members of a group such as a family, a church, or a community.

interdependence, and group responsibility have been the central values of African traditions and have thus been important to the self-concepts of African peoples as well as of African Americans.

The sense of self as internal, individual, and set apart from others, from society, and from nature is not shared by many cultures. In fact, some would consider it bizarre. In present-day Morocco, an adolescent does not exhibit the level of rebellion or conflict many Americans assume is normal for this age. "One reason is the hierarchical nature of Moroccan society, fostering the great respect shown to the aged. Another is the importance of the family unit rather than the individual. Zawiya youth generally do what is best for the family rather than what is best for themselves as individuals; this eliminates many of the sources of conflict" (Davis & Davis, 1989).

Josselson (1988) states that most adolescents, but especially adolescent girls, develop a sense of self through their relationships with other people. Josselson explains that although adolescents may choose to do things differently than their parents, their parents are still quite important to them. In fact, they probably continue to care a lot about whether their parents accept their changing selves. Josselson believes that adolescent anxiety may result from fear that important relationships will be lost as the adolescent changes. If the sense of self is based on the ability to make and keep relationships, a loss of friendship can feel like a part of the self has been lost. As late adolescents develop new relationships, they do not throw away old ones. Instead, they bring new friends home. Josselson believes that the new sense of self must develop through a mixing of old and new relationships.

In recent research, Josselson (1996) suggests that there are four trajectories that can be identified in women's development all of which continually evolve throughout the lifespan around issues of connection and feelings of competence. The four trajectories are: guardian, pathmaker, searcher and drifter. Josselson describes guardians as high achievers who are committed to family expectations and tend to be rigid in outlook and resistant to change. Pathmakers are not afraid of risk or commitment. They try to balance their own needs with the needs of others. Searchers tend to be idealists who may have difficulty making commitments and are overwhelmed by choices. Finally, the drifters live for the moment and avoid having to make choices.

Josselson's views are consistent with the position taken by Carol Gilligan and her colleagues (Gilligan, 1988; Gilligan & associates, 1990) (see Chapter 5; see also the discussion of identity development later in this chapter). Gilligan, like Josselson, believes that relationships are important to the sense of self. When Gilligan and her research team asked adolescent girls to describe themselves, the girls presented themselves as involved in relationships with other people and not as isolated individuals. Although relationships with others were most important to their self-descriptions, the girls were also interested in how they were distinct from others (Stern, 1990). These adolescent girls appeared to be defining a sense of self while remaining connected with others.

Jean Baker Miller and her colleagues at the Stone Center at Wellesley College (Miller & others, 1997; Kaplan & Klein, 1985; Miller, 1976) agree with

self-in-relation
The theory that a clear understanding of our relationships with others is most important in the definition of the self, especially for women.

Josselson and Gilligan. They have developed a theory, which they call **self-in-relation**, that states that a clear understanding of our relationships with others is most important in the definition of the self, especially for women. Women ask themselves not just "Who am I?" but "Who am I in relation to others? Where am I going and how will that affect others? How am I going to get there and how will my getting there affect other important people in my life?"

Self-in-relation theory was formulated through interviews and discussions with women. Few empirical studies have attempted to test this theory. Ruebush (1994) studied 50 mothers and their 15- to 18-year-old daughters and found some support for the theory. Maternal empathy was significantly related to the daughters' successful individuation, in terms of being free from feelings of conflict and guilt with their mothers and not requiring excessive levels of

AN APPLIED VIEW

Self-In-Relation Therapy

Traditional psychotherapists believe that independence is quite important and would treat a depressed client by encouraging her to end unfulfilling relationships, become less dependent, and find fulfillment elsewhere. A therapist who follows the self-in-relation model, on the other hand, would emphasize the positive aspects of the relationship. The therapist would also help the client find ways of doing things for herself and of increasing her happiness without ending her relationships.

The therapist might help the client explore her personal and family history to find out why she feels she must sacrifice her own needs and help her change her

belief that attending to herself is wrong. As part of this process, the client might keep a record of when and how often she does things for herself and for others. She might keep a list of the activities that give her feelings of accomplishment. This would help her become more aware of how much energy she invests in taking care of others and how much satisfaction she gets from doing things for herself. The goal of self-in-relation therapy is to develop interdependent relationships through which the needs of all participants can be satisfied equally (Enns, 1991).

maternal support and approval. In other words, when mothers understood their daughters, the daughters were not likely to feel guilt and anxiety about separation. Kenny and Donaldson (1992) found that freedom from guilt and anxiety toward parents and feelings of positive parental attachment were associated with personal and academic adjustment for first-year college women. Allen and Stoltenberg (1995) looked at gender differences in college students separation from parents and found that female students reported establishing more kinds of supports, were more satisfied with their supports and saw their families as more cohesive than male students. Perhaps the societal focus on the affiliative skills of women help to prepare young women for developing social support outside the family. These studies also suggest, consistent with the theories of Gilligan, Josselson, and Miller, that supportive and empathic relationships with parents can foster healthy separation.

According to self-in-relation theory, the development of independence and self-esteem takes place *within relationships* in which all individuals are encouraged to adapt and change. This is what we meant earlier by "interdependence." Not all relationships are healthy, however. Some women (and some men) sacrifice their sense of self for others. They disregard their own needs and focus all their energy and efforts in pleasing and taking care of others. When that happens, they are likely to lose self-esteem and feel depressed.

The Self and Social Learning Theory

Social learning theorists such as Albert Bandura, who was introduced in Chapter 2, view the self as developing in a different way. They hold that the self is based mainly on the *expectations* people believe that others have for them and whether those expectations are reasonable. The self can be greatly influenced by whether you think you can succeed and whether you think other people think you can succeed. **Self-efficacy** is the name Bandura gives to your self-expectations, or beliefs about what you can accomplish as a result of your efforts.

self-efficacy
Our self-expectations or beliefs about what we can accomplish as a result of our efforts, which influences our willingness to attempt the task and the level of success we achieve.

These beliefs influence how a person behaves. If you believe you will be successful, you will try harder. This helps you to succeed. In turn you are more self-confident. Other people also see you as successful and will expect you to succeed. As a result, they might spend more time teaching you new things and helping you to succeed further. Similarly, if you believe that people will like you, you will approach them more positively. They in turn will respond to you positively. This reinforces your view of yourself. Because self-efficacy influences what a person will attempt and how well a person will actually perform, developing a strong and positive sense of self-efficacy is very important.

People who are depressed don't believe they will do anything right. They believe that people will not like them. Their behavior results in failure, which in turn reinforces their beliefs.

According to Bandura (1978), your view of yourself is learned. It develops from the way you have been treated by others, from your successes and failures, and from your view of those experiences. The self develops the same way during adolescence as it does during early childhood. By adolescence, however, it is more difficult to change the way you think about yourself. You have had many experiences that have taught you that you are either good or bad. Each of those experiences has made your view of yourself more rigid. This view affects future behavior and the reactions of other people. Bandura calls this process through which beliefs, behavior, and environment affect one another **reciprocal determinism.**

reciprocal determinism
According to Bandura, the process through which beliefs, behavior, and environment affect one another as an individual's view of himself or herself is learned.

Self-Concept and Self-Esteem

Whereas self-concept answers the question "Who am I?", self-esteem answers the question "*How do I feel about who I am?*" Self-esteem is related to self-concept. A well-defined self-concept leads to high self-esteem (Campbell &

Lavallee, 1993), which in turn often leads to successful behavior (Rosenberg, 1985). Persons with high self-esteem like and accept themselves. They do not feel that they are perfect or better than other people; rather, they are aware of their limitations and work toward correcting them. Self-esteem is not the same thing as conceit. People with high self-esteem are generally well-liked.

Research indicates that self-esteem is important to adolescent motivation toward success, achievement, and mental health. People with high self-esteem often do well in school (Bell & Ward, 1980). They feel they are in control of their lives (Rosenberg, 1985). They are likely to view their successes as resulting from their own efforts (as opposed to being lucky or well-liked). Failures are believed to result from bad luck or not trying hard enough. Thus, people with high self-esteem believe that if they work hard, they can succeed.

Adolescents with high self-esteem respond constructively to stress in their lives (see Chapter 11) and have positive ways of solving life's problems. However, increases in life stress can result in loss of self-esteem (Youngs & others, 1990). People with low self-esteem are more likely to have greater emotional and behavioral disorders such as anxiety, depression, delinquency, drug abuse, and eating disorders than are people with high self-esteem (Rosenberg, 1985).

Factors That Affect Self-Esteem

Harter (1990; 1993; 1999; Harter & others 1998) and Rosenberg (1979) have drawn on the works of historical scholars of the self, James (1892) and Cooley (1902), in identifying factors that affect self-esteem.

• *Doing well where it matters.* James (1892) theorized that global self-esteem was dependent upon the ability to do well in areas important to the individual. Harter (1993) has tested James's theory in research with children and adolescents and has found tremendous support for this conceptualization. In general, high self-esteem results when teens are doing well in areas where success is important to them. On the other hand, low self-esteem does not result if they are doing poorly in areas where success is not important. For example, if appearance and athletics are important to you, you must do well in those areas to feel good about yourself. If job success is not important to you, doing poorly on the job will not cause you to feel bad about yourself. Harter's research suggests that physical appearance, being liked by others of the same age, and doing well in sports are most important to adolescent self-esteem. Doing well in school and behaving well are rated as least important by the majority of adolescents. In recent work by Harter and her colleagues (1998) teens with higher feelings of self worth evaluated themselves more positively in domains that they considered important than peers with lower feelings of self worth. Adolescents with higher self worth were also more able to discount their weaknesses than were lower self worth counterparts.

social construction
The idea that self-esteem represents, in large part, the attitudes that significant others hold.

looking-glass self
Cooley's theory, which suggests that our self-concept is formed through the reflection of the attitudes of others back onto ourselves.

How important are the views of parents, teachers, and friends to adolescent self-esteem?

• *Earning the esteem of others.* For Cooley (1902), self-esteem was a **social construction**. In other words, self-esteem is the perception of what others think about us. The self is formed by taking on the attitudes of others. It is as if the adolescent is looking into the mirror and becomes the image that is reflected back. For this reason, Cooley used a mirror metaphor in explaining the concept of the **looking-glass self**. Adolescents who feel that their parents, close friends, and classmates support and approve of them will also like themselves. Adolescents who feel that they are not liked by important others will not like themselves. Harter (1993) has found that perceived support and approval from classmates and parents are central to self-esteem among adolescents. Rosenberg (1986) explains problems of self-esteem in adolescence as a function of adolescents' concern with gaining others' approval. Because adolescents are highly concerned and preoccupied with the impression they are making on others, self-esteem will be unstable. Efforts to please a variety of persons will result in frustration. Feedback gained from different individuals will likely be contradictory, contributing to confusion and self-doubt.

Positive self-esteem develops when parents, teachers, friends and other people who are important to you believe in you and expect you to succeed in areas that matter to you. Among children, parents' attitudes are most important to self-esteem. During adolescence, the opinions of classmates become more important, although parents are still very important. Furthermore, the influence of other people in general on self-esteem also decreases during adolescence. Whereas for early adolescents, self-esteem depends heavily on the attitudes of others, older adolescents develop a firmer sense of their own beliefs and attitudes and so are less dependent on the attitudes of others. Persons with low self-esteem seem to be most susceptible to the opinions of others in determining how they feel about themselves at a particular point in time (Campbell & Lavallee, 1993).

Although Harter originally considered *Doing well where it matters* and *Earning the esteem of others* to be separate factors affecting self-esteem, she has come to see these factors as related. Her research (1993) suggests, for example, that the domains of competence rated least important by adolescents, doing well in school and behaving well, are those rated as most important by parents. Since perceived support and approval from both classmates and parents are central to adolescent self-esteem, the conflict and threats to self-esteem for the adolescent become apparent. Her research demonstrates, additionally, that when adolescents forfeit competence or fail to meet expectations in areas deemed important by significant others, such as parents and classmates, the adolescent may perceive an accompanying loss of support from others. Thus, the adolescent who disappoints parental expectations for school success may feel a loss of parental support and approval, which in turn has a damaging effect on self-esteem.

Harter and Marold (1989) studied suicidal adolescents to see how these two factors influenced self-esteem. The suicidal adolescents were not doing well in areas that were important to them or their parents. They felt that they were letting their parents down. Furthermore, they did not feel that their parents or their friends would like or support them if they did not do well. These adolescents did not feel good about themselves, felt hopeless about their futures, and saw suicide as the only way out. Interestingly, Markstrom (1999) has found that religious involvement is related to positive self esteem. Perhaps, as discussed in Chapter 5, religious involvement provides some adolescents with a sense of connection and hope thus impacting self esteem.

DEVELOPMENTAL CHANGES IN SELF-ESTEEM

The quality of self-esteem changes during adolescence. It often drops at age 11 and reaches a low point between ages 12 and 13 (Rosenberg, 1986). Petersen

AN APPLIED VIEW

Parenting for Self-Esteem

It is clear that parents and other adults influence the self-esteem of adolescents. What specifically can they do to help adolescents feel good about themselves? Reasoner (1983) has given some suggestions. When parents possess high self-esteem, their children are also more likely to feel good about themselves. Parents who show interest in what their children are doing and expect their children to do reasonably well are likely to have children with high self-esteem. When parents are not interested in what their adolescents are doing, the adolescents are not likely to care much either. Adults often focus attention on the shortcomings of adolescents, which can make teens feel bad. Parents and teachers should point out to young people the things they are good at. Adolescents need to feel that they make a difference in the lives of other people, especially

their parents. Self-esteem will be higher even for teens who feel their parents have negative feelings about them but also feel that their parents are interested and concerned.

Parents also show interest in their children and provide feelings of security by setting limits. They provide the adolescent with a clear sense of what is acceptable behavior and make sure that those standards are followed. Giving adolescents responsibilities to fulfill can also give them feelings of importance. Allowing them to plan family activities such as playing games, picnics, or vacations can develop feelings of belonging. Helping them draw a diagram of the family's branches, composing a family history, interviewing the older family members, and visiting grave sites builds a sense of continuity. Finally, adults can help teens in setting and reaching their own goals.

and associates (1991) and Brooks-Gunn (1991) suggest that many early adolescents experience "simultaneous challenges" that can negatively affect self-esteem. These challenges include social events such as changing schools, changes in parent-adolescent relationships, and biological changes associated with puberty. Changing schools appears to be related to the decline in self-esteem (Simmons, 1987). Some younger adolescents are asked to make the change to junior high or middle school before they are emotionally ready. The change to a new school also breaks up old and comfortable friendships. New schoolmates may also value different things. For example, if an adolescent who values doing well in school joins a new group of friends who value physical appearance more, the adolescent's self-esteem may decline. These earlier findings are further reinforced by recent research. Eccles and associates (1999) found that both boys and girls experience a decrease in self esteem that coincides with the transition to middle or junior high school.

Physical maturation in early adolescence may also negatively affect self-esteem, especially in girls. Early adolescent girls may not be ready to deal with the new expectations people have of a person who has reached puberty (for example, an increased interest in the opposite sex). When young adolescents change schools, keeping friends who share similar values can be helpful to self-esteem.

Although problems in self-esteem are not unusual in early adolescence, the good news is that problems gradually decrease for most adolescents from grades 7 to 12 and throughout the college years (O'Malley & Bachman, 1983). The increased independence and personal freedom gained during the adolescent years are believed to benefit self-esteem. Petersen and associates (1991) point out, however, that for some teens the challenges of adolescence are over-whelming and self-esteem continues to decline. It seems that adolescence is a period during which the differences between the best adjusted and the most troubled adolescents increase. For some teens who experience difficulties prior to adolescence and for others who experience much stress during early adolescence, the entire adolescent period is a time of psychological distress.

Block and Robins (1993) studied changes in self-esteem among 47 girls and 44 boys from ages 14 to 23 and found decreases in self-esteem over that time period among 34 percent of the males and 57 percent of the females and increases in self-esteem among 64 percent of the males and 43 percent of the females. Girls who increased in self-esteem during adolescence were described at age 14 as moralistic, humorous, looked up to by others as a source of advice and reassurance, generous, and protective to those close to them. Boys who increased in self-esteem demonstrated different personality characteristics than their female counterparts. Males who were socially at ease, self-satisfied, calm, and relaxed were most likely to increase in self-esteem. Block and Robins explained these differences as a result of sex-role socialization. If girls are socialized to get along with others and boys are socialized to achieve competitively, the personal qualities associated with self-esteem appear consistent with sex-role expectations.

GENDER AND SELF-ESTEEM

The statistics reported by Block and Robins (1993) are consistent with other studies (e.g., Simmons & Blyth, 1987) reporting greater drops in self-esteem among adolescent girls compared with adolescent boys. A survey by the American Association of University Women (Bailey, 1992) revealed that at age 8, 60 percent of girls say they feel good about themselves. By age16, only 27 percent report feeling positive about themselves. In comparison, 67 percent of boys report being happy with themselves in elementary school. Almost half (46 percent) are still happy with themselves during high school. Thus, while both boys and girls of high school age report lower levels of self-satisfaction than elementary school children, the decline in self-satisfaction is greater for girls. Adolescent girls were also discovered to be less likely than boys to believe they will achieve their career goals. These findings are further supported by Eccles and associates (1999) who found that adolescence presents risks and opportunities for both boys and girls. While these researchers did find that girls' self esteem dropped more than boys', they found that both boys and girls suffer from decreases in self esteem. Interestingly, Eccles and associates (1999) further suggest that girls seem to be more negatively affected by failures and concern about failures than boys. This finding would suggest that girls might be less likely than boys to try new things or take risks which in turn could limit their future possibilities. Kling and others (1999) reviewed a large number of studies on gender differences in self esteem and found that male scores were higher on standard measures of global self-esteem but that the gender differences were relatively small.

A number of explanations have been given for adolescent gender differences in self-esteem. A girl may be "exposed from birth onward to the suggestion, whether conveyed brutally or delicately, of her inferiority" (Karen Horney as cited in Symonds, 1991). In addition, "it must be obvious even to a 4- or 5-year-old in some societies, if not all, that men dominate women" (Whiting & Edwards, 1988). Eccles and associates (1999) believe that puberty places young women at risk because of the narrow standards of beauty with which they become surrounded as they develop physically. In addition, they found that many adolescent girls still believe that high achievement related behavior conflicts with feminine behavior. Thus suggesting that gender role socialization and societal expectations play a crucial role in the development of self-esteem.
Myra and David Sadker (1991) suggest that many educational practices and teacher behaviors negatively affect the self-esteem and academic achievement of girls. Boys, for example, are often given more attention and instruction by teachers and parents. Boys seem to demand teacher attention by calling out in class more often. While teachers often respond or build on the boys' comments, girls who call out are more likely to be reminded to raise their hands and follow the rules. Girls often receive praise for the neatness and appearance of their work, while boys receive more praise for the quality of their ideas. Girls seem to be rewarded for silence and passivity. Boys conclude that adults hold males in higher regard. Girls, on the other hand, learn that adults assume females cannot achieve at the same level as males. Gender Gaps, a 1999 report published by the American Association of University Women (AAUW) is a follow-up to the 1992 AAUW report: *How Schools Shortchange Girls*. The 1999 report suggests that while there has been some improvement in terms of gender equity in schools, there are gaps that remain. For example, girls tend to get better grades in school than boys but boys generally score better on standardized tests, especially high stakes standardized tests (like the PSAT and SAT) that are tied to college admission and scholarships. It would seem that as long as these gaps persist, young women remain at high risk for decreased self esteem.

Carol Gilligan and her associates at Harvard (1990) believe that adolescent girls often conceal their true feelings, which results in loss of self-esteem. They interviewed girls attending a private school in New York state. At age 11, the girls expressed their opinions with much self-confidence. By age 15 and 16, the girls were much less certain and answered many questions by saying, "I don't know." Gilligan and her associates believe that adolescent girls hold back their real feelings, especially anger or resentment, in order to obtain approval, popularity, and attention at home and at school. During adolescence, girls begin to see that being assertive and outspoken is dangerous for women. They learn that they must be quiet, calm, and kind. The societal pressure on adolescent girls to be perfect and please others is believed to contribute to loss of self-esteem and, in some cases, psychological disturbance. In her best-selling book *Reviving Ophelia*, Mary Pipher (1994) shares cases from her therapy practice. She too suggests that the true selves of adolescent girls may become lost when young women are faced with the choice to be true to themselves or accepted by others.

Following her model of "multiple challenges," Petersen and associates (1991) point out that girls often mature earlier than boys and are more likely to experience puberty prior to the transition to middle school. Girls, therefore, are more at-risk because they experience both puberty and school change at the same time.

Instead of focusing on differences in self-esteem between adolescent boys and girls, Brooks-Gunn (1991) has been interested in identifying those girls who are most at-risk for loss of self-esteem and psychological distress during adolescence. Girls who experience high levels of conflict in the parent-daughter relationship and are early maturers appear to suffer the greatest loss in self-esteem. Early maturation means that these girls are experiencing weight gain earlier than their peers, thereby increasing dissatisfaction with physical appearance.

RACE AND SELF-ESTEEM

Research on the self-esteem of racial minorities has focused on African American youths. Early research assumed that poverty, low status, and perceived discrimination would result in low self-esteem. In an early study, Clark and Clark (1947) found African American children more interested in playing with white dolls than with dolls of their own color. This was interpreted as evidence of low self-esteem. Those results have since been interpreted as a reflection more of the racial prejudice of society and less of the children's feelings about themselves. Most recent studies suggest that there are no self-esteem differences among racial and ethnic groups (Phinney & Rosenthal, 1992).

Jenkins (1982) explains that the self-concept, as we noted earlier, has many components. A person's self-evaluation may be different for each of those components. Racial self-concept may be affected by the larger society's attitudes toward that race. When African American children choose white dolls, they may be reflecting those attitudes. This does not mean, however, that they devalue their friends, their sense of being loved by their families, or their own feelings of competence about academic or athletic achievements. Personal self-esteem or feelings of one's worth as a person are influenced more by the attitudes of important persons in one's life, such as family and peers. According to Cooley's model of the looking-glass self (discussed earlier in this chapter),

A SOCIOCULTURAL VIEW

Racial Self-Concept

One African American male recalls, "Much of my junior and high school years were difficult because, on top of the typical problems of this time period, I had to combine the struggle of being Black and having my race always looked down upon; expected to fail, expected to cause trouble, and expected to be unproductive. During this time, I had to fight to maintain my confidence. I did not know who I was . . . I was confused." He went on to describe how he was kicked off the football team for a failing grade, but when his parents spoke to the teacher, it was discovered that he actually had a C+. "He had given me an F not because I earned it, but because he expected me to deserve it" (John B. Diamond in Schoem, 1991).

A Mexican American male stated, "As I moved to junior high, the issue of my ethnicity became a problem. I remember thinking that I would be a great deal more popular if only I had Bobby's face and body and brains. I would look in the mirror and imagine what I would look like. The mythical Bobby was, of course, always white and popular with girls. This fantasy ate away at my self-esteem, and I found myself bitterly questioning why I had been born a brown-faced Mexican. . . . From this point on, all my energies were spent on the elusive quest for acceptance by my peers—and unconsciously, by myself" (Carlos Manjarrez in Schoem, 1991).

Do you think racial prejudice can make the development of self-concept and positive self-esteem difficult? Do you think the African American teen felt that he mattered to his teacher? What are the social expectations for him and how might they influence self-concept? What values of American society make the development of self-concept and positive self-esteem challenging for some minority teens?

African American children and adolescents are more likely to internalize the attitudes of important persons in their life (such as family members and friends) rather than the more distant attitudes of white society.

Despite the negative attitudes of the larger society toward racial minorities, the African American family has been able to develop positive feelings of personal self-esteem among their youth. African American families can help their children separate feelings of personal self-esteem from the negative views of society. Ward (1990) interviewed 51 minority adolescent women who were attending a prestigious private high school. Through her interviews, Ward learned how some African American families prepare their youths to survive the prejudices of society. Providing positive feelings and teaching youths how to deal with and understand racism seem to be most important. Ward explains, "when messages of white society say 'you can't,' the well-functioning black family and community stand ready to counter these messages with those that say, 'You can, we have, we will' " (p. 221).

Although personal self-esteem and racial self-concept reflect different aspects of the self, both are important to psychological well-being in adolescence. Recent research suggests that the self-esteem of older adolescents is related to the development of cultural identity (Phinney & Alipuria, 1990) (see discussion of cultural identity later in this chapter). Ward (1990) explains that a stable concept of the self both as an individual and as a group (African American) member are critical to the healthy growth of self. In recent years, the African American family and schools have also sought to build feelings of racial pride by helping youths learn about African American history and culture and the

Racial pride is enhanced by learning about one's cultural heritage and contributes to a positive sense of self.

achievements of African Americans. Phinney and Rosenthal (1992) suggest that the self-esteem of ethnic and racial minorities is influenced by a number of factors including the achievement of a committed ethnic identity, the presence of a strong and positive ethnic community with which to become involved, and gender-role expectations.

Recent research indicates that African American girls showed no evidence of decline in self-esteem at early adolescence (Eccles & others, 1999). In fact, theses researchers found that African American girls have higher self-esteem than European American girls and African American boys. African American girls scored higher on measures of confidence in their femininity, masculinity, popularity and physical attractiveness, while their European American counterparts scored higher on measures of worry about weight and social self-consciousness (Eccles & others, 1999). These differences suggest that the drop in self-esteem often linked with early adolescence is not inevitable but rather, is socially constructed.

As with white adolescents, the self-concepts of African American adolescents are influenced by relationships with parents, siblings, friends, and teachers. When African American adolescents feel liked by their family, friends, and teachers, they are likely to feel good about themselves. For this reason, racially desegregated schools can be difficult for African American adolescents. Students who are in the majority can make those who are different feel uncomfortable and out of place. During adolescence, students want to fit in with others around them, but having a different skin color makes a student stand out. Minority students may also be given few opportunities to be leaders in clubs and school organizations. All adolescents need teachers and peers who help them feel accepted (more on this in Chapter 8). Ward (1990) believes that African American student organizations can provide students with the social support they need to feel good about themselves and to learn about themselves and other African Americans. Teachers in predominantly white schools need to encourage African American students in their efforts to achieve while also encouraging them to maintain important sources of support with their families and peers of their own race.

African American and white adolescents probably value some aspects of the self-concept differently. African American adolescents generally describe peer relationships and athletic ability as more important than do white American adolescents (Epps, 1980; Harter, 1990). For many African American teens, participation in athletics has offered opportunities for recognition and advancement when other avenues have been denied. As a result, for some teens, athletics may have become more important to self-esteem.

Self-Concept and Social Competence

Although social competence has been defined in a variety of ways, the simplest definition is the ability to interact effectively with others. Socially competent people are able to make and keep friends and to accomplish their goals when relating to others. For adolescents, social competence is an important part of self-concept. Among the eight dimensions of adolescent self-concept identified by Harter (1990), social acceptance by peers together with physical appearance were most related to feelings of positive self-esteem. Adolescents who feel accepted by their peer group and their parents are likely to feel good about themselves.

Adolescents with low self-esteem are more likely to be disliked and rejected by peers. They are more likely to be shy and are less likely to be selected as leaders in clubs and social activities. Difficulties in social competence are not likely to disappear following adolescence. Growing research evidence suggests that adolescents who have poor peer relationships are more likely to have adjustment difficulties in adulthood (Reisman, 1985).

Adolescents' social relationships shape their self-concepts. They learn about themselves in relation to other people. As adolescents work out their social relationships, they sort out general and specific aspects of their own identities. A

WHAT'S YOUR VIEW?

Which Is Best—Looking Inward or Outward?

On the face of it, confiding in our friends would seem an excellent thing to do. However, several researchers have questioned whether adolescent sharing of personal feelings and worries, as described by Sullivan, is really beneficial. Mechanic (1983) believes that talking with friends about personal problems often makes adolescents feel *worse*. Conversation may lead to introspection (the self-examination of our innermost thoughts and feelings). Harter (1990) suggests that introspection is one of the greatest risks to the adolescent self-concept. Mechanic agrees with this, and believes that adolescent friendships are most helpful when they encourage looking outward rather than inward. Sharing exciting activities and adventures may be more healthy for some adolescents than sharing feelings and personal problems. Looking inward or outward—which is best? What do you think? Do you think that there are gender differences in styles of friendship during adolescence?

young adolescent's interest in the social behavior of peers is an attempt, in part, to learn about the self and how the adolescent is perceived by others. Long conversations between peers assist adolescents in understanding their social and emotional environment as perceived by others their age. Years ago, Sullivan (1953) described the importance of "chum" friendships. For early adolescents, chums increase each other's self-esteem and provide information, advice, and support in solving problems. By sharing personal thoughts and feelings, chums become more aware of the needs and desires of others. Sullivan believed that the understanding of self and others that comes out of chum friendships was important for the development of romantic and intimate relationships in later adolescence.

Generally speaking, adolescents find their friendships to be enjoyable. Together, they relax, joke, watch television, participate in sports, and talk (and talk and talk). These times give adolescents feelings of belonging with others who are liked and respected. Adolescents often feel they are best understood by their friends. Through these friendships, adolescents gain social skills such as empathy and understanding the point of view of others.

What adolescents learn about themselves from friends differs from what they learn about themselves from parents. The skills, values, and behaviors rewarded by one age group are quite different from those rewarded by the other. With parents, for example, adolescents may view themselves as polite and helpful. With friends, they may view themselves as loyal and talkative. They may act tough and aggressive on the basketball team, but be gentle and supportive when teaching Sunday school classes. Thus, the social relationship leads to differences in the self.

core self
According to Hart, a self that reflects our deepest values and is consistent across social relationships.

Hart (1988) suggests the existence of a **core self**. This reflects our deepest values and is consistent across social relationships. During early adolescence, parents have more influence on the core self. By late adolescence, relationships with a best friend are more important. About 75 percent of the self-concept, according to Hart, changes depending on the social relationship.

How adolescents view themselves in social relationships changes during adolescence. According to research completed by Smollar and Youniss (1985), early adolescents describe themselves as cooperative, sociable, and happy in their relationships with friends. Friendships allow the young adolescent to express the social self as helpful and considerate. The self-concept of older adolescents includes a view of the self as intimate, sensitive, and spontaneous. By late adolescence, social competence includes the ability to share personal thoughts and feelings and to be aware of the feelings and thoughts of others.

Social competence generally increases during adolescence. Dodge and Murphy (1983) believe that social competence is dependent on a person's ability to recognize cues in social settings. People who are socially competent are able to observe other peoples' behavior carefully and figure out what is acceptable. When unsure what to do, they model their behavior after someone who is more socially competent. Social competence, like social cognition (see Chapter 5) increases with age (Selman, 1980; Selman & others, 1997). As adolescents gain more cognitive skills, they are able to understand social situations better. They are less self-centered, better able to understand the views of other people, and more accurate in recognizing social cues.

Selman (1980, 1989; Selman & Schultz, 1990; Selman & others, 1997) maintains that healthy psychological functioning is related to an individual's skills in perspective taking (being able to understand from another's point of view) (see Chapter 5) and interpersonal negotiation (or working out differences with another person). Advances in perspective taking during adolescence can increase both self-understanding and social competence. The adolescent should be able to take a third-person perspective, to step outside the immediate relationship and view it from the perspective of another person. The adolescent is thus more self-reflective and aware of behavior and its effect on others. This increased self-awareness is accompanied by increased ability to control one's feelings and behavior and to understand the feelings of another person. During

mutual collaboration
In Selman's model, the ability to understand and respect one's own needs and those of other people.

pair therapy
A therapeutic technique developed by Selman in which two adolescents work together with a therapist toward the goal of developing more advanced interpersonal skills.

adolescence, according to Selman's model, the capacity for **mutual collaboration** in social relationships usually develops. Mutual collaboration involves the ability to understand and respect one's own needs and those of other people. It requires the capacity for intimacy, or sharing of experiences, and the capacity for autonomy, or ability to define one's interests and negotiate them with another person. Sometimes troubled adolescents do not develop expected skills in mutual collaboration. Selman and Schultz have developed a therapeutic technique called **pair therapy**. In this technique, two adolescents work together with a therapist toward the goal of developing more advanced interpersonal skills. The pair therapist plays the role of providing a third-person perspective and helps the adolescents work through conflicts in social interaction.

Thus far, we have examined the self from such standpoints as level of self-esteem and social competence. Another way to view the adolescent self is through Erik Erikson's concept of identity, which we described briefly in Chapter 2. In the next section, we look more closely at this extremely useful construct. As you read it, ask yourself if Erikson's ideas seem true for you.

Erikson's Concept of Self—The Identity

state of identity
According to Erikson, the general picture one has of oneself and also a state toward which one strives.

identity crisis
Erikson's term for the situation, usually in adolescence, that causes us to make major decisions about our identity.

repudiation
As described by Erikson, choosing one identity and therefore, repudiating (turning down) all other choices.

According to Erik Erikson, the main task of the adolescent is to achieve a **state of identity**. Erikson (1958, 1959, 1963, 1968, 1969; Erikson & Coles, 2000), who originated the term **identity crisis**, uses the term in a special way. In addition to thinking of identity as the general picture one has of oneself, Erikson refers to it as *a state toward which one strives*. If you were in a state of identity, the various aspects of your self-concept would be in agreement with each other; they would be identical.

Repudiation of choices is another essential aspect of reaching personal identity. In any choice of identity, the selection we make means that we have repudiated (given up) all the other possibilities, at least for the present. All of us know people who seem unable to do this. They cannot keep a job, they are not loyal to their friends, they are unable to be faithful to a spouse. For them, "the grass is always greener on the other side of the fence." They must keep all their options open and must not repudiate any choices, lest one of them should turn out to have been "the right one."

Erikson suggests that identity confusion is far more likely in a democratic society because there are so many choices. In a totalitarian society, youths are usually taught to have a uniform self-concept, which they are forced to accept. The Hitler Youth Corps of Nazi Germany in the 1930s is an example of a national effort backed by intense propaganda to get all the adolescents in the country to identify with the same set of values and attitudes—to accept a Nazi self-concept. In some societies, individual choices are limited because the family is valued above the individual. For that reason, most marriages are arranged in Muslim countries. In democratic societies, where more emphasis is placed on individual decision making, choices abound. As a result, some children feel threatened by this overabundance of options. Nevertheless, a variety of choices is essential to the formation of a well-integrated identity.

AN APPLIED VIEW

The Erikson Psychosocial Stage Inventory

The Erikson Psychosocial Stage Inventory (EPSI) was developed as a research tool to examine adolescents' resolutions of conflicts associated with Erikson's first six stages in psychological development. We remind you that these stages are concerned with basic trust, autonomy, initiative, industry, identity, and intimacy.

Because Erikson regarded adolescence as central to his theory of human development, an investigation of how the adolescent forms an identity is of value. EPSI has 12 items for each of Erikson's stages and was tested in a study of 622 adolescents (Rosenthal & others, 1981). On the basis of their extensive research, the authors concluded that the EPSI is a useful measure for studying early adolescence and for "mapping changes as a function of life events" (p. 525). This means that the test can be used to find out the relationships between a person's stage of development and his or her age, IQ, personality traits, and many other characteristics. Here are some sample items, which the respondent is asked to check true or false:

Item Number	Subscale/ Item
36.	*Trust*: Things and people usually turn out well for me.
13.	*Autonomy*: I know when to please myself and when to please others.
34.	*Initiative*: I'm an energetic person who does lots of things.
60.	*Industry*: I stick with things until they are finished.
10.	*Identity*: I've got a clear idea of what I want to be.
59.	*Intimacy*: I have a close physical and emotional relationship with another person.

From D. A. Rosenthal, et al., *The Journal of Youth and Adolescence*, 10(6):525–37. Copyright © 1981 Plenum Publishing Corporation, New York, N.Y. Reprinted with permission.

Further, it is normal for identity confusion to cause an increase in self-doubt during early adolescence (Seginer & Flum, 1987; Shirk, 1987). Shirk states that "such doubts should decrease during the middle-teen years, as social norms for self-evaluation are acquired through role-taking development" (p. 59). He studied self-doubt in 10-, 13-, and 16-year-olds, and found significant decreases with advancing age.

We believe that most psychologists would call Erikson the foremost theorist on adolescence today. He probably has done more research and writing on this fifth stage of identity formation than on all the others combined. This does not mean that all agree with his view that adolescence is a time of identity crisis. In fact, extensive cultural differences have been found in identity formation (Geertz, 1984; Rosaldo, 1984).

THE SEARCH FOR IDENTITY

Goethals and Klos (1976) argue that for well-educated youth, an identity crisis comes only at the end of adolescence:

❏ It is our opinion that college students do not typically have a firm sense of identity and typically have not undergone an identity crisis. College students seem to be in the process of identity seeking, and experience identity crisis toward the end of senior year and in their early post-college experience. A male or female's disillusionment with their job experience or graduate study, a female's disappointment at being at home with small children, is often the jolt that makes them ask what their education was for, and why they are not as delighted with their lives as they had been led to believe they would be (p. 129).

University of Michigan researchers (Bachman & others, 1978) studied changes in the attitudes and goals of 2,000 male adolescents. They conclude that contrary to the view "of adolescence as a period of great turbulence and stress, we found a good deal of consistence along dimensions of attitudes, aspirations, and self-concept" (p. 31). Few of their subjects gave any evidence of having experienced an identity crisis. A recent study by Kalakoski and Nurmi (1998) suggests that adolescent identity issues tend to be related to times of transition. For example, the transition from middle school to high school might trigger some identity exploration. These researchers found that the most common identity concerns for teens were school and career related.

Erikson, who himself had an extensive and rather difficult identity crisis in his youth, supposed that "My friends will insist that I needed to name this crisis in everybody else in order to really come to terms with it" (1975, p. 26). Born Erik Homberger, he seems to have rejected his difficult past. His Danish mother remarried a German Jew, and Erikson found himself shunned both by Jewish and Christian children.

His identity crisis was resolved by the creation of a brand-new person with a new name, religion, and occupation. Some biographers (e.g., Berman, 1975; Roazen, 1976) have suggested that the surname he chose, "Erikson," means he is the "son of himself." His theory of human development is no doubt colored by these experiences. At the same time, the intensity and degree of his identity crisis have made him extremely sensitive to the problems that all adolescents go through.

Perhaps the best conclusion we can reach, based on the available evidence, is that although the teen years are definitely a time of *concern* over one's identity, major decisions about it may be postponed by many until they reach early adulthood. This is probably truer today than ever because of the phenomenon that Erikson calls the **moratorium of youth**, which seems to be lasting longer and longer.

moratorium of youth
A time-out period during which the adolescent experiments with a variety of identities, without having to assume the responsibility for the consequences of any particular one.

THE MORATORIUM OF YOUTH

Erikson sees late adolescence as a period of moratorium—a "time-out" period during which the adolescent experiments with a variety of identities, without having to assume responsibility for the consequences of any particular one. We allow late adolescents this moratorium so that they can try out a number of ways of being, the better to come to their own particular identity. The moratorium period does not exist in pre-industrial societies. Some have suggested that only Western industrial societies can afford the luxury of a moratorium. Others say that only because the values in Western industrial societies are so conflicted do adolescents *need* a moratorium.

premature foreclosure
A situation where a teenager chooses an identity too early, usually due to external pressure.

Erikson stated that indecision is an essential part of the moratorium. Tolerance of it leads to a positive identity. Some youths, however, cannot stand the ambiguity of indecision. This leads to **premature foreclosure**. The adolescent who makes choices too early usually comes to regret them and is especially vulnerable to identity confusion in later life.

Erikson suggested that religious initiation ceremonies such as Catholic confirmation and Jewish bar and bas mitzvah can limit the young, forcing them into a narrow, negative identity. This can happen if the ceremony dogmatically spells out the specific behaviors expected by adults. On the other hand, such ceremonies can suggest to youths that the adult community now has more confidence in their ability to make decisions. The effect depends on the explanation of the goals of the ceremony.

Although some youths tend to be overly idealistic, Erikson believes that idealism is essential for a strong identity. Young people, in their search for a person or an idea to be true to, are building a commitment to an ideology that will help them unify their personal values.

IDENTITY STATUS

identity status
Refers to Marcia's four types of identity formation.

Erikson's ideas on adolescence have generated considerable research on identity formation. The leader in this field is James Marcia, who has made major contributions to our understanding through his research on **identity status**. He and his colleagues have published numerous studies on this topic (Marcia, 1994, 1980; Berzonsky & Kuk, 2000; vanHoof, 1999; Waterman, 1999).

Marcia believes that there are two essential factors in the attainment of a mature identity. First, the person must undergo several *crises* in choosing among life's alternatives, such as the crisis of deciding whether to hold to or give up one's religious beliefs. Second, the person must come to a *commitment*, an investment of self, in his or her choices. Since a person may or may not have gone through the crisis of choice and may or may not have made a commitment to choices, there are four possible combinations, or statuses, for that person to be in:

Table 6.1	Summary of Marcia's Four Identity Statuses			
	Identity Status			
	Confusion	**Foreclosure**	**Moratorium**	**Achievement**
Crisis	Absent	Absent	Present	Present
Commitment	Absent	Present	Absent	Present
Period of adolescence in which status often occurs	Early	Middle	Middle	Late

- *Status 1: Identity confusion*. No crisis has been experienced and no commitments have been made.
- *Status 2: Identity foreclosure*. No crisis has been experienced, but commitments have been made, usually forced on the person by the parent.
- *Status 3: Identity moratorium*. Considerable crisis is being experienced, but no commitments are yet made.
- *Status 4: Identity achievement*. Numerous crises have been experienced and resolved, and relatively permanent commitments have been made.

Table 6.1 shows the relationships among the statuses.

Erikson's eight stages (basic trust, autonomy, initiative, industry, identity, intimacy, generativity, and integrity—see Chapter 2) follow each other in a more or less unchangeable sequence. Research indicates that identity statuses have an orderly progression, but it's not as clear as in Erikson's stages. For example, Meilman (1979) studied males at the ages of 12, 15, 18, 21, and 24. They were rated on attitudes toward occupation, religion, politics, and, for the older subjects, sexual matters. In all these areas, as the group got older, fewer individuals were in the confusion status and more were in the achieved status.

GENDER AND IDENTITY

Which comes first—intimacy or identity? Why?

Erikson's theory of identity development was based on the study of men. Erikson suggested that a woman's identity may be left open at adolescence to accommodate her choice of a man. Career and ideological concerns were also considered less important to the identity development of women than of men. Some research (Douvan & Adelson, 1966; Marcia, 1980) suggests that the stages of identity development may differ for girls, with intimacy occurring either before or concurrent with identity development. For males, identity issues seem to be resolved before they begin working on intimacy issues. For females, intimacy and occupational identity are considered interrelated concerns.

Erikson's understanding of female identity development has been the focus of considerable debate. Researchers and theorists (e.g., Gilligan, 1982) interested in the psychological development of women argue that more accurate models of female identity development need to be developed based on the study of women. Josselson (1987; 1996) sought to expand Erikson's work on identity to include the experiences of women. She studied identity achievement in women by interviewing 60 women as college seniors. Thirty-four of the women were interviewed again as adults more than 10 years later. Based on these interviews, Josselson concludes that the pathway to identity achievement is different for women and men. "In short, the aspects salient to identity formation in women have been overlooked by psychological research and theory, which stresses the growth of independence and autonomy as hallmarks of adulthood. Communion, connection, relational embeddedness, spirituality, affiliation—with these women construct an identity" (p. 191). Women in diffusion appear to be lacking not only exploration and commitment, but also stable and positive relationships.

They may be estranged from parents or have no close friendships. In identity achievement, women continue to enjoy relationships that provided support and stability throughout the process of personal exploration. Josselson's work does not suggest that Erikson's stages need to be reordered for women, but rather that relationships are an important part of identity, at least for women. Unfortunately, Josselson's work shares the flaw of Erikson's work. Because Josselson did not interview men in addition to women, we do not know whether men would also describe close relationships as important to their identity.

After reviewing recent research in identity development, Cosse (1992) concludes that there are at least three pathways for identity development through adolescence. The autonomous pathway, involving the repudiation of earlier close relationships, is the predominant pathway for males and for some females. The interpersonal relatedness pathway, described by Gilligan (1982; 1988) and Chodorow (1978), enables a person to develop a unique sense of self while maintaining close relationships. This pathway is characteristic of individuals who have high levels of empathy for the feelings of others and serves to further increase feelings of empathy. The third pathway combines elements of the previous two. Cosse reminds us that adolescent identity development occurs in a specific cultural context. American culture rewards competitive competencies in school and work that enable someone to be an autonomous individual. Thus, some girls may choose the pathway of separation and autonomy but few boys will view the pathway of relatedness as culturally desirable.

Franz and White (1985) maintain that Erikson's stages explain the development of identity but not the development of intimacy. Whereas the prior stages of autonomy, initiative, and industry logically contribute to identity, their contributions to the development of intimacy are less clear. Franz and White propose that Erikson's framework be extended to include a second pathway, or series of stages, that leads to the attainment of intimacy. By attending to the development of both identity and intimacy, the model may be more appropriate and complete in explaining the development of men and women.

Archer (1992, 1993) believes that identity researchers have made a mistake by thinking of identity and intimacy as mutually exclusive characteristics. She argues that both men and women need to develop clear self identities and capacities for intimacy. Also, some experiences that are thought of as part of identity achievement also involve interpersonal relationships. Guidance from a mentor, for example, has been an important career development experience for young men and women. Archer thinks that identity researchers need to pay more attention to the social experiences that promote and inhibit identity development and that contribute to conflicts between the achievements of identity and intimacy. Identity development may be painful and confusing for the female adolescent growing up in a society in which adults continue to prefer boy babies over girl babies, in which one out of every four girls will be sexually abused by the age of 18, and in which women hold jobs of lower status and pay than their male counterparts. Conflict between intimacy and identity results when girls feel that the traditional female identity of being nurturing and helpful conflicts with career preparation or job entry. Theories of identity will not be complete, according to Archer, until more attention is given to the roles of stereotypes, wealth, and power based on gender, class, and race.

Grotevant (1992; 1997) explains that some aspects of identity are assigned whereas other aspects are chosen. People generally choose their occupational identities following a process of exploration described by Erikson. Gender, ethnicity, and adoptive status are components of identity that are assigned and not freely chosen. Assigned identities, especially those that are not valued highly by society, may limit choices related to other aspects of identity. For example, racial discrimination and gender stereotyping may limit the options available in developing an occupational identity. The challenge for the adolescent is to successfully combine areas of assigned and chosen identity while maintaining a positive and strong self-identity. We now turn to a discussion of another assigned identity—cultural identity.

THE DEVELOPMENT OF CULTURAL IDENTITY

cultural identity
That part of a person's self-concept that comes from the knowledge and feelings about belonging to a particular cultural group.

Forming an identity is complex for cultural minorities (sometimes called ethnic or racial minorities). Growing up in a society in which the dominant culture differs from one's culture of origin, ethnic minority youths need to integrate a positive ethnic identity with a strong and positive self-identity. Minority adolescents ask themselves not only, "Who am I?" but also such questions as "Who am I as an African American?" or "as a Latino?" **Cultural identity** is that part of a person's self-concept that comes from the knowledge and feelings about belonging to a particular cultural group. Cultural identity includes self-identification (whether I describe myself as Latino) and a sense of belonging (how close do I feel toward other people in my cultural group). It also involves an attitude toward one's cultural group (am I proud to be a member of my cultural group) and involvement (do I participate in the cultural practices of my group and do I have friends who are members of my cultural group) (Phinney, 1990).

Because the number of cultural minorities living in the United States is increasing, social scientists have become interested in the development of cultural identity. Many believe that a positive sense of cultural identity can help minority adolescents feel good about themselves in the face of discrimination and threats to their identity from the larger society. Some findings suggest that higher levels of cultural identity are related to better psychological adjustment and higher self-esteem (Parham & Helms, 1985; Phinney, 1989; Roberts & others, 1999). Ohye and Henderson Daniels (1999) believe that there is still much work needed to understand the complex issues of race, ethnicity and class. These researchers suggest that studies of adolescent females have been largely restricted to white females. Therefore, six million girls in the United States become adults without adequate representation in the developmental literature and research (Ohye and Henderson Daniels, 1999).

Many Americans believe in the melting pot philosophy. This is the idea that a new and superior race is developing from a mixing of many races and cultures. As these diverse peoples mix, they are expected to give up their customs and blend in with the larger society. Maintaining a sense of cultural identity may be

The melting pot concept holds that when you join a society (mainly through immigration), you should try to become as much like the members of that society as possible. Otherwise, you should not come. To this end, some states are considering making English the legal language, refusing to recognize any other language in school, on traffic signs, etc.

Other people argue that you can "maintain a foot in each camp." For example, you can speak your native language at home and English outside the home. You can celebrate the country's national holidays and those of your own culture as well.

What do you think? Is it possible to have a strong allegiance to a minority culture *and* the majority culture? Do people have to choose sides, especially where the two cultures clash? What's your opinion?

marginalism
Weak identification with both the majority culture and your own minority culture.

biculturalism
Strong identification with both the majority and your own minority culture.

seen as a barrier to blending in. Phinney (1990) presents a model suggesting four ways in which adolescents can identify with both their own minority culture and the majority culture. The adolescent may have a weak identification with both the majority and the minority group, may have a weak identification with one group (either minority or majority) and a strong identification with the other group, or may have a strong identification with both groups. Weak identification with both groups results in **marginalism**, which refers to a situation where a person feels isolated and alienated from both cultures. Strong identification with both groups is called **biculturalism**.

According to Vasqez and De Las Fuentes (1999), ethnicity and social class are critical socialization variables that have not been studied enough. Adolescents of color may suffer from institutionalized racism entrenched in systems developed by a dominant culture. For example, numerous programs developed to "help" Native American children remove these children from their homes and thus distance them from their families and their culture. Since family tends to be a source of support and security and a vehicle for ethnic identification, such programs are problematic. Acculturation may be the goal of these programs. However, acculturation can threaten family relationships as it can create conflict between parents and children around issues of traditional values (Vasqez and De Las Fuentes ,1999). We should consider this dilemma as we think about programs that "help" teens.

STAGES OF CULTURAL IDENTITY DEVELOPMENT

unexamined cultural identity
Early adolescents and others who may have given little thought to their identity.

cultural identity search
A process of exploring one's cultural identity.

The development of cultural identity may be thought of as a process that takes place over time, similar to Erikson's stages. Phinney (1989) suggests that the development of cultural identity takes place in a three-stage process. Early adolescents and others who have not been exposed to cultural identity issues are in the first stage, an **unexamined cultural identity**. People in this stage have given little thought to their identity. They may have absorbed their parents' attitudes without question, or they may prefer views of the dominant culture.

The second stage, **cultural identity search**, is a process of exploring cultural identity. It may begin after an important experience that forces awareness of choice. For example, the boy who endures racial slurs after changing to a school in which the dominant race is different from his may then become involved in reading about his race and participating in racial events. As a result of this process, people develop a deeper appreciation and understanding of their cultural background.

achieved cultural identity
When individuals find a way of resolving the differences between their cultural background and the dominant culture as well as the problem of the lower status of their cultural group in society.

The third stage, **achieved cultural identity**, results when people find a way of resolving two problems: (a) the differences between their cultural background and the dominant culture and (b) the lower status of their cultural group in society. At this stage, the individual has a clear and confident sense of cultural identity. Parham (1989) believes that the process does not end here, however. He suggests instead that the meaning and importance of cultural identity may be re-explored a number of times during a lifetime.

Achieving a clear and confident sense of cultural identity can be difficult in our society. Research shows us, however, that an achieved cultural identity is related to positive self-esteem and good mental health.

Can you think of additional challenges in promoting a positive cultural identity among teens?

Spencer and Markstrom-Adams (1990) have offered some ways to enhance cultural identity development. Parents need to be supported in their efforts to teach their culture to their children. Too often adults remain uncomfortable with issues of race or ethnic background and do not discuss these issues with their children. Parents need to be taught skills that build their sense of cultural pride and contribute to involvement in their children's schools. Teachers need to be better sensitized to the customs and traditions of cultural minorities. Negative stereotypes of minorities as presented in books, movies, and television must be challenged.

Boykin & Toms (1985) identified three challenges facing minority parents in raising children: the need to understand their own culture, the need to understand the mainstream culture, and the need to understand prejudice and racism. Phinney and Chavira (1995) studied African American, Japanese American, and Mexican American adolescents and their parents, and found that African American parents attended most to the ethnic socialization of their children. Parents emphasized the importance of achievement and taught their children that they would have to work harder than other adolescents to be successful. They stressed the importance of fitting into the mainstream, but also tried to prepare

Achieving cultural identity is an important development task for many adolescents. The ability to positively identify with one's cultural heritage and the majority culture is called biculturalism.

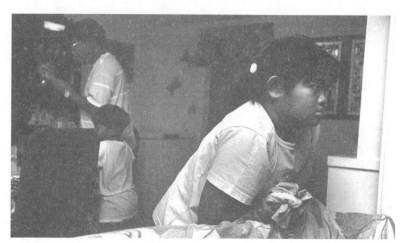

AN APPLIED VIEW

Forming a Strong Cultural Identity

Adolescents who are experiencing conflicts of cultural identity sometimes develop psychological symptoms like depression, anxiety, school failure, or delinquency. They may report conflicts with their parents, who represent the ways of the culture. Adolescents sometimes turn the hostility of the majority culture on themselves and suffer low self-esteem and depression. Sometimes they turn that anger outward and get into trouble at school or in the community.

Teachers, counselors, and others can work with youths to help them understand that developing a cultural identity can be difficult. Adolescents need to be guided into understanding their conflict by exploring what they feel are the positive and negative aspects of each culture. Then they can be encouraged to identify with what they consider the positive traits and to dissociate themselves from the negative traits. This helps them define their own identity. Counselors can make it clear that adolescents do not have to choose one identity over the other, for both have admirable qualities.

their children for the prejudice and discrimination they would face. Mexican American parents emphasized cultural pride more than the other two groups, and Japanese American parents emphasized achievement, with less emphasis on preparation for discrimination or cultural pride. Phinney and Chavira explain that African American parents respond to the reality of continued prejudice, especially in its subtle forms.

Sexual identity and gender role are also important aspects of self and identity. Beliefs about gender role have changed dramatically in the latter half of this century as a result of the feminist movement. This has had a profound effect on the ways in which many adolescent girls and boys (and adults) define themselves and their relationships with members of the opposite sex.

IMMIGRANT IDENTITY

There is a relatively large immigrant population in the United States and it is important to consider the unique experience of immigrant adolescents (De Las Fuentes & Vasquez, 1999). The immigrant experience can present additional challenges for teens beginning to grapple with issues of identity and connection. Even when the overall change brought about by immigration is positive the transition can trigger feelings of loss, alienation, and confusion. The desire to fit in, a hallmark of adolescence, can present a unique challenge for immigrant adolescents who may be trying to juggle the traditional cultural expectations and values of their parents and the expectations and values of their present surroundings in American society. Language barriers and socioeconomics can also be stressors for immigrant teens. The many stresses seem to be mediated by ethnic social support which makes sense in light of what we have just discussed in the previous section about cultural identity (De Las Fuentes & Vasquez, 1999).

sexual identity
Development of a cohesive sense of self as a sexual being in relation to culturally determined categories.

gender role
A pattern of behavior that results partly from genetic makeup and partly from the specific traits that are in fashion at any one time and in any one culture.

gender
Our conceptions of what it means to be male or to be female.

gender intensification hypothesis
Hill and Lynch's hypothesis that pubertal changes in early adolescence contribute to increased concern about conformity to gender stereotypes.

Sexual Identity and Gender Roles

❏ The problem lay buried, unspoken, for many years in the minds of American women. It was a strange stirring, a sense of dissatisfaction, a yearning that women suffered in the middle of the twentieth century in the United States. Each suburban wife struggled with it alone, as she made the beds, shopped for groceries, matched slipcover material, ate peanut butter sandwiches with her children, chauffeured Cub Scouts and Brownies, lay beside her husband at night—she was afraid to ask even of herself the silent question—"Is this all?"
Betty Friedan, 1963

This is the opening of the famous book *The Feminine Mystique*, which began the feminist movement in this country. Its early effects were mainly on late adolescent and early adult females, but today there is scarcely a female in the country (or a male, for that matter) whose life has not been affected by this movement.

   Contrast it with this statement made more than 2,000 years ago by the famed Greek philosopher Aristotle: "Woman may be said to be an inferior man." Most of us would probably disagree publicly with his viewpoint, but its underlying attitude is still widespread. People today are far less willing to admit to a belief in female inferiority, but many still act as though it were so. The influence of the women's movement as well as of science and other forms of social change, however, is profoundly affecting the way we view **sexual identity** and **gender role**.

AN APPLIED VIEW

The Young Adolescent's Rigid View of Gender

The term **gender** is used all the time. But what does it mean? Gender refers to our conceptions of what it means to be male or to be female. Ronald Slaby (1990) has explored gender as a social category system. By that he means a mental filing system: we use the categories of male and female to organize the information we receive. Slaby argues that gender categories are loaded with meaning. For example, when we hear of a person named Mary, before even meeting her we hold certain preconceptions of what she will be like. She will be female, so she can't be male, and will (probably) always be female. If you are female, then a person named Mary shares with you the things that you think all females experience. If you are male, then there are things that you think only males experience that you feel certain a person named Mary does not experience. In this manner, gender acts as a powerful organizing system.

   Most adolescents wrestle with what it means for them personally to be male or female. In doing so, they develop their gender categories. Studies of the adolescents' ideal men and women reflect very stereotyped

notions of the sexes. According to Slaby, this is due to the inflexibility of the adolescent's developing gender categories. Hill and Lynch (1983) explain this according to the **gender intensification hypothesis**. They believe that early adolescents experiencing puberty are concerned with the significance of their gender. When the meaning of gender is somewhat unclear, adolescents are more likely to use the categories of male and female in a rigid and stereotyped manner. That clarity helps adolescents solidify their understanding of gender. Once they become confident and comfortable in their ability to figure out maleness or femaleness, they can use the gender categories more flexibly. Then they are able to understand that even though Mary might fit their category of female in most ways, in some ways she may not.

   What about you? Are your concepts of gender role flexible or rigid? And what about your friends and relatives? It might be fun to design a questionnaire on gender roles and ask them to respond to it.

gender identity
Self-identification as a male or female based on biological characteristics.

gender-role behavior
The extent to which a person's behaviors, occupations, and interests are considered masculine or feminine according to cultural norms.

sexual orientation
Choice of a sexual partner of the same sex or opposite sex.

First, we should make a distinction between the two. Sexual identity comes from the development of a cohesive sense of self as a sexual being in relation to culturally determined categories (Savin-Williams & Rodriguez, 1993). According to the framework originally set forth by Money and Ehrhardt (1972), sexual identity includes **gender identity, gender-role behavior**, and **sexual orientation**. Gender identity is a person's self-identification as male or female and results largely from those physical characteristics that are part of our biological inheritance—the traits that make us males or females. Genitals and facial hair are examples of characteristics that affect gender identity. Gender role, on the other hand, develops largely from the specific traits that are in fashion at any one time and in any one culture. For example, in our culture, women appear to be able to express their emotions through crying more easily than do men, although there is no known physical cause for this difference. Gender-role behavior is the extent to which a person's behaviors, occupations, and interests are considered masculine or feminine according to cultural norms. Sexual orientation refers to a person's choice of sexual partner of the same sex (homosexual), of the opposite sex (heterosexual), or both sexes (bisexual). Sexual orientation will be discussed more fully in Chapter 9.

It is possible for people to accept or reject their gender identity and/or their gender role. Jan Morris (1974), the British author, spent most of her life as the successful writer James Morris. Although born a male she deeply resented the fact that she had a male gender identity and hated having to perform the male gender role. She always felt that inside she was really a woman.

The cause of these feelings may have been psychological—something that happened in her childhood, perhaps. Or the cause may have been genetic—possibly something to do with hormone balance. Such rejection is rare, and no one knows for sure why it happens. Morris decided to have a transsexual operation that changed her from male to female. The change caused many problems in her life, but she says she is infinitely happier to have her body match her feelings about her gender role.

gender-role adaptation
Defined by whether the individual's behavior may be seen as in accordance with her or his gender.

Some people are perfectly happy with their gender identity, but do not like their gender role. **Gender-role adaptation** is defined by whether other people judge individual behavior as masculine or feminine. If a person is seen as acting "appropriately" according to her or his gender, then adaptation has occurred. People who dislike their gender role, or doubt that they fit it well (e.g., the teenage boy who fights a lot because he secretly doubts he is masculine enough), may be said to have poor gender-role adaptation.

Research indicates that while gender roles may be modified by differing cultural expectations, gender identity is fixed early in a person's development. Money and Ehrhardt at Johns Hopkins University (1972) believe there is a critical period for the development of sexual identity, which starts at about age 18 months and ends at about age 4.

Once a child's gender identity has been established, it is unlikely to change, even when biological problems occur. Even in such extreme cases of chromosome failure (see Chapter 3) as gynecomastia (breast growth in the male) and hirsutism (abnormal female body hair), gender identity is not affected. In almost all cases, adolescents desperately want medical treatment so they can *keep* their gender identities.

Whereas gender identity becomes fixed early in life, gender roles usually undergo changes as the individual matures. The relationships between the roles of our two genders also change, and have altered considerably in the past few decades.

THE PSYCHOLOGICAL SIGNIFICANCE OF GENDER ROLES

Before the feminist movement began in the 1960s, most psychologists believed that conformity with traditional sex-role expectations was necessary for healthy psychological adjustment. The feminist movement, however, raised concern about the consequences of adhering to traditional gender-role expectations. One sex-role researcher, Sandra Bem (1975), argued that traditional American roles are unhealthy. She said that highly masculine males may be better adjusted psychologically than other males during adolescence, but as adults they may become anxious and neurotic. Pressures felt by American males to achieve in their careers and to hide feelings of sadness or weakness, for example, might contribute to anxiety and neuroticism. The highly feminine female would suffer from societal pressure to conceal her strength and competence.

androgynous
Refers to those persons who have higher than average male and female elements in their personalities.

Bem believed we would all be much better off if we were to become more **androgynous**. The word is made up of the Greek words for male, *andro*, and for female, *gyne*. It refers to those persons who are higher than average in both male and female characteristics. Such people are more likely to behave in a way appropriate to a situation regardless of their gender.

For example, when someone forces his or her way into a line at the movies, the traditional female role calls for a woman to look disapprovingly but to say nothing. The androgynous female would tell the offender in no uncertain terms to go to the end of the line. When a baby left unattended starts to cry, the traditional male response is to find some woman to take care of its needs. The androgynous male would pick up the infant and attempt to comfort it.

Androgyny is not merely the midpoint between the two poles of masculinity and femininity. Rather, it is a higher level of sex-role identification than either of the more traditional roles. Figure 6.1 illustrates this relationship.

A great deal of research has been done to assess Bem's claim that androgynous males and females, possessing high levels of both masculine and feminine traits, are better off psychologically. Although some argument in favor of the androgyny model has been found (Mullis & McKinley, 1989), the strongest

Figure 6.1

Relationships among the three sex roles.

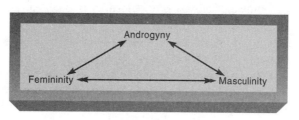

support has been for a positive relationship between masculine traits, such as independence, action, and mastery, and psychological well-being (Allgood-Merten & others, 1990; Cate & Sugawara, 1986; Lamke, 1982; Whitely, 1983). Examination of findings that supported the androgyny model revealed that the positive relationship between androgyny and psychological well-being was due to the presence of masculine traits rather than the combined presence of masculine and feminine traits (Whitely, 1983, 1984). In recent research, Allgood-Merten and associates (1990) found that high school boys and girls who described themselves as having high levels of masculine characteristics reported low levels of depression. Towbes and associates (1989) found that high school girls who described themselves as high in masculine traits such as a sense of personal control and feelings of mastery were better able to cope with life stress than were girls reporting low levels of masculine traits.

> What do you think? What kind of gender-role socialization is best for teens?

Are traditional male gender roles more healthy than traditional female roles? Critics (Gil & others, 1987; Nichols & others, 1982) say this is not necessarily so. Because measures of masculine traits such as independence, mastery, assertion, and self-control are so similar to psychological measures of self-esteem and freedom from depression, it is not surprising that masculinity is positively related to those measures. Critics suggest that psychological measures of well-being are biased because they list few traditional female characteristics (e.g., concern for others, a willingness to express feelings, and seeks help when needed). Measures of femininity also reflect societal bias and list characteristics that are not psychologically desirable—passivity, dependence, childishness, shyness, and gullibility. In contrast, few undesirable male characteristics are included on measures of masculinity. Critics also point out that traditional male characteristics may be helpful in certain situations, such as work and school, whereas traditional female characteristics are helpful in other situations, such as developing close friendships or intimate relationships.

Current findings concerning the psychological consequences of gender roles are not clear. Psychologists are particularly concerned about the effects of gender-role conformity for adolescents. As you will recall from the discussion of the gender intensification hypothesis (Hill & Lynch, 1983) earlier in this chapter, pressures to behave in gender-appropriate ways often increase in adolescence. Carol Gilligan and her colleagues (Brown & Gilligan, 1992; Gilligan & others, 1990) believe that this creates a dilemma for adolescent girls. At the same time that girls feel an increased need to conform to female gender-role expectations in order to be accepted by peers, they become aware that feminine interpersonal qualities are not valued by male society. Nolen-Hoeksema and associates (1991) suggest that adolescent girls who feel that they must hide their competence to win the acceptance of adolescent boys and adults also have feelings of hopelessness, helplessness, and depression. Pipher (1994) also suggests that there are risks to mental health when the true selves of adolescent girls become lost when young women are faced with the choice to be true to themselves or be accepted by others.

Recent literature and research looking at the development of boys (e.g. Pollack, 1998) suggests that societal expectations of gendered behavior limits boys as well as girls. Pollack (1998) describes what he calls the "Boy Code" as "the outdated and constricting assumptions, models, and rules about boys that our society has used since the nineteenth century" (p.6). Pollack believes that the "boy code" places boys in what he calls a "gender straitjacket" (p.6) and limits boys ability to be as connected to others as they would like and need to be. Pollack also describes a "mask" (p.5) behind which boys hide their true feelings. He believes that boys become so good at wearing the "mask" that they are not necessarily even aware of their true feelings.

Other writers also describe how pressures to conform to gender-role stereotypes negatively affect adolescent boys. Joseph Pleck and his colleagues (Pleck, 1983; Pleck & others, 1994) found that adolescent males who conform to traditional beliefs about male gender roles report more drug and alcohol use, more delinquency, more school problems, and higher levels of sexual activity than boys with less traditional gender-role beliefs. Other research suggests that failure to meet traditional gender roles is psychologically damaging for males. College men who described themselves as failing to meet traditional male standards of behavior reported more anxiety, depression, and social maladjustment than did college women who deviated from traditional female gender roles (O'Heron & Orlofsky, 1990). Many college men continue to think that

A SOCIOCULTURAL VIEW

A Guatemalan Indian Girl Assumes a Woman's Role

Rigoberta Menchú Tum, a Guatemalan peasant woman, has told the story of becoming a woman in her community. At age 10, girls there are taught about the practices of womanhood by their mothers:

❑ She [Menchú Tum's mother] taught us to look after and preserve our household things. Our cooking pots, for example. She had a lot of earthenware pots that she'd had for many years and they hadn't broken or been ruined because she knew how to look after her things. Well, she told us that if you are poor, you can't buy things all the time, nor must you only expect things from your husband. You yourself have to do your part to keep your little things, too. And she gave us examples of people whom she knew or that she'd helped to improve themselves: "That's what happens with women who don't look after their pots and then when they don't have them any more, they have to go and buy more."

She was like that with everything. Another of our customs she taught us was that you mustn't mix women's clothes with men's clothes. She told us to put our brothers' clothes on one side when we wash them. First you wash the men's clothes and then, at the end, our own. In our culture we often treat the man as something different—the woman is valued too, of course—and if we do things we must do them well first, because they are our men, and second, because it's a way of encouraging them, in the same way our ancestors did for their men. Not mixing the clothes was, I think, the order they respected. My mother said that we women have certain things that a man doesn't have, like our period for instance. So we keep all our clothes separate. It's the same for everything: we don't mix them, but most of all with our clothes. However, with kitchen utensils and all the things for the house, there isn't one for each (Burgos-Debray, 1984, p. 214).*

What is the real meaning of the washing of clothes? What does this say about female roles in Guatemala?

*From Elizabeth Burgos-Debray, editor *I, Roberta Menchú*
© Verso/NLB, London and New York, 1984, p 214.

expressing feelings is inconsistent with the male sex role. According to Cournoyer and Mahalik (1995), college men who are fearful about expressing feelings are also likely to experience anxiety, depression, and difficulty in establishing intimacy. It seems that recent changes in gender-role expectations have provided more flexibility for females than for males. Males continue to pay a psychological cost for adherence to traditional male gender roles.

How can gender-role socialization be changed to promote the psychological well-being of adolescent boys and girls? During the 1970s, Bem believed that androgyny was the answer. In response to criticism of the androgyny model and disappointing research findings, Bem (1981) then suggested that children and adolescents should be raised to be **gender aschematic**. Traditional socialization practices result in **gender schema**, or classification systems in which events and behaviors are labeled as appropriately masculine or feminine. This classification system leads to negative self-evaluation and low self-esteem when a person does not meet societal expectations for his or her own gender. Adolescent girls' intense dissatisfaction with their physical appearance, for example, can be understood as a failure to meet cultural standards for female attractiveness. Bem believes that optimal psychological health will follow when boys and girls are free of unrealistic standards. Rather than requiring high levels of both masculine and feminine traits (the androgyny model), gender schema theory suggests that psychologically mature individuals describe and use their talents without concern for culturally imposed gender-role expectations.

Gilligan and her colleagues (Brown & Gilligan, 1992; Gilligan & others, 1990), however, do not expect gender-role differences to disappear. They maintain that the psychological development of girls is different from boys. To promote the psychological well-being of adolescent girls, both the girls themselves and the larger society must learn to value and reward female qualities such as nurturance, interpersonal sensitivity, and concern for interpersonal relationships.

gender aschematic
Not using gender as a standard for evaluating or classifying behavior.

gender schema
A cognitive process through which behaviors are labeled as masculine or feminine.

THE IDEAL ADOLESCENT

What's your image of the ideal man? The ideal woman? If you presented those questions to a group of adolescents today, would you expect the answers to be the same as when you were an adolescent? The same as when your parents were adolescents? Researchers seeking to answer these questions have been surprised at the results. They have found overwhelming evidence that little has changed in adolescents' concepts of ideal males and ideal females (Doherty, 1998; Mills & Mills,1996; Snodgrass, 1992; Stiles, Gibbons, et al., 1987; Stiles, Gibbons, & Schnellmann, 1987).

Snodgrass (1992) surveyed college students and found that clear stereotypes of men and women prevail. To the students she studied, men represent forceful self-assertion and women represent nurturance and emotionality. The study did reveal one difference. In contrast to studies conducted some years earlier, Snodgrass found that these college students applied the personality traits of being independent, active, and self-confident to women as well as to men. In a

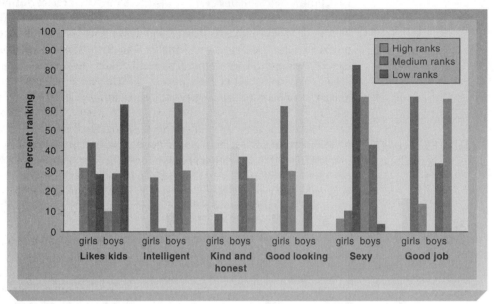

Figure 6.2
Significant differences between boys' and girls' rankings of the ideal woman.

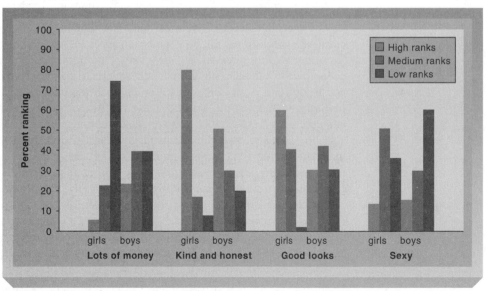

Figure 6.3
Significant differences between boys' and girls' rankings of the ideal man.

recent study of college students looking at conformity Doherty (1998) found that nonconforming females were evaluated more negatively than nonconforming males.

Younger adolescents describe men and women even more stereotypically by manner (Stiles, Gibbons, et al., 1987; Stiles, Gibbons, & Schnellmann, 1987), as can be seen in Figures 6.2 and 6.3. In this study, adolescents aged 11 to 15 were asked to describe and to draw pictures of what they thought to be the ideal man and ideal woman. Stiles and her colleagues found that:

> ❑ Girls depicted the ideal man as "the chivalrous football player" (kind and honest, fun-loving, smiling and bringing flowers). Boys depicted the ideal man as "the frowning football player" (fun-loving, frowning, and engaged in sports). Girls described the ideal woman as "the smiling hardworker" (kind and honest, smiling, intelligent and having adult responsibilities). Boys perceived the ideal woman as "the smiling sunbather" (good-looking, sexy, smiling and engaged in leisurely activities) (Stiles, Gibbons, et al., 1987, p. 411).

Overall, adolescent boys and girls differed considerably in their ideas about important traits for men and women. Looking at adolescents at transition time from junior high school to high school, Alfieri and colleagues (1996) found that gender stereotypes became more flexible at the transition and then progressively less flexible again as students settled into high school. Mills and Mills (1996) asked middle school teachers if they thought that their students had more flexible attitudes toward gender roles than previous generations of students. The students gender role attitudes were also measured. The researchers found that while 90% of the teachers believed that their students had more flexible attitudes than previous generations the measure of student attitudes indicated that they did not have more flexible attitudes toward gender roles.

Adolescents' views of the ideal man and woman are highly influenced by cultural expectations. Gibbons and associates (1991) found that adolescents from nations that emphasize collectivist or group values and are less wealthy economically (adolescents from South American countries, such as Venezuela and Brazil, and from European countries, such as Greece and Yugoslavia, were included in this group) express more traditional gender-role ideas than do adolescents from wealthier, individualistic nations, such as the United States, Canada, Germany, and Great Britain. In other research (Stiles & others, 1990), male and female adolescents from the United States described their opposite sex ideals as having lots of money and being fun, popular, good-looking, and sexy. In comparison, adolescents from Mexico described their opposite sex ideal as liking children, helping others, being very intelligent, and possessing inner qualities, such as goodness and honesty. It seems as though American adolescents value external characteristics of the opposite sex more than do Mexican adolescents, who generally describe internal qualities as more important.

CONCLUSIONS

A significant aspect of adolescent development is the formation of the self. The self is central to the personality. Cognitive changes that take place during adolescence influence the development of self-concept in positive ways, but also contribute to depression and anxiety for some adolescents. Traditional psychoanalytic theory views the self as emerging and stabilizing during adolescence through the process of separation—individuation. That view has been challenged in recent years by theorists interested in the development of women, who believe that interpersonal relationships are critical to a sense of self. Social learning theorists suggest that the development of self is a gradual process that takes place over time as a result of one's experience, evaluation of those experiences, and expectations of the self.

The self is defined by the self-concept, which changes with age. A critical ingredient of self-concept is self-esteem, which in turn depends on such factors as social contacts, parental behavior, developmental changes, gender, and race.

Erikson offers us another dimension of the self—one's identity. Although we each possess an identity, Erikson thinks of it more as an ideal state of personality in which there are no internal conflicts—all parts of the self are in concert. We strive for a state of identity, usually by passing through a state of moratorium in our teen years. Some adolescents become stuck in a foreclosed identity, one of James Marcia's four identity statuses (the others are identity confusion, moratorium, and achievement). Marcia suggests it is necessary to work one's way from identity confusion through the moratorium to an achieved identity, while avoiding foreclosure. Erikson's theory was developed through the study of males. Researchers have begun to study the identity development of females and suggest that female identity development may be both similar and dissimilar to male identity development.

Members of cultural minorities can find it challenging to define their own identity. It appears that there are three stages to this process, which may be called unexamined cultural identity, cultural identity search, and achieved cultural identity.

The advent of feminism has promoted major innovations in our thinking about the appropriate gender roles for males and females. Adolescents must wrestle with what it means personally to be male or female, and how this fits with societal gender-role expectations and changing conceptions of masculinity and femininity.

In summary, the adolescent's personality is undergoing many changes, but the changes are probably no more traumatic than at any other stage of life. The major concern is to begin forming an adult identity, which means choosing certain values and repudiating others. As Nightingale and Wolverton (1988) put it in their report to the Carnegie Council on Adolescent Development,

❒ Adolescents have no prepared place in society that is appreciated or approved; nonetheless they must tackle two major tasks, usually on their own: identity formation, and development of self-worth and self-efficacy. The social environment of adolescents today makes both tasks very hard. . . . We must change the view that many people hold of all youth as troubled and harmful to the rest of society (pp. 1, 16).

With this chapter, we conclude our consideration of individual and internal aspects of change during adolescence: puberty, cognition, morality, spirituality, and the self/identity. We now focus on the adolescent in interaction with the world: family, social relations, education, and work. The family, the peer group, the school, and the workplace are all important contexts or settings that influence the development of adolescents in critical ways.

CHAPTER HIGHLIGHTS

Cognitive Change and Adolescent Self-Concept

• A person's self-concept changes in four basic ways during adolescence: it becomes more abstract; differentiated; idealized; and subject to introspection.
• These changes contribute to depression and anxiety for

some adolescents. The vast majority of adolescents maintain positive self-concepts.

Individuation, Interdependence, and the Self

• According to psychoanalytic theory, there is a process of individuation in childhood and a second individuation during adolescence. The process of individuation is important to the development of the self as an independent person.

- Several psychologists, interested primarily in the development of adolescent girls, have suggested that rather than independence, interdependence should be the goal of personality development.
- Expectations and self-evaluations play a large role in the development of self-concept, or self-efficacy, according to social learning theorists.

Self-Esteem

- Self-esteem affects adolescent motivation, school achievement, and mental health. Adolescents with low self-esteem are more likely to experience emotional and behavioral problems.
- Self-esteem is greatly influenced by social, developmental, gender, and cultural factors.

Social Competence

- Social competence is an important aspect of self-concept among adolescents. Adolescents who feel good about their friendships generally feel good about themselves.
- Adolescents' self-concepts develop further as a result of their relationships with friends. Adolescents learn about and come to define themselves through these relationships.
- The cognitive developmental changes that take place during adolescence contribute to greater self-awareness and greater social competence.

Erikson's Concept of Identity

- Erik Erikson's psychosocial theory of development calls the adolescent stage a crisis in identity formation and repudiation versus identity confusion.

- An important aspect of Erikson's adolescent theory is the moratorium of youth.
- James Marcia has defined four identity statuses: confusion, foreclosure, moratorium, and achievement.
- Based on interviews with college women, Josselson suggests that the path to identity is different for women.

Cultural Identity

- A strong and positive cultural identity is related to psychological adjustment and high self-esteem.
- Because of the existence of prejudice in our society, minority adolescents may find it challenging to build a strong cultural identity.
- Developing a bicultural identity—feeling good about one's family background and having the cultural knowledge to participate in the main culture—is an ideal (although complex) goal for adolescents.

Sexual Identity and Gender Roles

- Gender identity results from those physical characteristics and behaviors that are part of our biological inheritance.
- Gender role, on the other hand, results partly from genetic makeup and partly from the specific traits that are in fashion at any one time and in any one culture.
- Views of acceptable gender-role behaviors have changed considerably during the past 20 years. Adolescents' views of acceptable gender-role behavior are often more rigid than adult views.
- Part of identity development in adolescence is defining oneself as male or female and determining what that means. That meaning is influenced by changing views of gender roles.

K E Y   T E R M S

achieved cultural identity 193
androgynous 197
biculturalism 192
core self 184
cultural identity 191
cultural identity search 192
gender 195
gender aschematic 200
gender identity 196
gender intensification
  hypothesis 195
gender role 195
gender-role adaptation 196
gender-role behavior 196

gender schema 200
ideal self 169
identity crisis 185
identity status 188
individuation 170
interdependence 171
introspection 169
looking-glass self 176
marginalism 192
moratorium of youth 187
mutual collaboration 185
pair therapy 185
personal identity 165
premature foreclosure 188

reciprocal determinism 174
repudiation 185
second individuation 170
self-concept 165
self-efficacy 174
self-esteem 165
self-in-relation 173
sexual identity 195
sexual orientation 196
social construction 176
state of identity 185
unexamined cultural identity 192

W H A T   D O   Y O U   T H I N K ?

1. What does individuation mean to you? Are you individuated?
2. In a "perfect" world, how would gender and race influence self-esteem?
3. Is it natural for females to care more about relationships than males?
4. Do you believe you have had an identity crisis? If so, what makes you think so?
5. What would you say are the main personality traits of the typical adolescent? Is there a typical adolescent?
6. Do members of minority groups have a different type of identity from those in the majority culture? If so, in what ways?
7. What is your attitude toward androgyny? Would you describe yourself as androgynous? If not, do you wish you were?

S U G G E S T E D   R E A D I N G S

Angelou, M. (1970). *I know why the caged bird sings.* New York: Random House. Angelou's childhood in rural Arkansas is recounted. Her strength and resilience model the building of a strong personal and cultural identity.

Hayslip, L. (1989). *When heaven and earth change place: A Vietnamese woman's journey from war to peace.* New York: Doubleday. Hayslip recounts the torture, deprivation, and survival instincts that developed her personality as a young woman.

Knowles, J. (1960). *A separate peace.* New York: Macmillan.

Pollack, W. (1998). *Real Boys.* New York: Henry Holt Co.

Raths, L., Merrill, R., & Simons, S. (1966). *Values and teaching.* Columbus, OH: Merrill. In a book filled with usable ideas, the authors explain and then demonstrate their approach to helping people come to terms with their own value systems. Especially appropriate for teenagers.

Salinger, J. D. (1951). *Catcher in the rye.* Boston: Little, Brown.

Wright, R. (1945). *Black boy.* New York: Harper and Row. This is a moving autobiography of the novelist's adolescence in the Deep South. Its insights are entirely relevant to today's world.

G U I D E D   R E V I E W

1. What is the emotional risk when an ideal self-concept develops during adolescence?
   a. Some adolescents develop unrealistic self-concepts.
   b. Some adolescents become very concerned with how they appear to others.
   c. Some adolescents are disappointed because they are different from the way they would like to be.
   d. Some adolescents have difficulty determining who is the "real me."
2. According to self-in-relation theory, the self develops through
   a. connections.
   b. separations.
   c. individuations.
   d. repudiations.
3. The domains of competence typically rated LEAST important by adolescents are
   a. doing well in sports and doing well in school.
   b. looking good and being accepted by friends.
   c. being liked by a close friend and doing well in a job.
   d. doing well in school and behaving well.
4. The ability to interact effectively with others is called
   a. social competence.
   b. perspective taking.
   c. interpersonal negotiation.
   d. individuation.
5. According to Selman, the capacity for mutual collaboration requires
   a. a core self.
   b. a committed identity.
   c. positive sense of self.
   d. capacity for autonomy and intimacy.

6. Self-esteem of racial minorities is most strongly affected by
   a. poverty.
   b. discrimination.
   c. family relationships.
   d. media.

7. According to Erikson, the main task of adolescence is the achievement of
   a. intimacy.
   b. identity.
   c. moratorium.
   d. foreclosure.

8. Vicki has been drifting in high school. Her grades are mediocre and she hasn't thought about college. She has no interest in world events or political issues. Her thoughts are focused on makeup, the opposite sex, and her friends. Marcia would describe her as
   a. confused.
   b. foreclosed.
   c. in moratorium.
   d. achieved.

9. Continuing to enjoy relationships that were important in the past is central to identity achievement according to
   a. Erikson.
   b. Marcia.
   c. Goethels.
   d. Josselson.

10. The part of a person's self-concept that comes from knowledge and feelings about belonging to a particular cultural group is known as
    a. identity status.
    b. biculturalism.
    c. cultural identity.
    d. marginalism.

11. The self-concept becomes more _____ (uniform/differentiated) as the person enters adolescence.

12. _____ (Self-efficacy/Self-in-relation) is the term coined by Bandura to define our self-expectations, or beliefs about what we can accomplish as a result of our efforts.

13. According to Harter, dissatisfaction with _____ (scholastic competence/physical appearance) is a prime cause of low self-esteem among adolescent girls.

14. Boys who increase in self-esteem during adolescence were viewed by others at age 14 as _____ (generous/self-satisfied).

15. _____ (Looking-glass self/Self-efficacy) is the idea that the approval of parents, friends, teachers, and classmates is important to positive self-esteem.

16. Marcia has made a major contribution to our understanding of identity formation through his research on _____ (identity status/cultural identity).

17. In identity foreclosure, a crisis is _____ (present/absent) and a commitment is _____ (present/absent).

18. According to Josselson, the pathway to identity development for women involves the _____ (maintenance/repudiation) of earlier close relationships.

19. A person who can identify strongly with both the minority and majority cultures is a _____ (biculturalist/marginalist).

20. A person who is _____ (androgynous/gender schematic) does not view events or situations as appropriately masculine or feminine.

Answers

1. C, 2. A, 3. D, 4. A, 5. D, 6. C, 7. B, 8. A, 9. D, 10. C, 11. differentiated, 12. Self-efficacy, 13. physical appearance, 14. self-satisfied, 15. Looking-glass self, 16. identity status, 17. absent, present, 18. maintenance, 19. biculturalist, 20. gender schematic

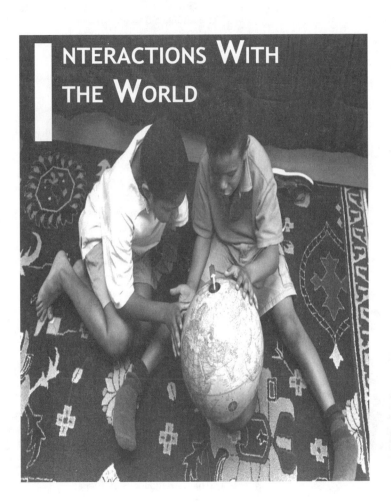

INTERACTIONS WITH THE WORLD

FAMILY RELATIONS

It is, after all, the simplest things we remember: A neighborhood softball game, walking in the woods at twilight with Dad, rocking on the porch swing with Grandma. Now, no one has the time to organize a ballgame. Our woods have turned to malls. Grandma lives three states away. How will our children have the same kind of warm memories we do?

B. F. Meltz, 'Saving the Magic Moments,' 1988

Adolescence is a period of biological, psychological, and social change for each individual teenager. Each adolescent is a member of a family and a society that are also changing. As a result of the social and economic changes of the 20th century, family life is now very different than it was in the past century when most families lived on farms. Since the 1960s, the family has experienced many other changes, caused by such powerful new trends as the altered status of women, the greater number of divorces and remarriages, and the increase in parenthood among teenagers. This chapter considers how the changes of adolescence influence family life and how family changes have affected the lives of adolescents.

After reading this chapter, you should be able to:

- Describe the family characteristics that promote healthy autonomy.
- Describe the role of conflict in early adolescent development.
- Describe the effects of adolescent pubertal changes and parental midlife concerns on the quality of parent-adolescent relationships.
- Discuss, according to three recent models, how adolescents are able to gain independence while remaining close to their parents.
- Explain Baumrind's parenting styles and an additional parenting style identified by Dacey.
- Explain how changes in the American family, including the high rate of parental divorce and remarriage, affect adolescent development, and identify factors that help adolescents adjust to those changes.
- Describe the importance of extended family, flexible social roles, and an ancestral worldview for many ethnically and racially diverse families.
- Discuss factors associated with causes and consequences of teenage parenthood.
- Discuss these issues from an applied, a sociocultural, and your own point of view.

ecological theory
A view of the growing child or adolescent as an active agent in a series of interacting systems.

The perspective on adolescent development and family relationships described in the introductory paragraph to this chapter is characteristic of **ecological theory**, as presented by Urie Bronfenbrenner. According to Bronfenbrenner (1979), human development is best understood as a series of changes and accommodations between an active, growing person and the immediate social environment, including the family, school, and peer group. The interactions between the person and the immediate social environment are also affected by the larger social environment, including the community, culture, nation, and world economic and social events. Ecological theory helps us understand the many contexts that interact to affect the development of a maturing adolescent. This chapter focuses on the development of the adolescent as part of a changing family context. Subsequent chapters in Part 3 will focus on the importance of peer and school contexts in adolescent development.

The Changes Brought on by Adolescence and their Effects on the Family

STRIVINGS FOR AUTONOMY

Adolescence is a time during which children typically gain in independence from their families. Independence involves gains in both physical and psychological autonomy. Adolescents, in comparison with younger children, for example, can do more for themselves and spend more time away from parents and with peers. (The influence of the peer context on adolescent development is discussed in Chapter 8.) As adolescents spend less time with the family and more time interacting with the larger world, they encounter a broader range of values and ideas. In conjunction with the cognitive changes occurring during adolescence (Chapter 4), the exposure to different ideas often contributes to an examination of parental values and teachings. As a result, adolescents begin to question and challenge parental views and develop ideas of their own. No longer are parents viewed as all-knowing authorities. Optimally, through the course of the teenage years, adolescents develop more mature and realistic views of their parents as persons with particular skills, talents, and knowledge, who deserve their respect but who can also make mistakes. Part of the process of gaining psychological autonomy requires adolescents to revise images of parents and develop personal ideas. In Chapter 6, we discussed this as it relates to individuation.

Few students of adolescent development would argue that the achievement of psychological autonomy is an important developmental task of the adolescent years. Disagreement exists, however, about the type of family environment that is most conducive to the development of autonomy. Traditional psychoanalytic theorists (Blos, 1979; Freud, 1958) have long argued that the development of healthy autonomy is achieved through a loosening of family ties. More recently, theorists and researchers (Grotevant & Cooper, 1986; Larose & Boivin, 1998; Sim, 2000; Youniss, 1989) have argued that healthy autonomy is promoted by positive and supportive parental relationships. Supportive parental relationships, they argue, allow for the expression of both positive and negative feelings, which foster social competence and responsible autonomy. Lamborn and Steinberg (1993) studied more than 2,000 high school students to examine the relationships between parental support, emotional distancing from parents, and psychological well-being among adolescents. According to their findings, adolescent strivings for autonomy seem to be accomplished best within family environments that simultaneously offer support and allow adolescents to gain emotional independence. Although adolescents who remain emotionally dependent on their parents may continue to feel good about themselves, they evidence less competence, self-reliance, pride in work, and academic achievement than do adolescents who achieve emotional independence.

ADOLESCENCE AS A STAGE OF CONFLICT

How much difficulty do adolescent strivings for autonomy cause the family? Many people think of adolescence as a time of storm and stress, rebellion, and conflict with parents. This belief is based mainly on early theories, popularized by G. S. Hall and Sigmund and Anna Freud (see Chapter 2). Anna Freud (1969a) studied troubled adolescents, noted that conflict with parents is common, and concluded that this is a normal and necessary part of adolescence. She believed that the biological changes of puberty set off this conflict, which was needed so that adolescents could become independent from their families. Adolescents who got along well with their parents were thought to be immature.

A different view has been suggested by research with normal adolescents (Arnett, 1999; Douvan & Adelson, 1966; Offer, 1969; Offer & Offer, 1975). The majority of adolescents admire their parents, turn to them for advice, and feel loved by them. Only about 25 percent of families report a lot of parent-child conflict. Most of those families had problems before their children reached adolescence. Ary and others (1999) found that families experiencing high levels of conflict were more likely to have low levels of parent-child involvement. Parents and teenagers in most families seem to argue about small things, such as chores, curfew hours, keeping one's bedroom clean, clothing, and choice of foods and snacks. Adolescents generally agree with their parents about basic values, such as the importance of education and work. It seems that there is much *continuity*, rather than great change, in parent-child relationships from childhood into adolescence (Rutter, 1987). That is, the relationship continues to be in adolescence what it was earlier.

Laursen (1995) suggests that some conflict is part of all significant interpersonal relationships in which mutual needs must be negotiated. He surveyed high school juniors and seniors to learn more about their conflicts with parents as well as with friends, siblings, romantic partners, and other adults. Adolescents reported the most conflict with mothers, followed by friends, romantic partners, siblings, and fathers. Conflicts with parents and friends focused on different issues, however. Parental conflicts were generally about responsibilities (e.g., cleanliness of room, chores), school (e.g., grades, homework), and autonomy (e.g., privacy, where you go and what you do), whereas peer conflicts involved friendships (e.g., sharing a personal problem, being ignored) and heterosexuality (e.g., intimacy of relationships, dating behavior). Sagrestano and colleagues (1999) found that parents reported more conflict than their children did which suggests that adults and teens may assess conflict differently. Laursen and colleagues (1999) also suggest that there are multiple dimensions to conflict: conflict rate, conflict affect (or intensity) and total conflict. These researchers found that conflict rate and total conflict decrease from early to late adolescence but that conflict affect increases from early to middle adolescence.

Minor conflicts and disagreements with parents may actually help adolescents to establish independence (Smetana & others, 1991). Conflict may help adolescents to develop interpersonal negotiation skills and independent thinking, as long as the argument takes place within a supportive and accepting family environment (Allen & others, 1994; Hauser & Bowlds, 1990). When parents are hostile, the adolescent is likely to feel rejected and avoid further discussion and interaction with parents. From this perspective, the critical issue is not that minor conflicts take place, but how adolescents and their parents interact to resolve these conflicts. According to research by Rubenstein and Feldman (1993), boys in mid-adolescence who were able to negotiate adaptive compromises (i.e., try to reason, listen to what parents say, and try to understand) to parental conflicts had parents who demonstrated consistency, support, positive mood, and emotional control when their sons were early adolescents. Sons were more likely to walk away from and avoid conflicts when their parents were inconsistent and unsupportive. When parents were hypercritical and rejecting and showed low levels of control over their own feelings, adolescent sons were likely to respond to conflicts by attacks (e.g., get mad and throw things, get mad and yell, get sarcastic, or say something hurtful).

Although most parents and adolescents do not fight a lot, there is evidence that conflict increases somewhat in early adolescence for a brief period of time. This increase seems to take place at the time of puberty.

THE EFFECTS OF PUBERTY

Early adolescence is a period of bickering and squabbling in many families (Steinberg, 1981, 1987, 1988). In general, early adolescents express less affection toward their parents, spend less time with them, and bicker more when they are together. Mothers and sons seem to argue the most. One recent study (Flannery & others, 1994), however, suggests that fathers may be very important to the quality of communication between adolescent children and parents: Adolescents' descriptions of parental communications as negative were related not only to the pubertal change but also to the number of negative feelings expressed by fathers when interacting with their adolescent children. Arguments between adolescents and parents generally decline toward the end of puberty. These conflicts support the psychoanalytic belief that the changes of puberty set off conflict. Conflicts are generally small, however, and are not like the storm and stress described by G. S. Hall and Sigmund and Anna Freud. Bickering takes place while adolescents are still quite close to their parents. Some theorists believe that this helps adolescents to make new close friendships outside the family without breaking family ties (Steinberg, 1990).

Petersen (1985), however, believes that puberty is not a direct cause of parent-child conflicts in early adolescence. She points out that other changes, such as a change of school, are taking place in the adolescent's life. When students enter middle school, they face many new expectations socially and academically. Meeting these new expectations can be stressful for early adolescents who are not ready. During puberty, early adolescents change physically, from looking like children to looking more like adults. Because the teens look more physically mature, parents, teachers, and other adults may suddenly expect more mature behavior. Petersen believes that these sudden changes in expectations are confusing for some early adolescents and result in more conflicts with

AN APPLIED VIEW

How Well Do Parents Know What Their Teens Are Doing?

The Who's Who organization recently (1997) surveyed 3,370 teenagers 16-18 years old, all of whom have an "A" or "B" average and are planning to attend college. Because these students are among the highest achievers in the United States, one might assume that their parents would be reasonably well aware of their activities. As the chart below reveals, there are some serious discrepancies.

Do you think your child:	Parental Myth	Teen Reality
Has contemplated suicide	9%	26%
Has cheated on a test	37%	76%
Has had sex	9%	19%
Has friends with drug problems	12%	36%
Has driven drunk	3%	10%
Worries about pregnancy	22%	46%

Source: Who's Who Special report, 1997

parents. These changes can also be difficult for parents. Dekoric and others (1997) found that parents and adolescents had consistently different expectations of developmental tasks and that level of conflict was associated with these differences in developmental expectations.

The Identity Crisis Versus the Midlife Crisis

The levels of stress and conflict experienced in families with adolescent children are associated with the life tasks faced by parents and their success in meeting those challenges. Quite often, for example, parents find themselves undergoing major life changes just at the time that their children are entering adolescence. The typical parents of adolescent children are in their middle adult years. Levinson (1978) has studied the adult development of middle-aged men. He described midlife as a time when men typically look at what they have done with their lives. They begin to realize that they may not accomplish all they had hoped to. All their dreams may not be met; at the same time, they are dealing with some loss of physical strength. This evaluation of one's life can cause feelings of anxiety known as the **midlife transition**. In recent research, Levinson (in press) found that much the same kinds of crises occur in middle-aged women's lives, although not necessarily at the same ages. How does the life stage of the parent fit with the life stage of the adolescent?

At the same time that adolescents are growing rapidly and maturing sexually, their parents are beginning to feel concerned about their own physical health, attractiveness, and sexual appeal. Whereas adolescents feel that their whole lives are ahead of them, parents are aware that the time left is limited. Teens are

midlife transition
Feelings of anxiety that sometimes occur at midlife as individuals examine what they have done with their life and think about what they would like to do in the remaining years.

beginning to date and to think about career choices. Their parents made these choices years earlier and must deal with the disappointment so many middle-aged people feel. The net result is: *parents with adolescent children report less satisfaction with their marriages than parents of younger or older children* (Anderson & others, 1983). This is bound to affect many families negatively. In many families, it is a time of crisis—in fact, a clash between two crises: the identity crisis of the children and the midlife crisis of the parents! However, in a recent study Seiffge (1999) found that marital communication, family closeness and opportunity for personal growth within the family are factors that need to be considered. Perhaps difficulties that appear to arise in marriages as children reach puberty are a reflection of earlier unaddressed marital dissatisfaction.

Adolescent children also bring expenses to the family. They plead for certain clothing and stereo equipment at the same time that many parents are trying to save money for college. Middle-aged parents may also be caught between two generations' needs—those of their adolescent children and those of their own elderly parents. Aging parents may need financial help, as well as support and attention, from their middle-aged children just at the time that youthful children want and need more attention.

Research (Koski & Steinberg, 1990; Silverberg & Steinberg, 1987) indicates some interaction between the midlife concerns of parents and the developmental changes of adolescents. Silverberg and Steinberg, for example, suggested that when young adolescents question their parents' values and rules as part of the process of individuation (see Chapter 6), parents may also begin to question themselves. Parents of early adolescents often report higher levels of stress. As adolescents begin to challenge parental authority, parents may feel less comfortable and less adequate in the parental role. Silverberg and Steinberg found that parents' midlife concerns were most affected by the growing independence of the same-sex adolescent. That is, fathers were most affected by the increasing emotional autonomy of their sons, and mothers were most affected by the increasing emotional autonomy of daughters. Mothers, but not fathers, reported lower levels of life satisfaction when their sons and daughters challenged parental authority.

family therapy
A form of psychological treatment in which all family members work together with a therapist to help the family as a whole to function better and to meet the needs of all family members.

AN APPLIED VIEW

The Role of Family Therapy

Perhaps the biggest problem with the clash between the adolescent identity crisis and the adult midlife crisis is that, typically, none of the parties involved is aware that this discord is taking place. Even when the parents are aware that their teenagers are struggling to define themselves, the children seldom realize it. And most adults cannot discern that they, too, are enduring a significant reexamination of their own lives until they nearly complete it.

That is why in virtually every case needing adolescent psychotherapy, the entire family must be involved. This approach is known as **family therapy** (as opposed to psychoanalytic or behavioral therapy, for instance). Sometimes those working with teens can help them and their parents understand the volcanic forces that are feeding their interpersonal struggles. In most cases where the conflict is serious and/or of long duration, a family therapist will be needed to help all the members recognize their roles in the problem.

Koski and Steinberg (1990) found that mothers who are experiencing many midlife concerns are likely to be unhappy in their role as parent. High levels of midlife concerns may reduce the amount of time and energy that a parent can devote to the challenges of an adolescent child, and may increase feelings of inadequacy as a parent. It seems, however, that a good marital relationship can contribute to satisfaction in parenting, even when midlife concerns are high.

It is clear that parents and adolescent children view their lives from different points in time. Those differences can cause conflicts. At best, they require adjustment and understanding between generations.

New Models of Parent-Adolescent Relationships

Because traditional views of parent-adolescent relationships being full of storm and stress have been challenged by recent research, theorists and researchers have turned to new models. These models view healthy adolescents as staying close to their parents and becoming independent at the same time. How is this possible?

One growing model of adolescent-parent relationships is **interdependence theory** (Grotevant & Cooper, 1986; Smollar & Youniss, 1989). According to this view, the relationship between parents and adolescent children is constantly changing. During childhood, parents have most of the control in family decisions. Typically, for example, the parent decides what time the child should go to bed. During adolescence, parents give their children more freedom and let them have more say in decisions. The typical teen might discuss bedtime hours and curfews with parents, and together they come to a decision. Older adolescents continue to seek their parents' advice and respect their views. At the same time, they have more freedom to make choices and to participate in making decisions. Older adolescents come to understand their parents not as all-knowing persons but as separate individuals with both positive and negative qualities. This change in the parent-child relationship helps adolescents become independent of, yet continue to feel connected to, their families. The research findings of Lamborn and Steinberg (1993) described at the beginning of this chapter are consistent with interdependence theory.

Another model that has gained increasing attention in understanding the role of parental behavior in child and adolescent development is **attachment theory** (Ainsworth & others, 1978; Bowlby, 1969). This theory has been popular in helping psychologists, educators, and other scientists understand parent-child relationships during early childhood. According to this view, a trusting relationship with one's parent or caretaker gives the child feelings of security and self-confidence. Parents who are sensitive, warm, and responsive to their young children and who support their independence help them develop *secure attachments*. In this model, independence results from closeness and secure attachment.

What can parents do to cope with the stress of parenting an early adolescent?

interdependence theory
A new model of parent-adolescent relationships in which adolescents gain independence not through rebellion, but through gradual increases in freedom and responsibility.

attachment theory
A theory proposing that caretakers who are consistently responsive and sensitive contribute to the development of secure attachment.

Such children feel confident in leaving their parent to learn about the world. They know that they can depend on their parent to be there giving help when it is needed. Through their experiences, children learn more about the world and other people. They become more socially competent (see Chapter 6). Children who do not have a trusting or secure attachment to their parent will be anxious. They will stay near their parent rather than explore the world and will be less socially competent.

In the last ten years, theorists and researchers have begun to think about how attachment works in adolescence. When teens develop more friends outside the family, they do not cut themselves off from their families. Their ties to parents do not make them dependent. Rather, when adolescents feel that their parents believe in them and will help them out if they need it, they may be more self-confident and more willing to try new activities. Research indicates that adolescents with secure parental attachment have higher self-esteem and are more socially competent than those with insecure attachments (Armsden & Greenberg, 1987; Kenny, 1987; Kenny & Donaldson, 1991; Kobak & Sceery, 1988). Secure parental attachments may contribute to positive views of self and lower levels of depressive symptoms among adolescent boys and girls (Kenny and others, 1993). Adolescents who feel detached from parents also feel rejected and lack self-confidence (Ryan & Lynch, 1989). Larose and Boivin (1998) found that teens with perceived security to parents at the end of high school maintained this feeling of security even after leaving home for college. In addition, these feelings of security were positively associated with emotional adjustment outside the family.

Consistent with attachment and interdependence theories, Allen and others (1994) maintain that parent-adolescent relationships that are characterized by both autonomy and support are most healthy for adolescent development. To test that hypothesis, those researchers observed interactions among 72 mothers and fathers and their adolescent sons or daughters when the adolescents were 14 years old and again when the adolescents were 16 years old. When parents were observed as encouraging their adolescents to think autonomously and providing support for the expression of independent views, adolescents reported high levels of self-esteem and ego development (see Chapter 6) at both ages. Adolescents who gained in self-esteem and ego development between 14 and 16 years of age were those whose fathers most actively challenged their children's thinking at age 14, but did so in an accepting and supportive manner. Allen and others suggest that the challenges posed by fathers in the context of a safe and supportive relationship may provide opportunities for the adolescents to comfortably test out their ideas and to begin to establish themselves as independent adults. Their findings also suggest that more attention should be paid to the importance of fathers' relationships with their adolescent children.

systems theory
A view of the family and other parts of the social environment, such as the school and peer group, in which the experiences of one individual are believed to affect other individuals and the system as a whole.

Many social scientists and family therapists view the family as a system. **Systems theory** views the family as a whole, not just as a group of individuals. When something happens to one person within the family, it affects all the family. When parents argue, for example, their anger affects every child.

Studying family systems has helped us learn how families adjust to change. Family therapists, such as Minuchin (1974) and Haley (1980), use that knowledge in helping troubled families. Minuchin believes that families need to fill two human needs—the need to belong and the need to be separate. That is, families should help children feel that they are accepted and loved. Families also need to allow children to feel separate or different. For example, all family members should not be expected to have the same ideas.

Families that do not provide a sense of belonging are called disengaged. A **disengaged family**, for example, is not concerned about how one of the children is doing in school. The child may react by dropping out of school. A

disengaged family
Families in which there is too little involvement or caring among family members.

AN APPLIED VIEW

What Is Your Family System Like?

Family therapists sometimes find it helpful to ask family members to produce a genogram. The genogram, which is a diagram of the family's relationship system, can help the therapist and the client learn about the family system. From the genogram, for example, you can see those family members who are close, those who are disengaged, and those who are in conflict.

Use the space below or a separate sheet to diagram your family system. Indicate the names, ages, gender, race or ethnicity, and relationship to you for each person in your family. Add additional circles as needed to represent other family members. Describe the quality of your relationship to family members by drawing lines connecting you and each family member, using the following system:

_____ (close)   /////// (conflict)
.... (distant)   ===] [=== (cut off)
O = female   + = male

- **Grandparents**
- **Great Aunts**
- **Great Uncles**

- **Parents**
- **Aunts/Uncles**

- **Your Generation**
- **You/Sisters/Brothers**
- **Cousins**
- **Stepsisters/Stepbrothers**

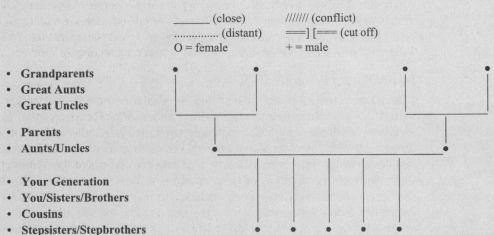

Do you feel closer to members of your own generation or to older family members? To family members of the same or opposite gender? To family members with similar or different ethnic or racial backgrounds? Where are sources of conflict in your family? Which family members would you like to feel closer to? What else can you learn about your family by examining your genogram?

For further information on genograms, see M. McGoldrick and R. Gerson (1985). *Genograms in Family Assessment*. New York: W. W. Norton.

disengaged family provides freedom but fails to provide feelings of belonging. Families that don't allow children to be separate enough are called enmeshed. An **enmeshed family**, for example, does not allow members to speak for themselves (family members are always interrupting to explain what the others mean). The adolescent in an enmeshed family may not learn to take responsibility for his or her actions.

Family systems feel a lot of pressure from changes that are taking place within and without. Outside pressures come from schools, jobs, and the community. Inside pressures come from changes in family members. Adolescents put a lot of strain on the family system. They bring new styles of dress and ideas into the family. They make many new friends, with whom they spend more time than with other family members. They may ask for help from parents one minute and reject that advice the next.

Minuchin believes that families need to be flexible to adjust to these changes. Problems in adolescence result when families do not adjust to meet the adolescent's need for independence. For instance, when an enmeshed family does not allow an adolescent girl to have her own ideas or to express her feelings, she may develop an eating disorder (see Chapter 11). When a disengaged family reacts to an adolescent boy's need for independence by giving up all control, the youth may feel that his family doesn't care, and when he needs support, will not go to them. He may also become depressed or act in antisocial (delinquent) ways.

Parenting Styles

Changes occurring within and among adolescents, parents, and the family system as a whole contribute to the challenges of raising adolescent children. In providing guidance for parents handling this task, psychologists have been concerned with what kind of parenting style is best for adolescents.

BAUMRIND'S MODEL OF PARENTING

Diana Baumrind's research identified three distinctly different styles of parenting. Some parents employ an **authoritarian style**. They respond to their teenagers' challenges by becoming more strict. They feel that if they clamp down right away, their teens will come home on time, keep their rooms clean and dress neatly. Other parents follow a **permissive style**. They decide to give their adolescent children lots of freedom and hope they will do what is best. A third group, the **authoritative style**, includes the majority of parents and tries to steer some course between these two. Baumrind (1971, 1989, 1991b, 1996) has studied children from early childhood through adolescence to find out the effects of these three different parenting styles.

The three parenting styles differ along two dimensions: *responsiveness* and *demandingness*. Responsiveness, including warmth and encouragement of independence, was mentioned as important in our earlier discussion of interdependence and attachment. Healthy families were described as providing warmth and closeness, yet allowing individuals to express opinions and become inde-

enmeshed family
Families in which members are very close and limit the freedom necessary for individual growth.

authoritarian parenting style
Describes parents who are demanding, give their children little autonomy, control their children by rules with little explanation, and fail to show warmth to their children.

permissive parenting style
Describes parents who give their children lots of freedom and hope they will do what is best.

authoritative parenting style
The most common and most successful style of parenting, in which parents hold high expectations, provide explanations for rules, and create an environment of warmth and caring.

pendent. Baumrind's model adds the dimension of demandingness. Demanding parents set high expectations and standards and supervise the activities of their adolescents.

- *Authoritarian parents* seek to control their children through rules. They use rewards and punishment to make their children follow the rules. They give orders and are not likely to explain them. Authoritarian parents are demanding but not warm. They give their children little autonomy.
- *Permissive parents* have little or no control over their children and have few methods of discipline. Permissive parents usually give their children too much autonomy. Some permissive parents choose this style because they believe children should have a lot of freedom and not be controlled by adults. These parents are called **permissive-indulgent**. They are responsive but not demanding. Other parents are permissive because they are just too tired or stressed to enforce rules. These parents are called *neglectful*, or **permissive-indifferent**. They are neither responsive nor demanding. Some authors (Maccoby & Martin, 1983; Lamborn & others, 1991) argue that combining both indulgent and indifferent styles in the permissive category overlooks important reasons for being permissive, and that we need to think about indulgent-permissive and indifferent-permissive as separate parenting styles.
- *Authoritative parents* are supportive, make their standards known, value self-control, and provide explanations for rules to their children. They believe that both parents and children have rights, but the parent has the final say in decision making. Authoritative parents are high in responsiveness and demandingness.

permissive-indulgent parenting style
Describes parents who are warm, but not demanding. They set few rules for their children because they believe that children should have a lot of freedom.

permissive-indifferent parenting style
Describes parents who have little control over their children and have few methods of discipline because parents are not interested or have other stressors in their lives.

According to the research of Baumrind (1971, 1989, 1991b, 1996), authoritative parenting is most likely to produce a healthy child and adolescent. Scaramella and others (1999) also found that supportive parents with good management skills contributed to positive mental health outcomes in their children. Parental responsiveness seems to develop social skills and a strong self-concept. Demandingness helps children to develop self-control and to be more cooperative with others (Maccoby & Martin, 1983). Baumrind (1991b) found that the adolescent children of parents who were demanding had few behavior problems and were not likely to use drugs. Adolescents whose parents used reason to influence their behavior and who supervised their lifestyle, health, friends, and school life were more competent and had higher self-esteem than adolescents whose parents used authoritarian methods of behavior control. Jacobson and Crockett (2000) found that high levels of appropriate parental monitoring of adolescents was associated with adolescent adjustment. Explaining rules is particularly important in parenting adolescents (Baumrind, 1968). Their cognitive development helps them to better understand their parents' reasoning. Low self-confidence and a lack of responsibility have been found among adolescents when their parents refuse to reason with them.

Authoritative parents explain rules and share their reasoning. Adolescents of authoritative parents generally show high levels of responsibility and self-confidence.

In a study of more than 4,000 adolescents between the ages of 14 and 18, Lamborn and associates (1991) found that adolescents who describe their parents as authoritative are more socially competent and better adjusted psychologically than other adolescents. Adolescents who describe their parents as authoritarian are likely to obey rules and keep out of trouble but have less self-confidence than other adolescents. This study suggests that distinguishing between indulgent and indifferent parenting styles is important. Adolescents from indulgent homes are likely to have positive self-concepts, but may not do well academically or may get in trouble for misbehavior at school. Adolescents who described their parents as neglectful scored lowest on measures of self-esteem, social competence, and problem behavior.

Parenting style may also influence the peers with whom high school students affiliate. Durbin and associates (1993) found that high school students who described their parents as authoritative were more likely to choose well-rounded peers such as jocks, normals, populars, and brains. In choosing these peers, the high school students were able to meet the norms and expectations of their peers without alienating themselves from their parents. Students who described their parents as indifferent or uninvolved were most likely to belong to peer crowds such as druggies or partyers that do not endorse adult values.

Baumrind's model was developed by studying European American families. The relevance of these parenting styles for more diverse families is therefore uncertain. Several researchers have begun to address this research gap. Consistent with research with European American families, Steinberg and associates (1991) found that authoritative parenting style was associated with higher self-reliance, lower psychological distress, and lower delinquency among adolescents across the lines of socioeconomics, family structure (single-parent versus two-parent families), and ethnicity (African American, Asian American, and Latino).

Chao (1994) argues, however, that Baumrind's model of parenting does not accurately reflect the parenting style of Chinese families. According to Baumrind's framework, Chinese parents are most often categorized as authoritarian. Authoritarian parenting is associated with poor academic achievement in European American children, although Chinese and other Asian American children often achieve well in school. Chao explains that Asian American children do well in school because of the training (*Chiao shun*) provided by parents in socially desirable and culturally approved behaviors. According to Asian culture, parental care, concern, and involvement includes control and governance (*Guan*) of the child. By not taking into account *Chiao shun* and *Guan*, Baumrind's model does not accurately depict Asian parenting. It is clear that further research is needed to understand optimal parenting styles across diverse cultural groups.

NURTURANT PARENTING

An additional parenting style was discovered in a study of 56 families in which at least one of the adolescents was highly creative (Dacey, 1989; Dacey & Packer, 1992; Dacey & Lennon, 1998). Dacey and his research team labeled

nurturant parenting style
A style of parenting in which parents, by modeling and family discussions, espouse a set of well-defined values and expect their children to make personal decisions based on those values.

Can any single style of parenting be recommended for all adolescents? What do you think?

this parenting method the **nurturant style**. Parents in these families are devotedly interested in their children's behavior, but make few or no rules to govern it. Instead they involve their children in family discussions about values, and model the behavior they value. They make it clear that they expect their children to make their own decisions based on those values. Recent research by Fletcher (2000) suggests that teen involvement in extracurricular and community activities reflects parent involvement which speaks to the importance of modeling desired behavior. Obviously this style is quite time-consuming: lengthy discussions and "practicing what you preach" demand a lot of energy. Therefore, this style is not for everyone. After children make decisions, nurturant parents give them feedback about what was done. Even when they disapprove, they rarely punish. Most of the teens in the study said that their parents' disappointment in them is enough to change their behavior.

In addition to helping children form values, nurturant parenting also helps children develop other qualities, such as a greater ability to take risks, to solve problems, to wait for rewards, and to be free of sex-role expectations. These qualities help the children make sound, insightful decisions.

The success of nurturant parenting is based on a well-established principle: *people get better at what they practice*. These parents give their children many chances to practice decision making, self-control, and creative thinking. They "serve as caring coaches as their children learn how to live."

SIBLINGS

Importance of sibling relationships is another understudied topic. However, current research (Seginer, 1998; Widmer & Weiss, 2000) suggests the crucial role that siblings can play in each others lives. Seginer (1998) found that positive sibling relationships contributed to an overall sense of social support. Looking at disadvantaged adolescents with older well-adjusted siblings, Widmer and Weiss (2000) found positive image of an older well-adjusted sibling was associated with adjustment in younger siblings. Further research on this important topic could offer new insights for mentoring opportunities.

Cultural Diversity and the Family

Many of our views of the family, of how it has changed, and of how it is supposed to be, come from research on white American middle-class families. Social scientists know much less about ethnic families of color, such as African American, Native American, Latino, and Asian American families. Research done with families of color has used methods that limited what we have learned. One research method has been to compare families of color with white families. When the research showed that ethnic families differed from white families, the differences were described as deficiencies in the ethnic families. Many social scientists now recognize the need to learn more about the cultural traditions of ethnic minorities (Allison & Takei, 1993).

We know that there are many differences within and between ethnic family groups. Obviously, not all African American families or Asian American families are the same. Harrison and associates (1984, 1990), however, point out several ways in which most families of color differ from the average white American family. Some of these differences have developed as ways of surviving economically and raising children under conditions of poverty and discrimination. The **extended family** is an important concept in families of color (Wilson, 1989). Family members see each other often and depend on each other for help. It is not unusual, for example, for an African American adolescent to go live with another relative when the family is having problems (Allen & Majidi-Ahi, 1989). Relatives who are better off financially might help out an adolescent by, for example, buying a class ring, senior pictures, and graduation cap and gown. While this would be unusual in many white families, the extended family of color has learned to cope by sharing housing, household tasks, child care, and other resources. According to Jarrett (1995), extended families frequently help poor African American adolescents gain access to friends, education, jobs, and other resources that enable them to improve their economic futures. Friends of the family may also be an important source of help. These friends are not blood relatives but are treated like close family. Native Americans may think of the family as including not only the mother, father, and grandparents, but the community or even the entire tribe as well.

Among families of color, social roles are often more flexible. Older siblings often help to parent or take care of younger siblings, while several adult family members, such as parents, aunts, and uncles, share the responsibility for financial support. In addition, families of color are often bicultural or demonstrate a knowledge and understanding of both the majority culture and their own ethnic culture. They teach their children the values, customs, and rules of social interaction of their own culture as well as the values and rules of the majority. They want their children to be successful in society while being proud of their own racial background. According to the family value system, the well-being of the group (family and others of the same racial background) may be more important than the achievements of any individual family member. Harrison and associates (1990) call this the **ancestral worldview**. In contrast to the white American emphasis on individualism, many ethnic minorities feel that loyalty to their own group is more important than individual achievement. Interdependence, or being able to help out others of the same ethnic/racial background, is valued more than personal advancement. The religious practices of Buddhism and Confucianism, as well as many Latino religious ceremonies, emphasize harmony and group loyalty. The church has also been an important source of comfort and support that has helped African American families unite and survive during hard times.

Theories developed in white American and European cultures cannot be used to decide whether an ethnic minority family or adolescent is psychologically healthy. Those theories view adolescence as a time of increasing independence from the family. This value conflicts with the extended family system, where people feel a stronger need to stay near each other rather than to become independent through moving away. Parents from some ethnic groups, such as Puerto Ricans, may not view independence as being as important as many whites do. In

extended family
A family structure in which grandparents, aunts and uncles, and other family members beyond the nuclear family play an important role in providing emotional, physical, and financial support.

ancestral worldview
The belief, common among ethnic minority cultures, that the well-being of the larger group, such as the family or the community, is more important than the success of a single individual.

Puerto Rican farming communities, for example, adolescents must help out with the farming and learn its technology from their parents (Inclan & Herron, 1989). Some ethnic minority children have to stay close to their families in order to survive economically. When ethnic minority teenage daughters seek freedom from the family, conflict between parents and daughters can become severe (Allison & Takei, 1993).

Conditions of poverty may also influence parenting practices. For example, parents in low-income inner-city African American neighborhoods where crime rates are high may restrict, closely monitor, and chaperone the activities of adolescent youths in ways that would be unusual in suburban neighborhoods (Jarrett, 1995). Parents sometimes enlist the help of younger brothers and sisters in "chaperoning" teens, as the company of younger brothers and sisters, rather than parents, is more easily accepted by the teens. It seems that even middle-class African American adolescents maintain closer relationships with family throughout their teenage years than do white adolescents (Cernkovich & Giordano, 1987; Giordano & others, 1993). They also describe themselves as closer to family members than to friends (Giordano & others, 1993).

For some ethnic groups, such as Asian Americans, obedience to parental authority and respect for elders and traditions mean that adolescents remain physically and emotionally close to their parents throughout adolescence and into adulthood (Huang & Ying, 1989). These practices also differ from Western ideas about adolescence and parenting. Some adolescents may experience feelings of conflict between parental values and traditions and what is considered the American way.

A series of recent studies (Chiu & others, 1992; Feldman & others, 1991; Rosenthal & Feldman, 1991) have focused on increasing understanding of families across cultures by examining adolescents and parents in Hong Kong and Australia in addition to Chinese American and European American teens in the United States. These studies suggest that parental values and expectations

A SOCIOCULTURAL VIEW

Immigrant Versus American Values

Adolescence can be a particularly difficult time for teens who are living in two cultures. Depending on how long an Asian American or Puerto Rican family has lived in the United States, the parents may have rules and expectations for their children that clash with the mainstream culture. Adolescents may not like it, for example, when they are not allowed to go out or stay out as late as other teens. Teens often acculturate more rapidly than their parents and may resent the Old World rules and values. The "Cholo" gangs in southern California have attracted some Mexican American youths. The gangs offer a quick route to assimilation, separation from the family, and a subculture rich in

slang, clothing, nicknames, and tattoo and graffiti techniques (Huff, 1990). Cultural values, rather than pubertal development, are the source of conflict between many ethnic minority adolescents and their parents (Inclan & Herron, 1989). Vasquez (1982) found that first- and second-generation Mexican Americans were more firmly grounded in the Mexican American culture and were less likely to be delinquent than were third-generation adolescents. As adolescents move away from the traditional culture, they are more at-risk for behavioral problems. Table 7.A documents the effects that intercultural clashes can have on family life.

continued

Table 7.A	Asian American Cultural Clashes	
Traditional Cultural Patterns	**vs. Immigration Factors**	**vs. American Society**
Close-knit family with strong father.	Father has to hold down more than one job to support family and is unable to provide the strong influence tradition demands.	More democratic approach to family structure. Mothers work. Increase of single-parent families.
High parental expectations.	Language difficulties; inability to make friends; difference in educational systems.	More tolerance of children who are not high achievers.
Shame associated with having a problem.	Immigration brings many adjustment problems.	More openness regarding problems, and more willingness to seek professional help when necessary.
Pride in one's worth and value to one's self, the family, and the community.	Inability to communicate, obtain a job, achieve a level of academic excellence.	Prejudice, stereotyping, both old and new; exotic, humble, inscrutable, studious, gang member or participant in Tong Wars.
Women are supposed to stay home and raise the children.	Women may have to enter job market to help support family.	Acceptable for women to enter job market and lead a more independent life-style.
Drinking is perceived in many cases as being an acceptable part of family and community life. Alcohol is a part of ritual and festivals.	Increased drinking because of pressure may not be considered a problem.	Drinking not perceived as being culturally important.

From *Drugs, Kids, and Schools* by Diane Jane Tessler. A GoodYearBook. Copyright © 1980 by Diane Jane Tessler. Reprinted by permission of Scott, Foresman and Company.

and adolescent behavior vary across cultural contexts. Hong Kong adolescents, for example, described their families as less accepting and less demanding. Their families place more value on traditions and prosocial behavior and less value on individual competence and success in terms of wealth, power, and social recognition. Hong Kong adolescents reported less misconduct in such behaviors as cutting classes, cheating on tests, swearing, smoking, and drug and alcohol use. Interestingly, however, the same parent characteristics—low levels of supervision and less emphasis on wealth, prestige, and power—were associated with increased adolescent misbehavior across all cultural groups (Feldman & others).

Comparisons of first and second generation Chinese American, Chinese Australian, Euro-American, and Hong Kong teens revealed no differences in emotional distress. These findings (Chiu & others, 1992) challenge popular beliefs (see Table 7.A) that immigrant adolescents face more adjustment problems than do other teens. Parents of Chinese American, Chinese Australian, and European American teens were described as more controlling than Hong

Kong parents. Ratings of parental warmth were similar across all cultural groups. Although the experience of immigration may change adolescents' perceptions of their parents' behavior, those changes do not always result in more emotional distress. Chinese immigrant students reported that they put more effort into school than did European American and Anglo-Australian adolescents (Rosenthal & Feldman, 1991). Parental and individual factors contributing to academic effort and school achievement were similar across all cultural groups. Having a demanding family environment in which conflict is low and possessing personal characteristics of self-control and industry seem to promote academic effort and achievement regardless of ethnic background or immigrant status.

> Should American parents become more firm? More demanding? Less accepting? Provide more supervision? Emphasize tradition?

Table 7.1	The Family's Changing Role
Former Family Roles	**Societal Elements That Perform Them Now**
Economic-productive	Factory, office, and store
Educational	Schools
Religious	Church or synagogue
Recreational	Commercial institutions
Medical	Doctor's office and hospital
Affectional	Family

The Changes in the American Family and Their Effects on the Adolescent

Let us now look at the ways in which family life has changed in recent decades. As more of their time is being taken up by work, many adults seem to have less time to spend with family members. Members of the extended family, such as grandparents, aunts, uncles, and cousins, are likely to live farther away and not in the same house or community like they might have a century ago. The typical American family no longer consists of father as breadwinner and mother as housewife and child caretaker.

THE LOSS OF FUNCTIONS

In 1840, the American family fulfilled six major functions (Sebald, 1977). Table 7.1 lists those functions and suggests which parts of society now perform them. Today the first five functions—economic-productive, educational, religious, recreational, and medical—have been taken over by professionals. The major function for the family today is to provide affection, support, and nurturance for its members. In the 19th century, parents and children were more interdependent in the following areas (Coleman, 1961):

- *Vocational instruction.* For both males and females, the parent of the same sex taught them their adult jobs. Most men were farmers and most women housewives. Parents knew all the secrets of work, secrets passed on from generation to generation. Today, nearly all men work at jobs different from their fathers, and an increasing percentage of women are not primarily housewives like their mothers were.
- *Economic values.* Adolescents were important economically on the farm; without children, the farm couple had to hire others to help them. Work was a source of pride to the children and their work made it very clear that they were important to the family. Today, children are not expected to help their families financially, but it's still costly to raise children, especially adolescents.
- *Social Stability.* When families stayed in the same town all their lives, parents were able to tell their children everything there was to know about how to live in that town. Today, the average American moves every five years. The adults are as much strangers in a new place as the children. In fact, with Dad, and now frequently Mom, driving out of the neighborhood to work, the children may well know the neighborhood better than their parents do.

social capital
The knowledge, values, and beliefs passed on from one generation to the next through time spent together.

Some social scientists express concern that the American family is no longer fulfilling its functions adequately. Coleman (1987), for example, believes that the family has been losing its influence as a source of **social capital**. Coleman has defined social capital as "the norms, the social networks and the relationships between adults and children" (p. 36). Social capital serves to connect the generations through a shared body of knowledge and values and prepares young people to carry out responsibilities as adults. The office complex, shopping mall, and rock concert have replaced the household, neighborhood store, and family gatherings as places of work, shopping, and leisure. As a result, children and adolescents have less contact with caring adults and less opportunity to obtain social capital. Coleman suggests that new institutions developed to take over family functions such as child care and recreation must consider their role in developing social capital or in shaping the attitudes, motivations, and self-concept of youths. Personal attention and interest, involvement, emotional closeness or intimacy, and continuity over time are some of the characteristics Coleman believes are necessary for any institution that hopes to connect with and transmit values to youths.

Youniss (1989) shares Coleman's concern that family changes such as parental divorce may interfere with the family's role in providing social capital. As the labor market increasingly demands workers with highly technical skills, Youniss wonders whether our families will provide youth with sufficient motivation to seek out and obtain the necessary training. In conjunction with this concern, Youniss is reassured by numerous studies indicating that for the majority of today's adolescents, parents remain an important source of social capital. Despite social change, most adolescents respect their parents, feel close to them, and take seriously the values held by their parents. Parents remain authorities and advisors to their adolescent offspring on many matters.

According to a *Boston Globe* poll of 400 parents with children 18 years old or younger, 78 percent of women and 62 percent of men stated that it is harder for parents to raise children now than it was when their parents were raising them (Matchan, 1995). In a U.S. public opinion poll, almost everyone surveyed felt that children were worse off today compared with 15 years ago because of a decline in family stability and parental care (National Commission on Children, 1991). The Council on Families in America (1995) claims that the decline of marriage in this country is directly responsible for many of the terrible hardships that children face. Furstenberg (1994) argues against this widespread belief. By presenting data comparing family and child-raising patterns with other nations, he suggests that maternal employment and family instability by themselves have a small impact on the well-being of our youth in comparison with the impact of more complex social and economic factors. Increases in suicide, crime, and substance abuse observed in the 1980s were as prevalent among young adults who had grown up in traditional family environments as they were among teens who had grown up in changing family environments. Furstenberg concludes that family change occurred simultaneously with an increase in problem behaviors but did not directly cause those problems.

> What do you think? To what extent are changes in family life responsible for an increase in teen suicide, crime, and substance abuse?

Although the family appears to be adequately fulfilling its role in providing affection, support, and nurturance for most adolescents, it is not adequately meeting that need for all teens. Some social critics (Popenoe, 1988; Skolnick, 1991) point to changes in the American family as a source of teenage problems. Other social critics (Furstenberg, 1994) maintain that family changes are here to stay and that society needs to find better ways to support the family in fulfilling its nurturant role. Lingxin and Bonstead (1998) found that high levels of parent-child interaction increased both parent and child's educational expectations which impacted achievement. This study highlights the importance of finding ways to help parents be involved with their children because the positive effects of this involvement extends far beyond the parent child relationship. Marian Wright Edelman, president of the Children's Defense Fund, described it this way. If, as according to the African proverb, "it takes a village to raise a child, that village has been torn asunder. And the question is: How do we begin to reinvent community?" (Matchan, 1995) Flexible working options for parents,

WHAT'S YOUR VIEW?

The Future of the American Family

A number of researchers have argued that in spite of the many changes in the typical American family, it is basically the same as it always was. They cite the very high rate of marriage and a relatively stable birthrate among other factors as indicating the hardiness of the primary family unit.

There are many others, politicians as well as scientists, who say that the family today is in serious trouble. There are even those who fervently believe that the family as we

know it is on the way out. They point to the professionalization of child care, the absence of the mother from the home in the daytime, the epidemic of unwed teen pregnancies, and many other instances as evidence for their standpoint.

Imagine it is 20 years from now, and you are a social scientist who specializes in family study. What do you think you will be finding? Will the family still be intact or not? What do you think?

nurturing after-school resources for children, and income security for all children are some of the suggestions offered by Boston University Professor of Pediatrics Dr. Barry Zuckerman (Matchan).

During the past several generations, family life has changed in many ways for many Americans. Economic, cultural, and technological changes have had a large impact on family life. Recreational activities are often shared with teenage peers and not with adults, including parents. Following a trend toward earlier marriage and a rise in the birthrate after World War II, many women are now both marrying and having children later in life. More women are working outside the home. The rate of divorce rose greatly from 1965 to 1979, although it has leveled off since that time. Many children live at least part of their childhood in a single-parent family, and many adolescents are gaining stepfamilies as a result of their parents' remarriage (Furstenberg, 1990). Many adults have less time available to spend with their children. How have these changes in family life affected the lives of adolescents—for better and for worse?

THE EFFECTS OF DIVORCE

A smoothly functioning family can provide support and nurturance to an adolescent during times of stress. But when the family itself is in trouble, such as during a divorce, the family can become a source of stress rather than support.

Divorce has become commonplace in American society. Even with the slight decreases in the divorce rate in recent years, there are still more than one million divorces each year, which is about half the number of marriages performed during the same time (U.S. Bureau of Census, 1994). Divorce occurs most often in families with a newborn and second most in families with an adolescent present. It is estimated that as many as one-third to one-half of the adolescent population is affected by divorce (Jurich, Schumm, & Bollman, 1987).

What, then, are the effects of divorce on the development of the adolescent? Researchers have had difficulty answering this question. Divorcing parents are often unwilling to participate or let their children participate in such studies (Santrock, 1987). Thus we cannot be sure how true our research findings are for all families in which a divorce has taken place.

Divorce almost always means change and usually adds stress to a family. One obvious effect is economic. For some families, less money forces the family to move. The adolescent then has to make new friends and adjust to a new school. When the parents are not living together, two rents or mortgages must be paid. This leaves less money for other needs. Most teens, particularly younger adolescents, worry about how they are viewed by their friends. They feel that they need to have the right clothes or stereo to fit in with their group. They may resent no longer being able to have everything their friends have.

For the older adolescent, divorce may mean that there is less money to pay for college. Furstenberg (1990) points out that children from divorced families have been less likely than children of two-parent families to go to college. He believes that this may be because some fathers of divorced children are unwilling or unable to pay for their children's education. Wallerstein (1987) interviewed 38 adolescents 16 to 18 years old whose parents had divorced 10 years earlier. Most of the adolescents were still in school, but about half were not doing very well. Of the students in two-year and four-year colleges, only half received financial help from their fathers, even though their fathers could afford it. The students were disappointed with their fathers and felt hurt that they were not given more support.

Another usual effect of divorce is the absence of one parent. Often, custody is given to one parent (usually the mother), and so the children are likely to lose an important source of support (usually that of the father). In addition, the children may lose the support of grandparents and other extended family members. Extended family and friends can provide emotional, financial, and household support for single parents. This attention can help stabilize a family following divorce and can offer additional support to adolescent children.

When parents continue to fight after the divorce, it is hard for the child to be close to both parents. One or both parents may attempt to turn the adolescent against the other. Buchanan and others (1991) found that adolescents who feel caught between their parents are likely to feel anxious and depressed following a divorce. Asking an adolescent to give messages to the other parent or asking for information about the other parent's life are ways in which parents make adolescents feel caught. When parents continue to fight after a divorce, they often pull their adolescent children into the conflict. Having a close relationship with both parents after a divorce is helpful for adolescents. Continued parental fighting has a negative effect on adolescents, especially when they are drawn into the battle.

In further research, Buchanan and associates (1992) assessed the emotional adjustment of 517 children 10 to 18 years of age more than four years after parental separation. Comparisons were made for adolescents living with their mothers, with fathers, and with dual residence. Overall, differences in adjustment based on residence were small, with adolescents living with their fathers sometimes doing less well than the other teens. Adolescents adjusted less well living with their fathers only when they had shifted residences more than once since their parental separation, when parents were still hostile toward one another, when the teen did not feel close to the father, and when the father did not provide close monitoring of the teen. Guttmann and Lazar (1998) also compared the social adjustment of children from divorced and intact families, paying special attention to differences in custody arrangements. Their findings

AN APPLIED VIEW

Substituting for Absent Adult Role Models

Teens are at an age when they are especially vulnerable to the influence of others, whether for good or for ill. They need to be able to look up to appropriate adults, but often such people are unavailable.

What can you do about the absence of adult models? One contribution you can make is to expose the adolescents you work with to good adult models through the media. There are books, magazine articles, movies, and videos in which adults are depicted as behaving admirably. You can be on the lookout for these role models in the media, and you can expose the youth to these models as effectively as possible.

suggest that children from divorced families were generally less well adjusted than children from intact families and that custody arrangements matter very little.

Divorce can diminish a person's ability to parent. If the father is emotionally upset, he will not always be able to give attention and guidance to his children. If a mother has started working more hours outside the home as a result of the divorce, she will have less time to spend with her children. Fatigue and feelings of helplessness plus practical life changes leave some parents with less emotional and physical energy to devote to parenting. Children and adolescents who are upset about a divorce may be more difficult for a parent to manage and increase the parents' feelings of stress and incompetence (Hetherington, 1991). Smetana (1993) found that divorced mothers felt that they had less authority over their adolescent sons than daughters. This is consistent with other findings that divorced mothers sometimes have more difficulty in controlling sons. Some divorced parents, on the other hand, feel that they are better parents after the divorce. That is, once an angry or estranged spouse leaves the home, the remaining parent can sometimes go about the business of parenting without the interference of marital conflict.

Hetherington and her colleagues completed a number of studies of the effects of divorce on children (Anderson, Greene, hetherington & Clinempeel, 1999; Hetherington, 1991; Hetherington & others, 1989). Almost all children and families experience distress at the time of a divorce. Feelings of anger, sadness, and anxiety are natural reactions to changes in family life. Within two to three years after the divorce, most children and parents adjust to their new family life. Although the majority of children are able to adjust, difficulties continue for some. How well children adjust depends on a number of factors, including:

- *Cumulative stress.* The more a child's life changes following a divorce, the more likely it is that the child will have difficulty. When there is a great loss in income, a change in methods of discipline, a move to a new neighborhood, or a decrease in the amount of support given by one or both parents, stress is likely to continue and adjustment will be difficult.

- *The child's temperament and personality.* Some people are more easygoing than others and have less difficulty adjusting to change. For well-adjusted children, adjustment to divorce may not be difficult and may even prepare them to handle other changes in life. For children who have a hard time adjusting to change, divorce may increase behavior problems. If children have had behavior problems before the divorce, those problems will probably worsen after the divorce, at least for a while.

- *Age.* Adolescents feel pain and anger when their parents divorce, but their cognitive abilities help them to understand what has happened better than younger children can. Adolescents are also likely to have more friends outside the family than younger children. These friendships can offer support and, to some extent, can make up for loss of parental support.

Although divorce is difficult initially, most children and adolescents adjust successfully with time.

Children in divorced, single-parent households may be asked to participate in family decisions and household tasks. They may grow up faster and rely less on their families. Growing up faster can be useful if the adolescent becomes involved with positive peers and activities outside the family, such as school or work. If, however, the adolescent becomes involved in less constructive activities, such as drugs, alcohol, or just hanging out, the outcome can be disastrous. If divorce occurs during middle childhood or adolescence, teens will remember the conflict more than if they were younger. Wallerstein (1987) found that 10 years after their parents' divorce, adolescents viewed the divorce as the major event of their lives. Adolescents also report a greater need for their fathers than younger children do. In a 25 year follow-up to a longitudinal study on the effects of divorce, Wallerstein and Lewis (1998) found that adults now 27-32 years old had memories of abandonment, terror and loneliness. Many participants reported that they engaged in early sexual activity and experimented with substances. Participants also reported fear of intimacy and strained relationships with parents. These findings suggest that the impact of divorce may be longer lasting than previously believed.

Because divorce has become more common, it is not unusual for children to experience more than one parental divorce by the time they reach adolescence. Early adolescents who have experienced multiple parental divorces are more likely to have school problems, including lower achievement and more disruptive behavior than their classmates. These young adolescents also describe their parents as providing less supervision, being less accepting, having more parent-teen conflict, and granting their children less independence than other parents (Kurdek & others, 1995). These parental characteristics may explain in part the difficulties these teens are having at school.

• *Sex differences.* Because it is the father who typically leaves a family during divorce, there are often more negative effects for males than females. Boys living with single mothers have more problems, both at home and at school, than boys living with both parents. Recent findings (Hetherington, 1991), however, indicate that many girls from divorced families develop behavior problems similar to those displayed by boys as the girls enter adolescence. Buchanan and others (1991) found that adolescent girls are more likely to feel caught between parents than boys. Other research data (Wallerstein, 1987) suggest that adolescent girls often have difficulty relating to males during adolescence. Wallerstein found that 10 years after their parents' divorce, adolescent girls were often afraid of being hurt by boyfriends and were afraid that their own marriages would not last.

Despite the negative aspects we've outlined, you should keep in mind that divorce is not always a bad choice. Most adolescents, children, and parents adjust to their new family structure. Divorce is often better than keeping a stressful, unhappy family intact. In fact, the few studies that have compared adolescents from the two groups have shown that adolescents from divorced families do better than those from two-parent families in which the parents are fighting (Hetherington, 1973). If the outcome of the divorce is to be positive, parents need to find a way to get along with each other as much as possible and see that their conflicts do not harm their children.

STEPFAMILIES

Recent estimates suggest that 25 percent of children will spend some time in a stepparent family before they become young adults (Hetherington & others, 1989). Because most mothers don't remarry until five or more years after their divorce, a large number of children entering stepfamilies are in their teenage years. A new stepfamily often means new stepsiblings as well as a new stepparent. Getting along with a new stepparent and stepsiblings can be difficult for some adolescents. In fact, it seems to take adolescents and older children more time to adjust to their new stepfamily than it does to their parents' divorce (Hetherington & Clingempeel, 1988). However, according to some recent research (Hetherington & Stanley-Hagan, 1998) the critical factors seem to be family processes and functioning rather than family structure. This findings seems to indicate that how the family members interact is much more important than who makes up the "family". How difficult it is to adjust to a stepfamily depends on the same factors that affect adjustment to divorce. Let's look at how age and sex differences affect adjustment to a stepfamily.

Some of the developmental changes that take place in adolescence can make adjustment to a stepfamily more difficult (Hetherington & others, 1989). For example, as adolescents become more independent of their parents and have more to say in family decisions, they begin to view their parents as persons with strengths and weaknesses rather than as all-knowing adults. As a result, many adolescents are less willing to accept rules or discipline from a stepparent. Older children and adolescents seem to accept a stepfather better when he does not try to discipline the children. A stepfather is more likely to be accepted if he develops a good relationship with the children and supports the mother as she does the disciplining (Hetherington & others, 1985). A combination of parental warmth, involvement, and monitoring of activities (rather than enforcing rules and imposing discipline) has been found to be positively related with adjustment among early adolescents (Hetherington, 1991).

The adolescent's concern with sexuality may also make it difficult to accept the parent's affection toward a new husband or wife. Some adolescents have such negative attitudes toward new stepparents that it is very difficult to develop a positive relationship, even after a number of years. Hetherington (1991) found that many stepfathers and adolescent stepchildren do not mention each other as family members, even more than two years after the remarriage. Acceptance of a stepparent may not be as difficult for the late adolescent because she or he is often getting ready to leave home.

As you recall, divorce is often more difficult for boys. The mother's remarriage, if the children are living with her, can be good for sons after the initial adjustment. They can be helped by an involved parent of the same sex. Stepfathers are more likely to benefit younger sons than they are adolescents, however. The adjustment to and acceptance of a stepfather seems to be difficult for both adolescent boys and girls (Hetherington, 1991). Divorced mothers and daughters often form close relationships. Stepdaughters then resent the time and attention that is taken away by the stepfather (Hetherington & others, 1985).

In summary, although some teens react well to a divorce, most find it a disruptive event in their lives, from which they need time and support to recover. Let's turn now to a review of the effects of the parents' work roles on the family.

THE EFFECTS OF PARENTS' WORK ROLES

Mothers of adolescents are more likely to be employed outside the home than are mothers of younger children (Armistead & others, 1990). This is likely due, at least in part, to the assumption that children at this age are more independent. This may also be related to increased family expenses. Because the proportion of married working women with children doubled from 1960 to 1990, researchers have been interested in finding out whether having a working mother is good or bad for children. More research attention has focused on the effects of mothers' work roles than on those of fathers. Heath and Orthner (1999) found no gender differences in parents' perceptions of their ability to successfully manage family and work. Orthner (1990) reviewed the adolescent research and concluded that most studies show little effect. Whereas a few studies have suggested that having a mother who works outside the home results in fewer family activities, more adolescent behavior problems, and a drop in academic performance, most do not. Orthner believes that this is because mothers, adolescents, and families as a whole have adjusted well to changing parental work roles. For example, families with working mothers have learned to share many household responsibilities or simply relax standards concerning meals and housework. Orthner suggests that these changes in role expectations (beliefs about what mom should do) have decreased the stress (found in early research) associated with maternal employment. Almeid and colleagues (1999) looked at the impact that work stress had on tensions in the home. When fathers experienced work stress they were more than twice as likely to experience tension at home, especially if their wives worked full time. This may be a more subtle manifestation of role expectations.

It seems that the effects of a mother's working depend on many factors, such as whether the mother likes her work, how much stress the job creates both for her and the rest of the family, and how much the father gets involved in helping out with household chores and child care. For many women, employment is less stressful than full-time homemaking (Baruch & others, 1987). The amount of stress fathers carry home from their jobs also affects young adolescents (Galambos and others, 1995). Fathers who experienced work overload were found to have more arguments with their adolescents, which led to more adolescent behavior problems. When mothers were stressed from their jobs, adolescent children were likely to view their mothers as less warm and accepting. Adolescents showed no ill effects from maternal job stress as long as mothers maintained accepting views of their adolescents. When reduced maternal warmth and acceptance followed job stress, adolescent behavior problems also increased. Mason and associates (1994) found that African

American parents' satisfaction with their work and perceptions of having autonomy in their work were associated with their parenting style and level of family conflict as well as with levels of behavioral problems among early adolescent children. Working parents, it seems, need to be cautious to keep job stress from affecting their relationships with their adolescent children. Employers also need to be aware that parental work conditions have an impact on the family life of their employees.

Early adolescence is a time when many parents decide that their children no longer require after-school care and begin to leave children on their own. Very little is known, however, about how this unsupervised time affects early adolescents (Orthner, 1990). Although there is much talk about fathers getting involved in parenting and household chores, research suggests that not a lot of change has taken place (Furstenberg, 1990). Having a mother who works can help many children and adolescents become more independent and self-confident, especially if their mother enjoys her work and has enough help with household chores and child care. These adolescents may learn to make decisions on their own and take care of many responsibilities around the house. Children and adolescents of working mothers also have less traditional sex roles. That is, they are more likely to believe that men and women can do a variety of things without being limited by their sex (Hoffman, 1989).

Teenage Parenthood

So far, we have considered the difficulties of single parenthood as a result of divorce. In this section, we consider the difficulties of single parenting when the parent is still a teenager. Although teenage parenthood was not uncommon in

AN APPLIED VIEW

A Young Woman's Dilemma

"You're pregnant," the doctor said, "and you have some decisions to make. I suggest you don't wait too long to decide what you'll do. It's already been seven weeks, and time is running out!"

"Look, it just can't be true!" I replied. I was trying to convince myself that the clinic doctor was lying. It wasn't supposed to be like this! I was tired of the bitter quarrel I had been having with the doctor. I resented him with every passion. How could I let myself be seen like this?

I had been fearing this answer. I suppose I knew the truth all along, but I really didn't want to face it. I didn't want an abortion, that much I was sure of. Besides, where would I get the money?

For ages now, I had been thinking my period would come any day. Now the truth was in the open! I walked out of the office and headed aimlessly down the street. I looked around and saw only ugliness. I thought about God and how even He had deserted me. It all hurt so much.

"How could this have happened to me?" I thought. "Good girls don't get pregnant!" All of the things my mother had told me were lies. According to her, only the "fast girls get pregnant." The ones who stayed out late and hung around with boys. I wasn't part of that category!

I looked down at my stuffed belly and thought about my family. Would they be understanding? After all, they had plans for my future. They would be destroyed by the news.

"I'm not a tramp," I said to myself. "Then again, I'm only 16 and who would believe that Arthur and I really are in love?"

Written by a 17-year-old single parent

the past, the number of unmarried teenage parents has increased dramatically in the past several decades. The feelings of the unmarried girl that are expressed in the Applied View Box titled "A Young Woman's Dilemma" are all too typical. In past generations, the young girl's dilemma would probably have been resolved by marriage. That is much less likely to happen today. As recently as 1960, 85 percent of teenage mothers giving birth were married. That percentage had dropped to 34 percent in 1988 (Furstenberg, 1991). Many teenage girls decide to have their children without marriage. Children born of these pregnancies often have a difficult life (Garn & Petzold, 1983). It is important to note that the vast majority of research and writing about teen pregnancy focuses on mothers. Little attention has been paid to teen fathers. Studying teen fathers seems to be an important step if we are to come to a full understanding of the complex factors that are involved in the very significant social problem of teen pregnancy.

In this section, we consider the frequency, causes, and consequences of teenage parenthood.

TRENDS IN BEHAVIOR

The number of live births per 1,000 married and unmarried females in the United States dropped during the 1970s and early 1980s. Recent statistics indicate that the number of pregnant women over 30 climbed during the late 1980s and early 1990s, reflecting the increasing number of women who are postponing childbearing to their 30s and even 40s (National Center for Health Statistics, 1994). Of concern to public health officials is the large increase in the birthrate among teenagers between 15 and 17 years. Health officials are alarmed because pregnancies among this age group carry increased health risks for both the mother and the child. After declining in the 1970s and 1980s, the birthrate among girls age 15 to 17 rose by 19 percent between 1986 and 1990 (Bass, 1992). Those trends are shown in Table 7.2.

Table 7.2	Birthrates by Age of Mother, 1970 to 1991												
Age of Mother	**Birthrate (per 1,000 women)**												
	1970	**1975**	**1979**	**1980**	**1981**	**1982**	**1983**	**1984**	**1985**	**1986**	**1988**	**1990**	**1991**
10–14 years	1.2	1.3	1.2	1.2	1.1	1.1	1.1	1.1	1.2	1.2	1.3	1.4	1.4
15–19 years	68.3	55.6	51.5	52.3	53.0	52.7	52.9	51.7	50.9	51.3	53.6	59.9	62.1
20–24 years	167.8	113.0	109.9	112.8	115.1	111.8	111.3	108.3	107.3	108.9	111.5	116.5	115.7
25–29 years	145.1	108.2	108.5	111.4	112.9	112.0	111.0	108.7	108.3	110.5	113.4	120.2	118.2
30–34 years	73.3	52.3	57.8	60.3	61.9	61.4	64.2	64.6	66.5	66.5	73.7	80.8	79.5
35–39 years	31.7	19.5	19.0	19.5	19.8	20.0	21.1	22.1	22.8	23.9	27.9	31.7	32.0
40–44 years	8.1	4.6	3.9	3.9	3.9	3.8	3.9	3.8	3.9	4.0	4.8	5.5	5.5
45–49 years	.5	.3	.2	.2	.2	.2	.2	.2	.2	.2	.2	.2	.2

Source: U.S. Bureau of the Census, Statistical Abstracts of the U.S.: 1994.

As Lancaster and Hamburg (1986) point out,

❑ Except for the very youngest adolescents, contraception and abortion have lowered the birthrates for adolescents since 1970 to levels that are somewhat lower than those in the 1920s and 1950s. However, the rate of adolescent childbearing outside of marriage has shown steep increases (p. 5).

Although parenthood among teens in general has not increased, there is today more parenthood among unmarried teens (Furstenberg, 1990). In the 1950s, teenage marriage was more common, and those who became pregnant were more likely to marry. During the 1980s, two-thirds of all white teen mothers and 92 percent of African American teen mothers with their first baby were single (Furstenberg & others, 1989). The number of white unmarried adolescents bearing children more than doubled between 1975 and 1988 (Jorgensen, 1993).

Several changes in American society can help to explain these trends. The age at which couples are getting married has increased. More women see higher education and work as an alternative to early marriage and have put off marriage until they are in their 20s or older. In 1960, 40 percent of all 19-year-olds were married, compared to only 14 percent in 1988.

Sex, marriage, and children are less closely linked than they were 25 years ago. In the past, boys were much more likely to have sex earlier than were girls. Since the belief that you should wait until marriage has become less common, girls are having sex earlier as well. In 1983, 60 percent of white males had intercourse by age 18, and the same percentage of females had intercourse by age 19 (Hofferth & Hayes, 1987). According to a 1990 survey by the Centers for Disease Control, 72 percent of all high school seniors and 40 percent of ninth-graders have had sex (Bass, 1992). As a result, many more teenagers are becoming pregnant outside of marriage. Each year, about one million teenage girls in the United States become pregnant. About half of those pregnancies result in live births; one-third result in abortions; and one sixth result in miscarriages (Annie Casey Foundation, 1998).

What this means is that more children are being born without the cultural approval and support that marriage brings. Forty percent of all teenagers who become pregnant end their pregnancy by abortion (Furstenberg & others, 1989). Another 46 percent give birth to a child, and the remaining 14 percent of adolescent pregnancies end in miscarriage or stillbirth (Jorgensen, 1993). Only 5 percent of unmarried adolescents put their babies up for adoption (Jorgensen). Whether a teenager decides to have her baby depends on many things—her interest in school, the number of her friends who have children, and her family's attitude about keeping the baby all influence her decision. Most teens do not think they are mature enough to marry. Other teenage girls do not think that their boyfriends have much to offer in terms of financial support and stable companionship. Job opportunities for adolescent males are limited in today's economy. Contemporary gender roles challenge traditional views of father as breadwinner and mother as caregiver. The social stigma associated with unwed parenthood has declined in recent decades. Economic, social, and cultural factors thus combine to reduce the urgency for marriage (Furstenberg, 1990).

Teenage Parenthood—Causes and Solutions

The high rate of unwed teenage parenthood has become a controversial topic in this country. The public is concerned about the large public expenditure that has been directed toward supporting teenage mothers and their children. Some wonder whether the public financial support available to unmarried women with children has been an incentive to young teenage girls to bear children. Others argue that these girls don't plan to become pregnant. Rather, they do not take adequate precautions to prevent pregnancy. Dr. Leon Eisenberg of Harvard Medical School maintains that if we could do something to reduce unintended pregnancies, many other social problems would be resolved (Kong, 1995). Without a promising educational future and career to look forward to, the incentives for avoiding childbirth are not strong. The incentives for marriage and stable family life are few. Teenage boys cannot find employment that will support a wife and child. Many do not provide satisfying emotional support to young mothers. According to the Council on Families in America

(1995), the institution of marriage has become devalued in our culture. They maintain that the value of enduring marital relationships should be regained in our society.

Controversy over the solution to the rising rate of teenage parenthood is as heated as the debate over its causes. Should public Aid to Dependent Children be eliminated? Should unwed mothers place their children in orphanages? Would such measures wrongfully harm innocent young children? Should educational and employment opportunities be improved so as to deter adolescent women from early motherhood? Should teenage boys be taught more about becoming better husbands and fathers and be helped into jobs so that they can support their families and be more desirable marriage partners for teenage girls? Is better contraceptive education the answer? More confidence in marriage?

What do you think about the causes and solutions to parenthood among unwed teens?

Some people point to earlier menstruation as an explanation for these trends; others talk of the crumbling morals of today's youth and the financial rewards for motherhood available through public welfare. We can say that the images of the fast and easy girl and the sex-obsessed boy are surely false (Elster & Lamb, 1986; Kinard & Reinherz, 1987; Klein & Cordell, 1987; Stiffman & others, 1987). That unmarried parents are usually from the lower socioeconomic class and from one-parent families are also untrue generalizations. However, significant relationships between pre-school and school age physical abuse and neglect and teen pregnancy have been found.

INTENTIONAL TEEN PREGNANCY

We often make the assumption that when teenagers become pregnant the pregnancy is accidental. However, it is important to recognize intentional teen pregnancy as a growing area of social concern that needs to be addressed. Many young teenage mothers view having a child as a way to meet their needs for love and affection (Corcoran and others, 1997). Pregnancy may also be seen as a way to maintain a relationship. Sadly, adolescents alienated from their families and desperately in need of support and connection may see having a child as a way to create a loving relationship that is otherwise absent in their lives. (Williams & Vines, 1999).

RACIAL DIFFERENCES IN TEENAGE PARENTHOOD

Many studies have shown distinct differences among racial and ethnic groups of teenagers in sexual behavior, pregnancy, and teenage parenthood (Brooks-Gunn & Furstenberg, 1986; Chilman, 1985).

African American teenagers begin sexual experimentation and experience their first pregnancy at an earlier age than do white teenagers (Cummings, 1983; Moore, 1985; Zelnick & Kantner, 1980). Recent statistics indicate that among African Americans nearly two out of every three births are to single women, compared with one out of five among whites, and one out of three among Latinos. Although birthrates among unmarried women have been rising fastest among white women, unmarried African American teenagers still remain four times more likely to have children than are their white counterparts (U.S. Bureau of the Census, 1994).

Cummings (1983) analyzed data on 3,568 white, 969 African American, and 524 Mexican American women attending a planned parenthood clinic. Most subjects were age 14 to 24 years, and most sought counseling for an unplanned pregnancy. There were distinct differences in marital status among the groups. The largest percentage of married clients were Mexican American. African Americans and Mexican Americans decided to give birth to their unplanned child more often than did whites.

In making sense of these statistics, try to sort out ethnic values from socio-economic factors. Complex economic, social, and cultural factors must be considered. Racial differences in adolescent pregnancy and childbearing are reduced when family income and parental education levels are first taken into consideration. Poor teens, whether African American, Asian American, Native American, white, or Latino, are three to four times as likely to become unwed teens than are economically advantaged teens (Children's Defense Fund, 1987). The higher rates of teenage parenthood among economically disadvantaged youths are understandable. Teens who are behind in school, who lack basic skills, and who see few opportunities for their future are more likely to become parents as teenagers. Lacking educational and job opportunities, parenting may be one of the few available ways of achieving adult status. Unfortunately, becoming pregnant as teenagers often makes their lives and the lives of their young children very difficult. In an ongoing study of 293 racially diverse adolescent girls ages 14 to 19, Adler (1991) identified self-esteem and academic plans as factors influencing girls' attitudes toward pregnancy. Girls with higher self-esteem and the intention of continuing their education placed less value on having a baby and were less likely to expect that they would become pregnant in the next year than were girls with low self-esteem and no school plans.

Williams and Kornblum (1985) studied more than 900 teens who were growing up in poverty in cities and rural areas throughout the United States. According to their interviews, another cause of pregnancy relates to the opportunities in a teenager's life. When girls are not doing well in school and see little other hope for their future, motherhood can offer an escape from school. Being a mother is an identity for a girl who sees herself going nowhere in life.

Having a child can be viewed as a symbol of maturity and as a source of affection and love.

The Children's Defense Fund (1987), a national organization that is working to prevent teenage pregnancy, believes that teens need hope, strong self-esteem, and a sense of positive options for their future. Teenagers need to believe that there are better choices for their future. They need to be helped in developing school and work skills. They need to be offered opportunities for success outside of school. The Children's Defense Fund believes that when more teenagers have hope and vision for their future, teenage pregnancy will decline.

In order to solve the problem of teenage parenthood, we have to solve problems of education and job opportunities for all teenagers, regardless of race. We must also consider ways to help teenage mothers provide their babies with the needed emotional, intellectual, and physical care while also enabling the mothers to continue their education. These factors often make a difference in the future of the mothers and their children. Support from family members, as we will soon discuss, can also help.

Ethnic and racial group differences in teenage parenthood also, in part, reflect different values about parenting and family. Ethnographic interviews with African American, Hispanic, and white adolescents in New York City suggest that poverty and cultural practices interact to influence decisions regarding abortion, marriage, and out-of-wedlock childbearing as resolutions to teenage pregnancy (Sullivan, 1993). Thompson (1980) also compared whites and African Americans on their beliefs, perceptions, and decisions related to having children. African American teenagers, both male and female, felt more strongly that having children promotes greater marital success, approval from others, and personal security. They also expressed stronger beliefs that couples should have as many children as they wish. Adler (1991) reported that African American teenage girls are more likely than white girls to think that something good will result from having a baby.

Gabriel and McAnarney (1983) compared the decision about parenthood in two groups in Rochester, New York: 17 African American, low-income adolescents (age 15 to 18 years) and 53 white, middle-class adult couples. Their observations showed that the decision to become parents was related to different subcultural values. In contrast to the white adults, the African American adolescents did not see marriage as a prerequisite for motherhood, nor did they view completion of schooling and economic independence as phases of maturation that should precede parenthood. Instead, they expected that becoming mothers would help them achieve maturation and acceptance as adults. Health-care programs that encourage birth control to avoid unwanted pregnancies may be ineffective because they do not address the needs of African American clients in terms of the values of their own subculture.

In other research, Brown (1983) studied 36 African American adolescent, unwed, expectant couples (females, age 12 to 17 years; males, age 16 to 21 years) to assess the quality of their commitment and concerns as couples. It was found that fathers were primarily concerned with financial responsibilities to the child, parenting skills, continued schooling, problems with the girl's parents, and their own future. These results contradict the popular notion that unwed African American fathers exploit their female partners.

Despite the expressed commitment of low-income African American males to their unborn children, the vast majority of teenage and older fathers live apart from their children. The commitment to mother and child typically drops over time. Thirty-two percent of African American children under the age of six live with their mothers only (U.S. Bureau of the Census, 1994). Furstenberg (1994) conducted in-depth interviews with 20 young parents over a two-year period to better understand the factors that influence father involvement. These low-income, young parents residing in the inner city were also the former children of teenage parents. Most of the young fathers wanted to "do for their child," and to be better fathers than their own had been. They felt pride and excitement about fatherhood. Those young men who felt excluded from schools, jobs, and the broader society viewed fatherhood as a symbol of maturity and an opportunity to do something worthwhile. Many, however, had difficulty fulfilling their promises and expressed feelings of shame and loss as a result. The reasons for these broken promises reflect a complex combination of social, economic, and cultural factors. Most had not experienced good role models for consistent fathering. Young women and their families did not expect the new fathers to be stable or reliable providers. Opportunities for stable employment were often unavailable. Life on the street pulled them away from their commitments. Involvement in new relationships and the fathering of more children brought additional responsibilities. Many fathers contributed financially and emotionally to the care of unrelated children belonging to their new partners. Being a "real daddy" often had more to do with the role fulfilled in children's lives than with biological ties. Those fathers who sustained commitment to their biological children over time had been taught the value of sacrifice and had learned the skills needed for success in the labor market. They were attracted to girls with similar values and economic resources. Furstenberg concludes that young men will be part of the family only when they have the material and emotional resources to invest in their children.

THE CONSEQUENCES OF TEENAGE PARENTHOOD

While young teenagers may become parents because they see little hope for their future or believe that having a baby will make them feel more mature, important, or needed, the realities of motherhood limit their opportunities even further (Brooks-Gunn & Chase-Lansdale, 1995). Teenage mothers are more likely to drop out of high school. Because of their limited education, they will have more difficulty finding work and are more likely to need public assistance. They are also less likely to get married. Many young mothers know very little about how to care for a new baby. They find that an infant needs a lot of time and attention. The teen who no longer goes to school may see little of her friends. In addition, she may have little money to care for her new baby. As a result, the adolescent mother may feel lonely with little control over her life.

Some teenage mothers never break out of the cycle of poverty. Furstenberg found, however, that the lives of many teenage mothers do improve during their

20s and 30s. Those who have few additional children, who return to school, and who later marry and stay married are better off financially. The children of these women also do better in school and are less likely to have behavior problems than are the children of teenage mothers who continue to have children and do not return to school (Furstenberg & others, 1989). O'Callaghan and associates (1999) suggest that cognitive readiness to parent was an important factor in successful teen parenting and better future outcomes.

As you can easily see, many of the problems of unwed teenage pregnancy are associated with socioeconomic class—most of the time, it is harder for the poor. Separating socioeconomic class from race is not always an easy matter. Furstenberg (1991) suggests that teenage pregnancy and parenthood are as much symptoms as causes of economic and social problems. Those teenage mothers who experience the most difficult social and economic futures following the pregnancy and birth of their children probably had the fewest social and economic resources prior to pregnancy. Poverty, poor school achievement, and low occupational goals, for example, contribute to bleak educational and occupational futures whether or not the teenager has a child. Teenage parenthood certainly is not the solution to these social and economic ills, but solutions do need to consider the variety of factors that contribute to teenage parenthood.

Do you agree? How would you try to accomplish this?

THE ROLE OF THE FAMILY IN TEENAGE PARENTHOOD

The amount of support that a teen mother receives from her family can make a difference in how well she adjusts to motherhood. Teenage mothers who feel they receive emotional support from their families are often better mothers. They are more loving with their own infants (Colletta, 1981), are less likely to be anxious or depressed (Barrera, 1981), and are happier with their lives (Unger & Wandersman, 1988). Family members can provide guidance on how to care for a child, can actually take care of the child for part of the time, and can help pay for the costs of raising a child. When this happens, the adolescent mother is more likely to finish high school, to get a job, and to stay off welfare (Furstenberg & Crawford, 1978). Teenage mothers are most likely to turn to their own mothers for help. Older sisters, grandmothers, and aunts are other important sources of support. Teen parenthood often has an impact on the entire family (Brooks-Gunn & Chase-Lansdale, 1995).

AN APPLIED VIEW

Do Parents Really Care?

Some researchers doubt that communication in and of itself makes much difference. On the basis of their study of 287 African American and white teens, Moore and colleagues (1986) concluded that

❑ most parents do not want to get directly involved and, certainly, most teenagers are reluctant to encourage involvement . . . [*Most parents*] are relieved to discover that their

teenagers are obtaining contraception. Beyond that, it seems that most are either willing or prefer to respect the adolescent's privacy (p. 241).

Think of your own parents. Imagine that you are a teenager contemplating becoming sexually active. Would you engage one or both of them in a discussion about the appropriate decision to make? If not, what are the reasons you would not?

Whether a teenage father remains involved with his child may also depend on support from his own family and the family of the teenage mother (Furstenberg, 1994). Family members typically give opinions as to whether the pregnancy should be terminated and whether they believe the relationship between the teenage parents will survive. The mother's family, believing the father will not be reliable, may be quick to point out his failings. They may be protective of their daughter and grandchild. The father's family may warn the father of the difficulties of parenting and suggest that he may not actually even be the father. When both families offer support, because they approve of the father or believe it is in the newborn child's best interest, fathers are more likely to remain involved.

C O N C L U S I O N S

Today, American families are smaller, serve fewer functions, and are less stable than they were 100 years ago, when most families lived on farms. Since the middle of the 19th century, large extended families have been replaced by smaller nuclear families. Whereas children used to be an economic asset to the family, they are now a financial burden for an ever-lengthening period. Other changes include the lack of adult role models, the increase in age-related activities, the effects of divorce on all members of the family, the special problems and growing number of stepfamilies, and the effects of the parents' work roles. Nevertheless, most of us are still raised by our families, even though those families are different in many ways from the families of 100 years ago. In spite of social change, the majority of adolescents adjust well, continue to feel close to their parents, and turn to them as a source of advice and support.

Adolescence has long been thought of as a stage of emotional conflict, largely brought on by the hormonal changes of puberty. The internal conflicts were seen as naturally disrupting to family life. Now we believe that adolescents are no more conflicted than other age group members. Some of the external transitions of adolescence, like changing schools, can cause disruption for early adolescents. Another particular difficulty is that while teens struggle to deal with pubertal

change, their parents are often enduring midlife changes that are at least as disruptive. Conflicts between adolescents' search for identity and their parents' midlife adjustments can cause family strife.

Social scientists have identified three parenting styles, known as permissive, authoritarian, and authoritative. Permissive parents may be either indulgent or indifferent. This book suggests that another style should be added to this list—that of the nurturant parent.

The growing number of people in minority cultures in the United States is changing the way we view the family. The extended family, role flexibility, and an ancestral worldview have helped many ethnic and racial minority families to survive difficult economic times.

Unmarried teenagers are becoming parents at an earlier age than ever before. There are many problems—physical, psychological, economic, social—inherent in teenage parenthood. A number of educators and other leaders are arguing that we must do something to change this often heartbreaking situation.

Just as the family is in a state of flux, so are teen peer relations. In the next chapter, we investigate the changing nature of adolescent social relations.

C H A P T E R   H I G H L I G H T S

The Changing Adolescent and Parent
- Changes that the adolescent brings to the family include pubertal changes and the search for identity.
- Many parents are dealing with midlife concerns as a child enters adolescence. This requires adjustment between the generations and sometimes contributes to conflict.
- The majority of adolescents feel close to their parents. Several models of adolescent development were presented that explain how adolescents can gain independence while remaining close to parents.

Parenting Styles
- Baumrind has described three parenting styles. Adolescents who were raised by authoritative parents seem to have few behavior problems and resist drug use.
- Another parenting style, nurturant parenting, should be added to permissive (indulgent and indifferent), authoritarian, and authoritative styles as ways in which children are raised.

Cultural Diversity
- Families of diverse ethnic and racial groups often differ from the typical white middle-class family.
- Reliance on extended family members, flexibility in family roles, and bicultural socialization are several of the characteristics that have helped minority families survive economic and political hardship.

The Changes in the American Family
- The modern American family is seen as having lost five of its original six functions—economic, educational, religious, recreational, and medical functions are provided mostly by professionals in the community.
- Other factors affecting the modern family are the increase in peer activities, the rate of divorce, the stress of stepfamilies, minority status, and parents' work roles (especially the change in the mother's). While these factors often add stress to family life, they can also benefit the developing adolescent.

Teenage Parenthood
- Teenage pregnancy and parenthood are on the rise, especially among younger teens. With rare exceptions, this situation causes a lot of heartache for teen parents, their parents, and their child.
- Teenage mothers who receive emotional support from their families, who continue in school, and who have no additional children while they are still teenagers are likely to have better adjusted children and avoid the cycle of poverty.
- When teens possess strong self-esteem, feelings of hope concerning the future, and job and academic skills for entry into the job market, they are less likely to become teenage parents.

K E Y   T E R M S

ancestral worldview 222
attachment theory 215
authoritarian parenting style 218
authoritative parenting style 218
disengaged family 217
ecological theory 209

enmeshed family 218
extended family 222
family therapy 214
interdependence theory 215
midlife transition 213
nurturant parenting style 221

permissive-indifferent
 parenting style 219
permissive-indulgent
 parenting style 219
permissive parenting style 218
social capital 226
systems theory 216

W H A T   D O   Y O U   T H I N K ?

1. As most of us would agree, the family is quite different today from its 19th-century counterpart. Which would you say better fulfills society's needs: the typical family of the 1890s or the typical family of the 1990s?
2. In what ways is your nuclear and extended family different from your mother's or your father's?
3. Do the families of minority teens differ all that much from those of the white majority?
4. If you think the answer to question 3 is yes, are the differences due to race, socioeconomic status, or something else?
5. What are some ways that parents and teens can improve their communications?
6. What is your attitude toward family planning? Which parts of your attitude are simply matters of personal preference, and which parts do you think should be matters of universal agreement?

SUGGESTED READINGS

Auel, J. (1981). *Clan of the cave bear*. New York: Bantam. Auel's wonderful imagination and excellent knowledge of anthropology make this book on the beginnings of the human family a winner. In fast-paced fiction, she describes the relationships between two types of primitive peoples—those who communicate by voice and those who do so with their hands!

Guest, J. (1993). *Ordinary people*. New York: Viking Penguin. A story of family life and how family members contribute to and respond to the mental illness of their teenage son.

Laurer, R. H., & Laurer, J. C. (1991). *The quest for intimacy*. Dubuque, IA: Wm. C. Brown. The authors of this unusually comprehensive book state that "we believe your personal happiness is crucially tied up with the quality of your intimate relationships. Our purpose is to provide you not only with a basic understanding of marriage and family life, but to show you how you can apply the knowledge you gain and enrich your life" (p. xix).

McCullers, C. (1946/1985). *The member of the wedding*. New York: Bantam. Twelve-year-old Frankie yearns desperately to join her brother and his bride on their honeymoon. She learns a great deal about the transition from childhood to maturity from the devoted housekeeper.

Moravia, A. (1958). *Two women*. New York: Farrar, Straus & Giroux. This moving, compassionate tale describes the relationship between a peasant mother and her daughter in war-torn Italy. It involves the struggles of the mother to deal with her adolescent daughter's needs under extremely trying circumstances.

Steinbeck, J. (1945). *The red pony*. New York: Viking Penguin. A boy has a father who criticizes practically all of his behavior and a hired man who gives support. How does this alternative family meet the boy's needs?

Walker, A. (1983). *The color purple*. New York: Harcourt, Brace, & Jovanovich. This story of an African American child's life in the South details many destructive relationships, which fail to kill her faith in life and in herself.

GUIDED REVIEW

1. An alcoholic father affects the entire family. This relates to which theory of parent-adolescent relationships?
   a. systems theory
   b. interdependence theory
   c. attachment theory
   d. None of the answers is correct.

2. Lori's parents encourage her efforts as a member of the school swimming team. However, they have set firm rules about acceptable academic grade levels and responsibility for family chores. When deviations occur, there is extensive discussion and well-understood consequences. This style of parenting is
   a. permissive.
   b. authoritative.
   c. nurturant.
   d. authoritarian.

3. When parents use an authoritarian style of parenting, adolescents are likely to be
   a. high in self-confidence.
   b. socially competent.
   c. cooperative.
   d. obedient to rules.

4. Chinese American children may do well in school because of traditional parenting practices that include
   a. use of reason.
   b. training in socially approved behaviors.
   c. encouragement of independence.
   d. warmth and closeness.

5. Based on research on divorce, what can we conclude about the effects of divorce?
   a. Divorce is always devastating for all involved.
   b. Divorce is always harder on males than females.
   c. Divorce is always harder on females than males.
   d. Divorce almost always means some immediate stress.

6. Louis is 17 years old. If his parents divorce, what factors can help his adjustment?
   a. Getting an after school job will help him to forget.
   b. Having good friends outside the family can be a support.
   c. Not seeing his father will help him to avoid conflict.
   d. Moving in with a cousin can help him to escape the household stress.

7. African American parents in inner-city neighborhoods sometimes encourage younger siblings to chaperone their teenage brothers and sisters because
   a. parents don't have time to watch the younger children.
   b. extended family members are not available to help.
   c. younger children can learn from older siblings.
   d. because it increases the adolescents' safety.

8. Which of the following statements is true concerning parents' work roles?
   a. Parental job stress has no effect on adolescents.
   b. Adolescents are always negatively effected by maternal job stress.
   c. Fathers who experience high job stress have more arguments with their adolescent children.
   d. Adolescent behavior problems increase when mothers are warm and accepting and experience job stress.

9. Susan is 13 years old and no longer wants to attend an after school program. She insists that she is old enough to stay home alone while her parents are at work. Her parents want her to continue in the after school program. What advice would you offer to her parents based on current research?
   a. Let Susan stay at home alone every afternoon.
   b. Keep Susan in the after school program.
   c. Let her stay home alone two afternoons each week.
   d. There is insufficient research data at this time from which to advise her parents.

10. What factors did Adler (1991) identify as influencing girls' attitudes toward pregnancy?
    a. self-esteem and academic plans
    b. moral code and peer pressure
    c. parental values and marital status
    d. pubertal status and cultural background

11. Mild conflicts with parents within a supportive family environment can help young adolescents gain _____ (peer respect/autonomy).

12. At the same time that adolescents are growing rapidly and maturing sexually, many of their parents are undergoing a _____ (financial crisis/midlife transition).

13. According to _____ (interdependence/attachment) theory, a trusting relationship with one's parents or caretaker gives the child feelings of security and self-confidence.

14. _____ (Systems/Interdependence) theory views the family as a whole, not just a group of individuals. When something happens to one person in the family, it affects the whole family.

15. Families that do not provide an adequate degree of _____ (separateness/belonging) among family members are called enmeshed.

16. Authoritarian parents give their children too little _____ (autonomy/demandingness) and permissive parents give their children too little _____ (autonomy/demandingness).

17. Teenage children of authoritative parents are likely to select peers who are _____ (well-rounded/delinquent).

18. _____ (Social capital/Ancestral worldview) is the belief that the well-being of the larger group is more important than the success of the individual.

19. The higher birthrate among American teens in comparison to teenagers in Western European countries can be explained by _____ (greater sexual activity/less use of contraceptives).

20. The amount of emotional support that a teenage mother receives from her family can make a difference in _____ (her skills as a mother/how much government financial support she receives).

Answers

1. A, 2. B, 3. D, 4. B, 5. D, 6. B, 7. D, 8. C, 9. D, 10. A, 11. autonomy, 12. midlife transition, 13. attachment, 14. Systems, 15. separateness, 16. Autonomy, demandingness, 17. well-rounded, 18. Ancestral worldview, 19. less use of contraceptives, 20. her skills as a mother.

PEER RELATIONS

Nothing is more important to me than
hanging with my friends—nothing!

Bernie Cordts, a ninth-grade student

Social relations gain increasing importance during the adolescent years. Friendships take on new meaning and the peer group becomes a more important source of influence. Nevertheless, parents and other adults continue to be influential in the lives of adolescents. The value and influence of peer and parental relationships during adolescence has often been debated. Recent research has helped to resolve some of these debates. The importance of parent and peer relationships and the ways in which peer relationships change during the adolescent years are the focus of this chapter.

After reading this chapter, you should be able to:

- Show how friends become increasingly important as the maturing adolescent begins to move beyond the immediate family.
- Explain why parents remain an important source of influence and support to adolescents.
- Describe the differences in friendship patterns that are characteristic of adolescent girls and adolescent boys.
- List positive influences that the peer group can have on an adolescent's growth.
- Discuss the structure of large peer groups and the subgroups within them.
- List the elements that make up a subculture, and relate these to the subcultures of adolescents in the United States.
- Explain the psychogenic, culture transmission, and behavioristic models of the origins of subcultures.
- Describe the common elements of teenage subcultures.
- Evaluate interactions in the classroom.
- Discuss these issues from an applied, a sociocultural, and your own point of view.

Friendships

THE ROLE OF PARENTS IN PEER RELATIONS

Children depend much more on their parents than on their friends for their emotional needs. In adolescence, however, sexual interests and the need to become an individual distinct from the family strengthens the influence of peers. The young person is now more likely to confide feelings and problems to close friends rather than to parents.

Progressively more time is spent in the company of friends than at home. Many parents find the increasing absences of their child from the home difficult, particularly when dealing with their first child. This shift highlights the importance of the parents' role in monitoring adolescent behavior (Flannery & others, 1999). Appropriate parental monitoring and guidance around peer issues can help to protect teens from some of the risks of increased peer influence (Bogenschneider & others, 1998; Flannery & others, 1999). Young adolescents can be quite discerning about whether their parents agree with each other on how to handle an adolescent (Johnson & others, 1991; Scaramella & others, 1999; Steinberg, 1987). Having an adolescent in the house requires that

It appears to be a natural tendency for adolescents, as they grow older, to spend more time with friends, and less with parents. Nevertheless, most teens find that they continue to care very much about their parents' attitudes toward them.

parents grow and change in order to accommodate the changing needs of the teen. Many parents are able to make these adjustments without difficulty and are able to manage the conflicts that occur in this area (Jacobson & Cropckett, 2000; Paikoff & Brooks-Gunn, 1991; Steinberg, 1987).

Older views of this time of transition, especially Erikson's, Anna Freud's, and Blos's theories, stress that **emotional separation** from parents should be the adolescent's goal. From these perspectives, identity is formed through detaching emotionally from the family and shifting affection to peers. Within the framework of these principles, an adolescent who is respectful of parents and emotionally attached to them is considered to lack maturity (Baumrind, 1991a). Several sociologists and psychologists have criticized this view because it emphasizes the value of separation from others and downplays the need in humans at all stages of life for connectedness and a sense of community (Bellah & others, 1985; Gilligan, 1982; Gilligan & others, 1990; Kenny, 1987, 1990; Kenny & Donaldson, 1991; Lasch, 1979, Pollack, 1998; Pipher, 1994).

emotional separation
Identity is formed at adolescence through detaching emotionally from the family and shifting affection to peers.

An influential contemporary view (LeCroy, 1988) stresses that although the need for **emotional attachment** to family will not be expressed in the same way that it was in childhood, it does continue during adolescence. Although the teenager is ready to give up the dependence of a younger child, it does not necessarily follow that emotional distance from parents or opposition to parental values is healthy. As was discussed in Chapters 6 and 7, interdependence with parents becomes the goal. In support of this view is a study by Larose & Boivin (1998) which showed that adolescents who felt a sense of security from their parents continued to feel secure after leaving home for college. These same adolescents also reported feeling greater emotional adjustment outside the family. Many psychologists believe that when adolescents experience their distance from the family as a lack of support and acceptance, they may have unusual difficulty in forming a healthy identity or positive self-esteem.

emotional attachment
When adolescents give up the dependence they had as a younger child and develop interdependence with their parents.

Why do parents remain an important source of influence and support to most teens?

Another important study (Armsden & others, 1990) examined the attachment to parents and to peers of 43 early adolescents who were in treatment for depression with that of 52 youngsters of the same age who were not in treatment and 12 who were being treated for other psychiatric disturbances. The results suggest that the depressed adolescents had less secure attachment to a parent or parents, and that those who were most emotionally distant from parents were also the most severely depressed. Those in the depressed group were also less securely attached to peers than those in the control groups, but the researchers mention that this was just as likely to be a result of their depression as it was to be a cause of it. That is, the child may have had satisfactory peer relationships that later became damaged as the child became depressed and difficult to live with.

LeCroy's (1988) study of 85 boys and girls in the 10th and 12th grades investigated the influence of closeness to parents on the self-esteem of the students and on whether the students indulged in problem behaviors, ranging from skipping school to hitting a parent or using drugs. Students were asked to rate each parent on how intimate they were with the parent. The students evaluated the relationship on a scale of 1 to 7 on such statements as, "We want to spend time together"; "we enjoy the relationship"; "we love each other." LeCroy found that a healthy attachment to parents was related to high self-esteem and to few or no problem behaviors in these young people, especially if the father was taking an active part as a parent.

Older adolescents continue to be concerned that their parents approve of their choices. A 1986 study (Leslie & others) of 159 students in the junior and sophomore years of college showed that they wanted their parents to approve of the person they were dating. Most of the students had tried to influence their parents to notice the good qualities of their chosen partner, and to like him or her. The more seriously they were involved with their dating partners, the more important it was for them to have their mothers and fathers see the partner in a positive light.

From the research just discussed (and on the basis of common sense), it is obvious that the presence and support of parents are essential for the adolescent's emotional security and ability to take appropriate steps toward eventual independence in early adulthood. A study of 6,000 teenagers from 10 different countries (Australia, Bangladesh, Hungary, Israel, Italy, Japan, Taiwan, Turkey, the United States, and West Germany) found that adolescents worldwide usually had great respect for their parents and wanted to act in ways that would make their parents proud of them (Atkinson, 1989). Teens whose parents are appropriately involved in their lives—available but not intrusive—will most likely have friends and a peer group whose beliefs and activities are in line with values learned in the family. Especially if the parents are authoritative or nurturing (see Chapter 7) rather than authoritarian or permissive, adolescents trust and accept their guidance. Responsiveness, a quality made up of such things as being considerate of the adolescent's feelings, may be the most important factor in helping the young person to choose the right group of peers and friends for themselves.

Teenagers who rely excessively on peers or who become involved with dangerous age-mates are often suffering from a lack of parental presence, interest, and regard or are dealing with conflicting cues given by parents who frequently disagree with each other. These problems are by no means confined to poor families (Seidman, 1990). Echoing Elkind's 1984 concerns, family researcher Diana Baumrind (1991b) states that, in general, social organization in the United States today is "unstable and not adequately protective of youth" (p. 59).

Parents who are experiencing difficulties such as marital conflict or alcohol abuse may find it hard to be available and responsive to their adolescent children (Baumrind, 1991b; Bower, 1989). Adolescents being raised by a struggling single parent often wish for more adult attention and guidance—a wish, they might be surprised to learn, that is shared by many children of high-powered, successful parents. Adults who are devoting as many as 70 hours a week to a job or career have little time or energy left to invest in their teenage child when they finally arrive at home (Brooks, 1989; Gelman, 1990; Hewlett, 1991; O'Reilly, 1990).

Lack of parental or adult supervision in the after school hours can have negative effects on adolescent behavior and peer relationships. Young teens who lack adult supervision are more likely to get into trouble (Bogenschneider & others, 1998; Flannery & others, 1999). A 1991 study of 206 boys and their families who were seen at age 10 and again at age 12 confirmed that those who were not watched over enough by their parents were likely to be involved with antisocial peers by middle childhood. This continued into early adolescence (Dishion & others, 1991). Many parents in a study of after school programs for low-income young adolescents believed that supervision is no longer necessary for children age 12 to 14, and would leave them alone in the afternoons if there were no program. More middle- and high-income parents (27 to 30 percent) than low-income parents (17 percent) were willing to leave young adolescents on their own after school (Marx, 1989). In another 1989 study, 112 seventh-graders from dual-earner families who were left unsupervised both after school and from eight to ten hours a day in the summer were rated on peer involvement, deviant peer associations, problem behavior, impulse control, and peer acceptance. The researchers found that those who were both unsupervised and far from home after school and those who hung out in places like shopping malls reported a higher degree of peer involvement, association with dangerous peers, and problem behaviors (Maggs & Kolaric, 1990). This finding is further supported by Jacobson and Crockett (2000) who found that appropriate parental monitoring was associated with adolescent adjustment.

In conclusion, it is simply not true that teens drop family relationships for friends. The balance slowly changes, but both aspects of adolescent social life are essential to growth.

FRIENDSHIP PATTERNS OF ADOLESCENT GIRLS

In recent years, a growing body of research has been devoted to studying friendships between adolescent girls (Apter, 1990; Frankel, 1990; Gilligan & others,

1990; Lees, 1986; Stern, 1990). As was discussed previously, separation from family and other early emotional ties was thought until recently to be the most important developmental task of the teen years. It should be remembered that several older theories considered the male development to be correct and "normal," and if female development is not the same, it was considered a sign of something lacking. For example, Freud noted that girls do not separate from parents and friends over the course of adolescence as completely as most males do. He considered this a sign that girls and women fall short of full development (Stern, 1990).

Today, many theorists believe that girls' relationships are strengths rather than weaknesses. The girls themselves usually value their friendships and their ability to be a friend very highly. The girls who gave Stern (1990) a description of themselves each year for three years during their secondary school careers consistently mentioned their relational style and skills as most important to their self-image. Gilligan (1982) noted that girls and women who are asked to tell about themselves include information about their relationships with others as part of their self-description.

However, the girls also value independence, although they are likely to describe it quite differently than boys do. A greater ability to look outside oneself, an ability to allow oneself to depend on others when necessary, and an ability to realize how much everyone does depend on others are all defined by these girls as part of their idea of independence. They believe that it is not necessary to break ties with others in order to be independent. One girl stated, "I think I have realized that I can always stay close to my mother, even if, you know, it's different depending on her and staying close to her . . . you can be independent and still stay close to somebody like that" (Stern, 1990, p. 79).

Adolescent girls share confidences about their deepest feelings and important personal decisions with their closest friends. Apter (1990) also observed that giving and receiving advice in managing emotions was even more prevalent than confidence-sharing among the girls she studied. Girls often asked their friends for advice on how to behave in particular situations. More importantly,

A SOCIOCULTURAL VIEW

Interracial Friendship

Lorene Cary, a bright high school student from Philadelphia, was one of the first black girls to attend a wealthy private prep school in New Hampshire that had been all-male and all-white for 125 years. During her years there, she made many friends, both boys and girls. In her book, Black Ice, she tells how one of these friendships began:

❑ I began to feel more confident in the inevitable racial discussions in classes, at Seated Meal, after visitors' talks. I took the offensive and bore my gifts proudly. What the discussions concerned specifically, who was there, where they took place, I do not remember. I do recall hearing the same old Greek-centered, European-centered assumptions of superiority. Might made right. I had my stories about Chaka Zulu from my Harvard evening course (and I knew they worshipped Harvard!). Nothing mattered. I was like a child again, trying to argue that I was still somebody—I am Somebody! as we shouted back to Jesse Jackson on the television—even though black people had been slaves, even though we hadn't had the dignity to

continued

A SOCIOCULTURAL VIEW

jump off the boats en masse or die from tuberculosis like the Indians. More facts. I wanted more facts to show that it wasn't all fair now, that the resources that kept them here, ruddy and well-tutored, as healthy as horses, had been grabbed up in some greedy, obscene, unfair competition years before . . .

That's how I felt the night I left a racial discussion with a girl named India Bridgeman. A group of black girls had once asked her to take the role of plantation overseer in a student-choreographed dance. I'd kept in my head the image of her as she danced around the slaves with a whip, her classical ballet training showing in every movement. She'd visited England as a member of St. Paul's varsity field hockey team. She was an acolyte who knew the rituals of high mass: where to walk, what to carry. I knew her through Janie, but mostly I envied her from a distance as a symbol, a collection of accomplishments that I did not possess . . .

India translated what I had been saying into different words, and I listened, dumbfounded to hear them. It was clear that she, too, knew how it felt to be an outsider. I had never suspected it. India told me about her life growing up in Manhattan, and her own estrangement from many of our schoolmates. We talked until we grew hungry.

'Isn't there anything to eat, anywhere?' India jumped up from the floor, where we'd been sitting, and walked across the room to her stash. 'All I've got is mayonnaise,' she said as if the world would end. 'Hold everything! I know I had some crackers, too. Do you think that's gross, just putting mayonnaise on crackers?'

'Are you kidding? I was raised on mayonnaise. And mayonnaise, not that cheap-ass salad dressing.' I cut my eyes to the little jar in her hand. She whooped with laughter.

'What would you have done if I'd been holding some 'cheap-ass salad dressing'?'

'I would have died. But really, that stuff—'

'I know, it's awful,' she agreed.

India and I talked often and late into the night after that. We raged together at St. Paul's School—at its cliques and competitiveness; its ambivalence toward its new female members; its smugness and certainty and power. We talked about families and boyfriends, girls we liked and girls we didn't. We laughed at how we had appeared to each other the year before. Our talk was therapeutic, private, and as intense as romance. It was for me the first triumph of love over race (Cary, 1991, pp. 196–199).

From *Black Ice* By Lorene Cary. Copyright © 1991 by Lorene Cary. Reprinted by permission of Alfred A. Knopf, Inc.

they asked one another for help in managing their own emotions—how to get over being attracted to someone who doesn't like them, how to handle being hurt by a friend, how to stop feeling depressed. While this kind of friendship is very common among adolescent girls, it is not at all characteristic of boys of the same age. Although some boys do discuss personal matters with friends, this is not as deep and intense as it is between girls, a pattern that carries over into young adult life.

PEER GROUP PRESSURE AND FRIENDSHIP

When you were a junior or senior in high school, how important was peer pressure in your life?

Much has been made of the importance of peer pressure and group conformity in influencing adolescent friendships in harmful ways, such as encouraging involvement in antisocial or delinquent behavior, beginning to smoke and use alcohol or other drugs, becoming prematurely sexually active, or limiting one's efforts in school to prevent becoming a disliked brain or nerd. How vulnerable do adolescents feel they are to pressure from their friends?

In a study by Brown, Clasen, and Eicher (1986), more than 1,000 teenagers were questioned about how willing they thought they would be to do something their friends wanted them to do, even if they knew it was wrong. In this study, 20 different situations were described and the students were asked what they would "really do" in the cases described if urged by "a couple of your best friends" to conform to what the group wanted. Ten of the questions described neutral situations like going to concerts or school events, and ten involved some kind of misconduct, such as drinking beer or liquor, stealing something, or committing vandalism. Many said they were willing to go along with what the crowd wanted in the neutral situations, but few said they would succumb to pressure to join in misconduct.

The students also answered a set of 53 questions designed to show how much pressure they felt from their peers to behave in specific ways. Most of them reported that they felt pressured to participate in peer activities fairly often, but rarely felt pressured toward misconduct. Many of the students said that, in fact, they felt more peer pressure *against* misconduct and found that friends often discouraged each other from becoming involved in bad behaviors. However, some researchers suggest that parent involvement and supervision is a key variable when considering the impact of peer pressure (Bogenschneider & others, 1998; Flannery & others, 1999)

Another study (Pombeni & others, 1990) confirmed that adolescents usually choose to associate and identify with those who are like themselves, and who have similar problems, values, and hopes for the future. This is usually because such friends are better able to understand their perspective. Selman has studied this phenomenon closely, and has developed a theory of the relationships between five levels of interpersonal understanding and kinds of friendship. (See An Applied View box titled "The Relationship Between Level of Interpersonal Understanding and Friendship.") Such peers and friends are often a valuable source of support and encouragement for one another. However, as Ryan and Lynch (1989) suggest, if adolescents feel they are not very emotionally connected to their parents, or that their parents actually reject them, they are

The increase in the number of persons starting to smoke cigarettes is greatest among adolescent females.

AN APPLIED VIEW

The Relationship Between Level of Interpersonal Understanding and Friendship

Level 0   3–6 years

Undifferentiated and Egocentric
Concept of Persons: Undifferentiated—does not separate physical and psychological characteristics of persons.
Concept of Relations: Egocentric—no accurate notion of relations.

Friendship

Friendship depends on physical closeness and functional similarity; admires strength and quickness.

Level I   5–9 years

Differentiated and Subjective
Concept of Persons: Differentiates physical and psychological characteristics—intentional acts recognized.
Concept of Relations: Seen as one-way.

Friendship

Someone does what child wants or child does what other wants—implies recognition of an inner self.

continued

AN APPLIED VIEW

Level II 7–12 years

2d Person and Reciprocal

Concept of Persons: Can look at self objectively and realize that others can, too.

Concept of Relations: Reciprocal in that children realize that others do what they do (i.e., I know that she knows that I know)—sees people this way but not relationships (i.e., not mutual).

Friendship

Interactions become desirable in themselves—a "meeting of the minds"—objectively but only for specific interests. Still sees interactions as helping self.

Level III 10–15 years

3d Person and Mutuality

Concept of Persons: 3d person—self and others as subjects and objects. Can have mixed thoughts and feelings about something (love and hate).

Concept of Relations: 3d person view of self, others, and system. Looks on interpersonal interactions as including self, others, and the relationship.

Friendship

Goal is mutual interest and sharing.

Level IV 12+ years

In-depth and Societal-Symbolic

Concept of Persons: Individual seen as complicated, many things going on inside.

Concept of Relations: Interactions and relationships become complicated because they may reflect deeper levels of communication.

Friendship

Realizes that complex needs can be met by different relationships. Relationships are seen as open and flexible—helps in own self-identity.

---

more likely to "go along with the crowd." It appears that teens who suffer the insecurity of scant or absent emotional support from parents are willing to pay a higher price for acceptance by peers. Given the importance of friendships during adolescence, Brendgen and colleagues (2000) compared teens with deviant friends, teens without deviant friends and teens with no friends. These researchers found that adolescents with deviant friends tended to be delinquent more often than teens without deviant friends. They also found that teens with deviant friends were as likely as teens with no friends to be depressed.

As teens move from early to later adolescence, most of them become members of peer groups. In the next section we will investigate this development.

> We often focus on the importance of friendship during adolescence but this study suggests an important question: Is having deviant friends better than having no friends?

Peer Groups

DEVELOPMENTAL PATTERNS OF PEER GROUPS

Peer groups are important in adolescent development. Although friendships are vital throughout life, there seems to be something special about the role of the peer group during adolescence.

The role of peers as a source of activities, support, and influence increases greatly (Dumont & Provost, 1999; Savin-Williams & Berndt, 1990). There is also a developmental progression in the role of friendship during adolescence (McNelles & Connolly, 1999). Perhaps it is for these reasons that adults and the media have been interested in and anxious about the role of the peer group. Brown (1990) describes four specific ways in which the peer group changes from childhood to adolescence.

- As previously mentioned, adolescents spend much more time with peers than do younger children. As early as sixth grade, the early adolescent begins withdrawing from adults and increases time spent with peers. During high school, middle adolescents spend twice as much time with their peers as they spend with parents and other adults.

- Adolescent peer groups receive less adult supervision and control. Teenagers try to avoid close supervision by parents and teachers and are more independent; they find places to meet where they are less closely watched. Even at home, teenagers seek privacy and places where they can talk to friends without being overheard by parents and siblings.

- Adolescents begin interacting more with peers of the opposite sex. Whereas boys and girls participate in different activities and friendship groups during middle childhood, the sexes mix increasingly during the adolescent years. Interaction with members of the opposite sex seems to increase at the same time that adolescents distance themselves from their parents.

- During adolescence, peer groups become more aware of the values and behaviors of the larger adolescent subculture. They also identify with certain crowds, which are groups with a reputation for certain values, attitudes, or activities. Common crowd labels among high school students include jocks, brains, druggies, populars, nerds, burnouts, and delinquents. Interestingly, although the adolescent subculture changes over time, these crowds seem to exist in some form across all periods in which the adolescent subculture has been studied.

Brown (1990) has also thought about why peer groups change in these ways during adolescence and suggests several explanations. He maintains that the biological, psychological, cognitive, and social changes of adolescence affect the development of a teenager's peer relationships. Puberty increases adolescents' interest in the opposite sex and contributes to a withdrawal from adult activities and an increase in time spent with peers. While adolescents are in the process of becoming less dependent on their parents, they tend to increase their dependence on peers.

An adolescent's definition of a friend is quite different from a child's. Although both a 5-year-old and a 15-year-old might say a friend is "someone who is close to you," the same words would mean very different things to each. If questioned more carefully about what is meant by "close," the younger child might say it means "someone who lives near you that you play with." "The adolescent would have a much fuller set of requirements, which would not necessarily include living close by. By the teen years, the young person includes many psychological dimensions in the definition of a friend. These would include such things as values and interests in common as well as the idea that a friend is someone to be trusted with very personal information (McNelles & Connolly, 1999; Selman & Schultz, 1990).

Tedesco and Gaier (1988) studied what 204 adolescents in 7th, 9th, and 12th grades appreciate most in their friends. The 100 female and 104 male students gave written replies to ten open-ended questions about friendship values. For example, two of the questions were: "What is it about your best friend that you like most?" and "What are the most important things to consider in judging people?" Three categories emerged from the students' answers: interpersonal qualities, achievement, and physical qualities. A comparison of the responses for the different grade levels revealed what the researchers call "an interesting developmental phenomenon." Students of all ages gave some answers that showed high regard for interpersonal qualities, but the older a student was, the more he or she valued these qualities and the less weight he or she gave to attributes of achievement or physical appearance and dress.

AN APPLIED VIEW

The Power of Peer Persuasion

A nurse friend of ours recognizes the importance of the power of peer relationships in early adolescence. When she wants to get a patient to do something she knows he will want to avoid, she seldom tries to influence him directly. Rather, she discusses the situation with his friends and persuades them to help get him to do the necessary procedure. She finds that if she can get the friends to help influence the patient, she is more likely to be successful than if she presents logical arguments or threatens him.

She must be careful, as she solicits the aid of the friends, not to induce them to feel pity for the patient. She should try instead to help the friends imagine that they, too, could be in the patient's place. Evoking the "golden rule," she encourages them to do for the patient what they would want done for them if they were in the same predicament.

FUNCTIONS OF PEER GROUPS

In contrast with the popular view that peers are a negative influence during adolescence, Hartup (1982) noted that peer influence serves important social and psychological functions. When adolescents do not have the chance to be part of a peer group, they miss out on important learning experiences. Kelly and Hansen (1987) described six important positive functions of the peer group. The group can help teens to:

- *Control aggressive impulses.* Through interaction with peers, children and adolescents learn how to resolve differences in ways other than direct aggression. Observing how peers deal with conflict can be helpful in learning assertive rather than aggressive or "bullying" behavior.
- *Obtain emotional and social support and become more independent.* Friends and peer groups provide support for adolescents as they take on new responsibilities. The support that adolescents get from their peers helps them to become less dependent on their family for support.
- *Improve social skills, develop reasoning abilities, and learn to express feelings in more mature ways.* Through conversation and debate with peers, adolescents learn to express ideas and feelings and expand their problem-

solving abilities. Social interactions with peers give adolescents practice in expressing feelings of caring and love as well as anger and negative feelings.

- *Develop attitudes toward sexuality and sex-role behavior.* Sexual attitudes and sex-role behaviors are shaped primarily through peer interactions (Hartup, 1983). Adolescents learn behaviors and attitudes that they associate with being young men and women.
- *Strengthen moral judgment and values.* Adults generally tell their children what is right and what is wrong. Within the peer group, adolescents are left to make decisions on their own. The adolescent has to evaluate the values of his and her peers and decide what is right. This process of evaluation can help the adolescent develop moral reasoning abilities.
- *Improve self-esteem.* Being liked by a large number of peers helps adolescents feel good about themselves. Being called up on the telephone or being asked out on a date tells adolescents that they are liked by their peers, thereby enhancing feelings of positive self-esteem.

Although the peer group gains influence during adolescence, adults continue to play an important role in adolescents' lives. Adult and peer relationships seem to fulfill different needs in adolescent development (Savin-Williams & Berndt, 1990). Adolescents, for example, often talk with adults about their school progress and career plans. Adults provide an important source of guidance and approval in forming values and setting future goals. With peers, adolescents learn and talk about social relationships outside the family (Boehm & associates, 1999). They talk about more personal experiences and concerns, such as dating and views on sexuality. Adolescents generally feel more comfortable talking with peers about these concerns. They believe that peers will understand their feelings better than adults. Also, teens are afraid that they may appear foolish to the adults whose approval they seek.

> Can you think of any other ways that peer groups can serve a positive function both psychologically and socially?

WHAT'S YOUR VIEW?

Teens, Chat Rooms and the Internet

The internet has created a new sphere for the development of social relationships. However, the internet also presents new risks for teens. The internet has augmented the telephone as a source of interpersonal communication. Many teens spend hours each day visiting chat rooms and communicating with friends in cyberspace. Although there are certainly many positive opportunities available because of the internet, including the possibility of making friends from all over the world, there are also risks involved. Unfortunately, some adults visit teen chat rooms pretending to be teens and take advantage of unsuspecting young people. There are many web sites that are not appropriate for young people but screening devices and the laws regulating the internet are developing much more slowly than the internet. Also, spending too much time on the internet may be limiting teens participation in other extracurricular and social activities. Most experts recommend that parents need to monitor adolescent internet use to ensure safety. What do you think about teenagers and the internet? Do you think that there are problems, benefits? Do you think there are problems with making friends on-line?

AN APPLIED VIEW

Training Teens to Be Opinion Leaders

At the Peer Institute in Boston, 15-year-olds are trained to be role models and opinion leaders for other teens. Through the use of discussion groups, they learn how to "celebrate differences," and how to teach their peers healthy alternatives to destructive practices. Topics include race relations, date abuse, birth control, differences in sexuality and gender roles, helping immigrants, and dealing with stress through humor. Margie Henderson, director of the Medical Foundation Prevention Center which sponsors the institute, argues that since teens have such an influence on each other, there is much to gain from teaching some of them to be a positive influence on others.

PEER GROUPS AND SELF-CONCEPT

Peer groups may also be viewed as important to the development of self-concept (see Chapter 6). The crowd that an adolescent associates with allows her or him to try out a particular identity (Brendgen & others, 2000; Kinney, 1999). Cognitive developments in adolescence are also believed to affect changes in peer relationships. Higher level social-cognitive skills help the adolescent evaluate different crowds on the basis of their behaviors, values, and styles of interaction. Finally, the importance of the peer group may increase because of changes in the teens' social world. With the transition into middle school, students no longer spend time with a single teacher. As relationships with adults decrease, the student is introduced to a larger and often more varied group of peers. Finding one's place in a specific peer group with a particular reputation can offer feelings of belonging and security in the larger, more anonymous setting of the middle school. Kinney (1999) studied a group of teens who resisted the dominant peer group culture of their high school. Through interviews and observations it became clear that there were three primary groups: jock/preppy; headbanger; and nerd. The teens who formed an alternative "hippie" group had been categorized as "headbangers" but actively worked to develop a new group identity that was: accepting of others wishing to join the group, friendly to teens in other groups and open to differences within the group. Kinney (1999) believes that the "hippies" were carving out their own identity by acting on the dominant peer culture to form a new group.

Peers can give young adolescents a clearer idea of how they appear to others at a time when they are very sensitive to the imagined judgments of others (Pugh & Hart, 1999; Talwar & others, 1990). Peers may also give much appreciated support and encouragement in times of difficulty, especially those times when parents are not made aware of the problem. In two recent studies (Vernberg, 1990a, 1990b), the researcher questioned 73 middle-class boys and girls in the seventh and eighth grades of two medium-sized public schools about their experiences with peers. About half the group were new students in their schools because their families had recently moved into the school district. Three areas of peer experience—amount of contact with friends, amount of closeness with a best friend, and amount of rejection by peers—were measured to determine how socially acceptable the participants felt and whether or not they were depressed. Measurements were taken twice, six months apart. Those who had more contact with a peer group of about six members, had a best friend they were close to, and had seldom, if ever, felt rejected by peers were not depressed and felt that they were likeable. The students who had recently moved into the area were more likely to have had difficult experiences with peers. In this relocated group, boys were more likely than girls to suffer rejection by the already-established groups in the school. The boys were sometimes hit or teased in a mean way, whereas girls were not, but all the new students had more experiences that they felt were rejections by their peers than did the long-term residents. These results suggest that a family move may be particularly hard on young adolescents, who must make a place for themselves among already existing groups of peers.

The quality of relationships established in the teenage years has been shown to continue to have an effect much later in life. Pugh and Hart (1999) believe that the social identities that adolescents construct through interaction with their peers influence the development of ego identity. Hightower (1990) reports on data drawn from two large longitudinal studies, one begun in 1928 and one in 1931, comparing the mental health of the participants at ages 13 and 50 as it related to their interpersonal relationships in early adolescence. Out of the original 174 participants, 141 (69 males and 72 females) were still available for the study. Not surprisingly, those who had "harmonious peer relations," "positive relationships with adults outside the family," and "reasonable parental control" as teenagers enjoyed greater mental health in middle age than those who did not have these advantages. Another characteristic that had proved to be an asset for many from their early adolescence onward was the ability to form intimate friendships.

social isolates
Persons who are avoided by others, and typically avoid each other because they have low prestige.

sociogram technique
Establishing a social grouping's range from most popular to social isolate by asking each member to write down the names of the three people in the group whom they like the most and totaling the score.

AN APPLIED VIEW

Using the Sociogram Technique

A camp counselor we know works with a group of 12-year-olds in a summer camp, most of whom are white. However, there are kids in the unit from a variety of other races. These children tend to be isolated from the others, and also have not made friends with each other. The latter is not unusual—**social isolates** typically avoid each other, because each knows the other has low prestige.

Bill employs the **sociogram technique** to help him with this problem (see also Figure 8.2). He passes out 3 x 5 cards to each member of the unit and asks them to write down the names of the three people whom they like the most. He then scores these cards, giving each person who was named first a 3, second a 2, and third a 1. He adds up the scores for each person in the unit.

This gives him a "popularity" score: those with more total points are more popular. As expected, the minority group members are isolates (people who were picked by no one). Several of the white members are also isolates.

From time to time, Bill organizes the unit in groups of four for the purpose of fulfilling assignments. In each of these groups, he places one of the most popular children as well as one of the isolates. Because popular campers are less concerned about the opinions of the other campers, they are usually more likely to be friendly to the isolates. As the kids at the summer camp enjoy each day's activities, it naturally occurs that the isolates are able to identify with and be more friendly with the popular campers. In this way, they gradually gain self-esteem.

STRUCTURE OF PEER GROUPS

close friendships
Friends viewed as being supportive and available to provide support when needed.

In order to understand the importance and influence of peer groups in adolescence, it is necessary to consider the structure of peer relationships. **Close friendships** are the basic building block of the larger peer group structure. Having close and satisfying friendships is positively associated with the social, psychological, and academic adjustment of adolescents. Research suggests that having friends is related to having positive self-esteem (see Chapter 6), not being lonely, behaving appropriately in school, and achieving high grades. Adolescents who have no friends and are disliked by their peers are more likely to drop out of high school, engage in delinquent behavior, and show mental illness in adulthood (Savin-Williams & Berndt, 1990). Having more than one friend can be helpful because over-dependence on one friend can cause problems. However, there is little evidence that having many friends is better than having a few (Berndt, 1989). More important than the number of friends is whether friends are viewed as being supportive and available to provide support when needed (Savin-Williams & Berndt).

cliques
Relatively small and intimate groups of close friends of similar ages, backgrounds, and interests.

Cliques are relatively small and intimate groups of close friends of similar ages, backgrounds, and interests. Members of the same clique hang out together and often enjoy the same activities. In order to become a member of a specific clique, adolescents have to conform to the standards and behavior of that clique (Guerney & Arthur, 1984).

crowds
Peer groups with a reputation for certain values, attitudes, or activities.

Crowds, which were mentioned earlier, are larger groups of cliques, known to others by their interests or reputation. Members of the same crowd may or may not spend time together. Teenagers generally have little difficulty assigning their adolescent classmates to membership in different crowds. They also tend to agree with one another about the crowd affiliation of their classmates.

The crowd that an adolescent is associated with tells a lot about how she or he is viewed by peers. Crowd affiliation is a positive experience for adolescents when they are associated with crowds whose image they accept. Crowd affiliation can be distressful, however, if an adolescent is associated with a disliked crowd (e.g., nerds) or with a troubled crowd (e.g., delinquents). Although some adolescents may be able to select their crowd affiliation, this is not always possible. Not all crowds (e.g., jocks and populars) are open to new members. Adolescents are often assigned to crowds by their peers on the basis of personality, social class, ethnicity, and activities. Adolescents may resist being labeled by others as part of one crowd (Brown, 1997).

During the adolescent years, close friendships, cliques, and crowds seem to vary in their importance and influence. Younger adolescents, for example, place more value on being popular and believe that conforming to peer expectations is more important than do older adolescents (Gavin & Furman, 1989). Belonging to a crowd seems to be most important to younger adolescents. Older adolescents become dissatisfied with the pressures for conformity that come from being a crowd member. For the older adolescent, a few close friendships are more important (Brown & others, 1986). During early adolescence, same-sex friendships are most common. Young adolescents' choices of friends are fairly unstable. By mid-adolescence, close friends become important to share ideas

Late adolescence

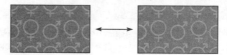

Stage 5: Beginning of crowd disintegration:
Loosely associated groups of couples.

Figure 8.1
The five stages of peer group structure.

Source: From D. Dunphy, *Sociometry, 16:* 235–236, 1963.

Stage 4: The fully developed crowd:
Heterosexual cliques in close association.

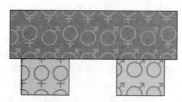

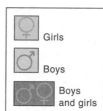

Girls

Boys

Boys and girls

Stage 3: The crowd in structural transition:
Unisexual cliques with upper status
members forming a heterosexual clique.

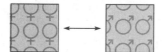

Stage 2: The beginning of the crowd:
Unisexual cliques in group-to-group interaction.

Stage 1: Pre-crowd stage:
Isolated unisexual cliques.

Early adolescence

and secrets with and as a source of information about the opposite sex. By late adolescence, opposite-sex friendships are common. Older adolescents tend to have fewer close friends, but those friendships tend to be long lasting (Guerney & Arthur, 1984; McNelles & Connolly, 1999).

Cliques and crowds change during the adolescent years in ways that increase the influence of the adolescent peer group (Brown, 1989). Dunphy's (1963) classic study of changes in the structure of peer groups through adolescence, even though completed three decades ago, remains the most complete effort to document these changes. Dunphy identified five stages of peer group structure (see Figure 8.1).

In your own words, how would you explain the reasoning behind the five stages of peer group structure and the order in which they occur?

WHAT'S YOUR VIEW?

Do Peer Groups Have a Negative Impact?

An ongoing debate in adolescent psychology is whether peers have a negative influence by pressuring adolescents to behave in undesirable ways. A substantial body of evidence suggests that peers sometimes do influence adolescents in negative ways. Hence the common mother's cry, "What happened to my wonderful child?" referring to marked changes in behavior that she has observed as her child moves through the teen years. Other evidence suggests that the peer group fulfills important functions, and that adolescents who are not members of a peer group are at risk for academic failure and emotional disturbance.

Although these two lines of evidence appear contradictory, a number of explanations have been offered to reconcile these findings. One explanation is that close friendships are necessary and important, but that the larger crowd influences are often negative (Cohen, 1983). Close friends, for example, might encourage an adolescent to work hard at school even though other classmates think that popularity or winning at sports is more important. Berndt (1989) argues that most adolescents do not distinguish between the values of their close friends and of other kids in their school. Friends, Berndt maintains, are just one source of influence. Other sources of influence are teachers, parents, and the media. All sources of influence can be positive or negative and can support or compete with each other in exerting their influence on the developing adolescent.

Another attempt to resolve the opposing views suggests that the values of parents, adolescents, and their friends agree more often than they conflict (Hartup, 1983). Adolescents, especially those who are well-adjusted, often select friends whose values are similar to those of their parents. Troubled adolescents, in contrast, are raised by parents who do not teach their children how to behave properly. As a result, the adolescents display mean and aggressive behaviors that cause them to be rejected by peers. In adolescence, they affiliate with those who share their antisocial attitudes. Deviance does not begin with the influence of antisocial peers, but with poor parenting and social failure (Hartup, 1989).

Think of two adolescents you know, one who is well-adjusted and one who has been experiencing social, psychological, or academic difficulty. Think about the influence of their close friends, peer subculture, and parents. Which of these explanations seems to best explain their successes and failures? Why?

Early adolescence is characterized by stage 1, in which isolated, single-sex cliques exist. The crowd has not yet developed, but emerges in stage 2 during mid-adolescence. The sexes are still isolated. Opposite-gender cliques begin to join to form heterosexual cliques in the 3rd stage, as members begin dating relationships. Once heterosexual activities become common, cliques begin to join with other cliques for social functions and form crowds. In stage 4, the crowd is fully developed. Crowds continue until stage 5 when, in late adolescence, the crowd breaks down into relatively isolated groups of intimate couples. Although it is uncertain whether the process Dunphy described is still accurate today, the transformation from simple classroom-based cliques in elementary school to independent cliques and crowds is evident in secondary schools throughout the country (Brown, 1989). In a recent study, Bukowski and colleagues (1999) found some support for Dunphy's description of early adolescent friendships. These researchers suggest that early adolescent girls show a preference for same-gender friendships. Early adolescent boys show this preference but not to the same extent. Bukowski and colleagues add that very popular and very unpopular young adolescents are more likely than their moderately popular peers to have opposite gender friends.

Savin-Williams and Berndt (1990) reported that before graduating from high school, the great majority of adolescents enter into dating relationships. This finding contradicts the notion that dating is becoming "outdated."

DATING BEHAVIORS

During elementary school, most boys and girls prefer friends of the same sex. By middle school, students show an interest in developing friendships with members of the opposite sex. Cross-gender friendships tend to develop with age. In earlier childhood, cross-gender friendships to the exclusion of same-gender friendships can be a source of social difficulty. However, by adolescence this difficulty is no longer apparent (Kuttler & others, 1999). Before graduating from high school, most adolescents have entered into dating relationships (Savin-Williams & Berndt, 1990). Montgomery and Sorell (1998) found that amount of dating experienced varied by age. Adolescents reported being "in love" about half of the time they reported being in an ongoing relationship. Interestingly, the researchers also found that boys reported falling in love earlier and more often than girls.

What do we mean by dating? Guerney and Arthur (1984) suggest that dating can be defined as follows: "Social activity that allows the opposite sexes to engage in social interactions with non-familial age-mates" (p. 97). Dating can refer to group activities or single-couple experiences. Dating has existed in the United States since the early 1900s and exists in few other countries. In many parts of the world, adolescents have less choice concerning their social and intimate companions and eventual mates.

Perceptions of dating and romance seem to be age and gender related. Young adolescents may confuse cross-gender friendships and romantic relationships, but this distinction is clarified by late adolescence (Shulman & Scharf, 2000). Craig and others (1999) found that descriptors used about relationships changed with age and experience. For example, references to intimacy increased with age and experience. Younger adolescents, especially girls, often develop crushes on an older adolescent or adult. Crushes seem to provide a means of expressing

A SOCIOCULTURAL VIEW

Cultural Influences on Peer Groups

An important theme throughout this chapter is the increasing influence of the peer group during adolescence. We have also emphasized that parents continue to have an important influence in the lives of teens. In fact, some studies have shown that parents have more influence over the academic achievement of adolescents than peers do (Davies & Kandel, 1981).

A recent study (Steinberg, 1990) examined whether the strength of parent and peer influence is the same for adolescents from different cultural groups. Steinberg and his colleagues studied 15,000 high school students from nine different high schools in Wisconsin and California. They found that white, African American,

Latino, and Asian American students showed fewer emotional problems, had higher self-esteem, and were in less trouble for behavior problems when they were raised by authoritative parents (see Chapter 7). Authoritative parenting was most related to academic success for the white students, however. For African American, Latino, and Asian American students, the peer group had a greater influence on school attitudes and behavior, including how much time students spent on their homework, whether they enjoyed school, and how they behaved in class. Fortunately for the Asian American students, their peers valued and positively influenced academic achievement. African American

continued

A SOCIOCULTURAL VIEW

and Latino adolescents had more difficulty finding and joining a peer group that rewarded academic success. Consequently, these youths—often experiencing conflict between the positive values of their parents for academic achievement and the negative values held by their peers—did less well in school.

Fordham and Ogbu (1986) found similarly that African American students felt that in order to be popular, they could not do well in school. When African American students of high ability attended school with only high-achieving students, they were no longer anxious about losing peer support and were more successful.

Steinberg (1990) found that high school peer groups were highly segregated across cultural lines. Most high school students did not know the crowd affiliation of students from other cultural groups. Instead of identifying students as jocks or populars, for example, classmates from other cultural groups were identified only by their ethnic affiliation. Steinberg believes that because of this cultural segregation, many African American, Latino, and Asian American teens have less choice about the peer group they can join. As long as African American and Latino youth are excluded from peer groups that value academic success, it will be harder for them to benefit from the good parenting they have received.

affection to someone outside the immediate family. They are generally intensely felt, but do not last long because other crushes take their place. During early adolescence, the object of affection (a crush or sweetheart) often is not even aware of this admiration. By mid-adolescence, both partners are generally aware of the other's affection. There is an increased openness and communication of affection between partners. Steady dating is most common among late adolescents.

Dating relationships range from informal casual dating to more involved steady dating. Steady dating is a more exclusive relationship with a higher degree of equality, commitment, caring, and intimacy between the partners. The degree of commitment felt by partners who are going steady, however, varies among couples. To some adolescents, going steady is the important step before marital engagement, while to others it is a common and passing arrangement.

Dating fulfills some of the same functions as friendships in addition to some functions unique to the dating relationship. Many social skills learned through friendships, such as showing intimacy and resolving conflicts, are developed further in dating and romantic relationships. Adolescents who date a particularly popular or physically attractive person may gain status in their group as a result of the relationship. Unfortunately, adolescents who do not date or feel they are unable to attract a desirable dating partner may suffer a loss of self-esteem (Skipper & Nass, 1966). Recent authors have added to the list of dating functions. According to Savin-Williams and Berndt (1990), dating also enhances identity development and the development of empathy. Adolescents are able to learn about themselves and explore who they are through dating relationships. Sensitivity and understanding of needs that are different from one's own can also be gained through dating. Because parents have little influence over the choice of dating partners, dating can help adolescents feel more independent and can provide a way of releasing sexual tensions (Guerney & Arthur, 1984).

Most of what we know about adolescent dating results from studies of heterosexual, white, middle-class adolescents. We cannot necessarily generalize their experiences to the dating relationships of lesbian and gay youths and youths from other racial groups and social classes.

DATING VIOLENCE

Dating violence, which has been called "The Hidden Epidemic" by Sousa (1999), can be defined as physically or emotionally aggressive behavior that occurs within the context of a single date or dating relationship. Dating violence often goes unreported and as a result, occurrence statistics are difficult to compile. Dating violence is much more common than most adults like to believe and unfortunately, is often ignored and/or minimized (Sousa, 1999). Sadly, many teen victims feel that they will not be believed or that they will be blamed for their victimization and therefore do not seek help. Some researchers suggest that teenage males and females are equally at risk for dating violence (Molidor & Tolman, 1998). However, females are much more likely than males to be victims of severe violence. Females also report more serious physical and emotional consequences of dating violence (Molidor & Tolman, 1998). Teen dating violence tends to mirror adult domestic violence and can include emotional and verbal abuse, physical assault, rape and murder (Sousa, 1999). There is a pressing need for more training of educators, parents and court officers around issues of teen dating violence.

The Adolescent Subculture

At the same time that they are forming peer groups, most teenagers are also becoming members of an adolescent subculture. This association is much larger and much looser than friendships and peer groups. We turn our attention to its formation in this section.

subculture
Any group that has its own customs (ways of dressing, for example) but is also part of a larger cultural group.

The word **subculture** refers to any group that has its own customs (ways of dressing, for example), but is also part of a larger cultural group. College students might be considered a subculture within the American culture. Some theorists have suggested that it is useful to consider adolescents as members of a subculture (e.g., Friedenberg, 1959).

ADULT ANXIETIES ABOUT THE ADOLESCENT SUBCULTURE

Thoughts of an adolescent subculture bring forth feelings of anxiety among many adults. Parents often worry that they will have less authority over the lives of their adolescent children as the subculture gains increasing influence. One of the worst fears is that under the influence of the youth subculture, teens will be pressured to be delinquent or join a gang (see Chapter 13).

Stereotypes of the adolescent subculture are easily found in the media, especially movies and television. For example, adolescents are portrayed as having styles of dress, language, and music all their own. They are viewed as spending most of their time in groups, whether at school, the mall, concerts, or on street corners. When they are not with their friends, adolescents are pictured as talking on the telephone to each other or listening to their headphone cassette and CD players. Adults sometimes have difficulty telling the difference between negative peer influences, such as delinquent behavior and premature sexual activity, and the more superficial signs of the adolescent subculture, such as taste in dress and music.

Why are images of the adolescent subculture a source of anxiety for many parents? Prior to the 1950s, parents were less concerned about peer influence. Although a spirit of freedom prevailed and many of the young flappers of the 1920s had unconventional attitudes toward adulthood, no one suggested that they had a culture of their own. It was not until the 1950s that social scientists began to suspect that young people were creating a new subculture. Among the earliest and most antagonistic to the cultural mainstream were the beatniks and the Hell's Angels.

Gilbert (1986), a social historian, believes that during the 1950s and 1960s, almost all adolescent problems were blamed on the peer group. He believes that there were a number of reasons for this. First of all, it was not until the 1950s that most American teenagers attended high school. As a result, adolescents began to mix with other teens from more varied social and economic backgrounds. Some parents feared the influence this might have on their children. In addition, teenagers began to listen to the same music and to dress alike, which made them different from their parents and more similar to one another, despite social class background. Gilbert believes that parents of the 1950s and 1960s were much more anxious about the negative effects of the peer group than was necessary. In fact, only a small percentage of adolescents dropped out of society to become beatniks or Hell's Angels.

In the 1960s, the writings of John Coleman (1961), Paul Goodman (1966), and Kenneth Keniston (1965, 1968) added to the public distrust of the adolescent subculture, which was described as challenging adult society. Adolescents were viewed as stubborn, irresponsible, and hedonistic (concerned only with seeking pleasure). They were described as rejecting the values of adults and selecting career goals, friends, and dating partners that they knew would not meet their parents' approval. About members of the adolescent subculture, Goodman concluded, "In principle, every teenager is a delinquent" (p. 19). In the late 1960s, the alienation was viewed by some researchers as so strong that they described adolescents as a "counterculture." Given these negative images of the adolescent subculture, parents' concerns are easy to understand.

How accurate are these views? Images of adolescent rebellion have changed since the 1960s. Certainly no one still believes that most teenagers are delinquents. During the 1970s and 1980s, for example, many adolescents were more concerned with career advancement and obtaining material possessions than with causes. Some writers feared that youths were losing their idealism and selling out to traditional values (Levine, 1980).

In the 1990s, however, an increased commitment to volunteerism and social causes has been observed among some adolescents. While the values of adoles-

AN APPLIED VIEW

The Football Field Versus the Math Field

Educator James Coleman has suggested that teachers and youth workers take advantage of a natural tendency of adolescents: finding joy in intergroup competition. Imagine yourself as a teacher at a high school on a cold, rainy November afternoon. You look out the window at the playing field and there are 50 students happily doing calisthenics. Why are they so highly motivated to do this obviously boring activity? Because they want to experience the ecstasy of being a member of a winning team.

Coleman suggests that we compare the playing field to the classroom, and he argues that we will see two major differences. On the playing field, if one individual does well, everyone on the team is happy, because the team benefits. Therefore, there is peer approval of success. In the classroom, on the other hand, success by one classmate may well mean that the others will suffer. This is especially true if the teacher grades on the curve. That means that there will be a certain percentage of A's, so many B's, and so forth. Therefore, if you help your classmate, you may be hurting yourself!

Another difference is with the rules in force in the two situations. The rules for most sports are reasonably clear, and most of the time, there is little argument as to when and how the rules apply. For example, you don't hear too often that "the referee lost us the game." In the classroom, however, hearing that "the teacher flunked me" is more common. This is because the rules for success are much clearer on the playing field than they are in the classroom. It is easy to believe that the teacher has unwritten rules and is willing to apply them differentially to the least- and most-liked students.

Coleman suggests that we emphasize **intergroup competition** in school, replacing **interpersonal competition**. These *academic* competitions can be between groups within the classroom, between classrooms within the school, or between schools, and they could be based on any school subject. We already have interscholastic competitions, mainly in the form of science fairs and math games, but these are not widespread at the school and classroom level. Coleman even goes so far as to suggest that under this system, teachers would earn a reputation and could be hired and paid on their merits, just as coaches are. In this way, he believes, principals, teachers, and students would all be motivated to help each other be winners.

intergroup competition
Applying the process and the rules of sports to the classroom so that students experience the joy of being a member of a winning team.

interpersonal competition
The traditional classroom model in which one classmate's success may well mean lower grades for others.

cents and their parents are often similar on important issues (Hartup, 1983), adolescents in the 1990s continue to differ from their parents in their choices of music, movies, and styles of dress. Adolescents also spend much more time with peers. According to one study (Nightingale & Wolverton, 1988), adolescents now spend only 7 percent of their waking hours with adults.

The adolescent subculture is most apparent not in political activism or in delinquent behavior, but in choice of hairstyle, clothing, and leisure activities (Fine & others, 1990). These public symbols make one's identification with the youth culture evident both to teens and to adults. The adolescent subculture is the target audience of many merchandisers, who seek to sell the symbols of youth subculture to teens. Adolescents have more leisure time and fewer responsibilities than most adults. Most of the money teenagers earn is spent on luxury products connected to leisure. They tend to save very little.

Merchandisers have taken symbols of youth leisure and rebellion, like blue jeans and black leather jackets, and have turned them into popular consumer goods. The adolescent consumer market is most apparent in the music industry, which has been supported by the cash of teenagers since the 1950s. Commercial music has been growing in importance since the 1920s in defining what it means to be young in America.

While it is clear that the adolescent subculture is alive and well in the 1990s, some social scientists believe it is viewed too simply. Brown (1990), for example, has documented the existence of many peer cultures in the United States. Not all adolescents share the same values and patterns of behavior or even listen to the same music. The subculture to which an adolescent belongs will be influenced by many factors, including the adolescent's gender, ethnicity, neighborhood, parents' education and income levels, and historical and social events. We have already noted how values of the adolescent subculture have changed from the 1950s to the present. Brown believes that the commonly held view of adolescents united in a hedonistic culture that opposes adult society is clearly false. He argues that social scientists need to study more closely the many peer cultures that exist, realizing that peer cultures change as history and society also change.

THE ORIGINS OF SUBCULTURES

How do subcultures get started? Why are there so many of them? Three major theories have attempted to explain the origin of subcultures: the psychogenic, the culture transmission, and the behavioristic theories (Sebald, 1977).

The Psychogenic Model

psychogenic model
A model explaining the rise of adolescent subcultures that assumes that teens are *psychologically disturbed* by the world they are living in and therefore form a subculture to escape or avoid reality.

All subcultures arise when a large number of people have a similar problem of adjustment, which causes them to get together to deal with the problem and help each other resolve it. Modern teenagers receive a much less practical and more abstract introduction to life than teenagers formerly did. They see the world as complex and ambiguous. They often feel it is unclear how they fit in and what they ought to be doing. Many try to escape from ambiguity into a more predictable and controllable world that they create with other teenagers. In the past, this way of creating an identity was used almost solely by delinquent youngsters who were unable to find a respected place in society. Today, escape or avoidance of reality is becoming a more common reaction to personal difficulty. This is known as the **psychogenic model** because it assumes that teens are *psychologically disturbed* by the world they are living in (not mentally ill, but disturbed or made uneasy).

The Culture Transmission Model

culture transmission model
A model explaining the rise of adolescent subcultures which posits that a new subculture arises as an imitation of the subculture of the previous generation.

According to the **culture transmission model**, a new subculture arises as an imitation of the subculture of the previous generation. This takes place through a learning process by which younger teenagers model themselves after adults in their twenties. Magazines, movies, and television programs aimed at teenagers have been effective mechanisms for perpetuating the subculture. Thus, though new forms of behavior may seem to evolve, in actuality most are only new versions of the solutions older people found for their problems when they were teenagers. Not surprisingly, this model argues that teenagers today are not all that different from those of previous decades.

The Behavioristic Model

behavioristic model
A model that sees adolescent subcultures starting out as a result of a series of trial-and-error behaviors, which are reinforced if they work.

The **behavioristic model** sees subcultures starting out as a result of a series of trial-and-error behaviors, which are reinforced if they work. It is like the psychogenic model in that a new group is formed by people with similar problems. It differs in that the psychogenic model views teen behavior as innovative, whereas the behavioristic model sees peer group members behaving the way they do because they have no other choice.

According to behaviorism, teenagers experience adults as "aversive stimuli"; that is, it is painful to interact with adults because in clashes over values, adults almost always win. In an attempt to escape from aversive stimuli, adolescents try out different behaviors with each other. They receive both positive reinforcement (their interactions with their peers make them feel better about themselves) and negative reinforcement (the pain they experience in interacting with adults *stops* when they stop interacting with the adult world).

Another factor in the perpetuation of youth subculture is inconsistent conditioning. For example, teenagers are expected to act responsibly in their spending, but on the other hand, they have to get parental permission for all but the smallest purchases, because they are not legally responsible for their debts. An example of inconsistent conditioning, one that no longer exists in the United States, is when teenagers are asked to fight and possibly die for their country, but are not allowed to vote and help influence their country's policies.

WHAT'S YOUR VIEW?

Which Model Is Best?

Which of these models best explains the origins of the youth subculture? Is more than one of them right? Can you think of a fourth explanation? If you were in charge of the world, what would you do about adolescent subculture?

ELEMENTS OF THE ADOLESCENT SUBCULTURE

Whether it has a similar or different origin from other subcultures, the teenage subculture clearly has a number of common elements, including the following.

- *Propinquity.* An obvious but often overlooked factor in the adolescent subculture is that the members of a group live near one another and know one another prior to joining the group. Gold and Douvan (1969) have suggested that physical closeness is the single most important factor in the makeup of teenage groups.

- *Unique values and norms.* All group members try to overcome limitations that they feel they would have if they did not belong to the group (Atkinson, 1989). Whatever the underlying reason for the group—race, age, politics, ethnic background—its members see a clear advantage in joining with others of similar values. In the teen years, adult domination creates a major motivation for attaching oneself to one's age-mates. In their study of parent and peer attachments, Armsden and Greenberg (1987) found that teens who are devoted members of their group "reported greater satisfaction with themselves, a higher likelihood of seeking social support, and less symptomatic response to stressful life events" (p. 427).

- *Gender differences.* There are some important gender differences in peer group values (LeCroy, 1988). Girls seek satisfaction of a wider variety of emotional needs in their friendships than boys do, but girls also tend to seek just a few close friends. Girls value loyalty, trustworthiness, and emotional support. Boys, on the other hand, seek friendships that help them assert their independence and resist adult control.

- *Peer group identity.* The youth subculture tends to force its members into a deeper involvement with one another. Because teenagers spend so little time with, and derive so little influence from, those older and younger than themselves, they have only one another to look to as models. Today, more adolescents continue their education for a longer period than in the past. As a result, they spend even more time with other youths and less time with adults.

A proposal by Newman and Newman (1976) recognizes the growing importance of this tendency. They suggest that we divide Erikson's identity stage into two stages: early adolescence (ages 13 through 17), called the group identity versus alienation stage, and later adolescence (ages 18 through 22), called the individual identity versus role diffusion stage. This division recognizes that it is necessary to identify with a group in order to achieve a strong personal identity.

Although we agree with their two-stage concept, we disagree with the ages they suggest. On the basis of our view of the research, early adolescence starts at about 11 for females and about a year later for males. Middle adolescence starts at about 14, and late adolescence occurs at about 17, leading into early adulthood at 19. We have chosen 19 because this is the age at which youths have usually been out of high school and into jobs or college for one year. We think this year represents a major turning point in the development of most individuals in the Western Hemisphere.

Evaluating Interactions in the Classroom

Social interactions in the classroom have considerable impact on learning. The class that is divided into small cliques and has many isolated students makes a poor environment for learning. The age of the students affects social groupings; junior high school students are more likely to develop cliques and strongly adhere to their rules than are younger or older students. Teachers can discover these patterns in the classroom and use the information to facilitate learning.

Figure 8.2
The sociogram.

The Sociogram: Questions

1. At which grade level do you think this class is?
2. What predictions can you make for the relationships among the students in cliques A and B?
3. What conclusions would you draw about students C, D, and E?
4. How many isolates are there in this class?
5. Which of the cliques in this class is the leading one?
6. If you were to form discussion groups for this class, which students would you be likely to put in one group?
7. For this sociogram, students were asked to say which students they considered to be good friends. What other questions might usefully be asked?
8. What conclusions would you draw about the learning environment of a classroom such as this?

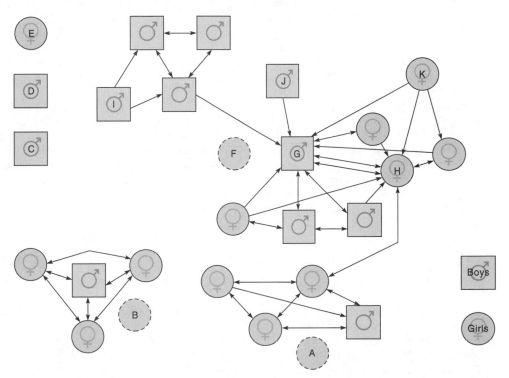

sociogram
Graphic picture resulting from sociogram technique.

One method of plotting social groupings is the **sociogram**, a schematic drawing that details the groupings in a classroom at any particular time. Figure 8.2 is a diagram of the social groupings in one classroom. Students were given 3 x 5 cards and asked to put their name on one side. On the other side, they were asked to list up to three students in the class whom they considered to be their good friends. If they did not feel that anyone in the class was a good friend of theirs, they could leave the card blank. The circles indicate girls and the squares, boys. The arrows between circles and squares indicate choices each student made. Arrows in both directions indicate mutual choice.

Try to answer the questions in Figure 8.2 about the sociogram. Answers to these questions and suggestions for using sociograms are found in the Applied View Box titled "Answers" (see also the Applied View Box titled "Using the Sociogram Technique").

AN APPLIED VIEW

Answers to Questions on the Sociogram

1. This is a seventh-grade class, which is indicated by the fact that the class is divided almost entirely into cliques and that the membership of each clique is primarily either male or female.

2. You can expect some friction and possibly hurt feelings among the girls in groups A and B. It is unlikely, especially in a seventh-grade class, that children of one sex who compete for the friendship of someone of the other sex are likely to maintain their own friendships for very long, and in fact they didn't.

3. These students are clearly isolates. No one picked them and they did not pick anyone else. Student C is Puerto Rican, and students D and E are Chinese. Each of these students recently joined the class and is bused to the school from another neighborhood. It is easy to understand why no other students have picked them yet, as this is a white, middle-class school. It is also typical that they have not picked each other. When a student has a characteristic that makes him or her an isolate from the rest of the class, the student is unlikely to pick another who has the same "negative" characteristic.

4. There are six isolates. Students C, D, and E are isolates; students I, J, and K also are isolates because, even though they have picked others, no one has picked them.

5. Clique F, the members of which are most often picked by students outside the clique, is the leading one.

6. Each of the discussion groups should include at least one of the isolates, as well as at least one of the most popular students. Students G and H, who were chosen most often by other members of the class, are the ones who can best afford to befriend an isolate and to help him or her become accepted by the group.

7. Some other questions might be: Which students do you most like to work with? Which students do you admire the most? Which students does the teacher like best? Which students would you least like to work with?

8. It is unlikely that there is a spirit of cooperation in a classroom that is so uncohesive and sharply divided as this one is. Learning in such a class must be especially difficult for the isolates, who have little chance of becoming accepted into the existing cliques. The teacher can have a considerable effect on this situation. The classroom in which the students feel psychologically safe is one in which the most effective learning is likely to take place. When a teacher establishes groupings that foster the breakdown of social isolation, and when the teacher is open to the contributions of the students, an atmosphere conducive to learning is created. Then, joining together in cliques for protection from perceived threats in the classroom is not necessary.

C O N C L U S I O N S

The first close relationships that children make outside the family are usually with age-mates of the same gender. We used to believe that this was the first step in emotional separation from the family. Now psychologists find that it leads to a different kind of emotional attachment. Girls in particular value these friendships. There has also been concern that new friendships will pressure teens to "go along with the crowd." It appears, however, that most often peer pressure is against misconduct.

It is completely natural that adolescents should form themselves into peer groups that give them the support and approval they need in this fast-changing period in their lives. For the most part, this clannishness does them no harm and has been typical of the age group for a long time.

Many social scientists agree that adolescents are differentiating themselves from adults more thoroughly than ever. Teens are a well-defined subculture that is influenced in its attitudes, values, and behavior mainly by the adolescents themselves and by those adults (musicians, manufacturers, etc.) who benefit from selling them things.

There is reason for concern, as we see a clear diminution of the influence of adults in general. It is, of course, still the case that adult role models such as teachers, clergy, and family friends provide meaningful and appropriate support to teens. Nevertheless, the effect of adult advice and moral leadership is clearly less today than in former times.

Is this decrease in adult influence having a negative effect on teens? If so, it is most likely to be seen in the area of sexuality. In the next chapter we investigate this increasingly important aspect of adolescent life.

C H A P T E R   H I G H L I G H T S

Friendships

- In adolescence, young people begin to depend on their friends more than on their parents for emotional needs.
- Parents may find the increasing time their adolescents spend out of the home with their friends difficult to accept.
- The support and encouragement of parents is still very important to growing teens, who continue to need some measure of adult supervision and concern.
- Friendship patterns of adolescent girls have been studied extensively in recent years. Girls value their skills in forming close friendships very highly.
- In spite of parental worries about the effects of peer pressure on their teenagers, studies have shown that peers often discourage misconduct.

Peer Groups

- Peer groups provide adolescents with a source of social activities and support, and an easy entry into opposite-sex friendships.
- The biological, psychological, cognitive, and social changes of adolescence affect the development of a teenager's peer relationships.
- Peer groups serve to control aggressive impulses, encourage independence, improve social skills, develop reasoning abilities, and form attitudes toward sexuality and sexual behavior. They may also strengthen moral judgment and values and improve self-esteem.
- Peer groups also aid in the development of self-concept, and allow an adolescent to try out a new identity.
- Moving to a new neighborhood and becoming accepted into a new peer group is harder for teens than for children or adults.
- Larger peer groups are structured into close friendships, cliques, and crowds.

The Adolescent Subculture

- A subculture is distinguished by having its own customs that differ in some ways from those of the larger cultural group of which it is a part.
- Adults may fear that the subculture of adolescents will have a negative influence on teens, and may interpret superficial factors of taste in dress and music as warnings of developing delinquency or premature sexual activity.
- In the 1960s, adolescents were depicted as highly rebellious, but this view has changed. Many teenagers of the 1990s are committed to volunteering in community and social causes.
- There is some evidence that adolescent subculture in America is not unitary, but consists of many different peer cultures.
- There are three major theories of the origins of subcultures: the psychogenic model, the culture transmission model, and the behavioristic model.
- Common elements of teenage subcultures are propinquity, unique values and norms, gender differences, and peer group identity.

Evaluating Interactions in the Classroom

- The sociogram is an excellent device for evaluating classroom interaction.

K E Y   T E R M S

behavioristic model 269
cliques 260
close friendships 260
crowds 260
culture transmission model 268

emotional attachment 248
emotional separation 248
intergroup competition 267
interpersonal competition 267
psychogenic model 268

social isolates 259
sociogram 271
sociogram technique 259
subculture 265

W H A T   D O   Y O U   T H I N K ?

1. In what ways is friendship among teenagers today different from the way it was 20 years ago?
2. What would you say to an adolescent who complained about being under peer pressure?
3. How is a teen leader different from an adult leader?
4. Why is the peer group so much more important to some youths than to others?
5. What are some positive and some negative aspects of the adolescent subculture?
6. What is the role of the professional person (teacher, nurse, counselor, etc.) in dealing with the teen subculture?

S U G G E S T E D   R E A D I N G S

Cary, L. (1991). *Black ice*. New York: Knopf. A young African American woman describes her experiences in the majority white culture. Educated in the most exclusive of schools, she manages to break down barriers and improve understanding between her and some wealthy white friends.

Takaki, R. (1987). *From different shores: Perspectives on race and ethnicity in America*. London: Oxford University Press. Provides an overall view of the contributions of Asian immigrants in the United States. We are familiar with the European immigrants' history, but much less so with the history of Asian immigrants.

Terry, W. (1984). *Bloods: An oral history of the Vietnam War by black veterans*. New York: Ballantine Books. African American soldiers constituted the vast majority of soldiers who fought in Vietnam. These are their own words.

G U I D E D   R E V I E W

1. When it comes to fulfilling emotional needs, children depend primarily on their
   a. parents.
   b. friends.
   c. classmates.
   d. teachers.
2. Erikson and Anna Freud believed that for the maturing adolescent, emotional separation from parents was
   a. unhealthy and disruptive.
   b. healthy and desirable.
   c. a sign of depression.
   d. temporary and inevitable.
3. Several older theories, including those of Freud, concerning the friendship patterns of adolescent girls held that a sign of weakness is seen in the girls'
   a. failure to separate from their parents.
   b. lack of competitiveness.
   c. separation from their peers.
   d. competitiveness.
4. Jake gives in to peer pressure on occasion. If Jake is like most teenagers, how often does he feel peer pressure to engage in misconduct?
   a. never
   b. rarely
   c. often
   d. constantly
5. Sarah is a typical 12th-grade student. If you asked Sarah what quality she appreciates most in her friends, she would probably say
   a. achievement.
   b. physical qualities.
   c. interpersonal qualities.
   d. dress.
6. All of the following are important positive functions of the peer group EXCEPT
   a. improving self-esteem.
   b. obtaining emotional and social support and becoming more independent.
   c. learning how to tell right from wrong.
   d. developing attitudes toward sexuality and sex-role behavior.
7. "I meet with my friends every day after school at McDonald's. We all get french fries and pop, and talk about what happened at school, who's dating who, and stuff like that." This person is describing her
   a. peer group.
   b. clique.
   c. club.
   d. crowd.

8. The adolescent subculture is most apparent in
   a. political activism.
   b. volunteerism.
   c. delinquent behavior.
   d. choice of hairstyle, clothing, and leisure activities.
9. Common elements of teenage subcultures include all of the following EXCEPT
   a. gender differences.
   b. propinquity.
   c. belief that there is power in numbers.
   d. unique values and norms.
10. By using a sociogram, teachers can get a better understanding of
    a. who are the isolates.
    b. who are the most popular students.
    c. who can best afford to befriend an isolate.
    d. all of the above
11. An influential contemporary view stresses that although _____ (emotional separation from/emotional attachment to) the family is not expressed in the same way as it is in childhood, it does continue during adolescence.
12. Parental _____ (responsiveness/authoritarianism), a quality made up of such things as being considerate of the adolescent's feelings, may be the most important factor in helping the young person to choose the right group of peers and friends.
13. Opposite-gender cliques begin to join to form heterosexual cliques in the _____ (third/fourth) stage of peer group development as members begin dating relationships.

14. _____ (Close friendships/Cliques) are the basic building block of the larger peer group structure.
15. Larger groups of cliques, known to others by their interests or reputation, are called _____ (crowds/peer groups).
16. The teenage members of a crowd are drawn to it because of _____ (similar activity interests/similar personalities).
17. According to _____ (Dunphy/Tedesco), there are five stages of peer group structure.
18. The _____ (culture transmission/psychogenic) model assumes that teens are psychologically disturbed by the world they are living in and therefore form a subculture to escape or avoid reality.
19. The _____ (behavioristic/inconsistent conditioning) model views subcultures beginning as a series of trial-and-error behaviors, which are reinforced if they work.
20. There are some important gender differences in peer group values. _____ (Girls/Boys) value loyalty, trustworthiness, and emotional support more than _____ (girls/boys), who seek friendships that help them assert their independence and resist adult control.

Answers

1. a, 2. b, 3. a, 4. b, 5. c, 6. c, 7. b, 8. d, 9. c, 10. d, 11. emotional attachment to, 12. responsiveness, 13. third, 14. Close friendships, 15. crowds, 16. similar activity interests, 17. Dunphy, 18. psychogenic, 19. behavioristic, 20. Girls, boys

Teenagers are talking about sex more than they used to, but they're not really doing anything more about it.

In the 1960s and 1970s the statement on the previous page was made by many experts in sexuality. Whether they were right or not, clearly this statement is no longer true. In this chapter, we turn to an aspect of adolescence that has perhaps changed more than any other.

The adolescent years are accompanied by increases in sexual arousal, interest, and behavior. Increased interest in sexuality results from both biological and social factors. Hormonal changes, for example, stimulate sexual interest and motivation. They also contribute to changes in physical appearance and attractiveness to members of the opposite sex. Changes in physical appearance and attractiveness indirectly affect sexual behavior as young adolescents suddenly find themselves becoming the objects of sexual attention. Social factors, including expectations and controls that limit opportunities for sexual behavior, also have an influence on the ways in which sexual interest is expressed. The timing and incidence of sexual expression is largely determined by sociocultural norms in the community, family, and peer group. Considerations, therefore, of what is normal sexual behavior and desirable types and levels of sexual expression at different ages are related to sociocultural norms and moral and ethical belief systems (Miller & others, 1993). Traditionally, culture has had a powerful limiting influence on the age, gender, kinship, and legal relationships among sexual partners. Cultural changes of the last 30 years, commonly referred to as the **sexual revolution**, have changed some cultural norms regarding sexual expression and behavior. In Chapter 3, we discussed the physical changes of adolescence. In this chapter, we examine the ways in which sexuality is typically expressed during adolescence and the importance of social and cultural factors as influences on sexual behavior. Although sexuality is a normal part of adolescent development, sexual behavior is accompanied by significant risks. In the final part of this chapter, we examine the risks of adolescent sexual behavior, including unintended pregnancy, sexual abuse, date rape, sexually transmitted diseases, and AIDS.

After reading this chapter, you should be able to:

- Describe a model of sexual well-being in adolescence.
- Specify the various concerns about adolescents engaging in sexual intercourse.
- Discuss theories concerning the development of homosexual orientation.
- Identify sources of stress associated with the revelation of a homosexual orientation.
- List the sources of sexual information that are available to adolescents; list suggestions for sex education programming.
- Discuss reasons why teenagers engage in premarital sexual activity.
- Discuss the prevalence and causes of unprotected sex in adolescents.
- Explain why some adolescents have the unfortunate experience of acquaintance rape.
- List the symptoms, consequences, transmission, and treatment for sexually transmitted diseases found in adolescents.
- Discuss these issues from an applied, a sociocultural, and your own point of view.

sexual revolution
The extraordinary change in human sexual behavior in the United States that occurred in the 1960s and 1970s.

Sexuality

There are few aspects of human behavior that have changed more in this century than sexual behavior. The situation has changed so much that it is reasonable to call it a sexual revolution. Although the percentage of teenagers who are sexually active is much larger than it was three decades ago, adolescent researchers know relatively little about healthy adolescent sexual development (Koch, 1993). Adolescent sexuality is studied most often from a problem perspective. We are concerned, for example, with finding ways to reduce sexual activity among adolescents and to avoid the negative consequences of sexual behavior, including pregnancy and sexually transmitted diseases. Given the magnitude of these risks, our preoccupation with the negative consequences of teenage sexuality is understandable.

Sexuality, however, is a normal part of adolescent development and contributes to healthy identity development (Brooks-Gunn & Graber, 1999). In addition to biological and social influences, sexuality is intertwined with a variety of developmental factors, including gender roles, self-concept, body image, emotional development, interpersonal relationships, capacity for intimacy, and spiritual beliefs. Developing a satisfactory sexual identity, including a positive view of oneself as a man or woman and as a sexual being, is also a core aspect of identity development (Brooks-Gunn & Graber, 1999; Koch, 1993). According to Selverstone (1989), healthy sexual development contributes to identity development by providing a sense of connectedness with another person, a feeling of being lovable, and a sense of power, joy, and hope. Healthy sexual development is thus a complex but important component of adolescent development.

SEXUAL WELL-BEING IN ADOLESCENCE

In order to promote healthy sexual development among teenagers, we need to first understand what that is. Brooks-Gunn and Paikoff (1993) provide a model of adolescent sexual well-being that emphasizes the interrelationships between sexuality and biological, cognitive, emotional, and social factors. The authors identified four developmental challenges that they believe are related to sexual well-being in adolescence.

- The first developmental challenge is faced during the first half of adolescence and accompanies pubertal change. Healthy sexual development requires a positive acceptance of pubertal changes, including satisfaction with body shape and size and feelings of physical attractiveness. As mentioned in Chapter 3, girls are frequently unhappy with the increase in body fat that accompanies puberty.
- A second developmental challenge is in the management and acceptance of feelings of sexual arousal and desire. Because increases in sexual feelings and arousal normally follow pubertal changes, adolescents need to view these as normal and healthy. Brooks-Gunn and Paikoff believe that this task is often

AN APPLIED VIEW

Television and Adolescent Sexuality

In a comprehensive review of the research on television and adolescents, Jane Brown and her colleagues (1990) found that:

- Adolescents today may have more access to their television sets than to their parents. One survey of an urban area found that two-thirds of the adolescents had their own television set. Most mothers of teenagers work outside the home, at least part-time. The average teen spends 12 hours a week with parents, mostly at meal times and while watching TV. Time spent alone with a father averages 10 minutes a day for adolescents; of that time, half is spent watching television together.

- Television greatly influences teenagers' standards of sexual attractiveness. Some people argue that because we don't have formal celebrations of "coming of age," teenagers create their own rites of passage from childhood to adulthood. An adolescent girl's rite may be trying to make her face and body match the ideals presented on television. The ideals of thinness are difficult for an adolescent to achieve (see Chapter 3 for the physiological explanation). More than one-third of the advertisements appearing on network prime time are "beauty" ads that sell sex appeal as well as products. It is estimated that the current standard of attractiveness portrayed on television is

the slimmest for women that it has been since the 1920s. Today, as in the '20s, the standard leaves adolescent girls at risk for eating disorders as they try desperately to achieve "the look."

- Watching television shapes adolescent expectations about sex and sexuality. The average teenage television viewer is exposed to 2,400 sexual references per year. Teens who watch soap operas and other dramas are exposed to even more. Studies of television have shown that intercourse between unmarried partners and prostitution are found more in action and adventure shows. Long kisses and married intercourse are more prevalent in situation comedies. These images affect adolescent conceptions of their own sexuality. Studies have shown that adolescents who chose to watch programs heavy with sexual content were more likely to have had sexual intercourse in the preceding year. (It is not clear as to whether one causes the other, however.) Research has also shown that adolescents are less satisfied with their own sexual experiences if they think television characters enjoy sexual experiences more than they do.

To what extent do images of sexuality presented in television undermine or promote the model of sexual well-being in adolescence promoted by Brooks-Gunn and Paikoff (1993)?

difficult for adolescents because the discussion of sexual feelings and desire is largely a taboo topic, especially for girls. While messages of sexual desire are widespread in the teen culture, including in music, videos, movies, fashion, and advertising, opportunities for open discussion of sexual feelings are often limited.

- The third challenge involves sexual behavior. Brooks-Gunn and Paikoff maintain that healthy sexuality includes feeling comfortable about choosing whether to engage in individual sexual behaviors, such as masturbation, and in sexual behaviors with a partner, ranging from kissing to intercourse. Healthy sexual development means that these behaviors are freely chosen and not practiced as a result of pressure or coercion.

- The fourth challenge is the practice of safe sex—including practices to avoid unwanted pregnancy and sexually transmitted diseases—among those who have chosen to engage in sexual intercourse.

Brooks-Gunn and Paikoff (1993) suggest that there is no single route to healthy sexuality. For some teens, healthy sexuality may involve abstaining from sexual intercourse but having positive feelings about one's body. Self-exploration without sexual intercourse may be a healthy choice for some adolescents, while engagement in sexual intercourse within a committed relationship during middle or late adolescence and using safe sex practices may be related to sexual well-being for other teens.

Efforts to develop a model of healthy adolescent development are complicated by religious, moral, and legal issues. Religious and moral perspectives may emphasize the importance of postponing intercourse until after marriage. Legal perspectives focus on the rights and responsibilities of minors and parents around issues such as access to contraceptives and abortion. Developmental perspectives examine the ability of teenagers to make informed choices about sexual behavior and its consequences. The developmental model developed by Brooks-Gunn and Paikoff is based on an acceptance of sexual feelings and sexual behaviors among adolescents in the context of variable community, family, and cultural standards, and focuses on developing the competencies to make informed choices.

On what assumptions would you base your model of adolescent sexuality?

Over the past 25 years, the age of first intercourse has dropped considerably and involvement in sexual intercourse by late adolescence is common enough to be considered normative (Brooks-Gunn & Paikoff, 1993). Sexual activity is now increasing among younger adolescents (Alan P. Guttmacher Institute, 1994). Some estimate that 600,000 pregnancies each year can be attributed to girls under 13 (Gershman, 1997). Although masturbation has been less common among adolescent girls than boys, the incidence of masturbation among teenage girls has increased over the past two decades, with about one-third of teenage girls reporting masturbatory behavior by middle adolescence (Chilman, 1983). Some authors (Brooks-Gunn & Graber, 1999; Brooks-Gunn & Paikoff, 1993; Fine, 1988; Koch, 1993) argue that cultural reluctance to acknowledge the normalcy and frequency of adolescent sexual behavior impedes healthy sexual development and increases the negative consequences of teenage sexuality in unintended ways. Because sex is widely practiced among teens, but not openly and positively discussed, teens may be poorly prepared to master the challenges of healthy sexual development. Other adults argue that healthy adolescent sexuality requires that more constraints and inhibitions be placed on adolescent sexuality and that attempts be made to reverse the changes of the sexual revolution. We now discuss the changes brought about by the sexual revolution, normative sexual behaviors in the 1990s, and the social factors that continue to influence the timing and practice of sexual behavior.

THE SEXUAL REVOLUTION

Seeing their elders flounder in a sea of confused values, adolescents have begun to consult one another more often on important matters like sex. Through media exposure, today's teens have access to sexually explicit material and information that was not available to previous generations (Carpenter, 1998). Edgar

Friedenberg, a far-sighted sociologist, saw the beginning of this "revolution" as early as the late 1950s. He described these new attitudes in *The Vanishing Adolescent* (1959). The yearning for love and world peace, perennially scorned by some cynical older adults, began to flourish among late teens and young adults in the 1960s. Many middle-age adults came to the disconcerting realization that they were beginning to admire and even emulate the values of their adolescent children. As the spirit of "love among brothers and sisters" grew, so did more open sexuality. And a great many adults were no longer sure this was wrong.

Although most teenagers are not ready for mature love, sexual feelings are unavoidable, and for many they are extremely frightening. They now face one of the most difficult decisions of life: Shall I say "yes" or "no" to premarital sex? Parents, clergy, teachers, police, and other adults used to be united in their resistance to it. But now, possibly for the first time in history, adult domination of the values of youth has faltered. As Williams (1989) expresses the change:

> ❏ Even as adults in America moderate their sexual activity in response to the threat of AIDS and shifting standards of behavior, teenagers in the last decade have developed a widely held sense that they are entitled to have sex (p. 4).

Evidence that the forces that traditionally kept the great majority of adolescents from engaging in sex are no longer powerful is presented in Table 9.1.

Examine this table. Notice that for the first four years, percentages increased for all age groups. The apparent decline in 1989 may be due to a growing concern about AIDS and other sexually transmitted diseases (STDs) (Koyle & others, 1989). The latest numbers suggest this decline may have been temporary. In a study of almost 50,000 high school students residing in small towns and cities of under 100,000 in the Midwest, 15 percent of 8th graders and 60 percent of high school seniors reported that they had engaged in sexual intercourse two or more times (Benson, 1993). Results of studies vary, but one recent report (Guttmacher, 1994) indicates that 56 percent of 18-year-old females and 73 percent of 18-year-old males are no longer virgins.

What makes a teen ready for sex? For mature love?

Table 9.1	Percentage of Never-Married Females Who Have Experienced Sexual Intercourse				
Age	1971	1976	1979	1983	1989	1990
15	14.4	18.6	22.5	NA	NA	43.0
16	20.9	28.9	37.8	40.2	34.2[a]	54.0
17	26.1	42.9	48.5	55.6	43.9	66.0

NA = not available
[a]Combines 15- and 16-year-olds.

Sources: Benson, 1993; Gullota & others, 1993; Rodman, 1989; Zelnick & Kantner, 1980.

THE JANUS REPORT

The largest, most scientifically designed study of sexuality since the Kinsey Report in the late 1940s was published in 1993. Compiled by Cynthia Janus, M.D., and her husband, Samuel Janus, Ph.D., it is entitled *The Janus Report of Sexual Behavior* (1993). The study covered a wide range of sexual topics, using questionnaire and interview approaches. Its sample of nearly 3,000 adults closely resembles the adult population described in the 1990 U.S. Census. Unfortunately, because of the legal problems involved in questioning children and adolescents, the researchers sought answers only from persons age 18 or older. A number of questions did involve the teen years, however, and among the researchers' most important findings are these statistics:

- Nearly one-fifth of men, but only 7.5 percent of women, reported they had had full sexual relations by age 14.
- Younger women responding to the questionnaire reported much younger ages at which they had their first full sexual experience than older women, thus indicating a continuing downward trend.
- Compared to Phase One of the study (1983 to 1985) 12 percent fewer men and women remained virgins until age 18 in Phase Two (1988 to 1992).
- The South has the earliest ages of sexual initiation and the most reported premarital sex (see Table 9.2).
- Asked whether they had had at least one homosexual experience, 22 percent of men (15 percent lower than Kinsey reported in the late 1940s), 17 percent of women, and twice as many career women as women who were home-makers answered yes.
- An amazing 11 percent of men and 23 percent of women reported having been sexually molested as children.
- Of the women who had had abortions, almost 20 percent had their first abortion before they reached 18 years of age.

Table 9.2	Age of First Full Sexual Experience by Section of the United States			
	Northeast	**South**	**Midwest**	**West**
N =	*558*	*928*	*661*	*573*
By age 10	2%	1%	0%	3%
11–14	12	16	7	11
15–18	44	65	51	51
19–25	39	17	37	34
25+	3	1	5	1
By age 14	14	17	7	14
Over age 18	42	18	42	35

From Janus & Janus, *The Janus Report on Sexual Behavior.* Copyright © 1993 Wiley-Liss, Inc. Reprinted by permission of John Wiley & Sons, Inc., NY.

The results of this study appear to indicate that several types of sexual experience are occurring even earlier than previous studies show.

We will now consider several aspects of sexual behavior during adolescence, including autosexual, homosexual, and heterosexual behaviors, in greater detail.

Autosexual Behavior

❏ If a boy in an unguarded moment tries to entice you to masturbatic experiments, he insults you. Strike him at once and beat him as long as you can stand, etc. Forgive him in your mind, but never speak to him again. If he is the best fighter and beats you, take it as in a good cause. If a man scoundrel suggests indecent things, slug him with a stick or a stone or anything else at hand. Give him a scar that all may see; and if you are arrested, tell the judge all, and he will approve your act, even if it is not lawful. If a villain shows you a filthy book or picture, snatch it; and give it to the first policeman you meet, and help him to find the wretch. If a vile woman invites you, and perhaps tells a plausible story of her downfall, you cannot strike her; but think of a glittering, poisonous snake. She is a degenerate and probably diseased, and even a touch may poison you and your children (Hall, 1904).

autosexuality
The love of oneself; the stage at which the child becomes aware of himself or herself as a source of sexual pleasure and consciously experiments with masturbation.

Psychologists have been debating **autosexuality** since this dire warning of G. S. Hall (see Chapter 2). There is still disagreement in the United States about masturbation, especially among females (Boston Women's Health Book Collective, 1989):

❏ As infants, touching and playing with our bodies, sometimes our genitals, felt good. Some of these experiences were explicitly sexual. Then many of us learned from our parents, then later from our schools and churches, that we were not to touch ourselves sexually. Some of us heeded their messages and some of us did not. But by the time we were teenagers, most of us thought masturbation was bad whether we did it or not. We felt guilty if we did masturbate, or we "forgot" it, or never discovered masturbation at all (p. 166).

Masturbation is probably universal to human sexual experience (Leitenberg & others, 1996; Meston & others, 1996; Weiderman & others, 1996). Although most people still consider it an embarrassing topic, it has always been a recognized aspect of sexuality, legitimate or not. Kinsey, in his 1948 study of male sexuality, found that 97 percent of all adult males masturbated. Jones and Barlow (1990) also found large gender differences in their study of the use of fantasy and masturbation in college freshmen. They asked the students to keep track of specific types of sexual fantasy and sexual behavior. They found great differences among men and women in their autosexual behavior. Jones and Barlow found that approximately 45 percent of the men masturbated once or twice a week compared to 15 percent of the women (see Table 9.3).

There were also gender differences in the use of fantasy with masturbation (see Table 9.4). Men were five times more likely to fantasize while masturbating than were women. It was also found that for men, the more sexually active they are, the more likely they are to masturbate. For college women, it was the opposite. The less sexually active they are, the more likely they are to engage in masturbation.

Table 9.3	Frequency of Masturbation[a]	
Frequency	**Male (%)**	**Female (%)**
Daily	0	0
Twice a week	26.5	4.3
Once a week	18.4	10.6
Once every 2 weeks	14.3	4.3
Once a month	12.2	8.5
Less than once a month	12.2	25.5
Never	16.3	46.8

[a]$x^2(5) = 20.89, p < 0.001.$

From J. Jones and D. Barlow, "Self-reported Frequency of Sexual Urges, Fantasies and Masturbatory Fantasies in Heterosexual Males & Females" in *Archives of Sexual Behavior*, 19(3):269–279, 1990. Copyright © 1990 Plenum Publishing Corp., New York, NY.

Table 9.4	Frequency of Fantasy During Masturbation			
	All subjects[a]		**Only those who masturbate[b]**	
Frequency	**Male (%)**	**Female (%)**	**Male (%)**	**Female (%)**
Always	53.3	9.8	58.5	12.0
Less than 75% of time	15.6	12.2	17.1	20.0
25–75% of time	15.6	17.1	17.1	24.0
Less than 25% of time	6.7	12.2	7.3	20.0
Never	8.9	48.8	0.0	24.0

[a]$x^2(4) = 25.66, p < 0.0001.$ For male, $n = 49$; for female $n = 47$.

[b]$x^2(4) = 20.57, p < 0.0004.$ For male, $n = 41$; for female $n = 25$.

From J. Jones and D. Barlow, "Self-reported Frequency of Sexual Urges, Fantasies and Masturbatory Fantasies in Heterosexual Males & Females" in *Archives of Sexual Behavior*, 19(3):269–279, 1990. Copyright © 1990 Plenum Publishing Corp., New York, NY.

These findings, however, must be taken with a grain of salt. They are based on the personal responses of college students to a questionnaire. It is difficult to know whether their practices are the same as the practices of college students who would not choose to participate in such a study.

Most 4- to 5-year-olds masturbate, are chastised for it, and stop, then start again at an average age of 14 (Masters & Johnson, 1966). If masturbation is so popular, why has it been considered such a problem? For one reason, it is believed that the Bible forbids it. Dranoff (1974) points out that the Latin word *masturbari* means "to pollute oneself." For generations, people have taken as a prohibition the passage in Genesis 38:8 in which Onan is slain by the Lord because "he spilled his seed upon the ground." Dranoff argues that Onan was not slain by the Lord for masturbating, but because he refused to follow God's directive to mate with his brother's wife. Instead, he practiced coitus interruptus (withdrawal from the vagina before ejaculation).

In addition to the biblical restrictions, for centuries the medical profession believed that masturbation caused disease. In 1760, Tissot asserted that a common consequence of masturbation is "locomotor ataxia and early insanity." There are many myths about masturbation: it causes one to go mad; it causes hair to grow on one's palms; it causes one to reject sex with anyone else. No research evidence shows that there are any intrinsically bad effects of masturbation. In fact, the American Psychiatric Association has stated that it should not be considered the sole cause of any particular psychiatric problem (American Psychiatric Association, 1985).

Although most psychiatrists feel that there is no intrinsic harm in masturbation and believe it to be a normal, healthy way for adolescents to discharge their sexual drive, some teens (mainly boys) feel such a sense of shame, guilt, and fear that they develop an "excessive masturbation" syndrome. In this case, masturbation is practiced even though the child feels very bad about it. These feelings are reinforced by solitude and fantasy, which lead to depression and a debilitating sense of self-condemnation.

Most psychiatrists argue that masturbation in childhood is not only normal but helpful in forming a positive sexual attitude. It cannot be obsessive at 4, so it should be ignored at that age. However, it can be obsessive at age 14, and if the parent suspects this to be the case, a psychologist should be consulted.

Homosexual Behavior

Not all sexual activity is between males and females. Same gender sexual activity often occurs as part of the adolescent process of sexual exploration. For most children from the age of 7 to about 13, best friends, the ones with whom they dare to be intimate, are people of the same sex. Feelings become especially intense between ages 10 and 12 when young people enter puberty and feel a growing need to confide in others. It is only natural that they are more trusting with members of their own sex who share their experiences. Occasionally these close feelings result in overt sexual behavior. One study found this to be true more than one-third of the time (Janus & Janus, 1993). In most cases, this behavior results from curiosity rather than from a homosexual orientation. According to other research (Remafedi & others, in press), fewer than 30 percent of high school students who report homosexual experiences identify themselves as homosexual.

homosexual
A person who prefers sexual interaction and intimate, interpersonal relationships with members of the same sex.

Sexual behavior thus differs from sexual orientation or identity. A **homosexual** is a person who prefers sexual interaction and intimate, interpersonal relationships with members of the same sex (Buunk & Van Driel, 1989). The majority of persons who come to identify themselves as homosexual have engaged in heterosexual sex, usually during adolescence (Savin-Williams, 1990). Some youths become aware of their homosexual orientation prior to engaging in homosexual sex or even in any sexual activity at all (Savin-Williams & Rodriguez, 1993). By late adolescence and young adulthood, sexual orientation determines the gender of a sexual partner on a much more consistent basis. Research suggests that one's sexual orientation as either heterosexual, homosexual, or bisexual develops gradually from childhood through adulthood, with approximately 8 to 10 percent of young people in the United States identifying themselves as homosexual or bisexual at some point in their lives (D'Augelli, 1988).

Male homosexuals are often called **gays** and female homosexuals are called **lesbians**. Unfortunately, gays and lesbians live in a society full of stereotypes about homosexuality.

Historically speaking, homosexuality has been surrounded by a number of myths:

gays
A name referring to male homosexuals.

lesbians
A name referring to female homosexuals.

- Male homosexuals are sissies and will never get involved in a fight.
- Boys with frail physiques and girls with muscular physiques have a strong tendency to become homosexuals.
- Homosexuality results from a mental disorder, usually caused by a hormone imbalance.
- Homosexual men have overprotective mothers and rejecting, inept fathers; in lesbians, the reverse is true.
- Homosexuals frequently attempt to seduce young boys. Since they cannot give birth to children themselves, this is the only way they can replenish their ranks.
- You can always tell the homosexual male because he "swishes" like a woman when he walks; looks at his fingernails with his fingers pointing away rather than toward himself; uses his hands in an effeminate way, with "limp wrists"; usually talks with a lisp; and crosses his legs like a woman.
- You can always tell a female homosexual because she has unusually short, cropped hair; refuses to wear a dress; hates all men; is unusually aggressive; and crosses her legs like a man.

What can you do to combat these myths?

Among the most difficult stereotypes confronting homosexuals is the belief that they are "sick." Until 1973, the American Psychiatric Association (APA) listed homosexuality among its categories of mental illness. In its decision to exclude homosexuality from that category, the APA Board of Trustees argued that because it cannot be said that homosexuality regularly causes emotional distress or is regularly associated with impairment of social functioning, it does not meet the criteria of a mental illness (American Psychiatric Association, 1985). Shortly after this pronouncement, however, one wing of the APA gained acceptance of a category called "Sexual Orientation Disturbance," established for those people, homosexual or otherwise, who suffer anxiety from the sexual choices they have made.

Clearly, some of the stereotypes about homosexuals are untrue and unfair. What generalizations, if any, do you believe can fairly be made about all homosexuals?

HOMOSEXUAL DEVELOPMENT

A number of suggestions have been offered about why people become homosexuals. The majority of Americans no longer assume homosexuality is an innate disease (Gallup, 1977). The three most often cited explanations are the psychoanalytic theory of homosexuality, the learning theory of homosexuality, and the biological theory of homosexuality.

psychoanalytic theory of homosexuality
Freud's theory that if the child's first sexual feelings about the opposite-sex parent are strongly punished, the child may identify with the same-sex parent and develop a permanent homosexual orientation.

learning theory of homosexuality
Theory that homosexuality is the result of learned experiences from significant others.

biological theory of homosexuality
A theory that postulates that the reaction of a fetus's brain to sex hormones may create a genetic tendency toward homosexuality.

- The **psychoanalytic theory of homosexuality**. Freud's psychoanalytic theory of homosexuality suggested that if the child's first sexual feelings about the parent of the opposite sex are strongly punished, the child may identify with the same-sex parent and develop a permanent homosexual orientation. Because researchers have noted many cases in which the parent's suppression of the homosexual's Oedipal or Electra feelings was not particularly strong, this theory is not held in much regard today.
- The **learning theory of homosexuality**. The learning theory of homosexuality offers another explanation: Animals that are low on the mammalian scale follow innate sexual practices. Among the higher animals, humans included, learning is more important than inherited factors. According to this theory, most people learn to be heterosexual, but for a variety of little-understood reasons, some people learn to be homosexual. Miller and Dyk (1993) suggest that a boy or girl may learn to respond homosexually through situations that produce homosexual arousal and create different sexual feelings.
- The **biopsychosocial theory of homosexuality**. No direct proof exists that people become homosexual because of genetic reasons. However, a number of studies (Diamonti & McAnulty, 1995; Hu, 1995; LeVay, 1993; Levay & Homer, 1994; Patterson, 1995) have offered some evidence of genetic predisposition (an inborn tendency). Some theorists argue that the fetus's brain reacts to sex hormones during the second through sixth month of gestation in a way that may create a genetic tendency (predisposition) toward homosexuality. Daryl Bem (1996) has found support for a biopsychosocial approach. He found that genetic inheritance influences temperament, which then interacts with a complex set of psychological factors that influence whether a person is heterosexual or homosexual. Because our current culture strongly favors heterosexuality, most people turn out that way. In another cultural setting, things might be different, he believes.

According to Boxer and others (1993), there is no known single explanation for homosexuality in adolescence. Considerable evidence suggests that there may be a biological predisposition (Miller & Dyk, 1993), but this is difficult to prove.

SELF-AWARENESS OF HOMOSEXUALITY

For a long time, psychologists believed that homosexuality did not emerge until adulthood. Recent studies of male homosexuals reviewed in the *Journal of the American Medical Association* (Remafedi, 1988), however, indicate that this belief was the result of interviews with teens, most of whom were ashamed or otherwise unwilling to tell about their feelings on the subject. Beaty (1999) found that gay teens with close relationships to parents came out sooner than those teens who did not feel close to their parents. This finding would reinforce the idea that lack of disclosure or later disclosure is not necessarily a reflection of later emergence of homosexuality. Boxer and colleagues (1999) studied self-identification of 202 gay and lesbian adolescents ages 14-20. These researchers found a mean age of 16.7 for males and 16.0 for females for self-identification. This finding suggests that many young people may self identify as gay or lesbian long before they disclose their homosexuality to others. Current studies are in remarkable agreement that at least one-third of all males have had "a homosexual experience that resulted in an orgasm" at least once during their adolescent years. About 10 percent "are exclusively homosexual for at least three years between the ages of 16 and 55" (p. 222). Janus and Janus (1993) found that 22 percent of men and 17 percent of women have had at least one homosexual experience. On the other hand, other recent studies have found that only 1 percent are exclusively homosexual throughout life (Muir, 1993). Clearly, the truth here is elusive.

The paucity of longitudinal research has left us with little knowledge about the developmental events and experiences that contribute to the realization of a gay or lesbian orientation. Retrospective studies have given us some clues. Most adult homosexuals remember feeling that they were "different" at about 13 years old, the age when most boys are beginning to notice girls. Some gay men describe themselves as more sensitive, more easily hurt, less aggressive, and more interested in artistic enjoyments than were other boys their age (Isay, 1989).

Following the early feelings of alienation and discomfort comes an awareness of sexual feelings (Savin-Williams & Rodriguez, 1993). According to one study of gay men (Rodriguez, 1988), an awareness of same-sex attraction emerged around 11 years of age. Acknowledgment of those feelings as homosexual emerged around age 16, and self-labeling as homosexual occurred at 20 years of age. Among lesbians, awareness of same-gender feelings were evident at age 16 with self-labeling occurring around age 21 (D'Augelli & others, 1987). These ages are averages, of course, and are likely to vary considerably across individuals. Savin-Williams and Rodriguez believe that the age at which youths become aware of their homosexuality is dropping because of the increased visibility of gays and lesbians in our culture. Although most gays and lesbians are aware of their sexual feelings and orientations during adolescence, most do not disclose this to others until the college years or later. According to one study of gay college men, fewer than half had disclosed that they were gay to family members (D'Augelli, 1993). Parents are often not the first people to whom a gay or lesbian adolescent discloses, nor are siblings (Savin-Williams, 1998). Mothers are often told before fathers and mothers report knowing about their children's sexual orientation (whether or not they have been told) more often than fathers (Savin-Williams, 1998). Savin-Williams (1998) believes that disclosure to family members is one of the most critical events in the life of a gay or lesbian teen.

Paul Paroski (1987) interviewed 120 gay and lesbian adolescents about the development of their sexual identity. He asked them where they learned about homosexuality. Whereas females had found out about lesbianism from television and other media, most males had learned about being gay through sexual experiences. When the homosexual teens of both genders described the process they went through in coming to terms with their sexual orientation, however, a striking pattern emerged. Almost all teens described the exact same sequence:

1. The realization of one's desire to have same-sex relationships.
2. The development of guilt, shame, fear of discovery of one's homosexuality, and a sense of engaging in abnormal behavior.
3. An attempt to "change" to heterosexuality through behavior and fantasy.
4. Failure to change sexual orientation and subsequent development of poor self-esteem.
5. Investigation of the homosexual lifestyle through various methods, including sexual activity.
6. Acceptance and development of a positive gay/lesbian identity.

People who achieve an early and positive acceptance of their gay/lesbian identity evidence high self-esteem (Beaty, 1999; Boxer & colleagues, 1999). It is not clear whether coming to terms with one's sexual identity enhances feelings of self-esteem or whether those with high self-esteem are better able to accept their homosexual identity (Savin-Williams & Rodriguez, 1993). It is clear, however, that developing a healthy sexual identity and maintaining positive self-esteem are often more complicated for the gay or lesbian adolescent, who must struggle with problems of stigma and stereotypes. The psychological distress that accompanies the process of achieving a gay/lesbian identity is less when the environment is accepting of homosexuality (Boxer & others, 1993). Adolescents who have difficulty accepting their own homosexuality or fear the consequences of revealing their homosexuality to others may lie or conceal their sexual orientation. This occurs at great cost to feelings of self-esteem and well-being. Communication of one's sexual orientation to significant others is important to the development of a positive gay/lesbian identity. Revealing a homosexual orientation is accompanied with significant risks, however. Loss of friendships, threats to physical safety, loss of parental emotional and financial support, and restriction of career choices are some of the possible consequences. Because of the risks of self-disclosure, and the importance of support it may not be wise to pressure adolescents to acknowledge and reveal their homosexual orientation before they are ready (Savin-Williams, 1998; Savin-Williams & Rodriguez, 1993). There is some evidence that younger gay/bisexual male adolescents (those less than 18 years old) whose parents have knowledge of their child's sexuality may suffer more negative psychosocial consequences—such as having higher rates of substance abuse, dropping out of school, or requiring psychiatric hospitalization—compared to their older (18 years and older) gay peers (Remafedi, 1987). Among older adolescents and young adults, self-disclosure of one's sexual orientation to parents and feelings of psychological well-being are associated with secure parental attachments (Beaty, 1999; Holtzen & others, 1995).

Homosexuality poses special problems for some teens. In response to societal stereotypes and prejudices, gay males and lesbians frequently express feelings of isolation from family, peers, and social/educational institutions (Boxer & colleagues, 1999; 19D'Augelli & Rose, 1990; Martin & Hetrick, 1988). Social isolation and a lack of identity with other people have been associated with suicide, for which gay and lesbian adolescents appear to be at-risk (Harry, 1989; Savin-Williams, 1994). Gibson (1989) found, for instance, that 30 percent of teenage suicides are related to gay/lesbian status. Trotter (1999) suggests that gay and lesbian youth are two to six times more likely to attempt suicide and that 30 percent of gay and bisexual men have attempted suicide. Lock and Steiner (1999) found that gay, lesbian and bisexual adolescents were at increased risk for problems of both mental and physical health. It is important to remember that those psychological difficulties experienced by some gay and lesbian teens are because of societal prejudices and stressors, not because homosexuality is a mental illness. There is little institutional support for homosexual youths. Schools often do not recognize the existence of homosexual adolescents (Boxer & colleagues, 1999). And the adult homosexual community is reluctant to provide services for youths because of the "lingering myth that associates homosexuality and pedophilia [sexual preference for children, a psychiatric disease]" (Gonsiorek, 1988, p. 116). Clearly teenage homosexuals are in need of support from professionals working with adolescents. Studies of gay and lesbian adults in non-clinical settings provide strong evidence that most gays and lesbians enjoy psychological health despite widespread prejudice and societal disapproval (Koch, 1993).

We now turn to a discussion of heterosexual behavior and the development of a heterosexual identity, which concerns the majority of adolescents.

Heterosexual Behavior

heterosexuality
Preference for intimate, interpersonal relationships and sexual interaction with members of the opposite sex.

Heterosexuality is the preference for intimate interpersonal relationships and sexual interaction with members of the opposite sex. Adolescent heterosexual behavior most often begins with less intimate behaviors and progresses to higher levels of intimacy. Findings from a number of studies (McCabe & Collins, 1984; Schwartz, 1999; Smith & Undry, 1985) reveal that non-coital behaviors, such as hugging, kissing, and necking, then fondling and petting, typically precede sexual intercourse. Holding hands and embracing are much more frequent first-date behaviors than are petting and intercourse. In general, boys report greater desire for sexual intimacy on a first date than do girls. As girls' commitment to a relationship increases, however, their desire for sexual intimacy generally increases, becoming similar to the desires reported by boys. However, some researchers (Tolman, 1999) believe that insufficient attention has been paid to desire in adolescent girls because the assumption is often made that adolescent boys are interested in sex and adolescent girls are interested in love. Adolescents who are involved in steady or committed dating relationships report the highest and most intimate levels of sexual activity (Miller & others, 1986). Looking at adolescent attitudes toward timetables for sexual activity, Feldman and others (1999) found that most participants believed that sexual activity should happen with serious rather than casual partners. Interestingly, however, most participants also reported that their own sexual activity had often been with casual partners. This finding raises important questions about the possible discrepancy between attitudes toward sexual activity and actual behavior. By the mid to late teens, most adolescents have engaged in a fairly predictable sequence of sexual experiences, usually including sexual intercourse.

First Coitus

Although sexuality develops throughout life, first intercourse is viewed by most as the key moment in sexual development. When this moment occurs is influenced by numerous factors (Lerner & Semi, 1997). For example, the more the adolescent engages in risk behaviors such as drug and alcohol abuse and delinquency, the earlier first coitus is likely to take place (Costa & others, 1995; Santelli & others, 1998; Savin-Williams, 1995). When do most Americans first experience intercourse? The statistics vary, but all research confirms that this experience occurs at a younger age than it did for previous generations (Annie Casey Foundation, 1995; Besharov & Gardiner, 1997; Carnegie Corporation, 1995; Rosenthal, 1996). By the end of adolescence, more than 80 percent of boys and 70 percent of girls will have been sexually active. In a study of adolescents in the Midwest (Benson, 1993), 30 percent of students in grades 6 through 12 had experienced intercourse two or more times. However, it is important to note that self-report is not always the most reliable way to collect data. In a recent study of self-report honesty, Siegel and others (1998) found that some middle school students admitted that they had not been honest when completing a sexual behavior questionnaire. The researchers found that 78 percent of middle school boys and 94 percent of girls reported honestly. Fourteen percent of the boys who admitted dishonesty overstated their sexual behavior and 8 percent of the girls admitting dishonesty understated their behavior.

Why are adolescents engaging in sex at earlier ages? Some theorists point to changes in social context (Walsh, 1989). They argue that today's youths learn about sexuality much earlier and from more sources than in the past. Sexually explicit magazines, rock music videos, advertisements displaying sexual situations, and movies depicting sexually graphic material are all part of the everyday culture of teenagers today. In the 1950s such materials weren't commonly available. The women's movement and its focus on double standards about sexuality also contributed to the social context of teens today. Early feminists questioned the **double standard**, the attitude that engaging in sexual relations was acceptable for males but not for females. Together these factors create a

double standard
The belief that engaging in premarital sexual relations is acceptable for males but not for females.

Table 9.5	Most Common Reasons for First Coitus			
	Women (N = 412)		Men (N = 261)	
Reason[a]	n	(%)	n	(%)
Love/caring	108	(27.0)	45	(18.1)
Partner pressure	97	(24.3)	17	(6.9)
Curiosity	90	(22.8)	46	(18.6)
Both wanted to	68	(16.7)	47	(19.0)
Alcohol/drugs	39	(9.9)	23	(9.2)
Sexual arousal	36	(9.0)	45	(18.1)
To "get laid"	8	(2.0)	37	(14.9)
Total	446[b]		260[c]	

[a]Respondents could give more than one reason. Infrequent motivations are not included on table.
[b]19 cases were missing.
[c]14 cases were missing.

From P. Koch, *Journal of Adolescent Research*, 3(3–4):345–362, 1988. Copyright © 1988. Reprinted by permission of Sage Publications, Inc., Newbury Park, CA.

Table 9.6	Evaluation of First Intercourse Experience			
	Women (N = 412)		Men (N = 261)	
Evaluation	n	(%)	n	(%)
A disaster	58	(14.0)	10	(3.9)
A disappointment	85	(20.6)	35	(13.6)
Neither positive or negative	69	(16.7)	15	(5.8)
Okay, fine, all right	101	(24.5)	64	(24.8)
Pleasant and pleasurable	85	(20.6)	89	(34.5)
Terrific, fantastic	14	(3.4)	45	(17.4)
Total	412		258*	

*3 cases were missing.

From P. Koch, *Journal of Adolescent Research*, 3(3–4):345–362, 1988. Copyright © 1988. Reprinted by permission of Sage Publications, Inc., Newbury Park, CA.

Research states that youths from stable environments form close attachments, but they are less likely to have sexual intercourse.

social context that provides the developing adolescent with information about sex beyond what is learned from peers and family. It is in this new social context that adolescents make decisions about when to first engage in sexual relations.

However, when adolescents are asked to explain how they decided to first have sex, they do not say "the social context." Instead they talk about relationships and their own sense of their developing self. Tolman (1999) believes that we need to pay closer attention to the relational aspects of sexuality in order to better understand teen sexual behavior. In one comprehensive study, researcher Patricia Koch (1988) asked college students to recall their reasons for first intercourse. The motivation for first sexual intercourse for women listed most often was the desire to express love or care (see Table 9.5). For men, their belief that both parties wanted to engage in sex was listed most often. Although pressure from their first sexual partner was the reason that many women gave for their first coitus (second only to love), the number of men that mentioned this was negligible.

Koch also asked college students to evaluate their first sexual experience. In recalling first coitus, men identified "pleasurable" in first place, while the largest percentage of women described it as "all right" (see Table 9.6).

FACTORS THAT INFLUENCE TEEN HETEROSEXUAL BEHAVIOR

While the experience of intercourse is fairly common by late adolescence, much variability exists among teens in the timing of this sexual experience. Brooks-Gunn and Paikoff (1993) consider intercourse at age 15 or before as early onset. Many studies have focused on factors that affect the timing of first intercourse, with less attention given to the timing of other sexual behaviors.

Hormonal changes increase teenagers' interest in sex (Brooks-Gunn & Furstenberg, 1989). The timing of puberty also affects the amount of freedom a teen is allowed. Girls who mature early are likely to be given more freedom by their parents in setting curfews and choosing friends than are less mature girls. They are also more likely to choose older friends, who are more likely to drink, smoke, and be sexually active (see Chapter 3).

The American Academy of Pediatrics Committee on Adolescence has learned that the fertility rate among girls under 15 years of age has been rising rapidly. Fertility rates among 10- to 14-year-olds are expected to increase further by 2010 (U.S. Bureau of the Census, 1994). Improved nutrition and health care have also contributed to an increase in the potential for young girls to become pregnant (Waltz & Benjamin, 1980).

Many studies have found that family as well as other environmental factors influence adolescent sexual activity (Lammers & others, 2000; Wyatt, 1989). Lammers and others (2000) found that dual parent families, higher SES, academic achievement, religiosity, feeling cared for by parents and other adults and high parental expectations were all associated with later onset sexual activity. It is important to remember that these are not causal factors but rather variables that have been found to be associated with sexual activity.

A number of studies have looked at the relationship between sexual communications among family members and sexual behavior (Dittus & others, 1999; Feldman & Rosenthal, 2000; Raffaelli & others, 1998; Whitaker & Miller, 2000). All have found that parents can have a powerful effect on their children's behavior, including those who are in their late teens, when the parent-child interaction is good and talk about sexuality is direct.

Many other social influences besides the family affect teenage decision making about sex. It has been found that when teens are in steady relationships, they are more likely to engage in sexual relations. "Having a girlfriend or boyfriend may provide opportunity for and pressure toward sexual activity" (Scott-Jones & White, 1990, p. 224).

Research has also shown that adolescents' sexual behavior can be influenced by their thoughts about masculinity and femininity. However, studies of this subject have contradictory findings (Fingerman, 1989; Scott-Jones & White, 1990). Scott-Jones and White found that teens who hold stereotyped views of men as aggressive and dominant and women as passive and submissive are more likely to engage in sexual relations. Females who hold these views often cite persuasion as the reason to begin sexual activity. Males with "old-fashioned" views often see sexual relations as conquests. However, the other study had completely different findings.

Fingerman found that adolescents who were raised in families that encouraged equality between the sexes were more likely to engage in premarital sex than were youths from non-egalitarian families. In egalitarian families, teenagers generally had mothers who were professionals (such as doctors or lawyers). Fingerman concluded that for an adolescent, the mother's life was more than an example of nontraditional sex roles; it was the living out of the values of equality between the sexes. These values, according to Fingerman, encouraged the teenagers in her study to engage in sexual relations before marriage.

Researchers followed more than 1,000 adolescents over a two-year period and found that when teens become sexually active, they become friends with peers who are also sexually active. By doing so, youths find support for their decision. Billy and his associates (1988) also found that for some teenagers, engaging in sex affects their schooling.

Reasoning ability may influence whether a teen becomes sexually active and/or becomes pregnant. When compared with adolescents in England, Sweden, and Australia, teenagers in North America show less understanding of where babies come from (Goldman & Goldman, 1982). This may be because there is less sex education in schools in the United States.

THE MANY NONSEXUAL MOTIVES FOR TEENAGE SEX

In recent years, researchers have begun to pay more attention to the notion that teens engage in sex for many reasons other than the satisfaction of their prodigious sexual drives. In one of the most enlightening articles on this subject, two therapists who specialize in adolescence (Hajcak & Garwood, 1989) conclude that many adolescents use orgasm as a quick fix for a wide variety of other problems. Among these alternative motives for sex are the desire to:

- *Confirm masculinity/femininity.* For some teens, having sex with more than one partner (sometimes called scoring) is taken as evidence that their sexual identity is intact. This is particularly relevant to teens (especially males) who consciously or unconsciously have their doubts about it.
- *Get affection.* Usually some aspects of sexual behavior include physical indications of affection, such as hugging, cuddling, and kissing. To the youth who gets too little of these, sex is not too high a price to pay to get them.
- *Rebel against parents or other societal authority figures.* There are few more effective ways to get even with parents than to have them find out that you are having sex at a young age, especially if it leads to pregnancy.
- *Obtain greater self-esteem.* Many adolescents feel that if someone is willing to have sex with them, others will hold them in high regard.
- *Get revenge or to degrade someone.* Sex can be used to hurt the feelings of someone else, such as a former boyfriend. In more extreme cases, such as date rape, it can be used to show the person's disdain for the partner.
- *Vent anger.* Because sex provides a release of emotions, it is sometimes used to deal with feelings of anger. Some teens regularly use masturbation for this purpose.
- *Alleviate boredom.* Another frequent motive for masturbation is boredom.
- *Ensure fidelity of girlfriend or boyfriend.* Some teens engage in sex not because they want to, but because they fear their partner will leave them if they don't comply.

Using sex for these reasons often has an insidious result. As Hajcak and Garwood (1989) describe it,

> ❏ Adolescents have unlimited opportunities to learn to misuse sex, alone or as a couple. This happens because of the powerful physical and emotional arousal that occurs during sexual activity. Adolescents are very likely to ignore or forget anything that transpired just prior to the sex act. Negative emotions or thoughts subside as attention becomes absorbed in sex. . . . The end result is that adolescents condition themselves to become aroused any time they experience emotional discomfort or ambiguity. . . sexual needs are only partially satisfied and the nonsexual need (for example, affection or to vent anger) is also only partially satisfied, and will remain high. . . the two needs become paired or fused through conditioning. . . . Indulging in sex inhibits their emotional and sexual development by confusing emotional and sexual needs and, unfortunately, many of these teens will never learn to separate the two (pp. 756–758).

This is not to say that adolescents don't experience genuine sexual arousal. They definitely do, but this arousal does not by itself justify sexual activity. These therapists argue that teens need to be taught to understand their motives and to find appropriate outlets for them. In fact, some experts are now recommending sex education that teaches alternatives to premarital sex.

How Do Young Adolescents Learn About Sex?

social learning
The theory that children develop their attitudes toward sexuality by modeling and conditioning.

Early adolescence is an important period for the development of sexual attitudes and the acquisition of sexual knowledge. Peers are usually reported to be the most common source of sex information for adolescents. Researchers studying adolescence argue that the passing on of information about sexuality can best be understood from a **social learning** perspective (Andre & others, 1989) (see Bandura, Chapter 2). According to this theory, children develop their attitudes toward sexuality by modeling and conditioning. Parents in American culture generally offer their children the model that sex should go unseen. Children rarely are given a sense of their parents' sexuality. Parents condition their children to feel ashamed of their own sexuality by discouraging their natural touching of genitals or masturbation. Social learning theorists feel that these actions give children the impression that sex is a taboo subject with their parents. Therefore, during adolescence, teens turn away from their parents for sex information and turn toward their peers. Young teens also learn about sex and sexuality through media exposure. Today's teens have access to a tremendous amount of sexually explicit material and information (Carpenter, 1998). Unfortunately, the media's portrayal of sex and sexuality is often incomplete with little focus on relationships, responsibility and consequences.

Many parents of adolescents today may be responding to pressures and concerns about teenage sexuality. These parents strive to model openness in talking about sex. However, even parents who want to be able to discuss sex with their children may not always communicate openly. Feldman and Rosenthal (2000) found that parents and adolescents evaluated parent communication about sex differently. They found that parents evaluated their communication more positively than their teens. Parents may believe that they are effectively communicating with their adolescents but adolescents may not feel that the communication is effective. This gap in communication may be due in part to a lack of practice in and discomfort with talking about sex.

Who is the *best* source of information about sex? Peers often do not pass on accurate information about sex (Treboux & Busch-Rossnagel, 1990). Teenagers are likely to view their friends' sexual attitudes as more permissive than their own and to act according to how they believe their friends are acting (Brooks-Gunn & Paikoff, 1993). Some programs that are led by peers seem to be quite effective, however, in teaching adolescents about their sexuality (Koch, 1993). Parents are not always clear in their sex information either. Handelsman and her colleagues (1987) found that teens had the same level of accurate knowledge regardless of whether they turned to their parents or to their peers. The researchers found that parents, like peers, often lack either accurate knowledge or the ability to communicate information effectively.

In your opinion, what is the best source of sexual information for teenagers?

AN APPLIED VIEW

Postponing Sexual Involvement

"Postponing Sexual Involvement" is an approach designed for use with 13- through 15-year-old adolescents. It is aimed at reducing pregnancy by decreasing the number of adolescents who become sexually involved. It was developed in Atlanta, Georgia (Howard, 1985).

This program does not offer factual information about sexual reproduction and it does not discuss family planning. Rather, the program concentrates on social and peer pressures that often lead an adolescent into early sexual behavior. Particular emphasis is placed on building social skills to help adolescents communicate better with each other when faced with sexual pressures.

One main difference between this curriculum and most sex education programs is that it starts with a given value, that is, teens should not be having sex at such a young age. Everything in the curriculum is designed to

support this argument. Traditional sex education programs invariably have the implicit goal of reducing teenage pregnancy, but they usually include information on birth control and reproduction so that if young adolescents choose to have sex, they can behave in a responsible manner. This curriculum avoids the double message implied in such traditional programs.

This series on "how to say no" was designed to provide young adolescents with the ability to bridge the gap between their physical development and their cognitive ability to handle the implications for such development. It was not developed to replace the provision of actual factual information about sexuality and family planning.

SEX EDUCATION

Schools can also be sources of information about sex and sexuality. Approximately 80 percent of American high school students have taken a sex education course at school (Dawson, 1990). At the junior high school level, sex education courses often focus on puberty, reproductive anatomy, and dating. Senior high courses include family planning, contraceptive use, and abortion (Blau & Gullotta, 1993). In response to the AIDS epidemic, much-needed AIDS education has been added to the sex education curricula, but often at the expense of more comprehensive and positively focused approaches to human sexuality (Koch, 1993). The development of full-service schools (see Chapter 10) containing school-based health clinics increases the role of schools as a provider of sex education and contraceptive information (Dryfoos, 1995).

Sex education curricula that focus on sexual responsibility and contraceptive use have been controversial due to fears that increased knowledge will promote sexual activity among teens. Contrary to these concerns, sex education does not seem to increase sexual behavior (Furstenberg & others, 1986). Unfortunately, however, the effectiveness of sex education in decreasing sexual activity, increasing contraceptive use, and promoting safe sex practices has not been clearly shown (Blau & Gullotta, 1993). Some good news was indicated by a survey of more than 3,000 high school students in Massachusetts that revealed that students who receive AIDS education are less sexually active and are more likely to use a condom than are students who have not participated in AIDS education (Wong & Hart, 1994). The percentage of teens who were sexually active was almost 10 percent less among teens who had AIDS education and the rate of condom use was 10 percent higher. Unfortunately, gains in knowledge do not always result in changes in behavior. Programs that are most effective do more than provide information. Some evidence suggests that the addition of role playing, assertiveness training, values clarification, goal setting, interpersonal communication, and discussions of decision making and personal responsibility may improve the effectiveness of sex education programs (Blau & Gullotta). Based on what we know about sex education, Koch (1993) suggests some guidelines for promoting healthy sexual development:

> Do you agree with these guidelines? How would you change them?

- Adolescents should be taught what they want to know about their sexuality. Knowledge of sexuality will not increase sexual activity.
- Sexuality should be integrated in family discussions and academic subjects. It should be dealt with as a natural and positive aspect of life.
- The wide range of sexual attitudes and experiences among adolescents should be recognized and efforts should be made to reduce prejudices such as sexism, heterosexism, and racism.
- Learning experiences should address cognitions (what teens know), affect (what they value and how they feel), and behavior (what they do, including communication, assertiveness, decision making, and problem solving).

WHAT·S YOUR VIEW?

What Should Be Taught in Sex Education?

What role do you think schools should play in educating adolescents about sex, contraception, and sexually transmitted diseases? This question has received more attention with the rise in the AIDS epidemic. A national survey (Roper, 1991) found that 96 percent of adults surveyed think that children should learn about AIDS in school (see Figure 9.A).

Do you think respondents would have endorsed school education for sexuality in general as well? Should love, decision making about first intercourse, homosexuality, and premarital sex be discussed in school? Why would some people endorse talking about AIDS, but not endorse talking about these other topics? What's your opinion?

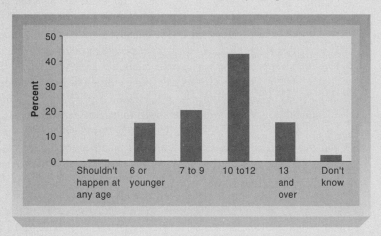

Figure 9.A

Age at which children should learn about AIDS in school

Source: The American Youth Poll, 1991, Roper Organization, New York, NY.

- Adults who work with adolescents need to become informed and comfortable with promoting healthy sexual development.

It seems that sex education alone is not enough. Studies have shown that teens need to learn about sex and sexuality from many sources at once (Andre & others, 1989; Handelsman & others, 1987; Saltz & others, 1995; Treboux & Busch-Rossnagel, 1990). This guards against misinformation and gives adolescents the opportunity to come to their own understanding about issues of sex, sexuality, and contraception.

ATTITUDES TOWARD FAMILY PLANNING

One of the risks associated with teenage sexual behavior is unintended pregnancy. Sexual activity is an obvious cause of pregnancy. Not all teens who are sexually active become pregnant, however. Attitudes toward family planning may help to explain, in part, why some sexually active teens become pregnant and others do not. American teenagers do not appear to be more sexually active than teenagers in some Western European countries (Brooks-Gunn & Furstenberg, 1989). Their greater rate of pregnancy is due partly to less use of contraceptives by American teenagers than by teens in Western European countries.

About one-half of American teenagers do not use birth control the first time they have sex. They say that this is because they do not plan to have sex, do not think about using birth control, do not have a method of birth control, and/or think that pregnancy is impossible (Zelnick & Shah, 1983). Most teens do not plan to become sexually active. They describe it as something that just "happened" to them (Chilman, 1983). Half of all first pregnancies happen within six months after sexual activity begins. Many teens continue not to use birth control after their first sexual experience (Roye, 1998).

It seems that many of the factors that influence whether teens use birth control are the same as those that influence sexual activity. Teens who do not get along with their parents, who are not doing well in school, and whose friends are teenage parents, are less likely to use birth control methods. Teenagers who are able to talk with their parents about sexuality and family planning are more

A SOCIOCULTURAL VIEW

The Most Important Value

Summer heat did not stop 100 members of Boston's Latino community from meeting inside the Huntington Avenue YMCA to discuss the economic and social problems facing Latino youth. Mayra Rodriguez-Howard, director of the Massachusetts Committee of Hispanic Affairs, placed the responsibility for sex education first on parents: "They should make the effort to educate themselves about sexuality so they can be in a position to educate their children."

Norma Wallace, mother of three Latino children and counselor to pregnant teenagers and single mothers, agreed that talking about sex in the home should not be a "sin." Because of the taboo that revolves around sex, many children learn about it in the streets and many times get the wrong information.

Nilda Rios, a Boston Juvenile Court probation officer, commented that peer pressure causes many children to abandon the values their parents have taught them. "When both parents have to work they can't give their children the attention they deserve, and also many homes have only one parent."

Rodriguez-Howard countered that the problems of youth should not be attributed to a loss of values, but to a lack of economic opportunity. "Youth today still possess the most important value—the desire to work hard to succeed."

likely to use birth control consistently and effectively (Jorgensen, 1993). Unclear understanding about how and when pregnancy occurs also results in poor use of birth control. Feelings of guilt and fear and concerns about the negative effects of contraceptives lead some adolescents away from their use (Furstenberg & others, 1983). Whether or not a girl has a steady boyfriend and how often she has sex also affects whether she uses birth control. Some adolescent girls need time to admit to themselves that they are sexually active. Having been taught that "good girls don't," they find that going to a doctor or a family planning clinic for birth control can be uncomfortable and creates emotional conflict. They have to admit that they *plan* to have sex (Children's Defense Fund, 1987). Because sexual desire among girls is generally not acknowledged, some adolescent girls have few strategies for handling it (Brooks-Gunn & Paikoff, 1993). Girls are frequently portrayed as the victims of male sexual desires. They are responsible for protecting themselves from the sexual advances of their male companions. Traditional social controls, such as supervision and chaperonage, no longer exist, but feelings about obtaining birth control remain ambivalent. As a result, girls continue to engage in unprotected sexual activity.

Most of what we know about teenagers' attitudes about birth control has been learned from teenage girls; we know very little about boys' attitudes. This is particularly unfortunate because 40 percent of teenage girls rely on their male partners to provide birth control (Brooks-Gunn & Furstenberg, 1989). Boys generally know little about reproduction, the menstrual cycle, and contraceptive methods (Blau & Gullotta, 1993). Given that a single male teenager has the potential to impregnate multiple teenage girls, and that both males and females are involved in sexual activity that leads to pregnancy, we cannot afford to ignore adolescent boys in our research and contraceptive education.

Unhealthy Sexuality

Adolescents are at risk for sexual victimization, including rape and sexual assault, from several sources—family members, friends, and strangers (Miller & Dyk, 1993). These unwanted experiences are clearly not a normal or desirable part of adolescent sexuality. Unfortunately, they are common (Elders & Albert, 1998). Sexual experiences among young adolescents are usually forced. At age 13 and younger, more than 60 percent of sexually experienced girls report forced intercourse. That rate drops to 25 percent at age 15 and 10 percent at age 16 (Alan P. Guttmacher Institute, 1994). Although the legal definitions of rape and sexual assault vary from state to state, they usually include non-consent of the victim, sexual relations, and the use of threat, force, deception, or intimidation (Davis & others, 1993).

Sexual Abuse

Adolescents are typically abused by someone they know and trust (Miller & Dyk, 1993). It is often just a continuation of abuse that started during childhood. The most common type of serious sexual abuse is incest between father and daughter (Alexander & Kempe, 1984) or stepfather and stepdaughter. This type of relationship may last for several years. The daughter is often manipulated into believing it is all her fault and that if she says anything to anyone, she will be seen as a bad person, one who may even be arrested and jailed. The outcome is often another adolescent statistic: a runaway or even a prostitute (Friedman & others, 1996; Gary & others, 1996; Jezl & others, 1996; Karp & others, 1996) (see Chapter 13). Sexual victimization prior to adolescence also contributes to the risk for delinquency as well as other risks for other emotional and physical difficulties in adolescence (Elders & Albert, 1998; Widom, 1994).

Russell (1995) believes that incest which is often covered up should be seen as a form of torture and treated as a human rights violation. Elders and Albert (1998) call for improved identification by health care professionals of young women who are at-risk of victimization and those who have been victimized, so that we may better understand and intervene to help victims receive professional attention earlier, which may alter the long-term effects of abuse.

Acquaintance Rape

"Beer goggle" sex, which refers to sexual relations that would not have occurred if the couple had not been drinking, is taking place among younger and younger teens.

beer goggle sex
Sexual activity resulting from drinking too much alcohol.

Acquaintance rape generally occurs within the context of adolescent social activities. The aggressor is typically a boyfriend or a more casual acquaintance of approximately the same age. Because this type of rape often occurs within a dating relationship, it is sometimes referred to as date rape. Being raped by a dating partner or friend is no less damaging psychologically than being raped by a stranger (Davis & others, 1993). In most cases, females are the victims, but there are incidents of male victimization as well (Miller & others, 1993). One survey (Koss & others, 1987) found that 54 percent of college women reported some form of sexual victimization, including rape (15 percent) and attempted rape (15 percent). Twenty-five percent of college males admitted to sexual aggression, including rape (4 percent).

Because heavy drinking has increased among adolescents in recent decades, a problem related to date rape has intensified. College students refer to it as **"beer goggle sex."** "Beer goggles" is another term for distortion of judgment by alcohol. Beer goggle sex is sexual activity that the person regrets the next morning and that would not have occurred if the person had not been drinking heavily. This phenomenon has not been studied sufficiently at this time, but we predict that scientists will be looking at it more closely in the years to come. One study (Canterbury & others, 1993) among recent high school graduates suggests

In your opinion, what beliefs and assumptions about males and females are used most often to justify date rape?

that those who drink during high school (two to four times per week) are more likely to experience date rape than those who do not drink or who consume less than one alcoholic beverage per week.

Acquaintance rape occurs with disturbing frequency among high school students. Davis and associates (1993) obtained questionnaire data from 237 9th- to 12th-grade high school students in three metropolitan Louisiana high schools, with 20 percent reporting that they had been involved in a situation where sex was forced. This was more common among girls (26 percent) than boys (11 percent). Only half of the teenagers reporting the experience of forced sex had told anyone of their experience. In another high school survey, Feltey and associates (1991) found similarly that 20 percent of high school students had engaged in sex when they did not want to. Because so many adolescents do not disclose their abuse experience, professionals working with teens should be alert to behavioral signs of sexual victimization. Some adolescents showing symptoms of depression, psychosomatic illness, irritability, avoidance of men, loss of confidence, nightmares, fears of going outside or inside, or anxiety may be reacting to an acquaintance rape.

Cultural beliefs about the sexes are partially blamed for the prevalence of acquaintance rape (Workman & Freeburg, 1999). In an Applied View box, we compare myths and realities about date sex. Many of these myths likely contribute to the incidence of date rape. Boys and men are more likely to agree with rape tolerant attitudes than are girls and women (Holcomb & others, 1991). Sixty percent of the high school boys surveyed by Davis and associates (1993) indicated that it is acceptable for a boy to force sex on a girl in some situations, such as when the couple had been dating for a while or had had sex in the past. In another study of adolescents in Los Angeles (Goodchilds & others, 1988), 12 percent of high school girls and 39 percent of high school boys reported that it was acceptable for a boy to pressure a girl into having sex if the boy had spent a lot of money. One-third of the males thought that pressure was acceptable when the male was highly aroused and could not stop, and 54 percent thought that

AN APPLIED VIEW

GHB (Gamma Hydroxybutyric Acid)- The Date Rape Drug

Recently, three teenagers in Detroit were convicted of killing a 15 year old girl who died of an overdose of GHB that had been slipped into her soda. GHB, known also as the date rape drug has been used by rapists to produce a sense of deep relaxation, happiness, lack of inhibition and unconsciousness in victims. GHB is tasteless and is usually slipped into a victim's drink. Drinking at social gatherings of high school and college students increases the risk that GHB can be used. Young people are being cautioned by law enforcement to be extra careful to avoid such risks.

pressure was acceptable when a girl first consented to intercourse and then changed her mind. Other common reasons given by high school students for justifying sexual coercion include the girl's behavior (she has done this with other guys; she is wearing sexy clothes; she says no but does not push him away), use of drugs or alcohol, and opportunity for the sexual relationship to take place (she goes to his house when his parents are not home or she invites him to her house when her parents are not home) (Feltey & others, 1991).

The beliefs used to justify sexual coercion reflect traditional sexual stereotypes regarding the uncontrollable sexual urges of males and an entitlement to sex based on length of dating relationship or amount of money spent. Girls are blamed for seducing boys and not recognizing the power of male sexuality. Among college-age men, almost one-half have reported that some women ask for and enjoy rape, one-third agreed that sexual aggression is closely tied to masculinity, and one-quarter agreed that rape is often provoked by women (Holcomb & others, 1991). Date rape prevention programs have been designed to alter the beliefs that perpetuate date rape.

Sexuality in the lives of late adolescents and young adults in the last decade of this century is very different from that in earlier decades (although perhaps not so different from several centuries ago). What is the relationship between this fact and the problem covered in the next section, sexually transmitted disease? This is a complex question.

AN APPLIED VIEW

Myths and Realities About Date Sex

Myths About Sex on Dates

1. A first date constitutes an intimate relationship.

2. A relationship means that the woman and the man should be having sex.

3. Two people involved in a physical relationship have the same motives.

4. A woman should say "yes" so a man doesn't dump her.

Realities About Sex on Dates

1. A first date often takes place between people who hardly know one another.

2. While a relationship may be an appropriate context within which two people can have a mutually caring physical relationship, there are no "shoulds" when it comes to having sex.

3. People engage in physical relationships for many different reasons and with different agendas.

4. Fear of "being dumped" is NOT a good reason to engage in sexual activity.

continued

AN APPLIED VIEW

Myths About Sex on Dates

5. It is a woman's responsibility to do what a man wants.

6. If a man takes a woman out, she "owes" him. Or if he pays, drives, is older/more experienced. Or just because he asks.

7. Drinking relaxes people, makes them less inhibited, better able to make decisions.

8. It is immature for a man or woman to say "no."

9. If a man does not score, there is something wrong with him.

10. If a woman says "no" to sex, it is a rejection of the man.

11. Once a man gets physically aroused, it is impossible for him to stop.

12. Once a woman engages in physical contact, she can no longer say "no."

13. It is never okay for a woman to say "yes."

14. Talking about having sex is not as romantic as just letting it happen.

15. It is okay to pressure someone for sex as long as there is no physical force.

16. It is okay to physically force someone to have sex if you think that they really want to.

17. Pressured/forced sexual contact is okay as long as it doesn't involve intercourse.

18. Forced sex and rape are different.

19. Rape does not happen between people who know each other.

20. If a woman says "no," she doesn't really mean it.

Realities About Sex on Dates

5. It is a woman's responsibility to herself and the situation to do only what she feels comfortable doing.

6. A woman never "owes" a man sex.

7. Drinking may make people feel more relaxed, but it impairs judgment. Decisions made with impaired judgment may not feel okay in retrospect.

8. Maturity involves the ability to take action on one's own behalf. It is not immature to say no to something that one does not want to do.

9. The pressure for men to "score" creates unnecessary expectations on the part of both men and women.

10. Saying "no" to sex is not an implicit rejection.

11. Though it may be unpleasant for a man to stop once he is aroused, it is NOT impossible.

12. Although thinking about and articulating one's intentions and desires in advance may help avoid misunderstandings, a woman can always say "no."

13. Women need to decide what they want to do about sex.

14. Just "letting sex happen" is not romantic, it is irresponsible and potentially dangerous.

15. Emotional pressure, even without physical force, is damaging.

16. Physical force is never okay; it is not up to one person to decide what another person wants.

17. Intercourse does not have to happen in order for sexual behavior to be coercive and therefore inappropriate.

18. Forced sex is rape.

19. Acquaintance/date rape is more common than stranger rape.

20. "No" means no. A women needs to be prepared to say it if she means it and a man has to be prepared to hear it and act accordingly.

From G. Hull, D. Margolis, and J. Dacey, an unpublished paper. Reprinted by permission of John S. Dacey.

Sexually Transmitted Diseases

AIDS

sexually transmitted disease (STD)
A class of diseases that are transmitted through sexual behavior.

AIDS (acquired immune deficiency syndrome)
A virus that attacks certain cells of the body's immune system, leaving the person vulnerable to fatal complications such as cancer and pneumonia.

Not long ago, when people thought about **sexually transmitted disease (STD)**, gonorrhea came to mind. In the 1970s, it was herpes. Today, **AIDS (acquired immune deficiency syndrome)** causes the most concern (Forstein, 1989).

AIDS was first diagnosed at Bellevue-New York University Medical Center in 1979 and has quickly approached epidemic proportions. What is known about AIDS is that a virus attacks certain cells of the body's immune system, leaving the person vulnerable to any number of fatal afflictions, such as cancer and pneumonia. In addition, the disease can directly infect the brain and spinal cord, causing acute meningitis.

More than 166,000 persons died from AIDS between 1982 and 1992 (U.S. Bureau of the Census, 1994). AIDS now ranks 15th among the leading causes of morbidity and mortality in children and young adults. It is first among 15- to 24-year-old males. Keep in mind that these are only reported cases of the full-blown AIDS disease. A combination of underdiagnosis and underreporting makes these estimates conservative at best. Studies suggest that about 50 percent of people who are infected with the virus will develop full-blown AIDS disease within 10 years and that 99 percent will eventually develop the disease (Lifson & others, 1989). According to U.S. government data, the greatest number of deaths is among those 30 to 39 years old. The number of deaths among 13- to 29-year-olds increased from 1,329 in 1985 to 3,809 deaths in 1992 (U.S. Bureau of the Census).

AIDS is reported to be increasing among intravenous drug users, women, children, the elderly, African Americans, Latino Americans, heterosexuals, and in small cities and rural areas (Catania & others, 1989; CDC, 1989a; Kirkland & Ginther, 1988). The only segment of society in which the incidence of AIDS is decreasing is homosexuals with no history of intravenous drug use, although this group still represents the single largest at-risk group (CDC, 1989b).

The AIDS virus is transmitted through the transfer of substantial amounts of intimate bodily fluids such as blood and semen. The virus is most often transferred through sexual contact, the sharing of hypodermic needles, and much less likely, through blood transfusions (a test for AIDS is now available at blood banks and hospitals). In addition, the virus can be transmitted from an infected mother to an infant during pregnancy or birth. In some central African countries, where AIDS is thought to have originated, the virus is found equally among men and women throughout the population.

Although there is no cure for AIDS, the disease can be effectively controlled through preventive measures. Use of condoms during sexual intercourse and clean, unused needles during intravenous drug use can drastically reduce the risk of contracting the disease. After a slow start, large-scale education efforts by grassroots organizations as well as by state and federal government agencies have begun to get these messages out, but the problems remain extremely serious.

The AIDS virus has been highly identified with a few select groups. If you're not gay or a drug user, you might think you don't have to consider preventive measures. However, a person exposed to the AIDS virus may not show any symptoms for up to 15 years! Further, this same person can expose other people to the virus during this incubation phase. Some people have reacted to this by becoming more particular about their sexual partners. Monogamous relationships have been on the rise again during the 1980s after the liberated days of the sexual revolution of the 1960s and 1970s. And the educational message seems to be getting through because condom use is increasing. But many people still ignore the dangers, and the consequences could be years away.

Some researchers believe that adolescents are a high risk group for AIDS because of the risky nature of teen sexual behavior (especially having multiple partners because of early onset sexual activity). Despite increased education and knowledge and intention to practice safe sex, many teens are still engaged in high risk behavior (Canterbury and colleagues, 1998).

OTHER SEXUALLY TRANSMITTED DISEASES

Often lost in the public focus on the burgeoning AIDS problem is a truly epidemic increase in the prevalence of other STDs. Because of its fatal nature, AIDS gets most of the press and the major funding. But STDs such as chlamydia, gonorrhea, pelvic inflammatory disease, herpes, syphilis, and hepatitis B are running rampant compared to AIDS, particularly among adolescents. The effects of such venereal diseases range from mildly annoying to life threatening.

Some of the more common STDs (other than AIDS) are:

chlamydia
The most common STD; it often has no symptoms and is diagnosed only when complications develop.

• *Chlamydial infection.* **Chlamydia** is now the most common STD, with about 4 million new cases each year (Sexually Transmitted Diseases Sourcebook, 1997). In one state, African American and Latino female teens have rates of chlamydia infection that are more than ten times higher than rates reported in white female teens (Massachusetts Department of Public Health, 1991). Chlamydia often has no symptoms and is diagnosed only when complications develop. It is particularly harmful for women and is a major cause of female infertility, accounting for 20 to 40 percent of all cases (Hersch, 1991). Untreated, it can lead to pelvic inflammatory disease (described later). As with all these diseases, it can be transmitted to another person whether symptoms are present or not. However, a seven day course of antibiotic treatment has been found to be very effective.

AN APPLIED VIEW

Sexually Transmitted Disease Prevention Education

The Massachusetts Department of Education (1994) completed a study of health education issues concerning sexually transmitted diseases (STDs). High school students' opinions concerning the strengths and weaknesses of STD education may be helpful in improving STD prevention education. As you read their comments, think about the approach you would use in teaching adolescents about STDs.

Perceived Strengths of STD Prevention Education

• I liked that the teacher treated us like adults. She was so straight with us: "You get it from this."

• It's comforting to have questions answered that are not even asked because it's like you want to know something but you don't know what questions to ask.

• One of the best parts was the chance to come after school for confidential conversations with the health teacher.

• It was real interesting to meet a person with full-blown AIDS, someone suffering from it: he told us how he got it and was real open; it touched me, made

me feel fortunate that I know what I know because he didn't know that then. I think this should go on in all schools.

Perceived Weaknesses of STD Prevention Education

• I remember last year, they handed us a sheet and it just listed all the STDs and symptoms and prevention, but we never talked about it.

• Our health teacher said "This is this and that's all. O.K., next subject."

• It doesn't mean anything that health's mandatory in my school because the assistant principal takes a lot of kids out. They just think: "Oh, health class, what are you going to do there? You're not going to learn anything."

• In our freshman health class they were able to bring in models of all the drugs that you could possibly have, but she wasn't allowed to bring in condoms, diaphragms, and she would rather have done that than drugs. She could only bring in a condom if she left it in the box . . . and she couldn't show us how to put it on or anything.

---

gonorrhea
Well-known venereal disease; the most common symptoms are painful urination and a discharge from the penis or the vagina.

pelvic inflammatory disease
Disease that often results from chlamydia or gonorrhea, and frequently causes prolonged problems, including infertility; symptoms include lower abdominal pain and a fever.

• *Gonorrhea.* The well-known venereal disease **gonorrhea** infects approximately 750,000 persons per year (Sexually Transmitted Diseases Sourcebook, 1997) and is common among adolescents and young adults. Gonorrhea is caused by bacteria and can be treated with antibiotics. When penicillin was introduced in the 1940s, the incidence of gonorrhea declined dramatically. The most common symptoms are painful urination and a discharge from the penis or the vagina.

• *Pelvic inflammatory disease (PID).* **Pelvic inflammatory disease** frequently causes prolonged problems, including infertility. It is usually caused by untreated chlamydia or gonorrhea. These infections spread to the fallopian tubes (see diagram in Chapter 3), resulting in PID. The scarring caused by the infection often prevents successful impregnation. There are more than one million new cases per year in the United States (Sexually Transmitted Diseases Sourcebook, 1997). Women who are most likely to get it are those who use an intrauterine device for birth control, have multiple sex partners, are teenagers, or have had PID before. PID is so widespread that it causes $2.6 billion in medical costs per year!

genital herpes
An incurable sexually transmitted disease with recurring pain.

Hepatitis B
Viral disease that is transmitted through sexual contact and through sharing infected needles; symptoms include high fever and aches; liver damage may result.

syphilis
A sexually transmitted disease that in its early stage has no symptoms; if not treated, the disease can lead to death.

- *Genital herpes.* **Genital herpes** is an incurable disease, with about 500,000 new cases every year. It is spread by a virus during skin-to-skin contact. Its major symptom is an outbreak of genital sores, which can occur as often as once a month. It is estimated that there are about 30 million people in this country who suffer from this infection. Unlike chlamydia, problems associated with herpes are mainly emotional and social rather than medical (Hersch, 1991). People with herpes often experience embarrassment and low self-esteem about their bodies.
- *Hepatitis B.* There are about 300,000 new cases of **Hepatitis B** in the United States each year (Sexually Transmitted Diseases Sourcebook, 1997). This viral disease is transmitted through sexual contact or through the sharing of infected needles. Although a preventive vaccine is available, those who are most at risk for Hepatitis B (intravenous drug users, homosexual men, and inner-city heterosexuals) usually do not have the vaccine readily available to them.
- *Syphilis.* Like gonorrhea, **syphilis** is no longer the killer it was before penicillin. However, this sexually transmitted disease still accounts for 134,000 new cases per year. It is caused by bacteria. Its first sign is a *chancre* ("shanker"), a painless open sore that usually shows up on the tip of the penis and around or in the vagina. This must be treated with antibiotics because the disease can be fatal.

Studies have shown that adolescents are at great risk for STDs. In addition, this is an age group that is particularly difficult to educate in any area concerning sexuality. The obstacles to education include individuals who refuse to take the information seriously and parents who won't let the information be taught.

The AIDS crisis and the STD epidemic have several features in common. On the negative side, misconceptions contribute to both problems. Many young people believe that only promiscuous people get STDs and that only homosexuals get AIDS. Having multiple sexual partners does increase the risk of contracting STDs, but most people do not view their sexual behavior, no matter how active, as being promiscuous. Recent research also suggests that machismo gets in the way of proper condom use, an effective prevention technique for all STDs. A "real man" doesn't use condoms. And finally, when people do contract a disease, strong social stigmas make accurate reporting difficult.

On the positive side, the preventive and educational measures are basically the same for AIDS and other STDs: dispel the myths, increase general awareness and acknowledgment of the problem, and encourage more discriminating sexual practices or abstention from sex altogether. Perhaps some of the educational efforts made on behalf of AIDS prevention and treatment will have a helpful effect on the current STD epidemic. Historically, the health focus on STDs has been on treatment, typically with antibiotics, but recently the Public

Not long ago, condoms and other birth control devices were never discussed in "polite society." The unfortunate spread of STDs has changed that practice greatly. In the larger cities of the United States, for example, there are now stores that sell nothing but condoms!

Health Service has shifted its focus for all STDs to prevention. Perhaps comprehensive efforts of this kind that emphasize all STDs will prove fruitful.

Major changes in adolescent sexual practices have occurred in recent decades. Many of them must be viewed with considerable alarm, especially when considering the tragic increases in AIDS, STDs, and pregnancies. The dangers of sexually transmitted diseases have become more well known with the media attention given to AIDS. However, much unprotected sexual contact still occurs. Among teens who are sexually active, 61 percent of sixth graders, 54 percent of ninth graders, and 47 percent of 12th graders reported that they did not use contraceptives (Benson, 1993). Which partner is responsible for contraceptive use? A study of adolescents found that when male teenagers first engage in sexual relations, they feel responsible for contraception and often use condoms. As they become more experienced, however, these same adolescents were more likely to see contraception as their partner's responsibility (Pleck & others, 1988). This could be because older adolescents may be more comfortable discussing contraception with their partner (Rickert & others).

Female adolescents making decisions about birth control find that it can be a complicated process. Durant and his colleagues (1990) have developed a model that illustrates the factors that may affect adolescent contraceptive use. They believe that female use of contraceptives is influenced by birth control knowledge, physical development, personality, peers, family, values, and the existence (or nonexistence) of an intimate relationship. Their model indicates that for female teenagers, making decisions about contraception can involve juggling commitments to values—their own and those of others.

C O N C L U S I O N S

For the four aspects of social interaction reviewed in this chapter—autosexual, homosexual, and heterosexual behavior and sexually transmitted diseases—there is one consistent trend with which all adolescents must deal: fast-paced change. Some of this change derives from groundswells in today's society, and some results from the nature of adolescence itself.

Perhaps nothing has had a more resounding impact on adolescent life than the recent changes in our attitudes toward sexuality. The greatest changes have been in the areas of homosexuality, sexually transmitted disease (including AIDS), and earlier and more widespread participation in sex by teenage females.

Some observers have suggested that the best way to deal with each of these problems is through the schools. In the next chapter, we acquaint you with the current situation in middle and high schools as well as in the adolescent workplace.

C H A P T E R   H I G H L I G H T S

Sexuality
• The sexual revolution has led many teenagers to become sexually active at increasingly younger ages.

Autosexuality
• Masturbation is believed to be a harmless and universal form of human sexual expression.

• It is of concern only when its practice causes emotional disturbance.

Homosexuality
• Homosexual behavior has been surrounded by many myths throughout history.

- Several theories suggest different origins of homosexual orientation: psychoanalytic, learning, and genetic.
- Many researchers now believe that homosexual orientation may already be set in adolescence, either by genetic factors, conditioning, or a combination of the two.

Heterosexuality

- Many teens still obtain a great deal of information and misinformation about sex from their peers.
- First intercourse is now occurring earlier than it did in past generations.
- A steady relationship with one member of the opposite gender is most likely to provide opportunity and pressure toward sexual activity.
- Television influences teenagers' standards of sexual attractiveness and behavior.
- There are many nonsexual reasons why teenagers misuse sex—to search for affection, to rebel against parents, to vent anger, and to alleviate boredom.
- Teens seem to benefit from learning about sex and sexuality from many sources at once. Parents and sex education programs can be most helpful.

Unhealthy Sexuality

- Adolescents are at risk for sexual victimization, including rape, from family members, friends, and strangers.
- Many adolescent runaways and prostitutes are the products of sexual abuse, often by someone they know, a family member, or parent.
- Acquaintance rape occurs within the context of adolescent social activities and is most common when social activities include heavy drinking. Cultural beliefs about the sexes are partially blamed for the occurrence of acquaintance rape.

Sexually Transmitted Diseases

- Today a very high prevalence of sexually transmitted diseases (STDs) is found in sexually active adolescents.
- AIDS (acquired immune deficiency syndrome) causes the most concern, as it is incurable at present and usually fatal. As of yet, it is not very common in adolescents, but may lie dormant in large numbers of them.
- Other STDs, including chlamydia, gonorrhea, genital herpes, syphilis, and others that affect adolescents, are increasing in epidemic proportions. We cannot help but worry that AIDS will follow this trend.
- In spite of increased availability and information about contraceptive methods, many teenagers continue to engage in unprotected sex.

K E Y   T E R M S

AIDS (acquired immune deficiency syndrome) 306
autosexuality 283
beer goggle sex 302
biological theory of homosexuality 287
chlamydia 307
double standard 291

gays 286
genital herpes 309
gonorrhea 308
Hepatitis B 309
heterosexuality 290
homosexual 285
learning theory of homosexuality 287
lesbians 286

pelvic inflammatory disease (PID) 308
psychoanalytic theory of homosexuality 287
sexual revolution 277
sexually transmitted disease (STD) 306
social learning 296
syphilis 309

W H A T   D O   Y O U   T H I N K ?

1. Why do you think there has been a sexual revolution in the past few decades? Why would it happen at this point in history?
2. In what ways has your morality about sex changed since you started thinking about it?
3. Do you agree with the theorists who claim that there are three stages in the development of love and sexuality, and that this development is natural?
4. Which do you think is the most important of the widespread changes in the sexual aspects of our lives?
5. If you were the mayor of a medium-size city, what actions would you take to try to reduce the incidence of sexually transmitted diseases?
6. If you were the mayor of a medium-size city, what actions would you take, if any, to change the sexual practices of teens in your city?

SUGGESTED READINGS

Calderone, M. S., and Ramsey, J. (1981). *Talking to your child about sex*. New York: Ballantine. This book offers a creative interpretation of human sexuality in a family setting.

Capote, T. (1948, 1988). *Other voices, other rooms*. New York: Signet. Written when Capote was 23 years old, this book is considered by many to be his best work. It describes his relationships with many eccentric characters, and tells how he became aware of his own homosexuality.

Conroy, P. (1987). *The prince of tides*. New York: Bantam. A penetrating study of the conflicts between attraction and family responsibilities.

Fromm, E. (1968). *The art of loving*. New York: Harper & Row. By reading Fromm's book, you will understand better what love is, how it relates to sexuality, and how you can give and receive it in highly effective ways.

Jacoby, A. (1987). *My mother's boyfriend and me*. New York: Dial Books. Sixteen-year-old Laurie doesn't know how to handle it when her mother's 27-year-old, handsome, blue-eyed boyfriend starts making advances.

Leeming, F. C., Dwyer, W. O., and Oliver, D. P. (1996). *Issues in adolescent sexuality: Readings from the Washington Post Writers Group*. Boston: Allyn & Bacon. Adolescent sexual activity has attracted considerable media attention. This volume contains a selection of articles that have appeared in the Washington Post and provides insights into the views of experts, policy makers, and the public concerning adolescent sexuality.

Tannahill, R. (1980). *Sex in history*. New York: Ballantine. This lively book describes the role of sex down through the ages.

GUIDED REVIEW

1. What is the largest, most scientifically designed study of sexuality since the Kinsey Report in the late 1940s?
   a. The Hite Report
   b. The Masters and Johnson Report
   c. The 1987 University of Michigan Study
   d. The Janus Report

2. When gay and lesbian teens experience emotional difficulties, it is often a consequence of
   a. mental illness.
   b. guilt.
   c. isolation from peers and family.
   d. sexual pressures from homosexual adults.

3. In Koch's study on the motivation for first intercourse, the most common reason women gave for first coitus was
   a. both parties wanted to engage in sex.
   b. pressure from their partners to engage in sex.
   c. desire to express love or caring.
   d. curiosity of what sex would be like.

4. Adolescents engage in sexual activity for all the following reasons except
   a. to alleviate boredom.
   b. to ensure fidelity of their girlfriend or boyfriend.
   c. to rebel against parents.
   d. to use contraceptives.

5. The most common source for sexual information for adolescents is
   a. sex education programs.
   b. parents.
   c. television.
   d. friends.

6. "Acting out" behaviors, such as truancy, running away, sexual promiscuity, and damage to school performance and family relationships, may directly follow the experience of
   a. sexual abuse.
   b. first heterosexual experience.
   c. first homosexual experience.
   d. first autosexual experience.

7. Sexual abuse victims usually discuss their experience only with
   a. police.
   b. mental health professionals.
   c. friends.
   d. teachers.

8. The most common type of serious sexual abuse is
   a. incest between mother and son.
   b. rape by a date.
   c. rape by a stranger.
   d. incest between father and daughter.

9. Which of the following beliefs is NOT used to justify date rape?
   a. If a woman says "no," she really means it.
   b. Males have uncontrollable sexual urges.
   c. If a woman wears sexy clothing, she is asking for it.
   d. Some women ask for and enjoy rape.
10. Beer goggle sex refers to sexual activity resulting from
    a. cultural beliefs.
    b. peer pressure.
    c. heavy alcohol use.
    d. having fun.
11. There are few aspects of human behavior that have changed more in this century than sexual behavior. The situation has changed so much that it is called the _____ (double standard/sexual revolution).
12. Masturbation is probably _____ (rare/universal) to human sexual experience.
13. _____ (AIDS/Chlamydia) is now the most common sexually transmitted disease.
14. The _____ (majority/minority) of persons who come to identify themselves as homosexual have engaged in heterosexual sex.

15. For some teens, having sex with many partners is a means of confirming _____ (fidelity of their girlfriend or boyfriend/their sexual orientation or identity).
16. Some teens regularly use masturbation as a way of _____ (rebelling against parents/alleviating boredom).
17. Based on sex education research, Koch recommends that sex education should include _____ (decision-making skills/presentation of facts only).
18. Being raped by a date is _____ (less damaging than/as damaging as) being raped by a stranger.
19. Women who use an intrauterine device for birth control, have multiple sex partners, are teenagers, or have had this disease before are at risk for _____ (pelvic inflammatory disease/gonorrhea).
20. Those who are most at risk for _____ (syphilis/Hepatitis B) are intravenous drug users, homosexual men, and inner-city heterosexuals.

Answers

1. D, 2. C, 3. C, 4. D, 5. D, 6. A, 7. C, 8. C, 9. A, 10. C, 11. sexual revolution, 12. universal, 13. Chlamydia, 14. majority, 15. their sexual orientation or identity, 16. alleviating boredom, 17. decision-making skills, 18. as damaging as, 19. pelvic inflammatory disease, 20. Hepatitis B

EDUCATION AND WORK

Cognitive skills are not the only outcome of schooling. Educators claim schools teach virtues ranging from patriotism and punctuality to curiosity and creativity. Critics claim that schools teach an equally wide range of vices, ranging from competition and conformity to passivity and authoritarianism.

Seymour Sarason

The Education of American Adolescents

American adolescents spend many of their waking hours within the walls of our nation's secondary schools. This was not always the case. In the last century, only the children of the rich attended secondary schools. Continuation in formal academic education beyond the elementary level was considered unnecessary for those who were not entering the professions. Many youths learned their trades informally through watching their parents or other adults in the community. The first U.S. high school, the Boston English School, was founded in 1821. In 1900, just 10 percent of the U.S. population attended high school. Today, nearly all American teenagers between the ages of 14 and 17 are enrolled in our nation's high schools (Murphy, 1987).

Although high school attendance has become common in this century, controversy over the role and importance of schools has increased. By the 1960s, all states required school attendance to at least 16 years of age. At the same time, evaluators of our nation's schools (e.g., Coleman & others, 1966; Jencks & others, 1972) claimed that schools had little effect on the behavior and achievements of students. Other critics have seen the schools as the cause of many of society's problems (Lightfoot, 1978). Following the claims that schools were damaging or at best unimportant, other writers (Madaus & others, 1979; Rutter & others, 1979) maintained that schools can be effective and can make a difference in the development of the nation's children and adolescents. There are a number of factors that have been associated with success at the middle school level including positive teacher regard for students and emphasis on effort (Roeser & Eccles, 1998).

The kind of education that is best suited to young adolescents has been hotly debated in recent years (Hechinger, 1993). The junior high school was an invention of the early 20th century, but did not become commonplace until mid-century. Prior to that time, children from ages 5 through 14 often attended elementary schools. The junior high recognized that young adolescents between 10 and 14 years of age had different needs than younger and older students and modeled itself after the successful factory, striving for efficiency and specialization. By mid-century, tracking students and employing teachers who were specialists in their content areas provided a hope for accelerating student progress to meet the Soviet competitors, who had launched Sputnik. Junior high schools, which had offered hope for educational improvement, came to be criticized for their factory resemblance and inattention to the needs of young adolescents. Middle schools replaced junior highs in some parts of the country and sought to be more personal and less like high schools. The philosophy of the middle school involves team planning and teaching, active learning, and the development of problem solving skills among students (Heller, 1993). Many critics believe that middle schools have not yet succeeded in the mission of educating young adolescents (Carnegie Council on Adolescent Development, 1989).

Controversy concerning our secondary schools continues today. Many writers note that the schools are being expected to fulfill functions, like moral development and sex education, that were previously accomplished in the home or the church. Although our expectations concerning what schools should be able to accomplish are high, the public image of the American high school as shown in television and movies is not positive. What do Americans expect of their secondary education system and how successful is that system in meeting those expectations? The first half of this chapter examines the functions of schools in American society, criticisms of the American middle school and high school, and suggested models for improving our schools. The second half of this chapter focuses on the importance of work and career development during adolescence.

After reading this chapter, you should be able to:

- List three functions of the American schools and evaluate how well secondary schools are fulfilling these functions.
- Describe the developmental needs of early adolescents and how well those needs are met by the structure and organization of the middle school.
- Discuss recommendations for improving secondary schools as suggested by the Carnegie reports and current psychologists.
- Describe four reasons for dropping out of high school and three factors that motivate improved performance from students at-risk for dropping out.
- Describe four ideas suggested for improving the schools.
- Explain matching, developmental, and sociological theories of career choice and how they can be applied in counseling adolescents.
- Explain the ways in which family, gender, and race influence and limit vocational choice.
- Discuss the advantages and disadvantages of employment for high school students.
- Discuss these issues from an applied, a sociocultural, and your own point of view.

FUNCTIONS OF THE SCHOOLS

According to Busch-Rossnagel and Vance (1982), schools are expected to fulfill three functions.

- *Literacy.* All American children, regardless of social class, are expected to learn basic literacy skills in reading, writing, and arithmetic. Prior to the Industrial Revolution, academic skills were not important to many trades. With the growth of industry, however, a more educated work force was needed. The technological changes of recent years have further raised the level of education needed for many jobs.
- *Transmitting social values.* Schools are expected to teach children the dominant social values of the society. Through school, they learn self-discipline, social skills such as how to get along with other children and adults, and acceptable standards of behavior. The social values taught within our schools largely reflect the standards of the middle-class society.
- *Improving society.* By teaching all children literacy skills and social values, regardless of socioeconomic class, the schools are viewed as a way of curing

the negative effects of poverty. School lunch programs and medical care offered through the schools are other ways in which the schools hope to reduce social and economic inequalities.

THE CRISIS IN THE AMERICAN SECONDARY SCHOOLS

Are American middle schools and high schools fulfilling the functions described above? According to many critics, American secondary schools are falling short in a number of areas.

Many reports have evaluated the status of American secondary education, but two of the most influential have been sponsored by the Carnegie Foundation. Middle schools were evaluated through the report *Turning Points: Preparing American Youth for the 21st Century* (Carnegie Council on Adolescent Development, 1989). Secondary education in the United States was evaluated by an earlier report of the Carnegie Foundation for the Advancement of Teaching entitled *High School: A Report on Secondary Education in America* (Boyer, 1983). Both reports identified serious challenges confronting our secondary schools, challenges that centered on improving the academic quality and the equity of our educational system. Other reports (Hechinger, 1993; Mullis & others, 1991; Secretary's Commission on Achieving Necessary Skills, 1991; U.S. Department of Education, 1994) have bolstered the positions taken in the Carnegie reports, which came to several disturbing conclusions.

• *Academic quality.* The world is changing rapidly in ways that are shaped by advances in science and technology. One consequence is that more highly skilled workers are needed. With rapid changes in science, adolescents growing up today may have to adapt their skills to a number of different jobs during their working years. If our nation expects to be able to compete economically with the other nations of the world, we need to prepare our youths with an education, especially in math and science, that surpasses that of other nations.

Although our economy demands a highly educated work force, statistics indicate that we are not doing a good job in meeting that demand. Results of national testing reported in April 1995 indicated that only one-third of high school seniors in the United States demonstrate proficient reading skills, and 30 percent of seniors failed to demonstrate basic reading skills (Associated Press, 1995). These percentages represent a significant decrease from scores reported just two years earlier. Only 28 percent of eighth-grade students were reading at proficiency level in 1994 (U.S. Department of Education, 1994).

American 13-year-olds have been found to be far behind adolescents of the same age in other industrialized nations in math and science achievement. The National Assessment of Educational Progress (NAEP), an assessment program sponsored by the federal government, found that just two-thirds of eighth-grade students demonstrated mastery of fifth-grade skills such as multiplication and division of whole numbers, and only 14 percent demonstrated success consistent with seventh-grade skills such as fractions, decimals, and simple algebra. Only 5 percent of high school seniors showed the understanding of geometry and algebra needed for further mathematics education (Mullis & others, 1991). Thus, many students appear to be graduating from high school without the level of mathematics preparation needed for college work or employment in areas of increasing growth (such as technology).

- *Social and economic equity.* By the year 2000, one-third of the students graduating from our nation's colleges and universities and entering the work force will be African American or Latino. By the year 2020, it is expected that half of all American schoolchildren will be nonwhite (Natriello & others, 1987).

WHAT'S YOUR VIEW?

Are the Schools to Blame?

According to statistics cited in this chapter, the American middle school does not appear to be fulfilling the primary function of education—to provide an adequate level of literacy to the majority of American students. While those statistics clearly tell us that we have to do a better job of educating our young people, some writers are less critical of the school system itself.

Entwisle (1990), for example, says that schools alone *cannot* make up for the pathologies of society, including poverty, discrimination, inadequate job opportunities, and poor parenting. All those factors are a strong influence in adolescents' lives. Fine (1986) suggests that even if all students graduated from high school, there still would not be enough good jobs for everyone. She suggests that in addition to reforming schools, we must improve job opportunities, housing, and health care services. Focusing just on the schools as a source of our problems, she believes, keeps us from looking at the broader range of social and economic problems.

Murphy (1987) points out that other nations are not necessarily more successful in educating all their students. Some nations, for example, separate and select out students based on test scores. Overall test scores in those nations look better because low-achieving students have been systematically tracked out of the higher academic courses. The democratic ideals of our school system may result in statistics that appear unfavorable in comparison with other countries.

Wehlage and Rutter (1986) argue that the schools are obligated to create a learning situation in which all youths can be successful. Schools should not be excused from their responsibility because some students are poor or do not speak English or come from single-parent homes.

How much are the schools to blame for the mediocre achievement of our middle school students and to what extent are other social factors responsible? How strong an influence is the school on the lives of adolescents? How does that influence compare with that of the family, the peer group, and the media? How do the democratic values of our educational system affect achievement? What is your opinion?

While the number of African American and Latino youths attending our high schools is increasing, the schools have been least successful in educating these youths. The gap between the quality of education offered at the best and the worst high schools has increased. Ten to 15 percent of American students receive an outstanding education in their high school. For another 20 to 30 percent, high school is an academic failure. For example, close to four out of every ten Latino youths drop out of high school. Half of those who drop out never even entered high school (U.S. Bureau of the Census, 1986). Although the dropout rate is higher among some minority youths, more white students drop out of high school than other ethnic/racial groups. The high school dropout rate for African Americans and whites is almost identical, at about 5 percent, with African American teenagers completing high school at the highest rates ever

(U.S. Bureau of the Census, 1994). When socioeconomic differences are controlled African American students are less likely to drop out than are white students (Carnahan, 1994). School dropouts are two times as likely to be unemployed as are high school graduates, and the jobs they do obtain often do not pay enough to support a family (National Center for Education Statistics, 1994; William T. Grant Commission, 1988).

As of 1993, 81.5 percent of whites, 70.4 percent of African Americans, 59.8 percent of Puerto Ricans, and 46.2 percent of Mexican Americans in the United States over the age of 25 had completed four years of high school or more (U.S. Bureau of the Census, 1994). Twenty-two percent of whites, 12.2 percent of African Americans, 8 percent of Puerto Ricans, and 5.9 percent of Mexican Americans had completed four or more years of college. Although African Americans continue to trail whites with respect to educational achievement, the differences have narrowed over the past decade (U.S. Department of Education, 1994). We must conclude, however, that American secondary education is not yet adequately fulfilling its third function—reducing social and economic inequalities.

The *High School Report* by the Carnegie Foundation for the Advancement of Teaching suggested a series of steps that should be taken to develop both excellence and equity in our high schools:

- Four essential goals are recommended for all high schools: to help students to think critically and communicate effectively, to learn about their human heritage and how to get along with other people, to be able to find work or enter higher education, and to fulfill their social and civic obligations through school and community service.
- Today's graduates should expect to change jobs several times in the future and must have a broad basic education on which to build. All high school students need to be taught to read and speak effectively using the English language. The high school should have a core curriculum for all students covering traditional subjects, such as history, math, and literature, as well as foreign languages, studies of non-Western nations, the meaning of work, and the importance of health. Electives during the last two years of high school would provide the opportunity to explore career choices or to take advanced academic subjects.
- Schools and teachers need to be taught to use technology and computers in the schools in ways that will improve students' learning. All students should learn about the social importance of technology, of which the computer is a part.
- Connections need to be made with higher education and the business world. These partnerships can help enrich the educational programs offered by the schools and can be used to better prepare students for the transition from high school to college or the world of work.
- In order to smooth the transition from school to adult life, class schedules need to be more flexible and allow for learning both in school and in the community. Greater flexibility in school size and use of time will help schools meet the needs of more students.

EARLY ADOLESCENCE AND THE TRANSITION TO MIDDLE SCHOOL

The crises of the American middle school and high school extend beyond issues of curriculum and academic achievement. The functions of the school, as outlined at the beginning of this chapter, include the social development of adolescents as well as the development of academic skills. In fact, as Entwisle (1990) points out, secondary schools affect the physical, emotional, moral, prevocational and political development of American teens.

As you will recall from previous chapters (see Chapters 6, 7, and 8), the transition to middle school (or junior high school) is a period of increased social and emotional difficulty for some early adolescents. Conflicts with parents, when they occur, are most likely to take place during the stage of puberty. Early adolescents typically turn their attention to peers and are exploring the world in ways that can be risky to their physical and emotional health. Exploration with drugs and alcohol during these years can put them at risk for later addiction. During this period of change, some adolescents feel isolated and confused.

As young adolescents move from elementary to middle school or junior high school, some students become less involved in school learning, and rates of absenteeism, drug abuse, and dropping out increase. Simmons and Blyth (1987) completed a large study on the transition from elementary to junior high school. In comparison with students who remained in an elementary school setting through grade 8, those who entered junior high schools had less positive attitudes toward school, achieved lower grades, and were less likely to participate in extracurricular activities. Girls displayed a drop in self-esteem and leadership skills following entry into junior high school and did not regain those losses throughout the high school years. Boys attending junior high schools were more likely to be robbed, threatened, or beaten compared to boys attending the seventh and eighth grades in elementary settings. Their findings further demonstrated that being robbed or beaten at school has a negative effect on self-esteem. Other authors (e.g., Lightfoot, 1983; Lipsitz, 1984) have also emphasized the importance of physical safety to student academic growth and emotional well-being. Simmons (1987) believes that students are forced to make the transition into larger and less-protected school settings before they are developmentally ready. She recommends that changes in school settings be more gradual (that is, junior high schools or middle schools should be more like elementary schools in structure), and students should not be expected to adjust to other life changes (e.g., changes in family and pubertal status) at the same time. In a recent study, Schiller (1999) found that a student's middle school academic success impacts the transition to high school. High achieving middle school students seemed to benefit from going on to high school with a majority of their middle school classmates. Whereas lower achieving students seemed to benefit from going to a different high school from the majority of their middle school classmates. This finding suggests academic success might create an educational hierarchy that impacts later educational outcomes.

The move from elementary school to middle school presents numerous social, academic, and emotional changes.

Seidman and associates (1994) studied 590 poor inner-city adolescents as they entered middle school. Declines in self-esteem, academic grade point average, and class preparation occurred for both boys and girls and for students of African American, Latino, and white race/ethnicity backgrounds. The declines were associated with changes in peer values, which were perceived as more antisocial, and with an increase in daily hassles at school. Students who experienced many daily hassles at school had lower expectations for academic success, prepared less for class, and achieved poorer grades.

Seidman and associates (1994) agree with the authors of the *Turning Points* report and other researchers (e.g., Eccles & others, 1993; Entwisle, 1990) that a mismatch occurs between the developmental needs of early adolescents and the structure and curriculum of our middle schools. This mismatch is thought to contribute to declining school involvement. At a time when early adolescents are confused and unsure of themselves, they move from smaller, more closely supervised school settings to much larger schools in which they feel no one cares about them. Middle school students suddenly have a large number of teachers, but feel that they know few or none of those teachers well. The teachers are often viewed as being less understanding and more controlling, just at a time when adolescents are seeking more independence.

The friendships that students enjoyed during elementary school are often disrupted, as different and larger groups are formed. Some early adolescents do not feel they belong to any of the new groups found in middle school. They are not finding the social recognition they need because in larger schools more students are overlooked and many do not participate in any school activities.

Eccles and Harold (1993) voice concern over the decline in parental involvement that generally occurs as children move from elementary to secondary school. Contrary to the popular beliefs that adolescents desire and need independence, a growing body of research indicates that parental involvement is critical to children's academic success at all grade levels. "Adolescents may indeed want greater autonomy, but they still need to know that their parents support their educational endeavors. They need a safe haven in which to explore their independence, a safe haven in which parents and schools are actively involved" (p. 575). Parents, teachers, and school administrators need to recognize the importance of parental involvement. Large and impersonal middle schools and high schools need to find ways to increase contact and communications between teachers and parents.

Not all adolescent researchers view the transition to middle school so negatively and some suggest that there are many variables to consider (Roeser & Eccles, 1998). According to Berndt and Mekos (1995), many early adolescents view their middle school and junior high school experiences quite positively. In fact, sixth and seventh graders reported more positive than negative events associated with moving to junior high both before and after their entry to junior high. Prior to entering junior high school, high-achieving students expressed fears about peer relationships in their new school. Girls were most concerned that they would be left out by friends or that their friends would not be faithful. Boys were fearful that they would be picked on by older students in eighth and ninth grade. After moving to junior high, few girls or boys mentioned peer problems or victimization by older students. High-achieving students found that their fears were not realized. Students who had behavior problems in elementary school expected that junior high school would be better. Once they entered junior high, however, their ratings were less favorable. It seems that there may be more continuity than discontinuity in student adjustment from elementary to junior high school. Berndt and Mekos believe that researchers need to look at which aspects of the junior high school experience cause stress, rather than assuming that the transition is detrimental for all early adolescents.

How stressful is the transition to middle school? What can be done to ease that transition?

According to recent research, school size does not seem to make a difference in academic learning as measured by standardized achievement tests. However, secondary schools (including middle, junior high, and high schools) that are relatively small (between 500 and 1,000 students) offer personal benefits for students. Almost all secondary schools have a student newspaper, band, student government, and other organizations. Within small schools, more students, and less-capable students, have the opportunity to participate in these and other activities. Participation in school activities increases chances for positive interactions with teachers and other school authorities. Although after school clubs and sports teams are often considered fringe benefits for youths and tend to be eliminated when school budgets are tight, recent research suggests that those activities are vitally important. In schools where participation in extracurricular activities is high, participation in at-risk behaviors such as alcohol use, sexual activity, attempted suicide, and delinquency is relatively low (Blyth & Roehlkepartain, 1992). It seems that these activities provide teens with a sense of belonging, help them develop skills and positive values, and provide a constructive way of contributing to the school and community. Large schools tend to have more behavior problems than smaller schools. This may be because smaller schools allow students to know their teachers well, and students feel responsible for their behavior. Small schools are often better able to meet and respond to the needs of individual students (Entwisle, 1990; Minuchin & Shapiro, 1983).

The authors of the *Turning Points* report believe that the following suggestions would enable middle schools to better meet the cognitive, emotional, and social needs of early adolescents.

- Create small communities for learning where relationships with peers and adults are close, stable, and respectful. In order to make larger schools more intimate, students and teachers would be grouped together in teams to form schools-within-schools.
- Teachers who are expert at teaching young adolescents and who are trained specifically for working with that age group should be hired.
- Families should be involved in their children's schools with a role in developing school policy. School staff should communicate with families about student progress and help parents to support their children's learning at home.
- Schools should be closely connected with the communities they serve. Community resources should be used to enrich educational programs. Students should provide service to their communities through placement in youth service programs.

ACADEMIC TRACKING AND ADOLESCENT EDUCATION

Middle schools and high schools differ from elementary schools in many ways, including academic tracking. Most secondary schools offer a number of "tracks," which students choose or are assigned to based on their ability. A student who is not interested in attending college can choose a general track or a vocational track. Although tracking makes sense in many ways, the practice has also been heavily criticized. Those who favor tracking argue that it assists teachers in meeting the learning needs of students. When students with different ability levels are grouped in the same class, it can be difficult for a teacher to prepare lessons appropriate for all students. Critics of tracking argue that it deprives some students of an equal education and reinforces privileges associated with social class. Existing research provides support for both views (Hallinan, 1991).

Some evidence suggests that ability grouping and exposure to many teachers benefit the economically advantaged, but not the disadvantaged, students. Students and teachers have higher academic expectations for high-track students, and high-track students are more likely to attend college. Furthermore, the most skilled and most experienced teachers are often assigned to high-track classes (Gamoran & Mare, 1989). Gamoran and Weinstein (1998) found that when with high quality instruction neither heterogeneous nor homogeneous grouping present obstacles to quality education. However, they (1998) also found that neither heterogeneous nor homogeneous grouping or tracking ensures effective teaching. These findings would suggest that it is quality of instruction for all students, rather than grouping that should be the focus of discussion. Economic resources in the form of new textbooks, better science laboratories, and counseling services are invested more often in the high-track rather than the low-track programs (Rosenbaum, 1991). Parents who have the most education seem to be more successful in advising their children and working with the school system so that their children are placed in higher track math courses (Useem, 1991). Economically disadvantaged students have been found to do well in some Catholic schools, which offer fewer tracking choices and in which all students are expected to achieve in the college preparatory track (Lee & Bryk, 1986). Students from economically disadvantaged families have also been found to do better when taught by a small number of teachers (Becker, 1987).

Rosenbaum (1991) suggests that many of the motivation and discipline problems observed among high school students are associated with tracking. Students who are not planning to attend college may have little motivation to study in high school. Rosenbaum believes that tracking systems need to be more flexible. That is, students in non-college tracks should be able to move into college tracks if they so desire. When tracking systems eliminate the possibility for students who are in middle school or high school to attend college, it is not surprising that those students have little motivation for academic study. Rosenbaum suggests, in addition, that employers need to reward work-bound students for getting good grades. Employers should consider school grades and teacher evaluations in hiring high school graduates. Simmons (1987) believes that secondary schools need to reward multiple talents. Because only a few students can achieve high levels of academic success, those who are less successful are likely to feel discouraged and alienated at school. They may turn to negative peer groups as a way of achieving status.

So far we have looked at the academic impact of tracking. However, tracking can also have an impact on social relationships. Students tend to make friends with classmates, so when there is tracking, students have friends within the same track (Hallinan, 1998). Although having friends within the same track is not necessarily problematic, being limited to having friends within a certain track because of the association may be socially limiting for some students.

Although more research is needed to resolve the tracking question, it appears that recent educational trends toward more academic tracking at an earlier age may not be meeting the needs of many adolescents and may be a source of bias in the schools.

GENDER AND RACIAL BIAS IN SECONDARY SCHOOLS

Critics of American secondary education point to the need for a work force highly educated in math and technology. Adolescent females and nonwhite males in general lag considerably behind white males in the development of math and technical skills (Asian Americans are an exception) and in completing higher education. Although high school girls have higher grade point averages than boys, they complete fewer years of post–high school education (Entwisle,

1990). Blame for the lagging performance of girls and African American boys can be placed on the family and the culture at large, but secondary schools have also been criticized as sexist and racially biased.

A report sponsored by the American Association of University Women, entitled *Shortchanging Girls, Shortchanging America* (Bailey, 1992), maintains that school curricula are biased against girls. Teachers provide more attention and encouragement and have higher achievement expectations for boys than for girls. The 1999 AAUW follow-up report, *Gender Gaps*, suggests that while there has been some improvement in terms of gender equity in schools, there are gaps that remain. For example, girls tend to get better grades in school than boys but boys generally score better on standardized tests, especially high stakes standardized tests (like the PSAT and SAT) that are tied to college admission and scholarships. In Chapter 6, we described how these practices can negatively affect the self-esteem of adolescent girls. Of additional concern are the ways in which educational practices reinforce gender stereotypes and discourage girls from achievement in math and computer science. Charges of gender bias in education are not new. Previous research indicates that males are more likely than females to be assigned to high-ability math groups and are more likely to be the object of praise, personal statements, encouragement, and jokes, as well as disciplinary comments (Brophy, 1985; Eccles, 1984; Hallinan & Sorenson, 1987).

It seems, however, that the positive attention and high expectations given to white males do not extend to African American males. Research suggests that teacher expectations are lower for African American males than they are for African American females (Ross & Jackson, 1991). African American males, who are independent and non-submissive, are viewed as least capable and most threatening by their teachers. According to Takei and Dubas (1993), adolescents are sometimes rewarded and punished by their teachers' grading practices for culturally related behaviors that have little to do with their actual academic performance. One college admissions officer described the situation in this way: "At any sign of wild behavior or rambunctiousness, that student (African American male) is automatically labeled as special and tracked into courses that aren't even going to prepare him for high school, let alone thinking of college" (Bloom, 1991). Fewer African American males are applying to college and those who do are often discouraged from entering challenging academic courses. One college junior who attended a vocational high school was told by his college advisor, "We don't think you can make it from that high school. You don't have the courses." This highly motivated young man, however, entered the college's remedial program and is ready to transfer into the engineering curriculum, to which he had initially been denied entry because of lack of preparation.

Recent research by Ferguson (1998) suggests that racial bias may be perpetuated by teacher expectations and perceptions. Looking at students with similar standardized test scores, Ferguson found that students of color often make less academic progress than their white peers with similar scores. Perhaps racial bias informs teacher perceptions so that less is expected of students of color. Steele

Do you agree? How biased are the schools in educating adolescent girls?

and Aronson (1998) believe that talented African American students may some-times suffer from a fear of being stereotyped that causes them to perform less well on tests. These researchers believe that merely being asked to record race in the demographic section of a test can negatively impact the performance of an African American student.

The quality and safety of the school environment often differs for teens of varying racial groups. Many African American teens, for example, attend schools that are not safe or orderly. When compared to white high school sopho-mores, African American high school sophomores are more likely to describe disruptions in the classroom by other students. They are also more likely to be injured with a weapon in school than are white students (U.S. Department of Education, 1994).

Criticism of education as sexually and racially biased has contributed to debate over the advantages of single-sex and racially segregated institutions. Although single-sex education at both the high school and college level became increasingly unpopular during the 1980s and 1990s, recent research and the popular press suggest that single-sex schools may offer girls more opportunity to hold positions of leadership and develop skills in math and technology without being labeled as unfeminine. Girls can concentrate on their academic learning without worrying what the boys will think. Advocates of coeducation point out that what is needed is a more equitable environment in coeducational institu-tions. Susan Bailey points out, "How else will boys learn of the strengths and abilities of girls if they can't learn with them?" (Graham, 1992).

THE HIGH SCHOOL DROPOUT

Critics of the American middle school and high school point out that the American schools are not meeting the educational needs of many youths, espe-cially those who are poor, Latino, and African American. Socioeconomic status and race/ethnicity are the two background characteristics most related to drop-ping out of high school. Dropout rates occur most often among Latino youths, followed by African American adolescents, and then white youths (see Table 10.1). There are more white students who are high school dropouts, however, than students from any other ethnic or racial group. *Poverty is the greatest determinant of who drops out of school.* Students from poor families are more likely to drop out of school than those from families with more money, regard-less of race (although many children from all social strata do not complete high school). High school dropouts are also more likely to come from single-parent homes, from large families, and from the South or large cities. Dropout rates of as much as 40 to 50 percent are reported for inner-city high schools. Poor school performance is also characteristic of dropouts. Dropouts generally had low school grades and low scores on standardized tests, had no plans for college, and were likely to miss school often or be in trouble for their behavior (Ekstrom & others, 1986; Wetzel, 1987). Those who drop out of high school are more likely to be unemployed, will have fewer job choices, and are more likely to be depen-dent on welfare in their adult lives (Gibbs, 1984). High school dropouts are also more likely to become pregnant while still high school age (Manlove, 1998).

Some urban adolescents leave high school because of responsibilities at home, such as caring for younger brothers and sisters.

The high school graduation rate in the United States rose from 4 percent of the population in 1900 to 75 percent in 1990, but the modern labor force demands educational skills that were not needed in the past (Carnahan, 1994). The need for unskilled labor has declined drastically.

This description of the high school dropout tells us little about *why* poor, inner-city youths from single-parent homes are more likely to drop out of school. To better understand this, Michelle Fine (1986, 1991) interviewed students who had dropped out or had been discharged from a New York City high school. Only 20 percent of the students who entered this school as 9th-graders actually graduated from 12th grade. Her interviews with former students identified four types of individuals who dropped out for different reasons.

Table 10.1   High School Dropout and Completion Rates

Dropout and completion measures	Total	White non-Hispanic	Black non-Hispanic	Hispanic	Asian/Pacific Islander
Percentage of youth ages 15-24 who dropped out of grades 10-12 October 1997 to October 1998 (event dropout rates)	4.8	3.9	5.2	9.4	(*)
Percentage of youth ages 16-24 who were dropouts in 1998 (status dropout rates)	11.8	7.7	13.8	29.5	4.1
Percentage of youth ages 18-24 who were high school completers in 1998 (completion rates)	84.8	90.2	81.4	62.8	94.2

* Sample size too small for reliable estimate.

SOURCE: U.S. Department of Commerce, Bureau of the Census, Current Population Survey, October 1998.

Many adolescents leave school because they do not believe that a high school diploma will give them a better future. They believe that there are few jobs available whether or not they have a diploma. Many youths know friends and relatives who have diplomas and are unemployed. They also know people who did not graduate from high school but are making a good living, often from illegal activities like selling drugs. Another group of students leaves because of other responsibilities. They need to earn money or help out their families. Sometimes this means watching younger brothers and sisters or caring for sick relatives. A third group of students leaves because they are overwhelmed by the poverty in their lives and are discouraged about their futures. They feel that it is already too late for them to improve their lives. Pregnant teenage girls may hope instead that their babies will have a better life. The content of the school curriculum seems to have little to do with the everyday realities of their lives. A fourth group of students leaves school because they feel they are not welcome there. These are the students who often cut class and talk back to teachers. These youths and their parents feel that school administrators are eager to get them to leave school as soon as legally possible. Rather than looking for ways to help students do better at that time, the schools feel that it is too late. School practices such as referrals to the dean, detention slips, and behavioral contracts are interpreted by these youths as signs that they are not welcome at school.

> Do you agree? Are high schools the vehicle for social mobility or do they perpetuate class differences?

After observing in the school and talking to teachers and administrators, Fine believes that the schools maintain the social class structure in our society. That is, rather than improving society (the third function of the schools discussed previously by Busch-Rossnagel and Vance), the schools contribute to a cycle of poverty among inner-city Latino and African American youths.

Out of concern about the increasing number of adolescents who are dropping out of high school, Wehlage and his colleagues (1989) sought to identify the characteristics of high schools that are successful in keeping and graduating their students. They conducted a national search and identified 14 schools that were effective in working with at-risk adolescents and preventing dropouts. The most outstanding characteristic of the successful schools was their ability to provide a sense of belonging among students. The schools were able to provide a supportive environment that helped the students overcome obstacles that usually would have led them to drop out. Teachers, for example, were highly committed to the education of their students and believed that they had the freedom and resources needed to develop programs that met the unique needs of their students. Instead of worrying about whether all aspects of the curriculum were covered, these educators tried to fit the curriculum to the needs of the students. Some students looked for supportive guidance from their teachers. Others sought vocational experiences that would prepare them for employment. Others, such as pregnant teens and mothers, responded best to courses in child development. The commitment and caring of the teachers and their ability to direct their teaching to the immediate concerns of the students appeared to be key factors in dropout prevention.

WHAT'S YOUR VIEW?

How Should Schools Deal with Problem Students?

Carnahan (1994) believes that specific and individualized programs should be developed for at-risk students. Dropping out of school can be explained by the poor fit between the student and the school environment. The type of program that meets the educational, emotional, and social needs of the pregnant adolescent is likely to differ from the type of program that works for inner-city male gang members. To encourage students to remain in school, school programs should function as "communities of support," in which students experience a sense of belonging.

Not all educational policy makers agree with Carnahan. Toby (1989), for example, argues that some students should drop out. If schools tolerate tardiness, truancy, and inattention in order to make schools a welcome place for all students, the school will be less effective for the majority students. Toby argues that school demands should be made tougher. Those students who want to learn will attend and those who are not interested will drop out.

Have the schools become too tolerant of deviant behaviors? Do a small number of students who don't fit in make learning more difficult for more cooperative students? Should the schools become tougher or more supportive of the individual needs of students?

THE ACADEMICALLY SUCCESSFUL AFRICAN AMERICAN ADOLESCENT

Much recent writing and research has focused on African American youths who are not being well served by our nation's schools, as evidenced by poor academic achievement and dropout rates. (Much less scientific research has been done on other nonwhites.) It is important to recognize that many African American youths succeed despite social and economic hardships (Hrabowski & others, 1998). Recent statistics reveal that the percentage of African Americans between the ages of 35 and 44 who completed at least four years of college doubled (from 8 percent to 16 percent) between 1980 and 1990. The percentage of African Americans who had graduated from high school also increased from 63 percent to 80 percent during the same 10-year period (Bovee, 1991). Between 1976 and 1992, the average SAT scores of African American students rose 20 points on the verbal section and 31 points on the math section. During the same period, scores of white students dropped 9 points on the verbal and 2 points in the math SAT (Henry, 1992).

Several researchers have decided to study African American youths who are academically successful even though their families are economically poor. Lee (1985) developed a profile of successful African American youths who were attending rural schools in the southeast United States. Although we do not know if this profile describes successful urban youths, it does suggest some ways in which the negative effects of poverty can be reduced by caring family, friends, and adults who provide guidance and encouragement and by positive school experiences.

Students in grades 9 through 12 who were described by their teachers as successful academically and socially described themselves as close to their families. They felt that their parents provided them with strong direction and guidance. Extended family members such as grandparents were often important in providing encouragement and listening to problems or concerns. These

A SOCIOCULTURAL VIEW

Making the Most of College

Much has been made of the educational problems surrounding the urban African American community. Dropout rates tend to be high. Students who do complete high school and enter college are often underprepared for the college environment because of the poor secondary school training common in urban settings and because of other social influences. Consequently, these men and women may have difficulty making the transition into college life.

Carroll (1988) examined a college discovery program designed to enhance the undergraduate experience for African American freshmen who arrive at college educationally underprepared. She discovered that the role of the counselor and the scope of the counseling services were very important in the retention of these students. In addition, it was found that if these students were encouraged to set their academic goals high, they were more likely to be satisfied with whatever results emerged.

It seems that a thorough and consistent counseling program that emphasizes motivations, attitudes, career choice, and effective study habits will assist African American integration into the college environment. It should also be noted that students benefit by being encouraged to pursue bachelor's degrees.

students also described close relationships with friends and adults outside the family, such as pastors, godparents, neighbors, and family friends. Parents, political and religious leaders (e.g., Dr. Martin Luther King, Jr.), sports figures, and entertainers were described as important heroes and heroines.

Most of the students also expressed strong religious beliefs and conservative moral attitudes about right and wrong. Successful students participated actively in school and church activities, but tended to limit their involvement in other community activities. These students had positive feelings about school and expected to continue their education beyond high school and be successful in future careers. They viewed themselves positively, had a realistic sense of their limitations, and felt that they had a reasonable amount of control in shaping their lives.

Studying a group of academically successful African American males, Hrabowski and others (1998) found that effective parenting was an important factor. Effective parenting was seen to include a loving parent-child relationship, limit setting, high expectations, open communication, positive identity: racial and male, and use of community resources.

Gregory (1995) conducted interviews with New York City high school students who had been identified by teachers as demonstrating the most academic improvement. Rather than studying students who decided to leave high school, Gregory was interested in finding out what motivated students at-risk for dropping out to improve their performance. Three factors emerged as common among these students. Despite many failure experiences in school, they had all experienced success at something. Additionally, they all identified a caring adult who had offered encouragement, and they all experienced some personal change that led them to reevaluate the direction of their lives. Gregory hopes that the schools will be able to help other adolescents turn around their lives by fostering those three conditions. Adolescents need opportunities for

A SOCIOCULTURAL VIEW

Top Students Are in Demand

The demographic composition of college classrooms is changing. The percentage of the African American population between the ages of 18 and 24 attending college increased form 16 percent in 1973 to 25 percent in 1993, and the percentage of the Latino population aged 18 to 24 attending college increased from 16 percent to 22 percent during that 20-year period (American Council on Education, 1995). Affirmative action policies may be credited for these increases.

Academically talented high school students, like Holly Johnson, are heavily recruited by the top colleges in the nation. Holly is African American, the daughter of a cab driver, the valedictorian of her class at an inner-city Chicago high school, and the recipient of a 1,200 score on her SATs. Eventually, she wants to be a doctor and return to Chicago to open a health center for poor children and families. With credentials like these, Holly receives daily phone calls from the top colleges, including Ivy League universities. She has received enough college catalogs and applications (without requesting them) to fill three file cabinet drawers, and

application fees have been waived for all the universities to which she is applying. She is being offered strong financial aid packages from top colleges.

Holly is clearly an exceptional student, but is receiving more attention from colleges than she might if she were European American. African American and Latino students with less impressive academic credentials are also being courted by many colleges. Many Americans believe that this is only fair. One college admissions officer explains, "These kids come from schools that don't prepare them well for college. There is something more in their hearts and souls that drives them to achieve beyond the limited opportunities of where they were born." Critics claim that the system of affirmative action has gone awry: An honorable idea has resulted in a flawed system where European Americans with equal credentials cannot compete for college entry. Debates concerning affirmative action policies are taking place nationwide. What's your view on this important and controversial issue?

success in learning, opportunities to think about and make sense of what they have learned, and the support of teachers, family, and peers in taking advantage of these opportunities. The importance of support from adults at home and at school has been confirmed by other research (Connell & others, 1995). African American middle school students who view their teachers and adults at home as supportive are more likely to view themselves as capable, independent, and fitting in at school, to be more involved in their schoolwork, and to still be in school three years later.

Scathing criticism of the public schools has contributed to demands for school reform and the restructuring of curriculum and student support services. We now review several of the initiatives that have been suggested to improve our secondary schools. As you read, evaluate the goals of each initiative in terms of the functions of education described at the beginning of this chapter.

IMPROVING THE SCHOOLS: FULL SERVICE SCHOOLS

full service schools
A model of educational reform in which a variety of educational, health, and psychological services are provided to children and families in a school setting.

Concern for meeting the educational, health, and mental health needs of economically disadvantaged youths and families has contributed to a growing interest in **full service schools** (Dryfoos, 1994, 1995). This concept recognizes that many youths and families, especially those in economically disadvantaged communities, are in need of several services, but obstacles limit their ability to obtain these services. Sometimes families and youths who are most in need are not identified by the social service system, or when identified, do not come to community centers to obtain services. Those who do obtain services often are confused by an inefficient system in which they meet with nurses, teachers, psychologists, employment counselors, and public safety and housing providers in different settings with different rules and regulations. These service providers do not communicate about client needs and may duplicate services or overlook some critical needs of the youths or their families.

Full service schools seek to reduce these problems by providing a variety of services in a single setting, the school. Since almost all children have contact with the school system, it is believed to be an ideal location for identifying youth and family needs. When services are available at the school site, youths do not have to travel to other locations to obtain needed services. Schools provide regular and continuing access to children from preschool through high school. The integration of services at one location provides the opportunity for a streamlined and seamless method of service delivery, in which all professionals communicate and collaborate in working with youths and their families. Although teachers, counselors, and school administrators are an important part of the full service school, school staff are not responsible for all these services. Community agencies funded by local, state, and national government and private sources provide services at the school setting.

Dryfoos (1995) described two types of full service schools that focus on adolescents. The school-based health clinic provides comprehensive medical, mental health, social, and health education services at the school site with 24-hour backup by a medical facility. Services are provided for illnesses and accidents, physical examinations, and mental health counseling. Family planning, nutrition counseling, substance abuse treatment, treatment for sexually transmitted diseases, and prenatal and postnatal care may also be offered. The number of school-based health clinics grew from just 10 in 1984 to more than 600 nationally in 1995. Youth centers also provide services to adolescents in full service schools. These centers provide services such as recreational activities, employment counseling and job placement, family counseling, drug prevention programs, and referral for health services. School doors are open before and after school, on weekends, and throughout the summer, with gyms, classrooms, and computers available for community use.

You may be wondering what these services have to do with education. Advocates of full service schools state that adolescents who are in poor health or who are worried about personal and family problems do not learn well in school. Teachers do not have the time or training to address the health and mental health needs of all students. Providing these services within the school should allow teachers to concentrate on the job of teaching. The full service school also advocates curriculum reform. "Restructured schools attend to individual differences, give staff a wide range of choices regarding teaching methods, organize curricula that are stimulating and relevant, and eliminate tracking and suspensions" (Dryfoos, 1995, p. 152). Schools are also expected to prepare students to become effective citizens (Haggerty & others, 1994), and thus the social and emotional development of students is also central to the schools' mission.

Although full service schools are too new for us to know how well they work, initial studies of these schools are very positive. These programs appear to have promise for reducing school dropout, substance abuse, and teen pregnancy rates and for improving school attendance and achievement (Dryfoos).

IMPROVING THE SCHOOLS: COMMUNITY SERVICE LEARNING

community service learning
A method of educational reform that involves students in service activities in community settings in conjunction with related academic study in the classroom.

Some schools have sought educational reform by building partnerships with community agencies and involving students in service learning activities. An educational philosophy of reform, **community service learning** engages students in active participation in organized service experiences that meet real community needs. The service experience is integrated into the academic curriculum and involves time for talking, thinking, and writing about what the student learned, gained, and lost through the service activity. The opportunity to apply academic skills and knowledge to real life is believed to be valuable because it enables young people to see the relevance of their learning and their power to make a difference in the world. Service learning is taught through problem solving and interactive and student-centered educational experiences, which may be more interesting and motivating for the student who has become bored and disinterested in school (Alliance for Service Learning, 1993). To implement these types of learning experiences and to build partnerships with community agencies, many high school administrators would have to drastically change the ways they think about and structure education.

As a method of educational reform, service learning focuses strongly on the social and emotional development of students and their preparation as future citizens (Perrone, 1993). At a time when adolescents are often viewed negatively by the community, service learning redefines youths as a source of solutions rather than as a cause of problems. The Carnegie report *A Matter of Time* (1992) describes how many young people have become alienated from adults and social institutions and have few opportunities to work closely with supportive adults. Young people and adults can learn respect, trust, kindness, caring, and ways to work together effectively as they collaborate in community service activities. Yates (1995) suggests that community service helps adolescents develop feelings of effectiveness, relationships to others, and awareness of moral and political issues, and thus it stimulates identity development. Kirkpatrick and colleagues (1998) found that adolescents who were involved in volunteer work had higher academic aspirations, better grades, higher academic self-esteem and more intrinsic motivation toward school work. Although much

WHAT'S YOUR VIEW?

Do Full Service Schools Fit the Mission of Public Education?

The concept of the full service school greatly expands the kinds of services provided through the public schools. Are these services consistent with the mission and functions of the public school system? Will these services improve academic learning or distract educators and students from the central mission of the school? Will more resources be brought into the public schools or will existing resources be spread out more thinly?

work needs to be done in formal evaluation of community service learning, reports from middle schools and high schools that have implemented such programs are very positive and promising (Anderson & others, 1991; Gonzalez & others, 1993; Kelliher, 1993; Kirkpatrick & others,1998).

IMPROVING THE SCHOOLS: THE SCHOOL-TO-WORK TRANSITION

Just as the Carnegie reports called attention to limitations of the public education system in developing the academic skills of youths, reports prepared by the William T. Grant Commission on Work, Family and Citizenship (*The Forgotten Half*, 1988) and the Commission on Skills of the American Workforce (*America's Choice*, 1990) have called attention to the failure of the public schools in preparing non-college-bound youths for the world of work. High school guidance counselors typically serve large numbers of students by compiling college catalogs and applications. In comparison with the large public expenditures spent on state-supported colleges, minimal public funding is invested in helping non-college-bound youths prepare for and enter the world of work (Smith & Rojewski, 1993). According to Glover and Marshall (1993), the United States is far behind all other industrial nations in its approach to the school-to-work transition.

Our poor investment in non-college-bound youths is problematic for a number of reasons. As indicated by the title of the William T. Grant report, many youths are forgotten. Only 20 percent of high school graduates receive a college diploma within five years after their high school graduation (Smith & Rojewski, 1993). As a result, we are failing to provide many youths with the technical skills needed for employment in our increasingly technological society. This results in a shortage of skilled labor and a high level of underemployment and unemployment among high school graduates. Many high school graduates obtain jobs in food services and retail, which require no training and offer no opportunities for advancement (William T. Grant Commission, 1988). Large numbers of late adolescents who do not attend college are unemployed for a period of time. Poor and minority students typically have the most difficulty finding employment because middle-class parents often have contacts to assist their children in finding jobs (Glover & Marshall, 1993). D'Amico and Maxwell (1994) believe that the unemployment of adolescent African American males explains the continuing wage differences between African American and white males. Whereas the earnings of older African American men have risen substantially in recent decades, the gap between the earnings of younger African American and white males continues to be large. This, they believe, is due to the high levels of joblessness among younger African American men who have not been prepared for or assisted in entry into the job market. The failure of high school studies or the diploma to be helpful to obtaining or advancing in a job decreases student interest and motivation in high school. As you will recall from our discussion of high school dropouts, many adolescents say that they leave school because it will not improve their future. Achievement in school must be connected to rewards in the world of work (Glover & Marshall).

As a method of educational reform supported by the School-to-Work Opportunities Act of 1994, **school-to-work transition** programs seek to better integrate high school learning and job entry through school studies, work experiences, and activities that connect school and work. Programs combine school and work in specific occupational preparation. Through the public school, for example, students engage in the process of career exploration beginning in the elementary grades and leading to the selection of a career major by 11th grade. Students also master a core of strong academic skills, which provide preparation for employment and for higher education, to ensure flexibility for the adolescent who later decides to attend college. Work-based learning includes a planned program of job training that is coordinated with school learning and leads to a skills certificate. Connecting activities include matching students with employers, student mentorship, communication between schools and employers, and providing postprogram services such as job search, counseling, or further training to students. Apprenticeships, cooperative education, community service, and vocational academies along with new innovative programs may contribute to the school-to-work learning process (Glover & Marshall, 1993).

IMPROVING THE SCHOOLS: ESSENTIAL SCHOOLS

Theodore Sizer (1984, 1992, 1996), former dean of the Harvard Graduate School of Education, suggests that there are a number of imperatives for better schools:

1. Give room to teachers and students to work and learn in their own, appropriate ways. Teaching and learning should be personalized as much as possible.
2. Insist that students clearly exhibit mastery of their schoolwork. The diploma should be based on a successful demonstration of central skills rather than courses or credits completed.
3. Get the incentives right, for students and for teachers. There should be a set of common goals for the school, but the means for achieving these goals will vary for each student.
4. Focus the students' work on the use of their minds.
5. Keep the structure simple and thus flexible.
6. Each student should master a number of essential skills and be competent in certain areas of knowledge.
7. The school should communicate expectations of trust and high performance, but not by threat (I expect much of you, but will not threaten you to deliver that).

Giving teachers and students room to take full advantage of the variety among them implies that there must be substantial autonomy for each school. For most public and diocesan Catholic school systems, this means the decentralization of power from central headquarters to individual schools. For state authorities, it demands the forswearing of detailed regulations for how schools should be operated. It calls for the authorities to trust teachers and principals—and believe that the more trust placed in them, the more their response will justify that trust. This trust can be tempered by judicious accreditation systems as long as these do not reinfect the schools with the blight of standardized required practice.

❏ The purpose of decentralized authority is to allow teachers and principals to adapt their schools to the needs, learning styles, and learning rates of their particular students. The temptation in every school will be to move toward orderly standardization: such is the instinct, it seems, of Americans, so used as we are to depending on structure. Good schools will have to resist this appeal of standardization: the particular needs of each student should be the only measure of how a school gets on with its business. Greater authority is an incentive for teachers, one that will attract and hold the kind of adults which high schools absolutely need on their staff (Sizer, 1984, pp. 214–217).

Sizer maintains that most American high schools are modeled after a factory. "You parade a kid through the assembly line slapping on various parts: History is the fender, French is the steering wheel. . . . None of the parts are related" (Richardson, 1991). Instead, Sizer believes, students should be helped to master a limited number of essential skills. Rather than covering a great deal of material superficially, they should focus on a smaller amount of material in greater depth. Students should be taught to actively think about what they are learning through debate with their teachers and other students. Curriculum should be shaped more by the students' attempt to gain mastery rather than by the need to cover content. Sizer believes that most teenagers are very interested in learning when their minds are stimulated, rather than numbed, by the schools.

essential schools
Sizer's model of education that maintains that every student should master a number of essential skills.

Sizer has tried out some of his ideas in the schools. Now a professor at Brown University, Sizer directs the Coalition of Essential Schools, which acts as an advisor and mentor to 52 different schools. Some of the **essential schools** have been remarkably successful. In the period of just one year, one Providence, Rhode Island, high school increased the number of its graduates attending college from 6 percent to 90 percent (Richardson, 1991).

Sizer (1997) believes that the recent proposals for national standardized testing contradicts many of the suggestions for effective schooling by excluding parents and local educators. The exclusion of parents and local educators minimizes the authority of teachers, administrators and parents to create schools that meet community needs.

Work and Vocational Choice

Which approach to educational reform do you recommend—full service schools, community service learning, school-to-work transition, or essential schools?

❏ Employment is a joke for most people and it's a joke for me. I'm growing up in a poverty-stricken area. It's hard trying to find a job. Day in and day out I'm looking in the want ads. Usually you have to be 18 years old. Most of the time the job is in some community I've never heard of. Every time something turns up that I'm qualified for, it's way out of my district. Once I went to a Wendy's because they had an ad in the paper. When I got there, I was told that there weren't any more applications in the store and to come back tomorrow. The next day I came back and I was told the ten positions had been filled. Then I asked the man was he prejudiced?

A student reported in Williams & Kornblum (1985, p. 33)

Adolescents who are preparing to enter the world of work have many possible career choices from which to select. The surge in high technology and internet opportunities present new challenges. Their great-grandfathers may have entered the business or trade of their fathers, giving little consideration to other choices. Few adolescents today automatically follow in their fathers' occupation, however. Although their great-grandmothers probably believed that few career options were open to them, young women today choose from almost all occupations. In some ways, however, adolescents' choices continue to be limited by available educational and economic opportunities, such as where you live, how much money your parents earn, and what kinds of jobs are available to people with your level of training. Although almost half of all high school graduates enter the job market directly following graduation, high school counselors and teachers are primarily concerned with assisting college-bound youth in the transition to college (Glover & Marshall, 1993). While college enrollment should be encouraged for many adolescents, Glover and Marshall and others (William T. Grant Commission, 1988; Commission on Skills of the American Workforce, 1990) express great concern that American high schools have no systematic procedures for helping high school graduates locate and maintain appropriate employment. "Most high school graduates not going to college are left to sink or swim—without advice or career counseling and without job-placement assistance" (Glover & Marshall, p. 588). This gap in career services most seriously affects poor and minority students, who have few connections for obtaining jobs with good prospects for advancement. Glover and Marshall believe that comprehensive school-to-work transition programs must be developed in our secondary schools (See discussion of the School-to-Work Transition earlier in this chapter).

To better understand how adolescents make career choices and to be better able to guide them in that process, we look now at current theories of vocational choice and influences on career development.

THEORIES OF VOCATIONAL CHOICE

Parsons's theory
A theory of vocational choice that suggests that people can make better choices if they match their skills and interests to the requirements of the world of work.

The first theory of vocational choice, **Parsons's theory**, was offered by Frank Parsons, who believed that if people would choose a vocation or career instead of just hunting for any job, they would be happier and more successful in their work. Parsons (1909) developed techniques to help people identify their career interests and abilities so they could "match" these traits to the requirements of different career fields. Parsons's belief that people can make better career choices by matching their skills and interests to the requirements of the world of work remains central to some of the contemporary theories of vocational choice.

These matching models offer little explanation of how peoples' interests and abilities develop over time, however. Developmental theories of vocational choice help us to better understand how a person's developmental level is important to the process of career choice. We now look more closely at several theories of vocational choice: those of John Holland, Donald Super, and the sociological views.

AN APPLIED VIEW

Identifying Your Personality Style

To identify your personality style, we recommend that you use a well-known interest inventory such as the *Self-Directed Search*, the *Strong Campbell Interest Inventory*, or the *Vocational Preference Inventory*. If you take these inventories, review them with a counselor who has training and experience in interpreting these instruments. A number of brief exercises like the one that follows have also been developed to give individuals an idea of their personality-related interests.

For this exercise, imagine that you are flying your airplane over the desert. You are running low on gas and are forced to land your plane on a desert oasis. There are six groups of people in the oasis; each group has the personality characteristics described in the following sections. You have to select a group to stay with until outside help arrives.

- _____**R Realistic type**—people who like athletics and mechanical activities, who like outdoor activities, who like to work with tools, machines, plants, or animals. They generally prefer working with their hands rather than with people. (An example of a realistic occupation is carpentry.)

- _____**I Investigative type**—people who like to watch and learn about the world around them, who are generally quiet, who spend a lot of time thinking in order to solve difficult problems. (Scientists are examples of the investigative type.)

- _____**A Artistic type**—people who like to think up new ideas, who are very creative, artistic, imaginative. (Musicians and interior decorators are artistic types.)

- _____**S Social type**—people who like to be with other people, who are talkative, friendly, and interested in helping or teaching others. (Teachers and social workers are social types.)

- _____**E Enterprising type**—people who like to influence, lead, and persuade other people, who are sociable and outgoing. (Salespeople, lawyers, and business executives are involved in enterprising careers.)

- _____**C Conventional type**—people who like to work with numbers and are good at carrying out details, who are neat, orderly, organized, and practical. (Typists, file clerks, and financial experts are conventional types.)

Which group of people would you prefer to be with until outside help arrives? Place the number 1 on the line next to the description of that personality. Assuming you could not join your first choice group, what would be your second choice? Your third choice? Write the numbers 2 and 3 next to each of those choices. Look back at the descriptions of the groups you chose. The description of each group fits with one of Holland's personality types. Did you choose to be with people who are similar to you or different from you? How strongly did you feel about your choices? Was it difficult for you to make each choice? What do your choices tell you about the kinds of activities you enjoy? About the kind of work environment you would prefer?

Adapted from Borchard and others (1988).

Holland's theory
A theory describing vocational choice as the matching of an individual's personality type to the type of environment in which the work takes place.

- *Holland's theory of careers*. John Holland (1985a) has developed a theory of careers that, like Parsons's approach, involves the matching of an individual to a work environment. In **Holland's theory**, the matching is between an individual's personality type and the type of environment in which the work takes place. Holland believes that most people can be categorized as one of six personality types, and every job can be classified as one of six work environment types. The categories for the personality types and the work environments are identical, because the work environments are defined by the type of people who work in them. Each of the six types are described briefly in the Applied View box titled "Identifying Your Personality Style."

realistic type
The Holland vocational personality type who prefers working with tools, machinery, animals, and the outdoors.

investigative type
The Holland vocational personality type who prefers working with ideas and is generally analytical, methodical, and precise.

artistic type
The Holland vocational personality type who is creative, expressive, and nonconforming.

social type
The Holland vocational personality type who enjoys working with and helping other people.

enterprising type
The Holland vocational personality type who enjoys business activities directed toward making money or achieving organizational goals.

conventional type
The Holland vocational personality type who enjoys working systematically with numbers, filing clerical records, and copying printed information.

Super's theory
An influential theory of career development presenting a series of developmental stages or tasks essential for the development of career identity and maturity.

Some part-time jobs help teens explore their career identities by testing out their interests and abilities in the world of work.

According to Holland, every person (and work environment) can be classified according to several types. An "SAE type" would describe a person who is primarily social, but also has characteristics of artistic and enterprising types. The best job choice for an SAE personality would be an SAE work environment. When an individual's personality and occupation are a good match, Holland says that they are "congruent."

Holland's theory has had an enormous impact on the field of career guidance. His career assessment instrument, the *Self-Directed Search* (SDS) (Holland, 1985b), is quick and easy to use and very popular. The SDS, however, has been criticized as sexist because most women fall into the social and artistic types, which fit the stereotypes of women as sensitive, caring, and expressive. Although this is probably not the fault of the measure since these types reflect the influence of socialization on girls and women, the results are not helpful when counselors are trying to help young women broaden the range of occupational choices they are considering. Adolescent girls, for example, may receive low scores on the realistic type because they never tried working with tools or machines and did not learn these skills from their fathers. Receiving a low score on the R (realistic) scale may further discourage an adolescent girl from experimenting with tools and considering occupations in which tools are used.

• *Super's developmental theory.* Donald Super has been one of the most influential figures in advancing theories of career choice and development during recent years. Super (1957, 1983, 1990) developed a life-stage theory of career development to explain how career identity develops over time and to determine a person's readiness to make a career choice. **Super's theory** describes five career stages, which he originally associated with different developmental periods. In more recent revisions of his theory (Super & Thompson, 1981), Super suggests that we recycle through each of these stages several times during our lives.

exploration stage
According to Super's theory, the stage of career development that most teens are at.

The **exploration stage** is associated primarily with ages 15 to 24. Through school, leisure, and part-time work activities, adolescents and young adults are exposed to a wide variety of experiences. Through these experiences, they further define their self-concepts and have the opportunity to test their abilities and interests. Early in this stage, initial work-related choices are made by assessing interests, abilities, needs, and values. Through the course of exploration, adolescents test out their interests and abilities in the real world and gradually become more realistic about a career choice. By the end of this stage, a beginning full-time job is often selected. Super believes that most adolescents are not ready to make definite career choices because they have not yet had the chance to adequately explore available opportunities.

A number of writers (Blustein & others, 1989; Neimeyer & Heesacker, 1992; Vondracek, 1993; Wallace-Broscious & others, 1994) have described the similarities between Super's theory of career development and Erikson's theory of identity development (see Chapter 6). According to Erikson (1963), a child's basic attitudes toward work are formulated during the elementary school years as the child develops either a sense of industry or one of inferiority. The growth stage is similarly important in Super's theory as providing the foundation for work attitudes. For Erikson (1959), forming an occupational identity is a critical aspect of identity achievement. Role diffusion is common as the adolescent initially explores unrealistic and unattainable career choices. Exploration preceding eventual commitment to a more realistic career choice are common to both Erikson's model of identity and Super's theory of career development. Research (Blustein & others; Wallace-Broscious & others) suggests that career maturity, including career exploration, career decidedness, and career planning, is high among adolescents and college students identified as identity achieved and low among those identified as identity diffused.

- *Sociological views of career choice.* Matching and developmental theories of career choice were developed by psychologists. Sociologists view career choice somewhat differently. Whereas psychological theories assume that the individual has at least a moderate degree of control in shaping career development, sociological theories view individual choice as limited by the social status of one's parents and the structure of social organizations (Hotchkiss & Borow, 1990).

The career development of an individual is related to the social status of parents. The educational level, type of job, and amount of money earned by parents influence the amount and quality of education the children will receive, which in turn affect the occupational level the children will realize as adults. When speaking of the structure of organizations, sociologists are referring to the unwritten rules in our schools, businesses, and communities that determine who has the most power and authority, who has the most opportunity to advance, and who receives the most money. An individual's race, ethnic background, gender, and school attended are a few of the unwritten rules that sometimes influence who gets hired and who gets promoted.

AN APPLIED VIEW

Theories of Vocational Choice

The vocational choice theories we have reviewed can be useful in guiding adolescents with educational and career concerns. Each theory offers a unique perspective on Karen's concerns.

❏ Karen is an eleventh-grade student at a small-town high school. She lives with her parents and two younger brothers. Karen's mother is a home-maker; her father is an advertising agent and commercial artist. Both parents are high school graduates. Karen is in the academic track in school, earns mostly As and Bs, and says she enjoys most of her classes. She has many friends at school, most of whom are planning to go to college. Karen expresses the following concern to the school counselor, "I always thought I'd go to college after high school, but now I'm not so sure. I really have no idea what I'd like to take up there or what I'd like to do once I have a college degree. I'm afraid I'll end up wasting four years of time and money on college and still not know what to do with my life when I'm through."

A counselor using Parsons's theory would work with Karen to assess her career interests, values, and abilities. Vocational interest and aptitude testing might be completed. From this Karen would learn more about the kinds of careers in which she would probably be successful and happy. Identifying her career interests and abilities and matching those with work opportunities would prevent Karen from "*wasting four years of time and money on college.*"

The counselor using John Holland's theory would also try to match Karen with the work environment. Karen's personality type would be identified through

administration of one of Holland's personality type inventories, such as the *Self-Directed Search*. The counselor might help Karen identify a college major that would be a good match with her personality type.

Applying Donald Super's theory, the counselor might wonder whether Karen is ready to make a career decision or choose a college major. The counselor would explain to Karen that adolescence is a time for exploring interests and developing the self-concept. Karen should not be concerned that she does not yet know what her major will be. Karen's readiness for career decision making might be evaluated through a test like the *Career Development Inventory* (Super & others, 1981). Based on the results of that test, the counselor would help Karen develop a plan for further exploring her interests, learning about the world of work, or gaining decision-making skills.

Sociological theory would lead the counselor to consider how Karen's social class and gender may be affecting her college and career plans. Do Karen's parents have the money to send her to college? Does Karen believe that her parents' money should be saved for her younger brothers' education? If financial concerns are an obstacle to Karen's college plans, the counselor might help Karen explore sources for obtaining financial aid.

As you can see, offering educational and career guidance to adolescents is complex. Theories provide some direction for the counselor. The counselor must complete a careful assessment and listen closely to the adolescents' concerns. It is important not to offer simple advice to complex career issues.

Sociologists have added to our understanding of career choice and development by making us aware of the obstacles to occupational mobility. They remind us that we cannot just blame the individuals who fail to achieve in our society. Instead, sociological principles remind us that we must work to reduce ethnic, racial, and gender barriers, which limit access to higher-level positions for many members of our society.

INFLUENCES ON VOCATIONAL CHOICE

In this section, we explain several factors that influence career choice: the family, gender, and race.

The Family

Several writers (Bratcher, 1982; Lopez & Andrews, 1987) have considered the ways in which the **family system** influences career selection. Families develop certain rules and boundaries that influence the roles of all family members, including career roles. The idea that some families always produce farmers or doctors or businesspeople is a reflection of family rules. Family systems also seek to maintain a status quo—keeping things the way they have always been. When a family member tries to do something different, the family system may be upset, putting pressure on the individual to conform to family tradition. Adolescents may be afraid that their career choices will disappoint their parents. Families also have boundaries that can be either too rigid or too weak. When the boundaries between the family and the outside world are rigid or the boundaries between parents and their children are weak (family members are too close), the adolescent will find it difficult to separate from the family and achieve her or his own career identity.

During adolescence, family rules and boundaries need to become somewhat more flexible, allowing children to explore and form their own vocational identities. Sometimes one parent is very close to an adolescent child and takes sides with that child against the other parent. In this case, one parent generally supports the child's career choices while the other parent is critical. The adolescent is caught between the parents' conflict and has a hard time moving ahead with career planning. When late adolescents experience anxiety about career decision making, problems in the family system may be at fault.

Gender

Although career opportunities have increased dramatically for women over the past 20 years, women continue to be underrepresented in certain school subjects, college majors, training programs, and occupations. Women are less frequently found in higher-level technical and managerial occupations and are more often found in less well paid clerical, social service, and educational positions.

Sundal-Hansen (1984) believes that women are less likely to be successful in the career world because of the **gender-role system:** the attitudes and beliefs concerning the ways in which the abilities and personalities of women are different from men. Based on the stereotypes of women as dependent and passive and the stereotypes of men as aggressive and independent, labor has been traditionally divided into what is recognized as "women's work" and "men's work." Women's work has historically been of lower status and lower pay.

Sundal-Hansen (1984) describes how women acquire gender-role beliefs. Little girls begin to learn stereotypes at an early age. Those stereotypes are reinforced by parents, teachers, peers, and the media. Because of what they view as appropriate for boys and for girls, girls are less likely to become involved in activities, such as athletics, math, science, and leadership positions, that would prepare them for higher-level employment. Gottfredson (1981) found that children limit their view of career options as early as elementary school based on what they believe is appropriate for their gender and social status.

family system
Family rules and boundaries that help to maintain the status quo or keep things the way they have always been.

gender-role system
The set of attitudes and beliefs about the ways in which the abilities and personalities of men and women differ and are suited to different career choices.

self-efficacy
Our self-expectations or beliefs about what we can accomplish as a result of our efforts, which influences our willingness to attempt the task and the level of success we achieve.

outcome expectations
Beliefs learned from teachers, parents, friends, and past experience about one's abilities and expectations for success and failure in areas such as school and career.

Hackett and Betz (1981) have borrowed the concept of **self-efficacy** from Bandura's model of social learning (see Chapter 2) to explain why women are generally less successful than men in the career world. Self-efficacy refers to a person's beliefs about whether he or she will be successful in certain tasks. Those beliefs influence whether an individual will attempt a new task. For example, an adolescent girl who does not believe she will be successful in math will be less likely to attempt math courses or consider a math-related occupation. She might also believe that even if she was good at math, she might not get hired because she is a girl, or that she would be disliked by others because she is good at math. These beliefs or **outcome expectations** are learned from parents, teachers, and peers and reflect the system of gender-role stereotypes. Hackett and Betz believe that self-efficacy and outcome expectations help explain how gender-role stereotypes influence the career development of women.

Although research concerning gender and career choice has focused on women, it is important to understand occupational choice for adolescent and adult males, too. Just as women have suffered from gender-role stereotypes, so have many men. Males are generally expected to be leaders, to develop mathematical and mechanical abilities, to be physically strong and athletic, and to have little interest in caring for others. Skovholt and Morgan (1981) described some of the ways in which gender-role stereotypes have limited the career choices of men. Men are generally expected to be the breadwinner in the family and are expected to choose an occupation that pays enough to support the family. While women have more freedom today in choosing to work inside the home or in other employment, most men continue to feel that choosing a home-making role is unacceptable.

Male identity and self-esteem often depend on career success. Men who do not advance in their careers or are unemployed often have difficulty feeling good about themselves. They are more likely to have health problems and are more likely to become angry or abusive in the home because of the depression, frustration, and humiliation they feel. In periods of economic decline or recession, many men are likely to feel anxious and defeated because possibilities for advancement are limited to few workers. Because of the traditional emphasis on male career success, many men have had to spend so much time at work that they have had little time to enjoy their families. In working with adolescent males, it is important to help them think of ways to feel good about themselves beyond career success and to be aware of the physical and emotional costs of job stress.

Adolescents' attitudes about work and gender-roles have changed in recent years in ways influenced by cultural values. In 1970, Sundberg and Taylor investigated 9th graders in the United States and the Netherlands and assessed their awareness of possible occupations and leisure activities. In both countries, girls were slightly more aware of occupations than were boys of the same age, but viewed fewer occupations as likely choices for themselves. Some interesting cross-cultural differences were also found. Dutch adolescents were aware of more occupations, but American adolescents were aware of more leisure activities, leading Sundberg and Tyler to hypothesize that U.S. adolescents might be

To what extent do the career choices of teens continue to be limited by gender-role socialization?

more pleasure oriented than Dutch adolescents are. Stiles and others (1993) decided to find out how the work and leisure attitudes of Dutch and U.S. adolescents have changed since 1970. Girls continue to demonstrate more awareness of occupational choices than boys do and to consider fewer occupations as actual possibilities for themselves. U.S. students linked work with success, achievement, and wealth and viewed work as more important than the Dutch adolescents. U.S. adolescents depicted the ideal man as ambitious, achieving, and earning a high income. The Dutch adolescents indicated greater concern for the quality of life, emphasizing the importance of sports, relaxation, and humor. Dutch girls were more likely than U.S. girls to value internal qualities in the ideal man, such as being caring, kind, happy, and making time for children.

Race

American youths have not benefited equally from the economic opportunities our nation has to offer. Within our labor force, dead-end jobs, offering little opportunity for advancement, are often filled by racial and ethnic minorities. African American males fill a greater proportion of R (realistic) jobs, such as the trades, and a much smaller proportion of E (enterprising) jobs, such as sales (Gottfredson, 1978). At every level of education, including college graduates, more African American youths are unemployed and underemployed than whites (Wetzel, 1987). In 1993, unemployment rates stood at 12.9 percent for all African Americans, 38.9 percent among 16- to 19-year-olds, and 22 percent among 20- to 24-year-olds. In the same year, rates for whites were 6 percent overall, 16.2 percent among 16- to 19-year-olds, and 8.7 percent among 20- to 24-year-olds; and rates for Latinos were 10.6 percent overall, 26.2 percent among 16- to 19-year-old teens, and 13.1 percent among 20- to 24-year-olds (U.S. Bureau of the Census, 1994). According to recent statistics, African American men 25 years of age and older who had completed four or more years of college earned an average of $31,380 per year in comparison with white men of equal education, who earned an average of $41,090 (Bovee, 1991). Earnings of African American men are only 86 percent of what white males earn for the same job (U.S. Bureau of the Census, 1994).

As of March 1992, unemployment rates stood at 5.9 percent for white women 16 years of age and older and 7.8 percent for white men. Among African Americans, the rates stood at 12 percent for women and 16.3 percent for men (U.S. Department of Labor, 1992). Unemployment rates among young adults are even higher. Among workers between the ages of 20 and 24, unemployment rates were 20.3 percent for African Americans, 11 percent for Latinos, and 6.8 percent for whites (William T. Grant Commission, 1988). Those who are most at risk for futures of chronic unemployment include school dropouts, pregnant teens, and youths who are performing poorly in school, who have a history of drug or alcohol abuse, or who have a criminal record. When teenagers experience long periods of unemployment and idle time before the age of 20, they are often headed for a lifetime of unemployment.

In discussing the influence of race in career development, it is essential to recognize that race and social class are often confused. The statistics given here describe those African Americans who are economically disadvantaged. Many

African Americans hold high-level jobs and earn large sums of money. The fact that a larger percentage of our racial minorities fares less well in our labor market leads us to ask why.

Researchers who explain high rates of unemployment among disadvantaged minorities focus on the characteristics of the unemployed, on characteristics of the society, or on a combination of both. When people who are employed are compared with those who are not, we often find that the employed have higher-level skills, better work attitudes, and are more highly motivated to succeed. These explanations suggest that individuals are to blame for their own unemployment—"if they only worked harder they would succeed." These explanations are thus sometimes referred to as "blaming the victim" (Ryan, 1976). Social explanations (Wellman, 1977), in contrast, blame the racist characteristics of our schools, political system, and economy.

job ceiling
The belief among minority youth that no matter how hard they work, they are limited by their race in how much they will be allowed to achieve.

Ogbu (1983) maintains that minority youth believe that a **job ceiling** exists in this country. That is, they believe that no matter how hard they work and how well they do in school, there is only so high they can go in the job market because of their race. Children learn about the job ceiling early in life by observing unemployed and underemployed parents, relatives, friends, and neighbors. Youths believe furthermore that only those who are willing to adopt an "Uncle Tom" attitude or to act as if they are white are likely to succeed. Fordham and Ogbu (1986) identified an "anti-academic" achievement ethic

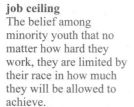

AN APPLIED VIEW

Career Development: Interventions for Economically Disadvantaged Youths

Your understanding of the causes of youth unemployment will influence your approach if you are to work with economically disadvantaged teens. If you believe that qualities of the individual are most important, you might focus on increasing teens' awareness of the importance of work behaviors such as punctuality, loyalty, and pride in one's work. You might help students learn more about their career abilities and interests and how they might be applied in future careers. You should also emphasize the development of academic skills. Your focus will be to help students adjust and adapt to existing conditions.

If you believe that our society is at fault, you might take the role of a social activist (Atkinson & others, 1977), trying to change the system of teaching and rewards in our schools and places of work. You might believe that schools need to be more sensitive to the cultural differences among students. Your focus will be to change the school and not expect the student to adjust to the existing school climate.

A third position recognizes that racial barriers exist in our society, but also recognizes that social change is

likely to be slow. According to this view, youths need to recognize that racism exists, but need to learn how to cope with it. You might help youths develop effective survival skills to cope with the stress of attending a predominantly white school or entering a high-powered predominantly white occupation. Developing a positive racial identity and feelings of racial pride might be important in helping minority youths feel good about themselves in predominantly white settings. You should also be concerned that students prepare themselves academically to meet the challenges ahead. It is important to maintain high expectations for students and not try to make up for past discrimination by lowering expectations (Hawks & Muha, 1991). You might also want to advocate for government-supported youth employment and training programs or for private programs, which provide financial support for higher education for economically disadvantaged youths.

Any of these programs may help youths take advantage of available opportunities. Which approach do you think you might take? Which specific strategies would you use? Why?

among African American and Latino high school students. Academic success was described by the peer group as "acting white," and many students knowingly limited their achievement by cutting classes and not studying. Ogbu and Fordham maintain that minority youths become distrustful of schools and businesses and adopt the attitude of "What's the use of trying?" What might be viewed as lack of motivation by some is viewed by Ogbu as a reasonable response to limited opportunities. Ogbu believes that programs such as Headstart, compensatory education, and special college admission, recruitment, and financial aid programs are necessary to undo the effects of past discrimination and help minority children to believe that America will give them an equal chance.

The job ceiling theory proposed by Ogbu and his associates (Fordham, 1988; Fordham & Ogbu, 1986; Ogbu, 1983) has become popular in understanding achievement motivation among African American youths. Taylor and associates (1994) found some support for the theory in their study of 344 African American and inner-city youths. Adolescents who are most aware of racial discrimination and perceived limited opportunities for achievement were likely to be uninvolved in their schoolwork. Schoolwork was not viewed as important to them. Perceptions of discrimination were not related to ethnic identity, however. Those teens who reported positive ethnic identity were likely to be doing well in school. Thus, it seems that adolescents' perceptions of external barriers to good jobs and career success are likely to limit their achievement. Adolescents do not, however, seem to limit their achievement out of allegiance to their ethnic/racial identity or to not "acting white." Consistent with our discussion of ethnic identity in Chapter 6, a positive sense of ethnic and racial identity is associated with healthy functioning in many contexts, including school.

THE WORKING TEEN

More high school students than ever before seem to be employed. In 1987, 41 percent of high school students reported that they were currently working and 75 percent of high school seniors said they were working or had been working very recently (Bachman & others, 1987; Fine & others, 1990). Yamoor and Mortimer (1990) found that the majority of teens between the ages of 11 and 16 reported paid work outside their home. Many adolescents spend numerous hours in the workplace.

Whether or not it is good for adolescents to work has been a topic of recent debate. Fine and associates (1990) summarized some of the arguments in favor of teenage employment. Through part-time employment, adolescents can build feelings of self-confidence, develop a sense of responsibility, increase positive work values and behaviors like following directions and being dependable, and learn more about the world of work and the value of money. Super (1957) viewed part-time employment as an important experience for adolescents in exploring the world of work, defining the self-concept, and testing out one's interests and abilities in the real world. Through part-time work experiences, teens might also develop skills that can be useful in later employment. When

Some part-time jobs teach few marketable skills and offer little opportunity for advancement. Part-time employment can have negative effects when teens work long hours, neglect other responsibilities, and learn few skills.

work experiences are properly arranged and supervised and provide meaningful experiences, they can help the adolescent develop a sense of industry and a vocational identity (Vondracek, 1993).

The circumstances of teenage employment are much different today than they were years ago. As a result, teenage employment may offer fewer benefits than it did in years gone by. Very few teens contribute their earnings to help out their families (Greenberger & Steinberg, 1986). Working adolescents spend most of their money on clothing, records, tapes, sports equipment, and entertainment (Bachman & others, 1987). Most of the jobs held by adolescents today are in retail and service areas, especially the fast-food industry, that offer little or no chance for advancement. They have little interaction with adults, working mostly with other teenagers, and learn few skills that can be used in future employment (Greenberger & Steinberg, 1981; Greenberger & others, 1981). In previous eras, young people worked closely with adults to learn trades and skills that they would use in their adult work.

Some research (Mortimer & others, 1992) suggests that part-time employment can be beneficial for teens when job responsibilities are complex and when the work tasks have some relevance to school learning and provide job skills that will be useful in the future. Other findings suggest that work can also have negative consequences for teenagers. According to Mortimer and associates, job stress contributes to depressive feelings, self-blame, and decreased internal control among ninth-grade boys. Greenberger and Steinberg completed a series of reports in the 1980s that indicated that working more than 15 to 20 hours per week was related to poorer school performance, less involvement in school activities, decreased closeness to parents, increased drug and alcohol use, and more negative work attitudes (Greenberger & Steinberg, 1981; Greenberger & others, 1981; Steinberg & others, 1982). It seems that the kinds of low-level jobs available to many teens contribute to poor work attitudes—e.g., that a person is foolish to do more work than he or she has to (Greenberger & Steinberg, 1981).

Should teens work— what do you think?

Students who work 10 hours or more per week have been found to do less well academically, report more psychological symptoms (depression, anxiety, tiredness, difficulty sleeping) and somatic symptoms (headaches, stomachaches, colds) than do students working less than 10 hours a week or not at all (Steinberg & Dornbusch, 1991). They are also more likely to use drugs and be involved in delinquent activities such as theft and vandalism. Students who work less than 10 hours per week reported self-esteem equal to students who did not work at all. Self-esteem was lowest among adolescents working more than 20 hours each week.

It seems that working a small number of hours can be positive for teens. As the number of hours increases, however, other areas of the teen's life suffer. Teens who work long hours have difficulty paying attention at school and are more likely to cheat, cut classes, and copy other students' assignments. Working long hours adds to the stress of being an adolescent, resulting in more psychological distress and more drug use. Working teens also feel free from their parents' control. This may be because work keeps them outside the home for many hours or makes them less dependent on their parents for financial support. It is also possible that teens who do not get along with their parents choose to work longer hours as a way of staying out of the home. Whatever the reason, working long hours increases stress for teens and can deprive them of important sources of family guidance and support.

CONCLUSIONS

Concern about our educational system seems to occur in cycles—every 20 years or so. We are clearly in a period of deep concern now as we are asking the schools to shoulder more and more of the burden of child rearing. This is due in large part to the declining role played by such institutions as the family and religion. The number of African American and Latino youths attending our high schools is increasing, yet our educational system seems to be failing to meet the educational needs for many of these youths. At a time when our labor market needs workers with strong skills in math and technology, schools are being accused of gender and racial bias, which is limiting the achievement of girls and racial minorities in technological areas. Although improving educational opportunities for adolescent girls and racial minorities is of great concern, it is also important to keep in mind that academic achievement for both groups has risen steadily in recent years.

A number of suggestions have been made to resolve the current educational crisis and enable the schools to better meet the social and emotional needs of adolescents. Among these are the development of a core curriculum strong in academic skills, less academic tracking, the creation of smaller communities within larger schools, increased parental and community involvement in local education, and providing teachers with greater freedom and flexibility in designing programs to meet individual student needs.

Choosing a career is an important concern of adolescents. Adolescents have many possible career choices, but options can be limited by educational and economic resources and available job opportunities. Theories of career choice and development can be helpful in guiding adolescents and helping them to identify and remove obstacles to career choice. Part-time work experiences can provide a way for adolescents to explore career interests and learn about the world of work. When students attempt to work long hours while attending school, however, the consequences for emotional, social, and academic development are often negative.

CHAPTER HIGHLIGHTS

Crisis in the American Secondary Schools

- The results of several influential reports indicate that American secondary schools are failing to provide a quality education and improve social and economic equality for all adolescents.
- Researchers from the Carnegie Council for Adolescent Development maintain that the developmental needs of early adolescents are not adequately met by most middle schools.
- The practice of academic tracking may benefit more capable and economically advantaged students but limit the motivation and achievement of economically disadvantaged students.
- Secondary schools have been accused of adding to the gender and racial bias in our society.
- Students drop out of high school for a variety of reasons related to family responsibilities, poverty, poor job opportunities, and not being understood by school teachers and administrators.
- Although many minority youths drop out of school before graduation, the number who manage to succeed academically is usually underestimated by the public. Caring family and friends, positive school experiences, and feelings of personal control seem to contribute to academic success.

Improving the Schools

- Full service schools seek to reduce problems that interfere with student learning by providing a variety of health and social services in the school setting.
- Community service learning initiatives focus on preparing students to be better citizens and to apply academic skills to real-life problems.
- The school-to-work initiatives are concerned with the preparation and entry of noncollege youth into the labor force.

- Sizer believes that high school students need to be encouraged to think critically and to master a limited number of essential skills. Teachers need to be given more freedom in designing a curriculum to meet the needs of individual students.

Theories of Vocational Choice

- The most influential theories of vocational choice include Parsons's theory, Holland's matching theory, Super's developmental theory, and sociological theory.
- Holland's theory postulates six vocational personality types: realistic, investigative, artistic, social, enterprising, and conventional.
- Super's theory presents five stages of vocational development. Exploration of career interests is the central task for adolescents.
- Sociology theory helps us understand how career choices and advancement can be influenced by socioeconomic status, gender, ethnicity, and race.

Influences on Vocational Choice

- Characteristics of the family system, such as family rules and boundaries, need to be flexible to allow adolescents to explore their vocational identities.
- Although career opportunities for women have increased dramatically over the past 20 years, adolescent girls continue to be exposed to gender-role stereotypes. Career success may be limited by lower self-efficacy and outcome expectations.
- Males are also affected by gender-role stereotypes. Self-esteem is dependent on career success.
- Racial and ethnic minorities have not achieved as high a level of career success as the white majority in this country. Racial barriers, poverty, and discrimination must be considered in understanding the unemployment and underemployment of minority youths.

The Working Teen

- Although several writers have described adolescents as having an aversion to work, in fact many teens work so many hours while also attending school that it is injurious to their health and learning.

K E Y   T E R M S

artistic type 339
community service learning 333
conventional type 339
enterprising type 339
essential schools 336
exploration stage 340

family system 342
full service schools 331
gender-role system 342
Holland's theory 338
investigative type 339
job ceiling 345

outcome expectations 343
Parsons's theory 337
realistic type 339
school-to-work transition 335
self-efficacy 343
social type 339
Super's theory 339

W H A T   D O   Y O U   T H I N K ?

1. What are the main problems of secondary schools? What causes these problems?
2. Should our schools have a narrow focus (improving cognitive skills) or a broad focus (nurturing mental health, teaching appropriate values, feeding hungry children, implementing safe sex, etc.)?
3. What effects do modern electronic media (radio, TV, cassettes, and CDs) have on the intellectual development of youth?
4. How should schools interact with these media to better educate teens?
5. What are some of the approaches suggested by current psychologists for improving our schools?
6. What are some suggested alternatives to today's system of high school education?
7. Does our society need "rites of passage" to accompany high school experience?
8. Should secondary schools be preparation for life? For the world of work? For training the intellect?
9. In your opinion, which theory of vocational choice is most useful?
10. Do gender-role and sexual identity affect vocational choice? How?
11. Do teens work too much? If so, what should be done about it?

SUGGESTED READINGS

Douglas, M. C. (1983). *Go for it! How to get your first good job: A career-planning guide for young adults.* Ten Speed Press. Besides giving sound advice about selecting the right job for you (or your teenage client), this guide includes information on filling out forms, taking tests, and writing letters. Especially good are the 90 job descriptions, which include requirements and benefits. Labor laws that concern minors are also presented.

Orenstein, P. (1994). *Schoolgirls: Young women, self-esteem, and the confidence gap.* New York: Anchor Books. This compelling book was written in association with the American Association of University Women. It describes experiences of adolescent girls in their schools and how these experiences contribute to a decline in self-esteem.

GUIDED REVIEW

1. Sal benefits from school because he receives school lunches and free medical care. This function of the school is called
   a. literacy function.
   b. transmitting social values function.
   c. improving society function.
   d. human heritage function.

2. Rosenbaum argues that in order to be successful, tracking programs need to be more
   a. vocationally oriented.
   b. flexible.
   c. academically challenging.
   d. college oriented.

3. Adolescent girls and nonwhite males lag behind white males in the development of
   a. physical agility.
   b. social studies skills.
   c. foreign language skills.
   d. math skills.

4. The greatest determinant of who drops out of school is
   a. gender.
   b. grades.
   c. age.
   d. poverty.

5. According to recent research, all of the following are important in motivating students who are at risk for dropping out to improve their school performance, EXCEPT
   a. freedom from responsibilities at home.
   b. opportunities to think about and make sense of what they have learned.
   c. support from teachers and parents.
   d. opportunities to be successful at something.

6. Concern about the large numbers of noncollege-bound youths who lack the technical skills for entry into promising jobs has contributed to interest in which of the following educational reform initiatives?
   a. essential schools
   b. community service learning
   c. full service schools
   d. school-to-work transition programs

7. Roxanne, 17, is exposed to a wide variety of experiences and has the opportunity to test out her abilities and interests. She begins making her initial work-related choices. This stage is what Super describes as the
   a. formative stage.
   b. exploration stage.
   c. commitment stage.
   d. maintenance stage.

8. The idea that some families always produce farmers or doctors is a reflection of family
   a. rules.
   b. myths.
   c. choices.
   d. structures.

9. Leon believes that no matter how hard he works and how well he does in school, there is only so far that he can go in the job market because he belongs to a minority group. According to Ogbu, minority youths believe that there is a job ceiling because of the
   a. high rate of teenage pregnancy.
   b. high rate of drug abuse in the inner city.
   c. poor academic skills displayed by friends.
   d. high rate of unemployment and underemployment in their neighborhood.

10. Work experiences that are most likely to be beneficial for teens possess all of the following characteristics, EXCEPT
    a. little interaction with adults.
    b. tasks that are relevant to school learning.
    c. job skills that will be useful in the future.
    d. job responsibilities that are complex.

11. Schools are expected to fulfill three functions: literacy, improving society, and _____ (incorporating multiculturalism/transmitting social values).

12. The authors of the Turning Points report and other researchers believe that there is a _____ (good match/mismatch) between the developmental needs of early adolescents and the structure and curriculum of our middle schools.

13. For students entering into adolescence, parental involvement in the schools _____ (is no longer important/remains important) to academic success.

14. After observing in the schools and talking to teachers and administrators, Fine believes that the schools maintain the social class structure in our society. That is, the schools contribute to a cycle of _____ (poverty/violence) among inner-city Latino and African American youths.

15. According to Gregory, teachers who are interested in motivating students at risk for dropping out might _____ (leave them alone/offer more support).

16. An essential component of full service schools is _____ (family planning services/collaboration among a team of professionals).

17. _____ (Self-efficacy/Self-concept) refers to a person's beliefs about whether she or he will be successful in certain tasks.

18. Holland believes that most people can be categorized based on six _____ (personality/environment) types, and every job can be classified based on six _____ (personality/environment) types.

19. The exploration stage in Super's theory of career development has been compared to _____ (Erikson's model of identity development/Kohlberg's model of moral development).

20. Based on existing research, high school students should work _____ (more than/no more than) ten hours per week in jobs that are unrelated to their academic studies or career preparation.

Answers

1. C, 2. B, 3. D, 4. D, 5. A, 6. D, 7. B, 8. A, 9. D, 10. A, 11. transmitting social values, 12. mismatch, 13. remains important, 14. poverty, 15. offer more support, 16. collaboration among a team of professionals, 17. Self-efficacy, 18. personality, environment, 19. Erikson's model of identity development, 20. no more than

TEENS WHO HAVE SERIOUS PROBLEMS

STRESS
AND
MENTAL DISTURBANCE

Young teen-agers today are being forced to make decisions that earlier generations didn't have to make until they were older and more mature and today's teen-agers are not getting much support and guidance. This pressure for early decision-making is coming from peer groups, parents, advertisers, merchandisers and even the legal system.

David Elkind, 1989

CHAPTER 11

CHAPTER OUTLINE

Dealing with Stress

stress
Emotional tension arising from life events and feelings of threat to one's safety or self-esteem.

What is **stress**? Stress can be understood as emotional tension arising from life events, or as feelings of threat to one's safety or self-esteem (D'Aurora & Fimian, 1988).

In Chapter 1, we cited the oft-repeated warning of many parents that adolescence is the last chance to have fun, because becoming an adult means taking on the heavy responsibilities of maintaining a job and a family. Well, this is not to say that infancy, childhood, and the teen years are free of stress—far from it! In this chapter, we discuss sources of stress in the lives of teens and the ways in which teens successfully and not so successfully cope with stress. Teens who experience much stress and are not able to cope effectively sometimes face mental disorders. In the second half of this chapter, symptoms and causes associated with those disturbances are presented.

After reading this chapter, you should be able to:

- Define stress and identify common sources of stress during the teenage years.
- Describe the relationship between stress and physical illness and mental health.
- Identify effective and ineffective ways of coping with adolescent stress.
- Identify the symptoms associated with anorexia nervosa and bulimia nervosa.
- Describe four factors that contribute to the development of eating disorders.
- Identify the symptoms associated with depression and depressive equivalents.
- Explain the causes for depression based on three theoretical models.
- List the warning signs of adolescent suicide.
- Explain how individual, family, and social factors contribute to adolescent suicide.
- Discuss these issues from an applied, a sociocultural, and your own point of view.

As you are aware from your readings in this text, the adolescent years present many challenges for young people. Among the challenges the adolescent faces are coping with the bodily changes of puberty, adjusting to cognitive changes, developing new ways of relating with peers and family, dealing with sexual and other moral dilemmas, completing academic requirements, and planning for a future occupation. Although these are normal developmental tasks, they are sources of stress for most adolescents. In addition to developmental stressors, many adolescents experience additional conflict because of changes in the family such as parental divorce, illness, or death. Hartos and Power (1997) believe that parents are not necessarily aware of the stress that adolescents feel. Studying mothers and adolescents they found that teens reported more stress than their mothers reported for them. As we have discussed previously, open parent-child communication is an important factor when considering adolescent stress.

Many more families today exist in a state of *ongoing, unending crisis*—families with only one parent, who may be unable to work; families living below the poverty level; families with one or more disabled or highly disruptive children (Smith, 1990).

Sometimes stress may be from childhood trauma that begins to manifest itself only during adolescence and adulthood. This may be due to psychological defense mechanisms such as denial (see Chapter 2). Brown and associates (1975) studied depression in working-class English women. Among those women who suffered with depression, there was a high incidence of separation from or loss of their mothers by death in early childhood. Many of these women were young children during World War II and, as a result of the Battle of Britain, were sent to the countryside or to other countries to protect them from the German bombing of the major British cities. Being abused or neglected as a child increases the risk of delinquency in adolescence. Widom (1991) followed into adolescence children who were abused or neglected. In comparison with a matched control group (children who were similar in race and socioeconomic status, but had not been abused), the abused children were twice as likely to have a juvenile record.

Knowing how stress works and how to handle it effectively are necessary building blocks in the process of maturation. To learn more about adolescent stress, several researchers (Compas & others, 1987; Newcomb & others, 1981; Stark & others, 1989) have developed checklists to identify life stressors among adolescents.

Newcomb and associates (1981) identified seven major sources of adolescent stress. They are:

- *Family/parents* (e.g., divorce, money problems, fighting, alcohol abuse, physical and sexual abuse).
- *Accident/illness* (e.g., family illness, death in the family, medication used).
- *Sexuality* (e.g., fell in love, got pregnant, started dating, was dating regularly, had a gay experience).
- *Autonomy* (e.g., found a new group of friends, joined a club, decided about college, started making money, took a vacation without parents).
- *Deviance* (e.g., got in trouble with the law, stole something valuable).
- *Relocation* (e.g., parent changed jobs, changed schools, family moved).
- *Distress* (e.g., face broke out with pimples, thought about suicide, ran away from home, gained a lot of weight, got poor grades in school).

You probably recognize some of these events as obvious sources of stress in adolescence. You may be surprised, however, to see some positive events. Finding a new group of friends, beginning to date, and deciding about college are generally positive events for young people. Nevertheless, these events require adjustments in the individual's life and for that reason can also be stressful. Many of the stressors reported by adolescents are related to the developmental changes they are experiencing (Compas & Wagner, 1991). Finding a new group of friends, for example, may be important as early adolescents move from elementary to middle school and spend increasing amounts of time with peers. For these reasons, finding a new group of friends may be more stressful during adolescence than it would be at other ages. Larsen and Asmussen (1991) believe that the biological and cognitive changes of adolescence are not stressful in themselves. Rather, they contribute to stress because they change the adolescent's social relationships with peers and parents.

Newcomb and others (1981) found that adolescent boys and girls reported a similar number of stressors. The *kinds* of stress most frequently reported did differ by sex, however. Adolescent males reported more events involving getting into trouble and adolescent females reported more events related to accident/illness distress. Byrne and Mazanov (1999) found that similar numbers of boys and girls reported that school was a source of stress. Middle adolescents reported the highest number of negative life events, suggesting that middle adolescence may be a peak period for life change. Late adolescents reported a higher number of life events in the autonomy area. Other researchers (Stark & others, 1989) also found that adolescents of different ages were concerned with different life stressors. Middle adolescent problems centered around school and parents, while late adolescents reported increasing concern with money and the future. Compas and Wagner (1991) found that negative family events were related to distress among junior high school students, whereas negative peer events were most distressful for high school students, and negative academic events were most distressful for college students.

The life events questionnaire developed by Newcomb and associates (1981) focuses on major sources of stress rather than on daily and ongoing problems such as conflict with a brother or sister, feeling pressured by friends, or taking care of brothers and sisters. Bruce Compas and his colleagues (1987) developed a life events questionnaire for adolescents, the *Adolescent Perceived Life Events Scale* (APES). The APES measures daily hassles as well as major life events. Research using the APES suggests that daily hassles are important in determining how adolescents respond to major life events (Wagner & others, 1988). Daily hassles are sources of annoyance, irritation, and unhappiness. Adolescents who experience both a major negative event, such as parental death or divorce, and many daily hassles are at higher risk for psychological difficulties. They are more likely to have emotional problems than adolescents who only experience a major life event. It is important to recognize that the daily stresses of adolescent life can be serious and harmful.

Some stressful life events are as much a consequence as a cause of behavioral and adjustment problems (Masten & others, 1994). For example, the stressful life events of getting a poor grade in school, getting in trouble with the law, and becoming pregnant may be a consequence of antisocial behavior and learning or adjustment problems. Those events might also be explained by impulsive and aggressive personality characteristics, which are symptoms of mental disorders. When adolescents who are already troubled experience more negative life events, their stress levels and problems only increase. In a longitudinal study of 10th-grade students, McFarlane and associates (1994) found that stressful life events were both a risk and consequence of depression.

You can see by now that you encounter stress from many sources regardless of your age. Stress is primarily due to change. It is the nature of human development to produce inexorable change in every aspect of our existence. This situation is difficult enough when we are young, but at least then we have the support of parents, teachers, and other adults, as well as a more resilient body.

AN APPLIED VIEW

The Adolescent Life Change Event Scale

Below is a list of common life events and the stress scores assigned to them. To evaluate the amount of stress in your life, check the events that happened in the past year. Total your score; it will be explained later in the chapter.

Adolescent Life Change Event Scale (ALCES)

1 A parent dying	98	18 Hassling with parents	64
2 Brother or sister dying	95	19 Trouble with teacher or principal	63
3 Close friend dying	92	20 Having problems with any of the following: acne, overweight, underweight, too tall, too short	63
4 Parents getting divorced or separated	86		
5 Failing one or more subjects in school	86		
6 Being arrested by the police	85	21 Starting a new school	57
7 Flunking a grade in school	84	22 Moving to a new home	51
8 Family member (other than yourself) having trouble with alcohol	79	23 Change in physical appearance (braces, glasses)	47
9 Getting into drugs or alcohol	77	24 Hassling with brother or sister	46
10 Losing a favorite pet	77	25 Starting menstrual periods (for girls)	45
11 Parent or relative in your family (other than yourself) getting very sick	77	26 Having someone new move in with your family (grandparent, adopted brother or sister, or other)	35
12 Losing a job	74		
13 Breaking up with a close girlfriend or boyfriend	74	27 Starting a job	34
		28 Mother getting pregnant	31
14 Quitting school	73	29 Starting to date	31
15 Close girlfriend getting pregnant	69	30 Making new friends	27
16 Parent losing a job	69	31 Brother or sister getting married	26
17 Getting badly hurt or sick	64		

Figures rounded to the nearest whole number.

From R. C. Yeaworth, et al., "The Development of an Adolescent Life Change Event Scale," Spring 1980 in *Adolescence*, 15, 93. Reprinted by perssmission of Libra Publishers, Inc.

We do seem to experience more stress as we get older. As we move into adolescence and adulthood, we must rely more on knowledge and insight to avoid having a stressful life. What follows is a description of just what stress is and how humans try to deal with it.

CHANGE AS A SOURCE OF STRESS

❏ If the last 50,000 years of man's existence were divided into lifetimes of approximately 62 years each, there have been about 800 such lifetimes. Of these 800, fully 650 were spent in caves. Only during the last 70 lifetimes has it been possible to communicate effectively from one lifetime to another as writing made it possible to do. Only during the last six lifetimes did masses of men ever see a printed word. Only during the last four has it been possible to measure time with any precision. Only in the last two has anyone anywhere used an electric motor. And the overwhelming majority of all the material goods we use in daily life today have been developed within the present, the 800th, lifetime. (Toffler, 1970, p.148)

With the incredible amount of change in this current "lifetime," it is not surprising that the 20th century has been called the most stressful in which humans have ever lived. These changes have resulted in so much stress that sociologist Alvin Toffler (1970) has labeled it a new disease: **future shock**. Future shock is the illness that results from having to deal with too much change in too short a time. Toffler compares it to culture shock, the feeling we get when arriving in a foreign country for the first time: We become disoriented and anxious, but inwardly we know that if this discomfort becomes too great, we have only to get on the plane and go back to our own culture where we can feel safe again. Future shock causes the same kind of stressful feeling except there is no going home to escape from it.

Why is change so stressful? Toffler suggests that stress results not from the direction or even the kind of changes that we are faced with in this century but rather from the incredible rate of change in our daily lives. He suggests that there are three major aspects of rate of change, each of which is rapidly increasing.

Transience is a lack of permanence of things in our lives. Toffler documents in great detail how much more transient (fast-changing) our lives have become in this century. Some changes are more novel than others. The **novelty** of new situations in our lives is more dissimilar from old situations than it used to be and, therefore, far more stressful. It also matters what percentage of our lives is in a state of change at any one time. People used to maintain stability in most of their lives, allowing only a few aspects to change at any particular point. With greater **diversity**, this stable proportion is now much smaller for most of us.

Although the escalating rate of change in our lives has increased the amount of pressure we feel, research is increasing our understanding of it and helping us deal with it better.

The Effects of Stress

❑ A merry heart doeth good like a medicine.
Proverbs 17:22

We are all aware of short-term physical upsets such as fainting, rapid heartbeat, and nausea caused by the strains of everyday life. Many of these reactions are impairing. Only recently have we come to understand more about the relationship between long-term emotional stress and illness.

Much of the research looking at the relationship between stress and disease has been completed with adults. Some studies have been completed with adolescents and the results are similar to adult findings. Compas and his associates (1987) found that stressful life events are related to psychological and behavioral problems in adolescents. Not surprisingly, behavioral and psychological problems of adolescence are also related to parents' stress. When parents experience a lot of stress, their children are more likely to have emotional difficulties as well (Compas & others). For African American adolescent females, a greater

future shock
The illness, according to Toffler, that results from having to deal with too much change in too short a time.

transience
Toffler's term for the lack of permanence of things in our lives, which leads to increased stress.

novelty
Toffler's term for the dissimilarity of new situations in our lives, which contributes to stress.

diversity
Toffler's term for the increase in the percentage of our lives that is in a state of change at any one time, which increases stress.

AN APPLIED VIEW

Measuring the Relationship Between Stress and Physical Illness

Psychologists and psychiatrists have long attempted to measure accurately the relationship between stress and disease. In the early 1900s, Adolph Meyer introduced the concept of *psychobiograph*, which emphasized the importance of biographical study in understanding the whole person. He attempted to relate the biograph of the person to the likelihood of that person getting a variety of diseases.

The most successful attempt in this area was by Holmes and Rahe (1967). They developed the **social readjustment rating scale**, which measures the relationship between events in one's life that require considerable adjustment and the likelihood of getting sick as a result of these crises. The scale is composed of life events that require coping, adaptation, or adjustment. The adolescent adaptation of this scale, the **adolescent life change event scale** was presented earlier in this section. Preschool and elementary adaptations also exist.

The events in the scale are ranked according to the relative degree of adjustment required by the average individual. A numerical weight, called the *life change unit* (LCU), is assigned to each event. The greater the degree of the life change, the higher the unit number.

Holmes and Rahe designed the original adult scale by developing a list of events that could affect psycho-logical well-being. The list was submitted to a sample of adults who rated each item according to the relative amount of adjustment required. The results of this study were then mathematically interpreted and a numerical value (LCU) was assigned to each event. For each LCU that you checked as having happened to you during the past year, give yourself the number of points indicated. Then total up your score.

Research has shown that a score of 200 or more makes some kind of illness likely. Colligan (1975) reports that 86 percent of those who experienced more than 300 LCUs in a year developed some serious health problem.

This does not mean that every person who has a high number of LCUs in one year will definitely get sick. It only means that the likelihood of getting sick is considerably increased. One caution is necessary here: there is the danger of a self-fulfilling prophecy. That is, persons who know they have a high number of LCUs in a year may believe they are going to become sick, and therefore they do. Nevertheless, people under a high level of stress should take especially good care of their bodies and should be ready to check with their doctors quickly if they do develop symptoms of illness.

---

social readjustment rating scale
A scale developed by Holmes and Rahe to measure life events that require considerable adjustment and to determine the relationship between those events and physical health.

adolescent life change event scale
A psychological measure developed to evaluate the amount of stress in an adolescent's life based on the number of stressful life events experienced in the last year.

number of negative life events was found to be linked to depression, conduct disorder, post-traumatic stress symptoms, and physical illness (Brown & others, 1989).

Other research (Daniels & Moos, 1990) found that depressed youths reported more major stressors and daily hassles than healthy youths. Biederman and Spencer (1999) found that a history of stress is common with depressed young people. However, the researchers believe that the relationship between depression and stress is still unclear. Stress may precipitate depression in some people, but depression itself may be a stressor. Youths with behavioral problems reported more parent and school stressors than healthy youths. At school, highly stressed adolescents are more likely to get into fights, talk back to teachers, play the class clown, and get headaches and stomachaches (Fimian & Cross, 1987).

What makes one person handle difficult life stress better than another person? One answer is practice. Success with similar situations leaves a person with some experience and confidence to draw on in coping with a new stressful situation. The person is less rattled, and is able to think more clearly and respond more logically to the situation.

Social support has also been found to be an important factor in a person's ability to remain composed and to adapt successfully to stressful situations. According to Bowlby (1973), these social supports are created during infancy when a person learns the essential base of security that will carry him or her throughout life. Being able to turn to a close and confiding adult was found to be very important to healthy functioning adolescents whose parents suffered from severe depression (Beardslee & Podorefsky, 1988). Close family and friends seem to provide a cushioning effect during the stressful times in our lives. The demands of living with a chronic physical illness pose special challenges for adolescents. Parental support has been shown to alleviate the negative consequences of stress for teens suffering from juvenile rheumatic diseases such as arthritis, especially when there were few additional stressors (Timko & others, 1995).

Relationships with family and friends can be sources of stress rather than support when those relationships are high in conflict (Gottlieb, 1991). When parents sexually abuse their children, the relationship becomes a source of distress rather than comfort. Feelings of isolation, an inability to trust others, anxiety, low self-esteem, substance abuse, depression, eating disorders, and

AN APPLIED VIEW

How Teens Feel When They Learn They Have Cancer

There is a wonderful book of quotes from adolescent cancer patients entitled *What It Is That I Have, Don't Want, Didn't Ask for, Can't Give Back and How I Feel About It* (Ohio Cancer Information Service, 1991). Below are some quotes reflecting how these teens felt when they first learned they had cancer:

- I kept some of my feelings to myself. I knew that my parents were upset, so I tried to be sorry for them. (Kathy)
- Mom cried and Dad got mad and I just felt numb inside. (Marsha)
- My doctor told me I had cancer, I didn't know what to think. I told him that someone had mixed up the report and had made a mistake. I wish they had. (Dan)
- I was very angry. I thought, "Why me?" Everybody reassured me it wasn't my fault. (Joyce)
- I panicked. I didn't understand. I didn't know who to go to or what to do. Fortunately, my parents were there to help me. (David)
- I was frightened. (Ellen)
- I couldn't believe it! I thought I would wake up in the morning and everything would be just as it was before I got sick. (Kay)

- It helped to talk to the doctors. They answered all of my questions and gave me a lot of support. (Paula)
- I had my future all planned and it didn't include getting sick. Guess I need to change my plans a little, or postpone them a few years. (Michelle)

If you read these quotes carefully, you will get a picture of how teens often react to severe stress of all kinds. How should you respond to a person who is undergoing a period of extreme distress, such as learning that he or she has cancer? The introduction to this book offers a good suggestion:

❏ "Your experiences and the feelings you have about your illness are unique to you. No one has ever felt exactly the way you feel. Sometimes it can be hard for you and others to understand what is happening to you. Other patients have gone through similar experiences. Some of them have shared their feelings in this book. Remember since no two people are alike, their experiences, treatments and feelings might be different than yours" (p. 1).

This book is available from the Ohio Cancer Information Service, 101A Hamilton Hall, 1645 Neil Avenue, Columbus, OH 43210.

suicide are among the long-term consequences of sexual abuse (Browne & Finkelhor, 1986). A growing body of research links stress with many different types of physical and mental disturbances. As a result of this research, we are getting a much better understanding of how stress can cause disease.

COPING STRATEGIES

coping strategies
Methods people use to handle stress.

In other efforts to find out why some people handle stress better than others, researchers have been looking at the methods, the **coping strategies**, people use to cope with stress. Differences in coping may determine whether stress has negative psychological effects. Coping is defined as any effort used in response to stressful events (Compas, 1987). Those efforts can be behavioral (what people do) or cognitive (what people think). Lazarus and Folkman (1984; 1985) have studied the coping responses of adults and developed the *Ways of Coping Scale* to measure coping strategies. They describe three general kinds of coping behavior:

- *Problem-focused coping*: a direct action to reduce a problem or increase resources to solve a problem.
- *Appraisal-focused coping*: thinking about a problem in a new way so that it can be more easily solved.
- *Emotion-focused coping*: what people do to handle feelings of tension

The *Ways of Coping Scale* has been used to assess the coping responses of adolescents. Stern and Zevon (1990) found that adolescents used problem-focused coping most often. Examples of problem-focused coping include making a plan of action to follow and coming up with different solutions to a problem. Another frequent coping strategy used by teens was seeking social support. Examples of social support are talking to someone about your feelings and asking someone for advice. Stern and Zevon also found that a positive family environment was related to problem-focused coping. Adolescents reporting a negative family environment were more likely to use emotion-based coping styles, such as blaming themselves, daydreaming, and reducing tension through eating or drug/alcohol use. Emotion-based coping is directed toward handling stressful feelings rather than solving the problem that is causing the stress.

Another way to think about coping strategies is whether they are *approach-oriented*, that is, dealing directly with the source of stress or its consequences, or *avoidance-oriented*, denying or withdrawing from the source of stress and its consequences. According to Ebata and Moos (1995), problem-focused coping efforts are generally approach-oriented. Some emotion-focused coping strategies, such as trying to see the bright side of a problem, are approach-oriented, whereas other emotion-focused strategies, such as trying to forget the situation, are avoidance-oriented. Approach-oriented strategies generally produce better solutions or outcomes than do avoidance-oriented strategies. Ebata and Moos also found that older adolescents use more approach-oriented coping than

Most teens have learned a number of ways for coping with stress. Discussing a problem with a friend is a common way of reducing stress.

younger adolescents. Adolescents who use approach-oriented strategies also believe they have some control over the stressors and have good sources of social support.

What other coping strategies are used by adolescents? Patterson and McCubbin (1987) interviewed high school students to find out. The most common coping strategies mentioned were relaxing (listening to music), developing independence, being optimistic, and making close friends. Females more often than males mentioned seeking social support, turning to family members to solve problems, making close friends, and being self-reliant. Ventilating feelings (swearing, blaming others for what's going wrong, and getting angry and yelling at people) and investing in close friends (being with a girlfriend or boyfriend and being close to someone you care about) were the most frequent coping strategies of adolescents who use drugs and alcohol. Friendships were probably related to substance abuse because most teens use drugs and alcohol with friends. Friends may also be important in persuading a teen to experiment with drugs. Turning to family to solve problems, seeking spiritual support, and engaging in demanding activity appear to be coping strategies that help adolescents keep away from drug use. Recklitis and Noam (1999) found that avoidance and ventilation were associated with more behavior problems, while problem solving and the use of interpersonal skills were associated with fewer problem behaviors. They also found that girls were more likely to use interpersonal coping strategies and boys were more likely to engage in physical activity as a coping strategy. Frydenberg and Lewis (1999) also found some gender differences in coping with girls using more self-blame and boys using physical activity to avoid dealing with a problem. McFarlane and associates (1994) found that teens who viewed their parents and siblings as sources of support were at low risk for the development of depression in adolescence.

It is apparent that some coping strategies are "healthier" for adolescents than others. Increasing evidence suggests that good coping strategies may help adolescents deal with the stress and tension of their lives. Consequently, some psychologists argue that parents and schools must help young people develop better coping skills (D'Aurora & Fimian, 1988). Adolescents can also be helped to develop and use sources of social support, such as close friends and supportive adults. If we are able to do this, we may be able to prevent social, psychological, and behavioral problems before they develop.

Walsh (1992) interviewed 20 homeless children and adolescents to learn more about how they cope with the stressful life events of becoming and then remaining homeless. The chronic daily hassles of life in a shelter or motel were often a greater burden than the single event of becoming homeless. Lack of privacy, blasting television sets, and no place to do homework are just a few of the numerous daily hassles experienced by these youths. Some children and adolescents used problem-focused coping and attempted to change the source of the problem (i.e., I'm going to get us out of this place). Most realized that they could not solve the problem directly. Adolescents were better able than younger

AN APPLIED VIEW

Coping with Homelessness

The following are excerpts from an interview (Walsh, 1992) with a 17-year-old boy who is a senior in high school. T. J. is the oldest of six children and has been living in a shelter for the past several months with his mother and siblings. What stressors does T. J. describe as part of shelter life? How has he tried to cope with this stress? How is his life similar to and different from that of other adolescents?

❏ I came here a few months ago. My mom has been here longer. . . . She did the best she could. She tried to do what was the best. Coming here was the best for all of us, because I guess she wants help.

The hardest part of living here is probably the way I am. I'm so set in the way I do things. Like I'm not conscious of other people sometimes. There's just so many different people. They're all ladies too. That's another thing. It's different with men and ladies, because ladies are different, have different attitudes than men do, I guess. . . . I'm used to my sisters and the way they are and they are used to me, but when you come here people have different ways, it's hard to explain. I mean nothing's really hard for me to do. It's just stuff I don't like. Since I'm seventeen they give me more freedom than the little kids. And so there's not a lot put on me. I know I have an attitude problem, but that's just the way I am

and I always probably will be. I probably get upset so easily. It's just different here. It's like having a bunch of little sisters, that's what it is. . . . To get away I usually go into my room and just listen to the radio, my tapes. I have a room of myself. . . .

In school, I usually just describe the shelter as an apartment house to the other kids. I just use that. It isn't a lie. It is a house. Five of my friends know I live here. They don't know it's a shelter. It's embarrassing. Not embarrassing but I don't like using the word shelter really. When they ask questions I try to avoid them and avoid the questions and stuff. I don't like to get too in-depth into the question. . . . It's like being in between houses. That's how I think about it. In between houses. I don't know what other kids think. . . . So I just say I'm living in an apartment house. . . .

What I like here is that I don't have to watch kids all the time. Where we lived before, I had to watch them while my mother went to work. And, I get a little more freedom here than I used to have. But I'm older now. It would probably be the same way if we didn't come here (pp. 165–166).

From Mary E. Walsh, *Moving to Nowhere: Children's Stories of Homelessness.* Copyright „ 1992 Greenwood Publishing Group, Inc., Westport, CT. Reprinted with permission.

children to understand the complexity of their situation and to recognize that the events that brought them to the shelter were beyond their control. With that recognition, emotion-focused strategies such as going outside and finding someone to play with were more feasible ways to cope. Although homelessness and poverty had understandably negative impacts on the lives of these youths, many of their stories revealed an unusual capacity for survival through difficult conditions. Recognition of the capacity for survival in the midst of adversity has contributed to a growing interest in resilience.

RISK AND RESILIENCE

resilience
Quality possessed by individuals who deal well with stress and who have few psychological, behavioral, or learning problems as a result.

risk factors
Stressors that place an individual at risk for academic or psychological difficulties.

protective factors
Characteristics of resilient individuals that protect them from stress.

Individuals who deal well with stress and who have few psychological, behavioral, or learning problems as a result of it are said to have **resilience**. In recent years, researchers have become interested in studying the characteristics of resilient individuals. The stressors that individuals experience are called **risk factors**. They place a person at risk for academic failure and psychological difficulties. Risk factors include poverty, chronic illness, racial inequality, homelessness, parental mental illness, alcoholism and drug abuse, exposure to violence through war or tragedies in the inner cities, physical and sexual abuse, neglect, lack of supervision at home, and the family experiences of divorce and teenage motherhood. Researchers have been interested in identifying **protective factors** (characteristics of resilient individuals that protect them from stress) (Jessor, 1993; Jessor & others, 1998). Three kinds of protective factors have been found so far: family environments, support networks, and personality characteristics. Hauser and Bowlds (1990) describe the characteristics of these protective factors.

Supportive family environments that provide feelings of warmth and closeness are often a protective factor for at-risk adolescents. Parents of resilient adolescents combine support and understanding with supervision of their children's activities and *consistent* discipline. Most resilient adolescents have been able to establish a close relationship with at least one family member, either a parent, grandparent, aunt, brother, or sister. That family member is especially important in providing care, attention, and guidance when there is marital conflict or parental psychopathology (Werner, 1990; 1995). Werner and Johnson (1999) believe that we must make use of what we have learned about resilience to inform our practice with vulnerable children.

Resilient adolescents are also able to obtain support through the social network outside the family. A school in which teachers and counselors are closely involved with students can contribute to resilience. Relationships with caring ministers, teachers, older friends, and peers can provide the support needed by adolescents in coping with stress in their lives.

Personality characteristics and temperament can also reduce the negative effects of stress. A remarkably similar list of personality characteristics was developed from the study of academically successful Latino high school students who grew up in poverty (Mason, 1967), Asian youths on the island of Kauai (Werner, 1990), and competent African American adolescents from working-class families (Lewis & Looney, 1983) and inner-city ghettos (Hauser & Bowlds, 1990). These characteristics include positive self-concept and self-confidence, feelings of control, social responsibility, sensitivity, cooperativeness, good communication and social skills, an outgoing nature, and an easygoing disposition.

PREVENTION

primary prevention
Programs directed at eliminating problems before they begin.

secondary prevention
Intervention during the early stages of a problem to reduce its severity or the length of time it will last.

Psychologists have been enthusiastic about identifying risk and protective factors in the hope that this knowledge will contribute to the prevention of psychological difficulties. Successful prevention programs reduce the occurrence of maladjustment and clinical dysfunction by reducing risk factors (when possible) and enhancing protective factors and effective coping strategies. **Primary prevention** programs try to eliminate problems before they begin. **Secondary prevention** programs intervene during the early stages of a problem to reduce its severity or the length of time it will last.

Prevention programs vary in whether they are directed toward all children and adolescents in a particular school or city, or whether only students identified to be at particular risk are invited. Among adolescents, prevention programs are most often delivered in the schools or in a community agency. School-based programs have been effective in building social skills and reducing at-risk behaviors. Effective programs have been developed for reducing the onset, use, and abuse of cigarettes, marijuana, and alcohol, reducing teen pregnancy, curbing dropout rates, improving academic performance, and reducing delinquent behavior (Kazdin, 1993). According to research by the Carnegie Council on Adolescent Development (1995), families, schools, and community organizations must work together to reduce risks, enhance protective factors, and promote healthy adolescent development.

Unfortunately, prevention programs are still limited and are not able to prevent the occurrence of mental disturbance. Psychotherapy, psychiatric hospitalization, and medications are needed to improve dysfunction for those individuals already suffering from mental disturbance.

MENTAL DISTURBANCE

In my dream I saw the world
in a frame of imitation gold.
I heard fear pounding in my ears
And in the white light I could see only black.
Blinded by the sound of darkness
I saw invisible fingers
And heard nonexistent sounds.
I was a nonexistent person
In a nonexistent world.
God help me
As I stab myself with a
Rubber knife.
By a 16-year-old girl living in an adolescent residential center

A number of psychologists and psychoanalysts (most notably Anna Freud and G. Stanley Hall, but also Blos and Erikson) have suggested that it is normal in adolescence to have distressing, turbulent, unpredictable thoughts that in an adult would be considered pathological. Here is an example of this view:

❏ The fluidity of the adolescent's self-image, his changing aims and aspirations, sex drives, unstable powers of repression, and his struggle to adapt his childhood standards of right and wrong to the needs of maturity, bring into focus every conflict, past and present, that he has failed to solve. Protective covering of the personality is stripped off, and the deeper emotional currents are laid bare (Ackerman, 1958, pp. 227–228).

Just how common and how serious are these problems? To what extent are they associated with normal developmental processes or real emotional disturbance? Studies (Offer & others, 1981; Petersen, 1988) indicate that between 10 and 20 percent of adolescents experience some type of serious emotional disturbance. That percentage is very close to the rate of emotional disturbance among adults. Weiner (1992) cites numerous studies indicating that "adolescent turmoil" is not normal among adolescents. Signs of turmoil generally indicate a need for psychological help. Furthermore, when adolescents become disturbed and do not receive appropriate treatment quickly, the chances are slim that they will "grow out" of their problems (James & Nims, 1996; Tolbert. 1996). Weiner warns,

❏ An indiscriminate application of "adolescent turmoil" and "he'll-grow-out-of-it" notions to symptomatic adolescents runs the grave risk of discouraging the attention that may be necessary to avert serious psychological disturbance (1970, p. 66).

In sum, adolescence is not a time of turmoil and distress for most teens. Rates of mental disturbance among teens are very similar to rates of disturbance among adults. Rates of mental disturbance are much higher, however, in the adolescent years than in childhood. As you will recall from the first half of this chapter, adolescents (and adults) are experiencing increasingly high levels of daily stress. Hechinger (1992) warns that teenagers are in greater danger than ever from the risks of alcohol and drugs, unwanted pregnancy, sexually transmitted disease, depression, and violence. It is not normal for adolescents to be experiencing high levels of psychological distress. When society exposes youths to serious risks such as drugs and violence, psychological distress and mental disturbance are likely to increase. The social and technological changes of recent years provide adolescents with new opportunities, but also expose them to risks that seriously threaten psychological and social well-being (Carnegie Council on Adolescent Development, 1995).

TYPES OF MENTAL DISORDERS

What types of mental disorders are found among adolescents? Studies charting the incidence the types of mental disorders that afflict adolescents are relatively rare but are increasing (Reimer, 1996). Several studies (Dori & others, 1999; Haliburn, 2000; Grilo, 1999) have suggested that the prevalence of major

How would you respond to signs of adolescent turmoil?

depression and the incidence of suicide are increasing in adolescent populations, particularly among males.

Kazdin (1993) reports that approximately 5 million out of the 28 million adolescents between the ages of 12 and 17 living in the United States have experienced a significant emotional or behavioral disturbance. He explains further that many disorders, such as autism, mental retardation, attention deficit, conduct disorder, and learning disabilities, emerge in childhood and continue into adolescence and adulthood. Other disorders, including schizophrenia, eating disorders, and depression, emerge or become much more common during adolescence. Although only 1 percent of children referred for treatment between the ages of 1 and 6 are diagnosed with major depression, that rate increases to more than 13 percent by age 12. Many of the problems that emerge during childhood and adolescence, such as depression, conduct disorder, and attention deficit, have consequences that continue throughout life. As a result, it is important that interventions be made during childhood and adolescence to alter the course and reduce the lifelong negative impact of these disorders. Completed suicide is also rare in childhood, but increases sharply in early adolescence and continues to increase throughout the remainder of life.

We have a great deal of data on specific adolescent mental disturbances. Most common among them are eating disorders and depression, which sometimes leads to suicide attempts.

ANOREXIA NERVOSA

anorexia nervosa
A syndrome of self-starvation characterized by intense fear of gaining weight, disturbance of body image, significant weight loss, body weight below minimal normal, and amenorrhea, none of which can be accounted for by a known physical problem.

Anorexia nervosa is a syndrome of self-starvation that mainly affects adolescent and young adult females, who account for 95 percent of the cases (Larson & Johnson, 1981; Mintz & Betz, 1988). Some researchers believe that eating disorders are more prevalent in males than previously indicated but that the illness may be covered up (Nelson & others, 1999; O'Dea & Abraham, 1999). "The essential features of Anorexia Nervosa are that the individual refuses to maintain a minimally normal body weight, is intensely afraid of gaining weight, and exhibits significant disturbance in the perception of the shape or size of his or her body. In addition, post menarcheal females with this disorder are amenorrhiec. (The term anorexia is a misnomer because loss of appetite is rare.)" (American Psychiatric Association, 1994). Amenorrhea is the absence of menstruation.

According to the DSM-IV (1996), the specific criteria for anorexia nervosa are the following:

- Refusal to maintain body weight at or above a minimally normal weight for age and height (e.g. weight loss leading to maintenance of body weight less than 85 percent of that expected; or failure to make expected weight gain during period of growth, leading to body weight less than 85 percent of that expected).
- Intense fear of gaining weight or becoming fat, even though underweight.
- Disturbance in the way in which one's body weight or shape is experienced, undue influence of body weight or shape on self-evaluation, or denial of the seriousness of the current low body weight.

- In postmenarcheal females, amenorrhea, i.e., the absence of at least three consecutive menstrual cycles. (A woman is considered to have amenorrhea if her periods occur only following hormone (e.g. estrogen) administration.

Health professionals have seen an alarming rise in the incidence of this disorder in the last 15 to 20 years among young women (Nelson, 1999; Rosen & others, 1987). Estimates of the frequency of anorexia nervosa range from .2 percent to 1 percent of the adolescent population (Whitehouse & Button, 1988). It is important to note that some researchers believe that eating disorders are more prevalent in males than previously indicated (Nelson & others, 1999). The frequency of anorexia has been reported to be as high as 5 to 7 percent among adolescent ballet dancers and as high as 30 percent among adult dancers (Garner & Garfinkel, 1980; Hamilton & others, 1985). Whether there is an actual increase of anorexia nervosa or whether it is now being more readily recognized has yet to be determined.

Although anorexia may begin before the age of 10 or after the age of 40, it occurs most frequently in early adolescence (accompanying the changes of puberty) and in late adolescence (prior to or during the time of separation from the family) (Halmi & others, 1979). Anorexia is found most often among the upper socioeconomic classes, although in recent years it has been found more frequently among all socioeconomic groups (Andersen & Hays, 1985).

BULIMIA NERVOSA

bulimia nervosa
A disorder characterized by episodic binge eating and depressed mood, with low self-worth thoughts following the eating binges.

Bulimia nervosa is a disorder related to anorexia nervosa and sometimes combined with it. It is characterized by

❏ binge eating and inappropriate compensatory methods to prevent weight gain. In addition, the self-evaluation of individuals with Bulimia Nervosa is excessively influenced by body shape and weight (American Psychiatric Association, 1996).

Bulimia has been observed in women above or below average weight, as well as in those of normal weight. Binging and purging seem to increase when others in the environment are exhibiting that behavior. Young women who are living in college dormitories or apartments may be more likely to witness that behavior than are adolescents living at home (Attie & others, 1990). The specific criteria of bulimia are:

- Recurrent episodes of binge eating. An episode of binge eating is characterized by both of the following:

  Eating, in a discrete period of time (e.g. within a 2-hour period), an amount of food that is definitely larger than most people would eat during a similar period of time and under similar circumstances.

  A sense of lack of control over eating during the episode (e.g. a feeling that one cannot stop eating or control what and how much one is eating).

- Recurrent inappropriate compensatory behavior in order to prevent weight gain, such as self-induced vomiting; misuse of laxatives, diuretics, enemas, or other medications; fasting; or excessive exercise.

- The binge eating and inappropriate compensatory behaviors both occur, on average, at least twice a week for three months.
- Self-evaluation is unduly influenced by body shape and weight.
- The disturbance does not occur exclusively during episodes of Anorexia Nervosa.

Why do you think that eating disorders have increased in recent decades?

CAUSES OF EATING DISORDERS

Why do some adolescents develop eating disorders? Any attempt to explain the causes of eating disorders needs to consider why their frequency has increased so greatly in recent years and why the disorders occur most often among early and late adolescent women.

Contemporary theory and research suggest that eating disorders are best explained through a multiple-risk model (Garfinkel & others, 1987). That is, no single factor explains why an individual develops an eating disorder. Many factors, including developmental stage, culture, personality, and family, must be considered.

- *Developmental stage.* The changes that take place in early and late adolescence help explain why eating disorders are so common among adolescents. Bodily changes, the changing sense of self, and changing demands of the school and peer environment are believed to increase the risk for eating disorders in early adolescence. Body dissatisfaction may be a contributing factor (Rosenvinge & others, 1999). Challenges of late adolescence, such as the establishment of intimacy, developing an identity apart from the family, and pursuing new education and career choices, increase the risk for eating disorders at that time. Attie and associates (1990) maintain that adolescent girls who are vulnerable because of personality, family, or other characteristics may respond to the physical changes and developmental challenges of adolescence by attempting to control their weight and thus may develop eating disorders. Anorexia develops most frequently in early adolescence and seems to be a response to the stressors of the early adolescent transition. Bulimia develops most often in late adolescence and seems to be a response, in part, to the developmental stressors of leaving home for college.
- *Culture.* Cultural standards for female attractiveness have changed in recent history. In earlier eras, a full figure and curves (such as Marilyn Monroe's) were the standard of feminine beauty and sexual attractiveness. During the past several decades, magazine and television models, Playboy pinups, and winners of the Miss America Pageant have all adopted a much thinner and leaner look (Mitchell & Eckert, 1987). Whereas men tend to be evaluated more by their actions and accomplishments, women continue to be judged to a great extent by their physical attractiveness, including how thin they are. Adolescent girls may experience intense social pressure to be thin, especially when they are in competitive schools and social environments that emphasize weight and appearance (Hamilton & others, 1985; Striegel-Moore & Cachelin, 1999). The extent to which our culture emphasizes physical attractiveness and thinness for females helps explain why they are more likely to develop eating disorders and why eating disorders have increased in recent

years. Mouton and colleagues (1998) suggest that eating disorders may develop as a means of approval seeking. In a culture that values thinness above many other attributes, this is sadly not surprising.

After many years of experience as a therapist with adolescent girls, Pipher (1994) blames culture for many of the psychological ills, including eating disorders and depression, experienced by young adolescent girls. She puts it this way: "American culture has always smacked girls on the head in early adolescence. This is when they move into a broader culture that is rife with girl-hurting 'isms,' such as sexism, capitalism, and lookism, which is the evaluation of a person solely on the basis of appearance" (p. 23).

- *Personality.* The personality development of the child who is at risk for eating disorders has been explained by psychoanalytic theory. Bruch (1981), for example, believes that early problems in the mother-daughter relationship create problems in the development of autonomy and a healthy sense of self. The daughter feels ineffective and behaves in ways meant to satisfy other people and not herself. Anorexics are described as high achievers who strive for perfection. These characteristics are understood by psychoanalytic theorists as efforts to satisfy others and as reflections of an underdeveloped or "false self" (Winnicott, 1965).

  Other psychoanalytic theorists believe that when mothers are unable to provide their infants with enough comfort and nurturance, the infant feels frustration, anxiety, and emotional hunger. These infants never learn how to control their own tension or comfort themselves. The bulimic's compulsion to eat is believed to reflect a lifelong feeling of emptiness and deprivation as well as an effort to satisfy feelings of emotional hunger by eating (Humphrey & Stern, 1988).

- *Family.* Psychoanalytic theory views the origin of eating disorders in early parent-child relationships. Family systems theory (see Chapter 7) maintains that problems in family relationships during adolescence contribute to the development of eating disorders. Minuchin and others (1978) describe families with anorexic adolescents as overprotective, rigid, and conflict-avoiding. They are believed to be close and protective of family members in a way that keeps the adolescent from developing a psychologically separate and independent sense of self. Anorexic families are not comfortable expressing conflict. Children feel much pressure to conform to family expectations without expressing disagreement. Refusal to eat is seen as a silent kind of rebellion. Families of bulimics have been thought to resemble anorexics in these characteristics. Recent research, however, suggests that bulimic families are better able to express anger and conflict and are less likely to be overprotective (Humphrey, 1989). Some evidence suggests that parental relationships of bulimics are characterized by insecure attachments (Armstrong & Roth, 1989; Kenny & Hart, 1992). More specifically, bulimic adolescents describe parental relationships as affectively negative, emotionally unsupportive, and interfering with the adolescent's efforts to become independent.

A number of new approaches to treatment and therapy are currently being researched (Scott, 1988). Eating disorders are usually so complex and potentially hazardous that only qualified personnel should attempt to treat victims.

Depression

depression
A condition marked by a sorrowful state, fatigue, and a general lack of enthusiasm about life.

The term **depression** can have many different meanings and manifestations. Originally a word for a pathological symptom, it has found its way into common usage by the general public, and its meaning has been greatly broadened. *Depression* may be viewed either as a mood (situationally caused), as a syndrome (a complex of behaviors and emotions), or as a clinical disease (Petersen & others, 1993). It is considered to be a basic affective state that, like anxiety, can be of long or short duration, of low or high intensity, and can occur in a wide variety of conditions at any stage of development. In certain circumstances, such as in reaction to a death in the family, it is a normal and appropriate affective
response.

Depression becomes pathological when it occurs in inappropriate circumstances, is of too long duration, or is of such great intensity as to be out of proportion to the cause. Depression is harmful to a person's development when it interferes with the capacity to work, to relate to others, or to maintain the healthy functioning of essential physical needs for sleep and nutrition. Serious depressive conditions can upset a person's functioning in all of these areas and more.

Recent statistics indicate that many adolescents show some depressive symptoms. When specifically asked, more than one-fifth of the 1,000 inner-city adolescents seeking routine health services reported frequent depressive symptoms (Schichor & others, 1994). This does not mean that most adolescents are actually depressed (Petersen & others, 1993). Some depressive feelings and behaviors are transient responses to life changes. For other adolescents, depressive symptoms lead to more serious depression in adulthood. Gjerde and Westenberg (1998) found that adolescent depression, especially for females, was an important predictor of psychological difficulty in young adulthoood. Up to 5 percent of adolescents suffer from severe levels of depression, another 10 to 15 percent appear to be moderately depressed, and between 20 and 35 percent are mildly depressed (Brooks-Gunn & Petersen, 1991).

In comparison with childhood, adolescence is a time of increased risk for the development of depression. Recent findings suggest that depressive symptoms rise between late childhood and early adolescence, and continue to increase throughout the adolescent years (Clarizio, 1994; Rutter, 1986). Hankin and Ambramson (1999) found this to be especially true for females. Because depression increases in conjunction with the developmental transitions of early to middle adolescence, researchers have begun to examine the relationship between developmental change and depression. Pubertal events, including hormonal changes and feelings about sexual maturation and body image, an increase in the stressful life events (such as school change and parental divorce), a decline in protective factors (e.g., loss of family support), and cognitive changes in the ability to understand and express emotions and reflect upon self and the future have been identified as factors related to the rise in depressive symptoms (Hankin & Ambramson,1999; Petersen & others, 1993; Rutter, 1986). Because adolescents are cognitively able to view failure as stable and long-lasting, they may be more vulnerable to feelings of hopelessness (Clarizio, 1994). When puberty, school changes, and family or friendship changes occur at the same time (Brooks-Gunn, 1991; Petersen & others, 1991), risks for diminished self-esteem and depression appear to increase, especially among adolescent girls. Having a parent who suffers from depression is a known risk factor for the development of depression and anxiety (Compas & Hammen, 1994). This may be because depressed parents contribute to a more negative and less nurturing family environment. Increasing evidence suggests that there may also be a genetic vulnerability. Genetics are more likely a factor in more serious cases of depression, whereas environmental factors contribute more to milder forms of depression (Weiner, 1992).

Adolescence is also notable for the emergence of sex differences in the frequency of depression. Prior to adolescence, boys are more likely than girls to show signs of depression (Gjerde & Block, 1991; Hankin & Abramson; 1999). By adolescence, however, females more likely to be affected by depression than are males.

Symptoms of Depression

Beck (1967) identified four types of depressive symptoms, which have been useful in observing and diagnosing depression.

- *Emotional manifestation*: dejected mood, negative self-attitudes, reduced experience of satisfaction, decreased involvement with people or activities, crying spells, and loss of sense of humor.
- *Cognitive manifestation*: low self-esteem, negative expectations for the future, self-punitive attitudes, indecisiveness, and distorted body image.
- *Motivational manifestation*: loss of motivation to perform tasks, escapist and withdrawal wishes, suicidal thoughts, and increased dependency.
- *Physical manifestation*: appetite loss, sleep disturbance, decreased sexual interest, and increased fatigability.

Not all depressed individuals will show all these symptoms, of course, but they are likely to exhibit one or more symptoms from these four categories.

Much of what we know about depressive symptoms comes from the study of depressed adults. In recent years, researchers have become interested in learning more about how depressive symptoms differ for children, adolescents, and adults (Cantwell & Baker, 1991). Symptoms of depression in adolescence are quite similar to those specified by Beck as characterizing adults. As summarized by Clarizio (1994), common cognitive features of adolescent depression include low self-esteem, dissatisfaction with body changes, feeling ugly, and a negative view of the future. Depressed adolescent girls are particularly likely to express dissatisfaction with their bodies. Escapism, withdrawal, thoughts of self-destruction, and diminished school performance are common motivational symptoms.

Physical symptoms, including changes in appetite and weight, sleep problems, and fatigue, are also common among depressed teens.

THEORETICAL MODELS OF DEPRESSION

Several theoretical models, including psychoanalytic, cognitive, behavioral, and environmental, have been used to understand the causes for depression throughout the life span. These models have been used frequently in understanding and treating depression among adolescents.

The Psychoanalytic Model

psychoanalytic model of depression
A theoretical model that explains the causes of depression as a history of loss.

According to the **psychoanalytic model**, the causes of depression, both normal and pathological, can be best explained under the heading of "loss" (Cantwell & Carlson, 1983; Carlson, 1983; Crumley, 1982; Curran, 1984; Petzel & Cline, 1978; Seiden & Freitas, 1980; Shaffer & Fisher, 1981; Tishler & McKenry, 1983). Seriously depressed individuals have usually experienced a series of losses, which may include losses of loved ones through death or relocation.

Depressed adolescents will often recount a history of parental separation, death, or divorce; a series of moves; death or loss of pets; moving away from trusted friends; or they express the feeling that childhood was a far better state than adolescence. These lead to a feeling of hopelessness or despair at not being able to regain the lost objects or status. The anger born of this frustration is often turned against the self with harmful results.

The Cognitive Model

cognitive model of depression
The view that depression results from negative cognitions or thoughts.

Beck's (1967) **cognitive model** views depression as resulting from negative cognitions or thoughts. Beck discusses three kinds of negative cognitions: negative views of the self, negative interpretations of one's experience, and negative views of the future. These negative cognitions lead to feelings of hopelessness, helplessness, and depression. This model has been supported by some research. Adolescents who rate themselves as depressed were found to have negative views of themselves, of the future, and of their own performance (Hammen & Zupan, 1984).

According to Beck's model, some people are more likely to become depressed because of negative cognitive patterns that developed in early childhood. Early life experiences, such as parental loss or a poor parental relationship, make an individual more sensitive to similar experiences later in life. Those experiences set off negative thinking. For example, an adolescent who suffered the loss of a parent during early childhood may feel rejected by that parent. Later, the adolescent may be overly sensitive to rejection from peers or dating partners. The expectation of being rejected sets off negative patterns of thinking about himself or herself and the future (such as "no one will ever love me") and causes depression. Kenny and others (1993) found that young adolescents who have less secure parental attachments are likely to have negative views of self, which in turn increase their vulnerability to depression. The researchers reasoned that when parents are not responsive and caring, teens do not view themselves as worthy. Parents who are sensitive, responsive, and caring contribute to a worthy sense of self, thereby providing a protective buffer in dealing with the challenges of adolescence.

Behavioral Models

behavioral model of depression
The view that depression results from low levels of positive reinforcement or from an inability to escape from punishment.

Depression is understood according to two **behavioral models:** the learned helplessness model (Seligman & Petersen, 1986) and the loss of reinforcement model (Lewinsohn & others, 1979). According to learned helplessness theory, people become depressed because they cannot escape from a bad situation. Their efforts to improve their circumstances do not result in an improvement of conditions. Over time, they learn that trying is not worthwhile and give up their efforts. The theory was developed by observing caged animals who were repeatedly subjected to a painful stimulus. When they were later given the opportunity to escape their torture, they continued to allow themselves to be punished. An adolescent may show learned helplessness, for example, when efforts to find a part-time job are repeatedly unsuccessful because of economic recession. Over time, the adolescent may give up trying, even when a seemingly good opportunity is available.

learned helplessness model of depression
A behavioral model of depression proposing that people become depressed because they cannot escape from a bad situation.

The concepts of internal, global, and stable attribution have been added to the **learned helplessness model** (Abramson & others, 1978). Whether a negative event brings on depression depends on the causes to which a person attributes the event. Individuals who explain bad events by causes that are their own fault (internal to them), that are likely to persist or remain stable over time, and that are global or exist across many situations are more likely to respond to bad events by becoming depressed. For example, a self-blaming or internal attributional style may make a person see all problems as his or her own fault. Research suggests that depressed children and adolescents believe that their successes are caused by external factors and that their failures are caused by internal factors (Kovacs, 1989). In other words, they blame themselves for their failures and expect that they will continue to fail in many situations in the future.

loss of reinforcement model of depression
A behavioral model of depression proposing that people become depressed because events or situations no longer provide positive reinforcement.

According to the **loss of reinforcement model**, depressive feelings result from low levels of positive reinforcement (see Chapter 2). This may be because few positive reinforcers are available, because events that were formerly reinforcing are no longer valued, or because the person lacks the skills to obtain or make use of available reinforcers. In short, life seems to have lost its ability to provide pleasure. As a teen moves from childhood to adolescence, some reinforcers, such as parental attention or being a member of the Boy Scouts, may become less reinforcing. New sources of reinforcement, such as the attention of a dating partner or obtaining a driver's license and car, may be harder or impossible to obtain. For this reason, the adolescent may become depressed. Sometimes being sick or depressed can bring attention from others. Depression can thus be maintained by reinforcement such as attention, sympathy, or being excused from school or homework responsibilities.

Environmental Models

environmental model of depression
A theoretical model that describes depression as a consequence of environmental circumstances, such as negative life events, or high levels of life stress and few sources of support.

In the **environmental model**, an environment that lacks supports and is disorganized and stressful can contribute to depression. According to some researchers, certain environmental and sociological changes, such as increases in substance abuse, less cohesive families, and societal pressure to achieve and be successful, have resulted in an increase in the occurrence of depression in recent years (Cicchetti & Schneider-Rosen, 1984). An environment that offers strong sources of support through family, friends, teachers, or others can offset or buffer the effects of environmental stress.

Evidence indicates that African American teens may be more vulnerable to depression than white Americans. Freeman (1982) examined emotional distress among 607 urban African American high school students 15 to 18 years of age. Subjects reported high distress primarily about feelings of economic disadvantage, volatile anger, interpersonal sensitivity, and loneliness. It is likely that environmental factors contributed to the high levels of depression reported by these teens.

DEPRESSION AND GENDER

Biological, social, and cognitive explanations have been offered for the increasing prevalence of depressive symptoms during adolescence and for the emergence of gender differences. Explanation for gender differences include the greater likelihood for the simultaneous occurrence of pubertal change and social change such as changing schools (Brooks-Gunn, 1991), increased dissatisfaction among girls with their bodies (Allgood-Merten & others, 1990; Hankin & Abramson, 1999), and increased female acceptance following puberty of traditional feminine sex-role characteristics such as passivity and introspection (Gjerde & Block, 1991; Hankin & Abramson, 1999). The tendency for girls to internalize rather than externalize their feelings has also been linked with depression. Boys are often described as **externalizing** their depression, or expressing it outwardly and often aggressively. Girls, in comparison, more often express their depression by **internalizing**, or by worrying, becoming anxious, and keeping their feelings inside (Gjerde & others, 1988; Reinherz & others, 1990). Both tendencies clearly seem to reflect the gender roles that society has assigned to us. Girls are socialized to be passive, self-evaluative, and sometimes helpless (Gjerde & Block; Pipher, 1994), increasing their vulnerability to depression. They believe that their successes are the result of luck or other persons or circumstances. Nolen-Hoeksema and others (1991) have also proposed that adolescent girls are more likely than boys to develop a global, stable, and internal attributional style, whereby they blame themselves for their failures and increase their risk for depression. As a result of socialization experiences, girls may be more likely to ponder and ruminate over negative feelings rather than acting in ways that reduce negative feelings (Nolen-Hoeksema, 1987).

externalizing
Expressing depressive feelings outwardly, such as through aggressive behavior, delinquency, running away, or rebellion.

internalizing
Expressing depressive feelings by worrying, becoming anxious, and keeping one's feelings inside.

Girls have been described by relational theorists (Gilligan, 1982; Gilligan & others, 1990; Josselson, 1988) as more concerned with interpersonal relationships. While this orientation toward care and connection is a source of support, it is also believed to contribute to the risk for depression when interpersonal relationships are unsatisfying (Kaplan, 1991; McGrath & others, 1990). Some research suggests that adolescent girls are more sensitive than boys to changes in interpersonal relationships (Kennedy & Petersen, 1989) and are more dependent on others for support and positive social interactions (Baron & Peixoto, 1991). Gore and associates (1993) studied 1,200 adolescent boys and girls in 9th through 11th grades. Their findings suggest that high levels of interpersonal caring and involvement in the problems of others do contribute to depression among girls. When a high level of family stress is occurring and adolescent girls become highly involved in the problems experienced by their mothers, they are more likely to become depressed. Adolescent boys are less likely to experience depressive symptoms as a result of their concern for or involvement in family problems.

Leadbeater and associates (1995) propose that adolescent girls' vulnerability to depression can be understood by a link between their sensitivity to interpersonal life events and internalizing tendencies. Although this hypothesis has not been fully tested, the researchers argue that adolescents (mostly girls) who seek attention and nurturing and fear abandonment are more sensitive to stressful events involving others. The adolescents are likely to react to this stress with internalizing symptoms because those behaviors (somatic complaints, suicidal gestures, withdrawal) are likely to elicit desired concern and attention from others. Externalizing symptoms (aggressive, destructive, or violent behavior) would serve to destroy or alienate rather than preserve desired relationships.

Which explanations for adolescent gender differences in depression make the most sense to you?

Further understanding of adolescent gender differences can be gained by reading our discussion of gender-role and gender differences in self-esteem found in Chapter 6. Many of the factors that are believed to contribute to gender differences in self-esteem are also believed to contribute to differences between adolescent males and females in levels of depression.

DEPRESSIVE EQUIVALENTS

During the 1960s and 1970s, depression in childhood and adolescence was popularly viewed as "masked" or expressed in "depressive equivalents." The basic assumption was that children and adolescents do not express depression in the same ways as adults. Toolan (1975) stated that "especially in the adolescent we seldom see a clear picture of depression" (p. 407). Boys who "have a need to hide their true feelings, and particularly the softer, tender, weak sentiments" (Tishler & McKenry, 1983, p. 732) were considered more likely to express their depression in masked forms.

depressive equivalents
The expression of depression in adolescence through symptoms that are different from those displayed by adults.

Depressive equivalents were believed to allow adolescents to discharge and seek relief for their feelings. Activity of this type distracts teenagers from thinking of their problems and facing the unpleasant images they hold of themselves and their lives. Examples of depressive equivalents are difficulty in concentrating, running away, sexual acting out, boredom, and aggressive behavior.

Teens display depression in many ways. Some teens run away from home to escape an unpleasant and seemingly unchangeable situation.

- *Concentration difficulty.* Often difficulty in concentrating is the earliest, most frequently cited, and only symptom present. Typically it is the only one of which adolescents are aware. There is a defensive quality to poor concentration. As the mind seeks to avoid awareness of painfully sad thoughts and feelings, it may skip actively from thought to thought, unable to stay still for fear of being caught by the waiting depressive alternative. The effect on school performance can be devastating.
- *Running away.* Depressed teenagers sometimes run away from the family home, foster home, or other residential setting as a means of actively dealing with overwhelming feelings that often originate in family relations. Running away provides a temporary release of tension and gives the feeling that one is in control.
- *Sexual acting out.* The urgent necessity to ward off underlying feelings of being unloved and unwanted may push the adolescent toward promiscuous sexual behavior. Close physical contact with another person provides relief. Females are especially vulnerable.
- *Boredom and restlessness.* Depressed adolescents often manifest their condition by swinging between states of short-lived but unbounded enthusiasm and periods of intolerable boredom, listlessness, and generalized indifference. It is to avoid coming any closer to an awareness of depressive effects that the cycle of excited activity and restlessness is again renewed. "I'm bored" is often an unconscious code phrase for "I'm depressed."
- *Aggressive behavior and delinquency.* Depressed adolescents, especially boys, sometimes carry out angry and destructive behavior such as vandalism in place of the depressive feelings. These actions may be designed to counteract the poor self-image and feelings of helplessness by artificially inflating the youth's self-image as a strong, fearless, and clever person.

The concept of masked depression and depressive equivalents has been challenged in recent years. As explained by Clarizio (1994), criteria for distinguishing adolescents with masked depression from those with conduct disorders alone have not been provided. Factors that contribute to the development of depression alone and depression in addition to conduct disorder also appear to be different. In some cases, depressive symptoms may follow symptoms described as masked depression. For example, aggressive, delinquent, and hyperactive youths commonly experience rejection and criticism that make them feel unhappy and worthless.

What we do know is that compared with adults, adolescents are much more likely to experience another disorder in conjunction with depression. Cantwell and Baker (1991) report that depressed adolescents are likely to experience behavior and attentional problems, to show high levels of anxiety, or to have an eating disorder or a language or learning disorder. Girls are more likely to exhibit both depression and an eating disorder, and boys are more likely to have both depression and a disruptive disorder (Petersen & others, 1993). Consistent with these multiple diagnoses, Renouf and Harter (1990) found that depressed adolescents express feelings of both sadness and anger. Forty percent of the depressed teens described anger toward both themselves and others, and another 39 percent reported just being angry at others.

Sometimes adolescent depression is most easily recognized by symptoms associated with other disorders. Behavioral problems often come to the attention of parents and teachers sooner than depressive symptoms. Although the behavior disorder may be the most obvious problem, depression is frequently identifiable when diagnostic criteria for depression are applied (Compas & Hammen, 1994). Adults working with adolescents need to recognize that adolescents expressing behavioral problems may also be experiencing depression. The consequences of overlooking these symptoms and disregarding them as typical adolescent boredom or rebelliousness can be serious and can deprive adolescents of much needed psychological help.

SUICIDE

Suicide and attempted suicide among adolescents are a growing national problem (Garland & Zigler, 1993; Holinger & others, 1987) and an increasingly common response to stress and depression among young persons (Kienhorst & others, 1987). A survey of Massachusetts high school students reveals that the percentage of high school students reporting that they had attempted suicide increased from 6 percent in 1990 to 10 percent in 1993. The percentage reporting that they had made a plan to commit suicide increased in the same time period from 14 percent to 20 percent (Wong & Hart, 1994). According to a recent Gallup Poll, suicide touches the lives of many American teens. Six percent of American teenagers reported that they have attempted suicide, and another 15 percent said that they had come close to attempting suicide. Sixty percent of adolescents polled said that they personally knew a teenager who had attempted suicide and 15 percent knew a teen who had succeeded (Freiberg & APA, 1991). The suicide rate among 10- to 14-year-olds almost doubled from 1980 to 1991 and increased by approximately 30 percent among 15- to 19-year-olds (U.S. Bureau of the Census, 1994).

Teenagers have become not only more suicidal but apparently more reckless and self-destructive in general. Langhinrichsen and colleagues (1998) found that male adolescents endorsed more risk-taking behaviors than female adolescents. As the suicide rate has risen steadily over the past 20 years, so, too, have the rates for motor vehicle accidents, accidents of other types, and homicides (Bem, 1987). Although suicide rates for teenagers have risen, the rate for most other age groups has decreased. While the United States as a whole has become slightly less suicidal, teenagers and young people in general (age 30 and under) have become *dramatically more suicidal*. The increase has risen most steadily and most consistently among teenagers.

It is in the area of *attempted* suicide that adolescents are truly a distinctive population (Rotheram, 1987; Spirito & others, 1987). Estimates indicate that there are 200 to 300 suicide attempts for every committed suicide in the 15- to 19-year-old age group (Curran, 1984; McIntire, 1980). There are no official records of suicide attempts since data of this type is very difficult to gather and assess on a national scale. Holinger (1979) and Phillips (1979) have suggested that a significant number of fatal single-car accidents are in fact undetected suicides. It is a commonly stated suicide fantasy among teenagers to die in a car crash. Yoder (1999) compared suicide attempters, suicide ideators (those who think about but do not attempt suicide) and non-suicidal adolescents and found that those who attempted suicide were significantly more likely to have been physically or sexually abused by a caretaker and to have a friend who had attempted suicide.

THE INFLUENCE OF RACE ON SUICIDE

Suicidal behavior remains, as it has consistently for decades, a behavior in which whites (U.S. Bureau of the Census, 1994) and the middle class (Tishler, 1981) are overrepresented. The suicide rate for young African American males has increased dramatically in recent years, however. Between 1980 and 1991, the suicide rate among 10- to 14-year-old African American males quadrupled and the rate among 15- to 19-year-olds more than doubled (U.S. Bureau of the Census). Suicide is the third leading cause of death, following homicide and accident, among African American males between the ages of 15 and 24 (Berman & Jobes, 1991). Statistics indicate that Latino and Native American youths are also increasingly at risk for violent death, including suicide (Berlin, 1987; CDC, 1986). The suicide death rate for Native Americans is higher than for white, Latino, and African American adolescents (Wyche & Rotheram-Borus, 1990), although rates vary greatly among specific tribes.

Reasons for differences in suicide rates across different racial and ethnic groups are not clear. Several authors have analyzed social and cultural factors in efforts to explain some of these differences. It is important to consider that prevalence statistics by race may not be totally accurate. For some cultures (especially Hispanic and African American), suicidal deaths may be underreported, because death by drug overdose, accident, or homicide is viewed as more culturally acceptable (Wyche & Rotheram-Borus, 1990). Racial and social class biases may result in a more cursory investigation of the deaths among poor and minority youths. Rising suicide rates among African Americans may be attributed to the persistent problems of unemployment and underemployment, the movement of greater numbers of African Americans to urban areas (suicide rates among African Americans are higher in the Northern cities than in the South), weakening community and family ties among those who experience upward mobility, frustration between expectations for racial integration and mobility and what has actually been achieved, increased use of drugs (especially crack) among African American youths, and a departure from traditional religious beliefs (Rutledge, 1990). Traditional cultural and religious beliefs may help to prohibit suicide among many African Americans and Hispanics (Wyche & Rotheram-Borus). Whereas whites, for example, understand suicide as a reaction to stress, Mexican Americans have traditionally viewed suicide as a crazy act, a weakness, and an embarrassment to the family. The teachings of the Catholic Church and the close-knit family structure of Hispanic families have also discouraged suicide and encouraged reliance on extended family for help when needed. African Americans have traditionally endured much pain and

How is culture a factor in the act of suicide?

suffering and have not regarded that as just cause to die. Native Americans may be less fearful of death. The soul is considered immortal and death is a natural part of life, without the rewards of heaven or the punishment of hell. Persons who work with suicidal youths from different racial and cultural backgrounds will need to become familiar with each youth's cultural practices and their impact concerning the meaning of life, death, and acceptable ways of obtaining help.

THE INFLUENCE OF GENDER ON SUICIDE

Major gender differences exist in rates of completed suicide and suicide attempts. The completed suicide rate is four times higher for men than for women (U.S. Bureau of the Census, 1994). This is because of the type of suicidal behavior engaged in by males, the methods they use (Langhinrichsen & others, 1998), the lethality of the attempt, and the degree of psychiatric disturbance present (Brent & others, 1999). Males and females are two very different suicidal types. Universally, males are 3.6 times more likely to die of suicide than females are. For youths ages 15 to 24, the ratio of male to female completed suicides was 5:1, an increase from the 3:1 ratio found a decade earlier (Freiberg & APA, 1991).

Attempt rates also show dramatic gender differences, but in the opposite direction. Failed attempts at suicide among females are much higher than for males (Woodruff-Pak, 1988). The literature consistently has cited female-to-male ratios of at least 3 to 1. Jacobziner (1965), Weiner (1992), and White (1974) reported ratios of 4 to 1. More recent studies show a far greater number of females among teenage suicide attempters: 5 to 1 (Curran, 1984), 9 to 1 (Hawton, 1982b, 1982c; McIntire, 1980), 9.5 to 1 (Birtchnell & Alarcon, 1971), and 10 to 1 (Toolan, 1975). One reason for the high survival rate among females is the method used (Garland & Zigler, 1993). Whereas males often resort to such violent and effective means as firearms and hanging, females tend to choose less violent and less deadly means, such as pills. Male suicidals are considered to be significantly more disturbed than are female suicidals (Hawton, 1982a, 1982b, 1982c; Otto, 1972; Teicher, 1973). They may also be more committed to dying and therefore succeed far more often (Weiner, 1992). One study (Meyer & others, 1991) indicates that males express more hostility and have higher expectations for success than females. The authors conclude that males may be at

WHAT·S YOUR VIEW?

Why Does the Suicidal Behavior of Males Differ from Females?

Why do males use such violent means in their efforts to destroy themselves? One answer is, "Males are just naturally more violent than females." But this begs the question. It doesn't really answer why this difference exists.

Do males have more access to guns, ropes, and poison than females? Are males more intent on actually dying, rather than on getting the sympathy of others? Are males more interested in hurting those they leave behind by dying in such mutilating ways? What's your opinion?

greater risk for completing suicide because they hold high expectations and are more likely to express their frustration and anger in aggressive ways. To reduce the risk for suicide, adolescent males need to learn how to express their anger in less dangerous ways and to become more accepting of their personal strengths and weaknesses.

THE MEANING OF SUICIDE ATTEMPTS AMONG TEENAGERS

The relative infrequency of adolescent suicidal deaths compared to the number of attempts raises questions about the actual meaning and intent of these apparently self-destructive acts. Several studies have explored the lethal intent in adolescent suicide attempts. Bancroft (1979) reported that among a general population of adolescents admitted to a hospital emergency room because of self-poisoning, 42 percent stated that they had no intention of dying. Adolescents in Bancroft's study were considered to have the lowest level of suicidal intent, compared to the 21–35 and the 36-and-over age groups, as learned from the self-reports of subjects. Curran (1984) asked teenagers who attempted suicide if they thought that adolescents who attempt suicide intend to die. Only 16 percent named "wish to die" as the primary motive.

Self-poisoning (usually through drug overdose) is by far the most common mode of attempting suicide among female and younger adolescents. Self-poisoning, however, is rarely of high lethality. McIntire (1980) reported that only 12 percent of cases intended to cause death.

Many teenage injuries to self are not attempts to end life (although, unfortunately, attempts sometimes end in death, even when that is not the intent). What then is the actual meaning of and reasons for such dramatic acts? What are the hoped-for effects of the suicidal act of low lethality?

Considerable research points to the highly communicative quality of this type of suicidal behavior, particularly in younger and female populations. Further, teenage suicide attempts appear to occur within an interpersonal context (Hawton, 1982a; Topol & Reznikoff, 1984; Wenz, 1979; White, 1974). That is, the suicidal adolescent often hopes to regain a lost love or influence a lover's affection. Bancroft (1979) found that 45 percent of the 16- to 20-year-old suicide attempters studied were "seeking help" by means of their suicide attempt, while 35 percent sought to "influence someone."

The finding that teenage suicide attempts are usually of low lethality in no way diminishes the seriousness of the action. The adolescent who attempts suicide is a needy person whose act should be treated with the utmost seriousness. This is also true for those who "only talk about committing suicide." Their remarks should always be referred to qualified personnel.

Adolescent suicide and attempted suicide can derive from a variety of conditions. There is no typical suicidal adolescent. However, some common factors have been found. In every case, suicidal behavior occurs as the culmination of multiple, long-standing, significant problems, both within the person and between that person and the environment (Brent & others, 1999). Suicidal adolescents generally feel overwhelmed by stress and do not feel they have the personal or social resources to handle that stress (Mazza & Reynolds, 1998;

Weiner, 1992). Research suggests that certain personal, family, and social factors are frequently associated with adolescent suicide (Brent & others, 1999). It is important to remember that interpreting research on adolescent suicide is difficult. Sometimes, the research is referring to suicidal adolescents or those who have been labeled by professionals as at-risk for suicide because of suicidal threats, gestures, or thoughts. Other studies have assessed teens who have attempted suicide but survived the attempts. Studies of completed suicides generally attempt to develop a psychological autopsy by interviewing friends, teachers, and relatives and examining the writings and belongings of the individual. It is impossible to know how many deaths judged accidental were actually suicidal deaths. Research findings must be evaluated with these limitations in mind.

PERSONAL PROBLEMS AS RISKS FOR SUICIDE

Historically speaking, adolescent suicidal behavior has been viewed as behavior of an impulsive nature, often indulged in by relatively normal teenagers (Crumley, 1982; Jacobziner, 1965). It has become increasingly clear, however, that teenage suicide attempters are significantly troubled individuals whose emotional problems are impressive (Yoder,1999). The results of a 1990 poll of American teenagers revealed that teenagers who had attempted suicide or who came close to attempting suicide were experiencing many problems. The most commonly given reasons for considering or attempting suicide were problems at home (47 percent), depression (23 percent), problems with friends and social relations (22 percent), low self-esteem (18 percent), boy-girl relationships (16 percent), and feelings that no one cared (13 percent) (Freiberg, 1991).

Labeling of suicidal adolescents has proved difficult because of the myriad ways in which teenagers manifest their symptoms and hide or obscure their real feelings to the adult world. A variety of psychiatric diagnoses have been noted among teenagers who have attempted suicide. The three most frequently found diagnoses are substance abuse (drug and alcohol), conduct disorder, and depression. In addition, the presence of aggression, low frustration tolerance, and impulsivity substantially increase the risk of suicide (Berman & Jobes, 1991), as does parental psychopathology and abuse history (Brent & pothers, 1999). These personality characteristics are common in the before-mentioned diagnostic groups and certainly contribute to their risk for suicide. Conduct disorder, depression, and substance abuse often occur within the same individual. Teens with more than one of these diagnoses is at greater risk than if only one problem is present. Personality problems frequently associated with suicide include the following:

- *Depression.* The feature most often seen in the literature on adolescent suicide is depression, including unipolar and bipolar (manic-depressive) forms. Depression appears to be most related to suicidal ideation or thoughts about killing oneself. Studies (Cantwell & Carlson, 1983) suggest that as many as 83 percent of youths with suicidal ideation are also depressed. However, the majority of depressed youths are not suicidal (Berman & Jobes, 1991).

- *Substance abuse*. Histories of substance abuse have been found in 15 to 33 percent of adolescents who completed suicide. Substance abusing adolescents are three times more likely to attempt suicide than non-abusing adolescents (Berman & Schwartz, 1990).

- *Conduct disorders*. A large percentage of adolescent suicide completers have been involved in some form of antisocial or illegal behavior, ranging from shoplifting to drug selling to prostitution. The risk for suicide is increased when the teen is experiencing strong feelings of anger and aggression.

- *Overreliance on limited support*. Adolescents who attempt suicide tend to overinvest themselves in very few, but very intense, interpersonal relationships. They appear to have a limited capacity to support themselves emotionally or to cope with their lives by means of internal strength alone. Rather, they rely heavily on the support of others, usually peers (Topol & Reznikoff, 1984; Walch, 1976). Other research suggests that many suicidal adolescents have no close friends or confidants or have recently lost an important friendship (Berman & Jobes, 1991).

- *Communication skills*. Suicidal adolescents tend to express troubled feelings through behavior rather than internal or interpersonal dialogue. Adolescents who are aware and tolerant enough of their unhappiness to talk about it are at far less risk of suicide than those who have no other expressive medium available to them.

- *Lack of control over the environment*. Corder and colleagues (1974) cite the inability to change one's environment as a frequent cause of attempted suicide. This concept was first studied by Rotter (1971), and a review of the studies of personal control has appeared elsewhere (Dacey, 1976). People tend to fall into one of two categories in terms of their sense of control over their lives. "Internals" see control as self-derived; they have a sense that they can influence what happens to them by their own actions. "Externals" see control as imposed by outside factors; thus, they see life as a matter of chance or luck. Some externals really do not have much control; others only imagine they do not. In either case, external individuals are far more likely to commit suicide than internals.

- *Hopelessness*. Teens who attempt suicide express high levels of hopelessness or the belief that things will not get better (Spirito & others, 1988).

- *Hypersensitivity*. All adolescents occasionally overreact to situations, but the hypersensitive youth will have an extreme reaction to situations that would only mildly disturb most people. The disruptions caused by seemingly trivial events may come together in a suicide attempt. Withdrawn, lonely, and supersensitive are labels commonly used in describing suicidal adolescents (Berman & Jobes, 1991).

- *Limited problem-solving skills*. Adolescents who attempt and complete suicide often have more stressful life events and fewer coping skills. Youths at risk for suicide are generally not able to think of many ways to solve the problems they face. Suicide may be one of the few choices they see as available (Berman & Jobes, 1991).

- *School performance.* Poorer school performance and underachievement are often found among suicidal adolescents. This may be related to problems of depression, drug abuse, or antisocial behavior (Berman & Jobes, 1991). Less often, high-achieving adolescents commit suicide. When this happens, it is often startling to parents, teachers, and peers. Clarizio (1994) described several reasons why gifted youngsters may be prone to depression. They may, for example, seek to achieve very high or impossibly high standards, may be preoccupied with idealistic and abstract moral issues, may feel alienated from peers who have different interests, or may have experienced a failure or humiliation that they find impossible to bear.

FAMILY PROBLEMS AS RISKS FOR SUICIDE

Considerable research was devoted in the early 1980s to the constitution, dynamics, and histories of the families of suicidal adolescents (Angle, 1983; Brent & others, 1999; Crumley, 1982; Hawton, 1982a, 1982b, 1982c; McKenry & others, 1982). It has been shown that the families of suicidal adolescents experience significantly more dysfunction, disorganization, mobility, and loss than the families of normal teens. The following characteristics tend to typify the families of many suicidal teens:

- *Family instability.* Suicidal adolescents often grow up in disrupted and disorganized families, in which abuse, family and marital conflict, psychopathology (Brent & others, 1999), and physical illness are common (Weiner, 1992). Parental losses tend to occur at an earlier age for the suicidal adolescent than for comparison groups of disturbed, nonsuicidal adolescents. Other family members, including parents and siblings, have themselves previously attempted or completed suicide. The threatened loss of a parent during adolescence because of death or separation can bring on a suicide attempt for a vulnerable teenager. A high incidence of parental deprivation, both physical and emotional, has also been reported (Wade, 1987; White, 1974). Pierce and Schwartz (1992) found that suicidal adolescents describe consistently negative relationships and a lack of intimacy with both parents. Physical or sexual abuse in the home has been cited as a relatively common element (Yoder, 1999). Suicidal youths are also more likely to have parents with a psychiatric disorder such as substance abuse or depression (Berman & Jobes, 1991; Brent & others, 1999). Psychiatrically ill parents are sometimes unable to provide the care and guidance their children need.

- *The appearance of not being needed.* When we feel that no one needs us, we tend to become lonely and self-centered. Suicidal adolescents report less caring, less affection, less support, and less enjoyment in their family interactions (Berman & Jobes, 1991). This may result from the family dysfunction described earlier or from the teens' beliefs that they are not meeting parental expectations.

- *Communication.* Serious impairment of communication between father and daughter is increasingly being noted and treated as a factor in the dynamics of the female adolescent suicide (Angle, 1983; Hawton, 1982a, 1982b, 1982c; McKenry & others, 1982).

SOCIAL FACTORS AS RISKS FOR SUICIDE

- *Peers*. Peer problems are considered a critical factor in the development of adolescent suicidal behavior (Celotta & others, 1987; Jacobs, 1971; Rohn, 1977; Teicher, 1973; Tishler, 1981; Walch, 1976; Wenz, 1979; Yoder, 1999). Teens who feel that they do not have any friends and do not belong to a peer group are likely to feel lonely, isolated, and alienated, and to lack support when experiencing problems. This is especially true for disturbed, suicidal adolescents whose family life has often been inadequate.

- *Suicide clusters and the media*. Epidemics of teenage suicides and suicide attempts in a single locality recently have caused researchers and laypersons alike to wonder about the contagion of adolescent suicidal behavior. Some excellent research does suggest that well-publicized suicides bring out latent suicidal tendencies in adults and significantly increase the rate of suicide in the geographic area covered by the publicity (Ashton & Donnan, 1981; Bollen & Phillips, 1982; Phillips, 1979). It is reasonable to assume that adolescents are at least as readily influenced as adults. High school surveys suggest that suicide attempters and teens who are thinking of suicide are likely to have known a peer or family member who attempted suicide (Harkavy-Freidman & others, 1987).

  Curran (1984) has demonstrated that teenagers are quite familiar with suicide as a behavioral alternative to coping with life's problems. He reported that 87 percent of the female high school students questioned knew someone who had attempted or committed suicide. In 55 percent of the cases, it was a person known well by the teenager—a friend, close relative, or family member. Among the males questioned, 57 percent knew someone who had attempted suicide; 29 percent knew the attempter well.

Teenagers who are experiencing high levels of stress and are feeling lonely and rejected by friends and family may be at-risk for suicide.

- *Firearms.* The availability and accessibility of a means for completing suicide is an important component in evaluating whether a teen is at-risk for suicide and is also an important risk factor (Brent & others, 1999). Firearms have become more accessible in recent years. Simultaneous with that, the rate of suicide by firearms among 15- to 19-year-olds has increased faster than rates by other methods (Garland & Zigler, 1993). Teens who commit suicide are more likely to have had firearms in their homes than teens who made suicide attempts. The increased availability of drugs and alcohol among teens has only added to the suicidal risk posed by the presence of firearms.

- *Stressful life events.* The risk for suicide is increased when a number of negative changes occur in a person's life. Often these events are related to personal factors (achievement pressures, failure at school or work) or family factors (family suicide, parental divorce, change in caretaker or living situation). Some writers (Elkind, 1981) have blamed increases in adolescent suicide on social changes that have increased the amount of stress and responsibility experienced by teens who are left unsupervised for long hours. Mazza and reynolds (1998) found that daily hassles and negative life events were significant factors for suicidal ideation in adolescent males, whereas lack of social support and depression were significant factors for females. Completed suicide is often precipitated by an event that is experienced as shameful or humiliating, such as an arrest, rejection by a romantic partner, or sexual embarrassment (Garland & Zigler, 1993). Sexual or physical assault is a negative life event that contributes significantly to the risk for suicide among girls. Gay youths are particularly sensitive to concerns of acceptance and rejection as they attempt to form their personal and sexual identities while being keenly aware of societal prejudices. Gay and lesbian 18- and 19-year-olds, struggling with decisions about disclosure of their sexual orientation, are two to six times more likely to attempt suicide than heterosexuals of the same age (Trotter, 1999). Societal prejudices may thus contribute to the stigma, stress, and suicidal risk experienced by adolescents.

IDENTIFICATION AND TREATMENT OF SUICIDAL RISK

Given the seriousness of suicidal behavior, parents and professionals are very concerned that at-risk youths be correctly identified and referred for the appropriate treatment. We have described a number of risk factors for suicide in the previous sections. Many youths possess one or more of these risk factors without being suicidal. How then do you know whether a teen is in danger? Sometimes the identification of suicidal risk is easy. A teenager may directly tell a parent or friend about "wanting to die." Parents and other adults also need to be alert to the warning signs listed in the Applied View box on page 389. The vast majority of adolescents who kill themselves provide some verbal and nonverbal clues (Berman & Jobes, 1991). Brent and associates (1988) found that more than 83 percent of adolescents who had completed suicide had made a suicidal threat to others, often to friends or siblings, in the past week. Few adolescents seek help on their own. Thus, it is critical that suicidal threats be taken seriously and that suicidal adolescents be helped to get the treatment that

AN APPLIED VIEW

One Person's History of Suicide Attempts (Anonymous)

❏ My first psychiatrist told my parents that my psychological tests indicated that I was potentially suicidal. I was 14 then. At 22, I had made five suicide attempts and had been in six mental institutions, which add up to 29 months as a mental patient and five years of intensive therapy. My diagnosis was borderline schizophrenia, chronic depression, and sadomasochism. Why? How had I become so obsessed with suicide?

When I flash back on my adolescent days, I remember feeling ugly, socially awkward, stuck away in an all-girls' boarding school reading Camus and Hesse, unpopular, and stupid! In fact, I was not quite as dreadful as all that, but in my mind I was. I felt different. I once wrote, "I'm at the bottom of an upside-down garbage can and it's so ugly." The world was horrible, but I was the worst part of it.

Suicide was my escape. Unsuccessful suicide attempts put me in the care of others who delicately forced me to confront my feelings of sadness and anger. I had to learn to share with others and sometimes that was what I secretly wanted. Two of my attempts, however, were calculated, purposeful acts. Despite what shrinks may say, I wanted to be dead, not taken care of.

What did death mean to me? One of my earliest memories is sitting on moss-covered ground in a grove of pines, reading The Prayer for the Dead with my basset hound curled up beside me. Suicide meant escape from hell on earth. No other purgatory could be worse than this one. Even if I were reincarnated, I would end up being some "lowly animal"

with the kind of mind that could not plague me with frightening, lonely, depressing thoughts. I clung to my friends and family, but it only increased my anger and self-contempt. I treated those people as my keepers who temporarily saved me from being left alone with my tormenting mind.

The final blow hit in Boston. I gradually withdrew from the few friends I had, as well as my family. Death had grown so close that I no longer felt that I had much time. It was impossible to commit myself to anyone or anything. I was reserved, yet few people could sense how obsessed I was with death. Signs of affection terrified me because I knew I could not let anyone count on me. I needed death if life became too unbearable.

It finally did. I had become so passive that I no longer made contact with people. They had to call me. So much time had elapsed since I had felt close to someone that it seemed my "disappearance" would not really upset anyone. In addition to this, I was convinced that I was too stupid to handle academics or even a menial job (even though I had two jobs at the time). On a day when I knew no one would try and reach me, I took three times the lethal dosage of Seconal.

I was found 24 hours later and came out of a coma after 48 more. My arm was paralyzed. This time, I was placed in a long-term hospital. Another try at life began. With the help of an excellent therapist and the patient love of those whom I had thus far rejected, I have started once more. It has been two years since I took the pills. I think I know why people bother to live now.

What personal, family, and social risk factors for suicide are apparent in this account?

they need. Because siblings and friends may be reluctant to reveal this information, they must be taught that it is better to lose a friendship because of sharing a secret than to lose a friend because of suicidal death. Within the school setting, school counselors and psychologists may be the most appropriate and helpful persons with whom to share a friend's suicidal threats. Parents might seek out treatment through a community mental health center for a child expressing suicidal threats or exhibiting any of the risk factors listed in previous sections.

AN APPLIED VIEW

Warning Signs of a Potential Suicide Attempt

The following behaviors are signs that a person may be contemplating suicide. If you see these behaviors in a teen, you should notify someone who is professionally capable of dealing with the situation:

1. Change in school grades
2. Withdrawal or moodiness
3. Accident proneness
4. Change in eating or sleeping habits
5. Other significant changes in usual behavior
6. Talking about killing oneself
7. Talking about "not being" or not having any future
8. Giving away of prized possessions

Trained clinicians can then evaluate whether there is an immediate or more long-term danger that the adolescent is at-risk for suicide. If imminent risk is determined, the initial goal of treatment will be to protect the adolescent from self-harm. This protection may require hospitalization. Sometimes it involves a mutual problem-solving strategy in which the adolescent is provided with the caring and support that he or she perceived was lacking. Therapists need to make themselves accessible to the suicidal adolescent so that the teen can contact them as needed. Family and friends may help to provide a suicide watch so that the adolescent will not be left alone. Steps should be taken to remove any available means (i.e., firearms, pills) of self-harm from the environment. No-suicide agreements, or contracts, are often used by which the adolescent agrees not to inflict self-harm for a certain time period, typically until the next therapy session. Efforts are also made to reduce environmental sources of stress that are disturbing to the adolescent or to develop some problem-solving strategies for relieving the immediate crisis.

After the imminent danger of a suicidal threat has passed, much therapeutic work generally remains to be done. For about one-third of suicide attempters, no further treatment is necessary (Clarizio, 1994). These adolescents are generally well-adjusted and became suicidal in response to specific life events that have been resolved. Adolescents with long-term problems of depression, isolation, family conflicts, and behavioral problems require ongoing treatment. Goals at this point typically focus on reducing the underlying pathology (depression, substance abuse, or conduct disorder, if present), and developing coping, problem solving, self-esteem, and social skills, if they were lacking. The goals and methods of therapy will be determined by the training of the therapist and an assessment of the personal, family, and social risks that contributed to the crisis.

SUICIDE PREVENTION

The dramatic rise in adolescent suicide and suicide clusters, or series of suicides in a single community, has resulted in a variety of efforts to prevent youth suicide.

Crisis intervention and telephone hotline services are among the most common suicide prevention efforts, aimed at averting attempts among teens who are already depressed and suicidal. Crisis centers and hotlines offer immediate help with 24-hour availability to listen and resolve crises. For adolescents, the anonymity and comfort with the telephone make this form of contact easier than seeking face-to-face treatment (Berman & Jobes, 1991). Hotline services seem to work in reducing suicide when they are used. Hotline users are most often white and female. Thus efforts need to be made to advertise these services more effectively to teenage males and to other racial groups. Garland and Zigler (1993) also emphasize that hotline services must be backed up by mental health services, which can provide emergency services when needed.

School-based prevention programs are designed to raise awareness of the problem of suicide, to train participants to identify adolescents at risk for suicide, and to educate participants about referral methods and mental health resources (Garland & Zigler, 1993). The programs are intended for high school students, parents, and teachers, with the hope that they will be able to identify and seek help for teens who need it. Although school-based programs have become popular, there is little evidence to date that supports their effectiveness. Some evidence (Spirito & others, 1988) suggests that students do gain some knowledge, but demonstrate no change in their attitudes.

Do you think school-based suicide prevention programs are helpful or harmful? Why?

Garland and Zigler (1993) argue against the continuation of current school-based programs for several reasons. In efforts to provide a climate that destigmatizes suicide and supports the identification of self and peers as suicidal, school-based programs often describe suicide as a response to severe life stress but ignore the serious personal problems experienced by many suicidal adolescents. In order to heighten awareness and concern about suicide, the number of teens committing suicide is sometimes exaggerated in these presentations. Garland and Zigler fear that the normalization of suicide through these curricula will increase rather than decrease the acceptability and incidence of suicide. Garland and Zigler recommend that this type of school-based prevention program is appropriate for teachers and counselors who need to know more about the identification of at-risk students but will not be harmed by the message. School-based interventions for students should focus on the reduction of known risk factors for suicide, many of which are common to other adolescent problems, such as substance abuse and delinquency. Problem-solving skills training, social skills, and self-efficacy enhancement can be taught within school prevention programs. Because these skills-based programs are directed toward reducing the risk factors associated with suicide and other adolescent problems, they may be considered more truly preventative than programs directed merely at identifying suicidal youths.

C O N C L U S I O N S

We are living in a time of rapid social and technological change. Adolescents must cope with the stress associated with those changes as well as with the stressors that accompany the biological and emotional changes of adolescence. High levels of stress bring increased risks for physical disease and mental health problems. Fortunately, most adolescents have developed good coping strategies and possess a number of protective factors, which help them to maintain good physical and emotional health.

Mental health problems may develop when life stress combines with other risk factors. Mental disturbance is not a normal part of adolescent development. Society is exposing youth to serious risks, including drugs, alcohol, poverty, and violence, which add to the psychological distress of growing up. Signs of psycho-logical distress and mental disturbance among teens need to be taken seriously by adults. Cultural change, as well as developmental, individual, and familial factors, contribute to the risk for eating disorders, especially among adolescent women. Many adolescents show some symptoms of depression, but a much smaller number are severely depressed. Depression is one of the many risk factors contributing to adolescent suicide. In this final quarter of the century, we have been learning a great deal about the causes of mental distress and the stress factors that are so often related to it. Much progress has been made in the discovery of effective prevention and remediation in both areas.

In the next chapter, we examine another serious risk of the adolescent transition—substance abuse. Substance abuse may occur in conjunction with stress and mental disturbance or as an independent problem.

C H A P T E R   H I G H L I G H T S

Dealing with Stress
- Change, especially that caused by future shock, is the major factor in stressful situations.
- Many life events, including major events and daily hassles, contribute to stress in our lives.
- It is not unusual for disease to result from long-term and/or varied stressors.

Coping Strategies
- Coping strategies can be adaptive (healthy) or maladaptive. Adaptive coping strategies help adolescents reduce the negative effects of stress.

Risk and Resilience
- Protective factors can shield or protect adolescents from the effects of stress.
- Adolescents who are exposed to many risk factors, but develop few behavioral or psychological problems, are called resilient.

Mental Disturbance
- The idea that those who develop mental illness during adolescence will "grow out of it" is not supported by research.

- The rate of emotional disturbance among adolescents is very close to the rate of emotional disturbance among adults.

Eating Disorders
- Two of the most disruptive problems for adolescents are the eating disorders known as anorexia nervosa and bulimia nervosa.
- Adolescent girls develop eating disorders more than any other age group.
- Developmental, cultural, individual, and familial factors are associated with the development of eating disorders.

Depression
- Among the symptoms of adolescent depression are major changes in study, eating, and other behaviors, accident proneness, depressed talk, and the giving away of prized possessions.
- Teens often mask their depression through depressive equivalents.
- Teens who are depressed often suffer from more than one psychological disorder.
- Depression has been explained according to psychoanalytic, cognitive, behavioral, and environmental models.

Suicide

- Suicide is a response to stress and depression for some adolescents.
- Suicide is most common among middle-class whites, but the suicide rate for African American males has increased in recent years.
- Females are more likely to attempt suicide, but males are more likely to die from it.
- The causes of suicide include personality, family, and societal problems.

K E Y   T E R M S

adolescent life change event
    scale 360
anorexia nervosa 368
behavioral model
    (of depression) 375
bulimia nervosa 369
cognitive model
    (of depression) 374
coping strategies 362
depression 372
depressive equivalents 377

diversity 359
environmental model
    (of depression) 376
externalizing 376
future shock 359
internalizing 376
learned helplessness model
    (of depression) 375
loss of reinforcement model
    (of depression) 375
novelty 359

primary prevention 366
protective factors 365
psychoanalytic model
    (of depression) 374
resilience 365
risk factors 365
secondary prevention 366
social readjustment rating scale 360
stress 355
transience 359

W H A T   D O   Y O U   T H I N K ?

1. What is your score on the social readjustment rating scale? Is it a cause for concern? Why or why not?
2. What are five productive ways of dealing with everyday stress?
3. Suppose you realized that a close friend had developed an eating disorder. What would you do?
4. Think back to the last time you felt depressed. Now imagine that it was much worse than it was. How would you feel? How would you act? What would you do?
5. Why is there such a big difference in suicidal behavior between males and females?

S U G G E S T E D   R E A D I N G S

Benson, H., & William, P. (1985). *Beyond the relaxation response*. New York: Berkley. A stress reduction program that has helped millions of people live healthier lives. Includes his concept of the "faith factor."

Bethancourt, T. E. (1985). *The me inside of me.* New York: Lerner. When his entire family is killed in a plane crash, 17-year-old Freddie Flores is left rich and alone. Now he must not only learn how to handle his grief, he must learn to deal with his newly gained wealth as well.

Blume, J. (1987). *Letters to Judy: What your kids wish they could tell you.* G. P. Putnam's Sons. Judy Blume offers letters from young adults who confide their concerns with friendships, families, abuse, illness, suicide, drugs, sexuality, and other problems. In return, the author shares similar moments from her own life, both as a child and as a parent. She does not hesitate to reveal her own embarrassing situations to help us feel less alone. A special "Resources" section lists books for additional reading and addresses of special interest organizations.

Bunting, E. (1985). *Face at the edge of the world*. New York: Ticknor and Fields/Clarion Books. Jed should have been the first one to know, but instead he finds out from the headlines in the morning paper—Charlie, his best friend, has committed suicide. Why?

Curtis, R. H. (1986). *Mind and mood: Understanding and controlling your emotions*. Scribner's. According to Curtis, knowing more about emotions and how they affect the body can help in understanding and controlling them. This book has chapters on the nervous system and endocrine system, which address the physiological impact on emotions; a chapter on behavior modification; and a section with personality tests that you can take.

Gibson, M. (1980). *The butterfly ward*. New Orleans: Louisiana State University Press. This set of short stories tells what it is like to be between sanity and insanity. It is a sensitive look at the world of the mentally ill, both in and out of institutions.

Pipher, M. (1994). *Reviving Ophelia: Saving the souls of adolescent girls*. New York: Ballatine Books.

Plath, S. (1971). *The bell jar*. New York: Bantam. This famed book tells the story of Esther Greenwood's painful month in New York City, which leads eventually to her insanity and attempted suicide.

Rebeta-Burditt, J. (1986). *The cracker factory*. New York: Bantam. This novel humorously describes the difficulties of a young woman who takes to drinking because of the pressures in her life and is eventually institutionalized because of an attempted suicide.

Sechehaye, M. (1970). *Autobiography of a schizophrenic girl*. New York: New American Library. Written by a Swiss psychoanalyst, this book describes the method of therapy as it was applied to a case of schizophrenia. Offers many insights into this malady that often starts during adolescence.

Selye, H. (1974). *Stress without distress*. Philadelphia: Lippincott. Offers many suggestions on how to achieve a rewarding lifestyle.

Walsh, M. E. (1992). *"Moving to Nowhere": Children's stories of homelessness*. Westport, CT: Auburn House.

G U I D E D   R E V I E W

1. Emotional tension arising from life events or as feelings of threat to one's safety or self-esteem is known as
   a. depression.
   b. stress.
   c. conflict.
   d. mental disturbance.

2. Which of the following is an example of autonomy as a source of stress?
   a. getting poor grades at school
   b. deciding about college
   c. homelessness
   d. thinking about suicide

3. Which of the following is true of stressful life events?
   a. They may be only a cause of behavioral and adjustment problems.
   b. They may be only a consequence of behavioral and adjustment problems.

   c. They may be both a cause and consequence of behavioral and adjustment problems.
   d. They are usually unrelated to behavioral and adjustment problems.

4. Because Jack's father has to relocate every two years for his job, Jack lacks permanence in his life. This relates to the concept of
   a. distress.
   b. future shock.
   c. transience.
   d. autonomy.

5. All of the following have been identified as risk factors for psychological difficulties, EXCEPT
   a. easy-going temperament.
   b. sexual abuse.
   c. poverty.
   d. parental mental illness.

6. For the anorexic adolescent whose family is over-protective, rigid, and conflict-avoiding, refusal to eat is seen as a(n)
   a. silent kind of rebellion.
   b. effort to bring the family together.
   c. attempt to reorder family priorities.
   d. form of negative identity.

7. Which model explains depression as a consequence of "loss"?
   a. cognitive model
   b. environmental model
   c. behavioral model
   d. psychoanalytic model

8. The vast majority of adolescents who kill themselves provide
   a. some verbal and nonverbal clues.
   b. verbal clues only.
   c. nonverbal clues only.
   d. no clues whatsoever.

9. All of the following personal characteristics have been associated with adolescent suicide, EXCEPT
   a. feelings of hopelessness about the future.
   b. history of substance abuse.
   c. lack of a number of close friendships.
   d. belief that they can change their environment.

10. According to Garland and Zigler, suicide prevention programs should include the following components, EXCEPT
    a. problem-solving skills.
    b. self-efficacy enhancement.
    c. normalization of suicide.
    d. social skills training.

11. Parents of resilient adolescents combine support and understanding with supervision of their children's activities and _____ (strict/consistent) discipline.

12. _____ (Appraisal-focused/Problem-focused) coping refers to thinking about a problem in a new way so that it can be more easily solved.

13. _____ (Older/Younger) adolescents are more likely to use approach-oriented coping strategies.

14. _____ (Risk/Protective) factors may enable resilient persons to cope effectively with high levels of life stress.

15. Psychopathology is _____ (normal/abnormal) during adolescence.

16. Adolescents display _____ (similar/higher) levels of psychopathology than younger children.

17. Symptoms of _____ (anorexia nervosa/bulimia) include an intense fear of becoming obese, disturbance of body image, significant weight loss, refusal to maintain a minimal normal body weight, and amenorrhea.

18. _____ (Anorexia nervosa/Bulimia) develops most frequently in early adolescence and _____ (anorexia nervosa/bulimia) is more likely to develop in late adolescence.

19. One explanation of adolescent gender differences in depression is that boys have been socialized to _____ (internalize/externalize) their symptoms, whereas girls have been socialized to _____ (internalize/externalize) their symptoms.

20. According to Garland and Zigler, _____ (high school students/teachers) may benefit from learning to identify the signs of suicide most often displayed by teenagers.

Answers

1. B, 2. B, 3. C, 4. C, 5. A, 6. A, 7. D, 8. A, 9. D, 10. C, 11. consistent, 12. Appraisal-focused, 13. Older, 14. Protective, 15. abnormal, 16. higher, 17. anorexia nervosa, 18. Anorexia nervosa, bulimia, 19. externalize, internalize, 20. teachers

SUBSTANCE ABUSE

The tremendous fact for every one of us is that we have discovered a common solution. We have a way out on which we can absolutely agree, and upon which we can join in brotherly and harmonious action. This is the great news [for] those who suffer from alcoholism.

Alcoholics Anonymous

CHAPTER

CHAPTER OUTLINE

He always wanted to explain things
But no one cared.
The teacher came and spoke to him.
She told him to wear a tie like all the other boys.
He said it didn't matter.
After that they drew.
And he drew all yellow and it was the way he felt about the morning
And it was beautiful.
The teacher came and smiled at him.
"What's this?" she said. "Why don't you draw something like Ken's drawing?
Isn't it beautiful?"
After that his mother bought him a tie.
And he always drew airplanes and rocket ships like everyone else.
And he threw the old picture away.
And when he lay out alone looking at the sky
It was big and blue and all of everything.
But he wasn't anymore.
He was square inside and brown
And his hands were still
And he was like everyone else.
And the things inside him that needed saying didn't need it anymore.
It had stopped pushing.
It was crushed.
Stiff.
Like everyone else.

By a 16-year-old boy who later died of drug overdose

The transition from childhood to adulthood can be difficult, but the majority of adolescents are able to negotiate it successfully. In coping with the challenges of adolescence, teens may turn to drugs and alcohol for a variety of reasons (Feigelman & Feigelman, 1993). Mood-altering chemicals may seem to be a way of dealing with or perhaps denying the changes and expectations that accompany adolescence. The choice to use drugs may be an impulsive act, based on what feels good at the moment, or a way of temporarily relieving feelings of anxiety. For teens, who as a group lack the money, power, and freedom of adults, using drugs, tobacco, and alcohol may be an effort to take part in adult pleasures. Because alcohol is acceptable for adults and often glorified in the media, the message for teenagers can be confusing. Teenage peers are an important source of influence, and many peers expect their friends to engage in some alcohol and drug use. Teens who feel rejected by peers may also turn to drug and alcohol use as a way of dealing with their isolation.

For most teens, alcohol and drug use does not become a problem. Unfortunately, as the opening poem suggests, some teens suffer from a crushing of their spirits that can lead to substance abuse, mental disturbance, and/or crime. Having one of these problems increases the chance that you will suffer from the other two. Sad to say, illicit drugs and alcohol remain highly available to young teens and are a tempting, although dangerous, way to deal with life pressures. Hersch (1990) states that, in contrast to earlier generations of adolescents, "Today's kids, whether from inner cities or the suburbs or rural towns, must navigate a narrow course between drugs, alcohol, crime, and various forms of danger" (p. 23). No study of American adolescence is complete without taking a hard look at these distressing areas.

In this chapter, we examine the extent of adolescent use of alcohol and illegal drugs, types and stages of drug use, factors that influence drug use, and to what extent sexuality and crime are involved. The chapter will also help you become better aware of your own attitudes toward drugs and alcohol. In the final section, we discuss ways of combating substance abuse.

After reading this chapter, you should be able to:

- Define basic terminology and phrases related to substance abuse.
- Describe four stages or types of drug use.
- Identify at least three risk factors associated with drug abuse among adolescents.
- Take a stand on which factor has more of an impact on drug use—quality of family relationships or family structure (divorce, etc.)—and state why.
- State at least three reasons for the high degree of correlation between sexual activity and substance use, and describe which drugs are most related to sexual activity and why. You should also be able to distinguish between the myths and the realities about sexual enjoyment and drug use.
- Explain the difficulty involved in trying to examine the connection between ethnicity and substance abuse.
- Assess your current level of factual knowledge about drugs and arrive at a deeper understanding of your personal attitudes and/or experience with substance use.
- Describe the prevalence of drug use among different groups of adolescents and between different types of drugs, and state any major differences between groups.
- State why it is difficult to get adolescents into treatment programs.
- Compare some common drug abuse prevention methods as to their relative strengths and weaknesses.
- Describe several activities that could be effective alternatives to drug use.
- Discuss these issues from an applied, a sociocultural, and your own point of view.

Some Definitions

drug abuse
Use of a drug in such a way that the individual's physical, mental, or emotional well-being is impaired.

drug addiction
Historically has referred to dependence on narcotics.

drug dependence
When a physical or psychological need, or both, result from continuous drug use.

drug tolerance
A condition that develops from continuous use of a drug when a larger and larger amount is needed to produce the same effect.

abuse of inhalants
Sniffing substances such as airplane glue, paint thinners, and gasoline to get high.

drug overdose
Taking so much of a drug that it causes an acute reaction, usually extreme anxiety, which is sometimes followed by stupor, low breathing rate, and, in rare cases, coma.

victimless crime
The term given to drug abuse by some people who feel that drug abusers are the only ones who suffer from the use of drugs, so they should not be fined or imprisoned for their actions.

controlled drugs
Those drugs that have been limited in their distribution and manufacture under the Controlled Substances Act of 1970.

A drug is any chemical or vegetable substance that causes an emotional or behavioral change. Although some drug use is common among teens, abuse is less typical. Children and teenagers may be introduced to drug use at home, such as taking a sip of a parent's beer. This guided experimentation is usually short-term and is motivated by curiosity (Okwumabua, 1990).

Drug abuse is use of a drug in such a way that the individual's physical, mental, or emotional well-being is impaired. According to Newcomb and Bentler (1989), any regular use of a drug can be considered abuse.

Historically, **drug addiction** referred to dependence on narcotics. Today it has so many meanings that experts are now beginning to use the term *drug dependence* instead.

Drug dependence occurs when there is a physical or psychological need, or both, resulting from continuous drug use. Psychological need occurs when the person feels anxious, depressed, or irritable when she or he doesn't have the drug. Physical dependence, on the other hand, occurs only when negative physical symptoms result from drug withdrawal, such as vomiting, sweating, muscle tremors, joint pain, delusions and hallucinations, and almost always a strong sense of anxiety. About 5 percent of all American teenagers are physiologically drug dependent.

Drug tolerance is a condition that develops from continuous use of a drug. It occurs when a larger and larger amount is needed to produce the same effect.

Abuse of inhalants such as airplane glue, paint thinners, and gasoline is becoming more common. The most popular new inhalant is amyl nitrite, popularly known as "poppers."

Drug overdose is defined as taking so much of a drug that it causes an acute reaction, usually extreme anxiety, which is sometimes followed by stupor, low breathing rate, and in rare cases, coma. Hallucinogens, marijuana, and stimulants can produce an anxiety attack even when taken in small doses; users become agitated, frightened, suspicious, and think that people are "out to get them." While medical attention is always necessary in the case of drug overdose, it is also important that those who are first on the scene remain calm and that they reassure the person he or she will be all right.

Victimless crime is a term often applied to the use of drugs (see Chapter 13). Some people feel that abusers are the only ones who suffer from the use of drugs, so they should not be fined or imprisoned for their actions. However, drug abusers tend to be poor financial risks, dangerous drivers, and often resort to theft to support their habit; they are usually dependent on their families and often on society for support. So their acts are seldom victimless.

Controlled drugs are all those that have been limited in their distribution and manufacture under the Controlled Substances Act of 1970. This act empowers the Attorney General of the United States to punish those who use or sell drugs illegally. There are five levels of punishment under this law. Most of the drugs discussed in this chapter are covered by the act.

AN APPLIED VIEW

What Do You Know About Drugs?

Test your knowledge of drugs by answering the following questions.

1. During which time was drug abuse a problem in the United States?
   a. during the Civil War
   b. in the 1950s
   c. in the 1960s
   d. all of the above

2. How do most drug users make their first contact with illicit drugs?
   a. through "pushers" seeking new customers
   b. through their friends
   c. accidentally
   d. through the media

3. Which of the following is the most commonly abused drug in the United States?
   a. marijuana
   b. alcohol
   c. cocaine
   d. heroin

4. Which of the following is not a narcotic?
   a. heroin
   b. marijuana
   c. morphine
   d. methadone

5. Which of the following is not a stimulant?
   a. amphetamine
   b. caffeine
   c. mescaline
   d. methamphetamine

6. Which of the following drugs does not cause physical dependence?
   a. ethyl alcohol
   b. morphine
   c. mescaline
   d. secibarbital
   e. codeine

7. Which of the following is not a hallucinogen?
   a. MDA
   b. LSD
   c. STP
   d. MPA

8. Why is intravenous injection the most dangerous method of taking drugs?
   a. because of the rapidity with which the drug enters the system
   b. because nonsterile equipment and solutions are likely to cause serious medical complications (such as getting HIV [virus] and AIDS)
   c. because the amount of drug entering the bloodstream is likely to be large
   d. all of the above

9. When a person becomes physically dependent on drugs, what is the primary reason he or she continues to take the drug?
   a. to experience pleasure
   b. to relieve discomfort
   c. to escape reality
   d. to gain acceptance among friends

10. Which of the following drugs has never been used to treat narcotic addiction in the United States?
    a. cyclazocine
    b. naloxone
    c. methadone
    d. psilocybin
    e. heroin

11. Which of the following is an effective treatment method for drug abusers?
    a. maintenance
    b. detoxification
    c. abstinence
    d. psychotherapy
    e. all of the above

Source: Special Action Office for Drug Abuse Prevention, 1976.

AN APPLIED VIEW

Answers to "What Do You Know About Drugs?"

1. (d) All of the above. The use of drugs is as old as the history of humankind. The United States has witnessed special drug abuse problems in different periods. During the Civil War opium was used medically, and since its addictive properties were not clearly understood, many wounded soldiers became addicted. Following the Civil War, the practice of opium smoking became popular on the West Coast and spread to many urban areas. Throughout the century, there were periodic "drug scares" created by the use of cocaine at the turn of the century, heroin in the 1920s, marijuana in the 1930s, and heroin again in the 1950s. The 1960s saw a social explosion of drug use of all kinds, from LSD to heroin and marijuana.

2. (b) Through their friends. With the exception of alcohol, which is usually first used at home, most drug users are introduced to drugs by friends.

3. (b) Alcohol. Estimates are that about nine million Americans are alcoholics.

4. (b) Marijuana. In the past, marijuana was legally classified as a narcotic but it isn't now. Marijuana's effects are similar to stimulants, sedatives, or hallucinogens, and its actual effects depend on dose, frequency of use, set (personality and expectation of the user), setting (environment), and other factors. Morphine and heroin are legally and pharmacologically classified as narcotics. Methadone is a synthetic narcotic.

5. (c) Mescaline. All are stimulants except mescaline, which is a hallucinogen with effects similar to LSD.

6. (c) Mescaline. Physical dependence on mescaline (the drug derived from the mescal and peyote cactus) and many other hallucinogens has not been verified.

7. (d) MPA. MPA is not an acronym for any known drug. MDA, LSD, and STP are hallucinogens with similar effects. MDA (Mellow Drug of America) and STP (Serenity, Tranquility, Peace) are street drugs.

8. (d) All of the above. In particular, using nonsterile equipment is a serious hazard often overlooked by the drug user.

9. (b) To relieve discomfort. When people stop taking a drug that they are physically dependent on, they develop physical withdrawal symptoms (such as muscle spasms, vomiting, sweating, insomnia, and so forth). Taking the drug relieves the discomfort of withdrawal symptoms.

10. (d) Psilocybin. Psilocybin is a hallucinogen that has no accepted medical use. All the other drugs have at various times been used to treat narcotic addiction. When heroin was introduced in 1898, some people thought it had possibilities for treatment of "morphinism." Methadone, cyclazocine, and naloxone are used currently to block the high produced by heroin.

11. (e) All of the above. All these methods have been used successfully to treat drug abusers, and many have been used in combination.

---

possession
Refers to illegally obtaining drugs from someone not legally sanctioned to distribute them.

Possession, dealing, and **trafficking** in drug substances are also distinguished legally. Illegal possession refers to obtaining drugs from someone not legally sanctioned to distribute them. Dealing is the sale of drugs on a small scale, usually carried out by a friend of the purchaser. Trafficking involves the sale of much larger amounts of drugs. Each of these violations of the law carries a different penalty.

Stages and Types of Drug Use

dealing
The sale of illegal drugs on a small scale, usually carried out with a friend of the seller.

trafficking
The sale of large amounts of illegal drugs.

experimental use
Limited use of drugs out of curiosity or to have new experiences.

social use
Infrequent use of drugs that is limited to parties, dances, and other special occasions.

medicinal use
Use of drugs to reduce tension and anxieties.

addictive drug use
Use of drugs as a necessary means for feeling good physically and emotionally.

When does drug and alcohol use become a problem? At what point should parents become concerned?

Adolescents may become involved in drug and alcohol use for a variety of reasons. As previously mentioned, not all drug and alcohol use escalates to the level of abuse and dependency. Weiner (1992) describes four categories of drug users, which are helpful in distinguishing between drug use and drug abuse.

Experimental users try drugs out of curiosity or to have a new experience and then stop using them. **Social users** take drugs in order to join in activities with their peers. Social use is limited to parties, dances, and other special occasions, and involves only infrequent use. Although infrequent social use does not generally interfere with daily functioning, it can be dangerous if teens operate cars after they have used drugs or alcohol. **Medicinal users** take drugs to reduce tension and anxiety. They care more about the pleasurable effects of the drug than its social function. They use drugs more frequently than experimental and social users and are more likely to use drugs alone. **Addictive drug use** also takes place outside of social situations. The user depends on the drug not only as a way to relieve tension, but also as a necessary means for feeling good physically and emotionally. Users are more likely to be using drugs at times and in places where such use is inappropriate (e.g., at school) and to experience withdrawal symptoms when drugs are not available. The majority of adolescents who have used drugs and alcohol may be classified as experimental or social users. Medical and addictive drug users are likely to suffer from other emotional difficulties. Low self-esteem, depression, and suicide are serious concerns among teens who are preoccupied and dependent on drugs and alcohol (Feigelman & Feigelman, 1993). Delinquency is also a concern for youths with substance abuse problems. Pliszka and colleagues (2000) looked at adolescents who had been placed in juvenile detention and found that the vast majority used or were dependent on drugs. They also found that 60 percent of these adolescents met the criteria for conduct disorder and 42 percent for affective disorder (depression).

The consequences of drug and alcohol use certainly differ depending on the type of drug use. According to Newcomb and Bentler (1988), for example, adolescent alcohol use has some negative consequences for young adult functioning, including decreased college involvement, lower religious commitment, and lessened social conformity. Involvement with hard drugs results in more serious consequences, including more sexual partners, earlier parenthood, lower educational attainment, poorer jobs, more psychological problems, and more involvement in criminal activity. Newcomb and Bentler maintain that hard drug use prevents adolescents from accomplishing the important tasks that prepare them for adjustment in young adulthood.

Some adolescents progress from drug use motivated by curiosity, experimentation, and social activities to more serious levels, characterized by preoccupation and dependency, and others do not. All adolescents who drink will not necessarily try marijuana and then move on to hard drugs. The reasons why some teens progress from alcohol to marijuana to hard drugs is not clearly understood. However, teens who use the most dangerous substances have

AN APPLIED VIEW

Do You Have a Drinking Problem?

The following questions may help you to decide whether you have a drinking problem:

- Do you drink to escape from the pressures of college life?
- Do you sometimes skip classes because of hangovers?
- Do you drink more than your friends?
- Do you drink to escape from reality, boredom, or loneliness?
- Do your friends or loved ones express concern about your drinking?
- Do you drink and get drunk even when intending to stay sober?
- Do you drink when you are alone?
- Do you drink frequently to a state of intoxication?

- Have you had two or more blackouts (can't remember some or all of what happened while you were drinking) in the past year?
- Have you gotten into trouble with the police and/or college officials as a result of your behavior while drinking?

If you must answer yes to any of these questions, you may have a problem with alcohol, and perhaps you should seek advice from trained personnel. If you must answer yes to more than two of the questions, especially those on the second half of the list, you definitely have reason for concern, and should seek help.

Source: Adapted from Chebator, 1993.

usually used more mild drugs first, and teens who are the heaviest users of more mild drugs and alcohol are most likely to move on to more dangerous substances. The stage theory, developed by Kandel (1980), suggests that if you can prevent young people's use of a milder substance, you can reduce the subsequent use of more dangerous substances.

PERSONAL, PEER, AND FAMILY INFLUENCES

Personal, family, and peer factors all play a role in whether adolescents begin to and continue to use drugs. Teens who progress from lower to higher levels of drug and alcohol use have been described as unconventional, impulsive, rebellious, and pleasure seeking (Weiner, 1993). Adolescent substance abuse has also been associated with aggressive and antisocial behavior in elementary school, delinquent behavior in adolescence, and low self-esteem, depression, poor academic achievement, and low religious involvement (Okwumabua, 1990).

problem behavior theory
The theory that early problem behaviors contribute to substance abuse and that persons who have one deviant behavior are likely to have others as well.

These findings are consistent with the **problem behavior theory** proposed by Jessor and Jessor (1977). According to this theory, early problem behaviors, including shy and aggressive behavior in kindergarten, academic achievement in middle childhood, and feelings of alienation in the teenage years, contribute to adolescent substance abuse.

Personal factors, such as unconventional attitudes, poor school performance, and frequent drinking, often lead to marijuana use. Beginning to use hard drugs is associated with more serious emotional and behavioral problems, such as heavy marijuana use, depression, alienation, and social withdrawal (Weiner, 1993). Parental neglect and indifference can also contribute to the movement to hard drug use.

When parents and peers provide models for the use and abuse of drugs and alcohol, abuse is also more likely. Starting to drink hard liquor seems to have more to do with imitating the behavior of parents or friends than with individual personality characteristics. Grant (2000) indicates that one out of every 2.3 children in the United States is exposed to alcohol abuse in their immediate family. Grant (2000) also suggests that living with an alcoholic increases an adolescent's risk for alcoholism. Parents' levels of alcohol use when their children are in their early teens influence whether teens are drinking by mid-adolescence (Peterson & others, 1994). Good family management practices, however, offset the negative effects of parents' alcohol use. Involving children in the alcohol use of adult family members, however, contributed to earlier alcohol use by the teens. Mitic (1990) found in his survey of 1,128 students from grades 7 through 12 that those whose parents were attempting to teach them responsible drinking habits by introducing small amounts of alcohol taken with meals at home drank little in the presence of their parents. Unfortunately, this group drank as much when with peers as those whose parents gave them no guidelines on alcohol use. Teens who have peers that use drugs are likely to be drawn into drug use because their friends encourage their use and make drugs available. Peers have the strongest influence on the decision to use drugs during the middle school and early high school years. Crowe and others (1998) found that adolescents who were alcohol drinkers spent more time with peers and less time with family. In addition, family time was generally described negatively.

A number of current studies (Duncan & others, 1998; Kilpatrick & others, 2000; Swadi, 1999) suggest that there are multiple risk factors for teen substance abuse. These factors include parent-child conflict, inept parenting, peer deviance, life events and academic failure. A 1991 study compared the psychological well-being and substance use of 48 adolescents (24 males and 24 females) whose parents had divorced during a five-year period with 578 adolescents whose parents were still married. The researchers found that boys in disrupted families were especially likely to have problems with increased substance use, particularly after the divorce. In contrast, girls were most upset during the troubled time preceding the actual end of their parents' marriage and were much less involved with drugs (Doherty & Needles, 1991).

The importance of family unity and authoritative parenting as a safeguard against drug abuse is emphasized by Baumrind (1991b). Authoritative parents (see Chapter 7) were most successful in protecting their adolescents from problem drug use and helping them feel competent. The participants in Baumrind's study were from a university community, and none used highly addictive drugs. Many of the adolescents experimented with alcohol or marijuana, but this did not lead to later addiction. A similar result was also obtained by Shedler and Block (1990). Blyth and Roehlkepartain (1993) also found that family care and support, in conjunction with monitoring and discipline, contributed to lower involvement in at-risk behaviors including drug and alcohol use among teens.

What are the roles of parents and peers in influencing drug and alcohol use and misuse?

Searight and associates (1990) found that the 40 drug-abusing adolescents in their study scored poorly both in autonomy and intimacy within the family. The 446 Anglo and Latino youths who were studied over a five-year period by Coombs and associates (1991) showed a similar pattern. Although both users and nonusers in this group valued the opinions of their peers, emotional closeness with their parents appeared to protect these youths from drug abuse. Piercy and associates (1991) looked at the families of 151 adolescents who were in outpatient treatment for drug use and had experienced other various delinquent behaviors, such as running away. They were surprised to find that aspects of the adolescent's family structure—for example, the number of children and whether the parents were married or divorced—did not seem to be related to the teen's pattern of drug abuse. However, the *quality of relationships* within the family was found to be important in identifying amount and type of drug use. A milder pattern of drug use was most related to family closeness.

Barnes and others (1994) found that similar family factors (i.e., parental support, supervision, and parent-adolescent communication) influenced whether both African American and white adolescents abused alcohol. Whether or not peers used alcohol was a stronger influence for white adolescents. Strong religious beliefs were found to be a protective factor that discouraged alcohol abuse for African American teens.

The movement to more problematic drug and alcohol use may also be influenced by genetic factors. Alcoholism seems to run in families. Studies of twins and adopted children suggest that the familial prevalence of alcoholism cannot be explained solely by family climate and parental role modeling (Weiner, 1993). Baer and colleagues (1998) found that prenatal exposure to alcohol was more predictive of adolescent alcohol use than family history of alcoholism.

Hirschi (Gottfredson & Hirschi, 1990; Hirschi, 1969) developed a theory that partially explains the importance of family in influencing adolescent behavior. According to **control theory**, adolescents with strong bonds to society are less likely to deviate from conventional behavior. Attachment to family is the most important element in developing bonds to society. Adolescents with strong attachments to their families will not want to jeopardize that relationship by engaging in deviant behavior, including drug use and crime. In addition, because adolescents care about their parents, they will believe in their rules and develop commitment to activities valued by their parents. They are also more likely to select friends who share parental values rather than friends who abuse drugs. In this sense, parental relationships influence the choice of values, activities, and friends. Consistent with control theory, Turner and associates (1991) found that emotional detachment from parents was related to experimentation with substances among adolescents. In general, positive family relationships discourage initiation into drug use, even if parents are modeling substance use (Wenar, 1994).

Adolescent substance abuse can contribute to further deterioration in parent-adolescent relationships. One recent study (Stice & Barrera, 1995) revealed that teens with parents exhibiting low support and control were more likely to use

control theory
The theory that adolescents with strong bonds to society are less likely to deviate from conventional behavior.

substances. Adolescent substance abuse also contributed to further declines in perceived parental support and control.

In summary, personal, peer, and family problems often go hand in hand and contribute jointly to substance abuse problems. Teens engaged in destructive drinking and illegal drug use are apt to suffer from poor mental health, unsatisfactory peer relations, poor family relations, and low educational status (Duncan & others, 1998; Swadi, 1999; Kilpatrick & others, 2000; Wirtz & Harrell, 1990). According to Wenar (1994), parent and peer modeling of substance use increase the likelihood that a teen will begin experimenting with alcohol or drugs. Movement from use to abuse is fueled by the heavy drug involvement and abuse by a close friend and by conflictual relationships with parents.

WARNING SIGNS

It is sometimes difficult for parents to recognize early warning signs of drug use because they are similar to normal adolescent behaviors. For example, teens who are beginning to use drugs may avoid their parents, but it is not unusual for teens to spend more time with peers. It is critical, however, that parents recognize the signs that accompany the more serious substance abuse so that adolescents can get the help that is needed. Bogenschneider and associates (!998) studied parental awareness of teen alcohol use and found that most of the parents were unaware of their adolescent's drinking. Gold (1986) identified 13 clues to substance abuse: (1) chronic lying, (2) disappearance of money or valuable objects from the home, (3) substantial changes in mood, (4) abusive behavior toward oneself or other people, (5) frequent outbursts of anger, (6) auto accidents, (7) frequent school absence and truancy, (8) drop in school performance, (9) possession of and interest in drug paraphernalia, (10) hanging out with peers who use drugs, (11) being tired frequently, (12) sore throat or cough that doesn't go away, and (13) chronic conjunctivitis.

CRIME AND SUBSTANCE ABUSE

The relationship between drug and alcohol use and crime and delinquency have long been of interest. Consistent with problem behavior theory (mentioned earlier), teenage substance abuse is associated with a broad range of high-risk behaviors, including participation in deviant peer groups, interpersonal violence, poor school performance, vandalism, unprotected sexual intercourse, sexually transmitted diseases, and unintended pregnancy and parenthood (Duncan & others, 1998; Ketterlinus & others, 1994; Kilpatrick & others, 2000; Swadi, 1999). As noted by Gottfredson and Hirschi (1994), all these deviant behaviors offer immediate satisfaction or benefit, but have substantial risks for long-term harm. In an early study, Santana (1979) examined the relationship between substance abuse and antisocial behaviors. The study looked at 19 types of undesirable behavior and compared their occurrence among drug abusers to their occurrence among nonusers. The disturbing finding of this study was that, for 18 of the types, users were much more likely to commit undesirable behaviors than nonusers. Even for the single exception, arson, the behavior was evenly divided. This does not prove that drug use causes crime, but does indicate a strong relationship between the two.

Another study of the relationship between drug use and both minor and violent delinquency in three different ethnic groups of adolescent males was carried out by Watts and Wright (1990). These researchers found that frequent use of illegal drugs was the best predictor of violent delinquency among all the groups in the study. Although it is commonly thought that drug users, especially those who use hard drugs, are regularly involved in criminal activities, Federal Bureau of Investigation statistics (FBI, 1992) indicate that the drug most associated with crime is alcohol. In 40 percent of assaults and 35 percent of rapes, those convicted were "under the influence."

It is not clear whether drug use causes crime, crime causes drug use, or whether both situations result from common underlying problems (Milgram, 1993). Studies (Duncan & others, 1998; Kilpatrick & others, 2000) indicate that a common set of risk factors, including early antisocial behavior, family conflict, poor family management, drug use and criminal behavior among parents and siblings, school failure, economic problems in the home, low commitment to school, and association with delinquent peers, are associated with frequent drug use and chronic and serious delinquency. Low self-control is a common personality characteristic found among delinquents and drug users (Gottfredson & Hirschi, 1994).

Use of drugs and alcohol clearly increases the probability that acts of aggression and violence will occur. Alcohol has been associated with many adolescent deaths related to accidents and to homicides. Some deaths that are attributed to accidents may actually be suicides. Alcohol and drug use often accompany suicide. Kaeding (1985) estimates that more than half of adolescent suicides were completed under the influence of alcohol. Alcohol can impair judgment and increase impulsivity and depression. When firearms are available, this combination can be lethal for the already depressed teenager. College women are also more likely to experience rape when alcohol has been used. Data suggests that either the rapist or the victim had been drinking in 83 percent of unplanned campus acquaintance rapes (Milgram, 1993). When incidents of fraternity gang rapes have been studied, alcohol and/or drugs have almost always been involved. Drugs and alcohol not only decrease the controls of the perpetrator, but also impair the judgment of the victim as well as her ability to defend herself. For some perpetrators, intoxication of a date is used as an excuse for abuse (Milgram).

SEX AND SUBSTANCE ABUSE

Without question, there has been a sharp increase in both adolescent sexual activity and experimentation with illegal drugs over the past 20 years. Whether one causes the other is not known for certain, but these problem behaviors are often found together. Elliott and Morse (1989) state that typically, the young teen will begin by engaging in some delinquent behavior, progress to drug use, and then become sexually active. The frequency of intercourse is much higher for teens who are also involved in drug use. The fact that both sexual activity and use of injected drugs create a grave risk for becoming infected with the HIV virus makes understanding this relationship increasingly important. Rhodes and Jason (1990) note that the trend is for middle-class youth to be less involved with drugs today than was true a few years ago. Unfortunately, drug abuse, dealing, and violence are increasing among the poor and youths of color in the inner cities.

Rosenbaum and Kandel (1990) studied the 2,711 youngest participants in the *National Longitudinal Survey of Young Adults*—those who were 14 or 15 years old when they were first interviewed—to try to determine the nature of the connection between drug use and early sexual activity. The group included African American, white, and Latino teenagers of both sexes. The researchers found that those who began having sex before the age of 16 were much more likely to have already been using alcohol, tobacco, or illegal drugs than those who waited until they were older.

Other investigators have also found that involvement in sexual behavior at an early age is often at least partially a result of the use of alcohol and other drugs (Brooks-Gunn & Furstenburg, 1989; Flanigan & others, 1990; Rosenbaum & Kandel, 1990).

AN APPLIED VIEW

Sex and Alcohol Abuse

Mary Pipher works as a therapist treating adolescents and their families. She has written about her work with teenage girls and described the problems of Casey (Pipher, 1994, pp. 193–194).

❏ Casey, an 18-year-old female, came to therapy with her parents after her father had discovered diet pills in her purse. She had come home drunk several times during the past month, and was described as messy, dishonest, and irresponsible with money.

In therapy, Casey told Dr. Pipher about her first date with Stan during her sophomore year of high school. Instead of taking her to a movie as planned, her date took her to the country. After having a few beers, he made sexual advances. Casey thought that he would rape her and resisted his efforts. Stan called Casey a lesbian and Casey wondered whether he could be right. Later when Sam asked her out, Casey wanted to

prove her sexuality. She got drunk and offered to have sex before he asked. Since then, this pattern has been repeated with numerous dates. Alcohol became a way of deadening her anxiety so that she could have sex and reduce her guilt feelings afterwards. She felt badly about her appearance and was grateful that any guy wanted her. She was so eager to please that she never considered her own needs. Now, she knew she wasn't handling things well and was afraid of getting AIDS.

How would you categorize Casey's substance use (according to stage/type)? How do you understand the relationship between sexuality and substance use/abuse in this case?

More specifically, a number of researchers have looked at the relationship between the use of drugs and sexual functioning (Buffum, 1988; Buffum & Moser, 1986; Higgins & Stitzer, 1986; Solow & Solow, 1986; Zabin & others, 1986). The purpose of these studies was to evaluate the claims of many youths that drugs make them more sexually capable and increase their enjoyment of sex. It is alleged that drugs stimulate sexual activity through releasing inhibitions and through direct stimulation of desire.

Table 12.1	Controlled Substances: Uses and Effects			
Drugs	**Usual Methods of Administration**	**Possible Effects**	**Effects of Overdose**	**Withdrawal Syndrome**
Narcotics				
Opium	Oral, smoked	Euphoria, drowsiness, respiratory depression, constricted pupils, nausea	Slow and shallow breathing, clammy skin, convulsions, coma, possible death	Watery eyes, runny nose, yawning, loss of appetite, irritability, tremors, panic, chills and sweating, cramps, nausea
Morphine	Oral, injected, smoked			
Codeine	Oral, injected			
Heroin	Injected, sniffed, smoked			
Hydromorphone	Oral, injected			
Meperidine (pethidine)	Oral, injected			
Methadone	Oral, injected			
Other narcotics	Oral, injected			
Depressants				
Chloral Hydrate	Oral	Slurred speech, disorientation, drunken behavior without odor of alcohol	Shallow respiration, cold and clammy skin, dilated pupils, weak and rapid pulse, coma, possible death	Anxiety, insomnia, tremors, delirium, convulsions, possible death
Barbiturates	Oral, injected			
Glutethimide	Oral, injected			
Methaqualone	Oral, injected			
Benzodiazepines	Oral, injected			
Other depressants	Oral, injected			
Stimulants				
Cocaine	Sniffed, injected, smoked	Increased alertness, excitation, euphoria, increased pulse rate and blood pressure, insomnia, loss of appetite	Agitation, increase in body temperature, hallucinations, convulsions, possible death	Apathy, long periods of sleep, irritability, depression, disorientation
Amphetamines	Oral, injected			
Phenmetrazine	Oral, injected			
Methylphenidate	Oral, injected			
Other stimulants	Oral			

continued

Table 12.1	Continued			
Drugs	**Usual Methods of Administration**	**Possible Effects**	**Effects of Overdose**	**Withdrawal Syndrome**
Hallucinogens				
LSD Mescaline and peyote Amphetamine variants Phencyclidine Phencyclidine analogs Other hallucinogens	Oral Oral, injected Oral, injected Smoked, oral Injected Oral, injected, smoked, sniffed	Illusions and hallucinations, poor perception of time and distance	Longer, more intense "trip" episodes, psychosis, possible death	Withdrawal syndrome not reported
Cannabis				
Marijuana Tetrahydro-cannabinol Hashish Hashish oil	Smoked, oral	Euphoria, relaxed inhibitions, increased appetite, disoriented behavior	Fatigue, paranoia, possible psychosis	Insomnia, hyperactivity, and decreased appetite occasionally reported

Source: Drug Enforcement Administration, *Drug Enforcement*, Fall 1982. Washington, DC: U.S. Department of Justice.

These findings contain remarkable accord: only amphetamines actually enhance sexual performance, and they do so only for a limited time. Marijuana may stimulate sexual desire under some circumstances because it calms the fears some people have of sexuality. Alcohol also has the effect of creating a feeling of freedom from inhibition in small doses, but in moderate and heavy doses, it decreases the ability to function (Steele & Josephs, 1990). Barbiturates and psychedelics, the other two drugs investigated in these studies, appear to have no effect on sex. The studies all conclude that response is mainly dependent on the psychological makeup of the individuals involved and on the setting. Those experienced in the use of drugs with sex say that even when there is a good result, it is short.

Another interesting finding of several of the studies was that those who use drugs regularly are more sexually active than those who do not. Several decades ago, Goode (1972) discovered that drug users are not only more sexually active, but they start their sexual activity at an earlier age and with a more diverse selection of partners. Again, the studies seem to agree that the major reason that drugs are helpful in the sex act is because the drug users think the drugs are going to be. It may also be that high risk takers are more likely to indulge in both sex and drugs. Many studies have shown that both risk taking and poor impulse control are closely related to sexual activity (Rosenbaum & Kandel, 1990).

Of particular concern is the finding that people most likely to combine drug use (including alcohol) with sex are also those who are most likely to engage in unprotected sex (Buffum, 1988). They are also most likely to engage in activities that involve the highest risk of getting AIDS (see Chapter 9). Table 12.1 offers some essential information on the most-used controlled substances.

ETHNICITY AND SUBSTANCE ABUSE

Although not a great deal is known about the comparative abuse of substances by ethnic groups, one study did look at the number of arrests for drug use per 10,000 members of an ethnic group. Seven groups were considered (Asian-American Drug Abuse Program, 1978). The researcher found that African American, Mexican American, and Native American arrests outnumbered white arrests by three to two. Arrests for Japanese Americans and Chinese Americans were negligible. Of course, these data may reflect biases of the legal system rather than actual use.

More recently, the *National Longitudinal Survey of Young Adults* mentioned previously (Rosenbaum & Kandel, 1990) also noted ethnic differences in the prevalence of drug use and early sexual activity in teens younger than 16. Marijuana was the only widely-used illegal drug among African Americans of both sexes. White and Latino participants, male and female, were more likely to have used a wider variety of illegal drugs and alcohol.

African American teens begin using alcohol at an older age than white teens and are less likely than white teens to use alcohol during early and mid-adolescence (Barnes & others, 1994; Peterson & others, 1994). Research suggests that this may be because African American parents perceive alcohol use as more harmful, are more strict in prohibiting adolescent alcohol use, and are less likely to involve their adolescent children in adult alcohol use than are white parents.

Maton and Zimmerman (1990) found that the use of alcohol, marijuana, and hard drugs among urban African American male adolescents was influenced by several situations. If the adolescent was not enrolled in school and had no meaningful work, use or abuse of both alcohol and illegal drugs was more likely. Teens who had no religious beliefs and did not attend a church were also in more danger of becoming drug abusers. Adolescents who felt that their families and friends respected and cared for them were less likely to be drug users. These research findings are very similar to those in previous studies of white middle-class adolescents. Rosenbaum and Kandel (1990) also found that having high educational goals, being religious, and having a strong family life reduced the risk of substance abuse.

Estrada and others (1982) reported on the use of alcohol and drugs by a group of 107 Latino seventh- and eighth-grade junior high school students (ages 12 to 16) in Los Angeles. Findings suggest that for these teens, the strongest link is between the use of alcohol and marijuana.

Dembo (1981) studied what caused the drug involvement of 1,101 African American and Puerto Rican seventh-graders. He looked at five factors: their home composition, relationships with parents, attitudes toward school, machismo values, and/or identification with drug-involved peers. Data showed that identification with drug-involved peers probably provides the most likely prediction of drug use for these youths.

Let us now look at the overall prevalence of substance abuse.

Prevalence of Use

How widespread is substance abuse? Table 12.2 gives an overview of the situation. We will look at the evidence more closely, but first let's alert you to several cautions about substance abuse studies. Some authors have suggested ways in which living in areas where drug abuse is common has negative effects on adolescents, *including those who do not use drugs*. In poor urban neighborhoods, an adolescent's desire for a quick way out of poverty often leads the teen to becoming a dealer of illegal drugs (Williams, 1990). In these neighborhoods, police may assume that all adolescents are involved with illegal drugs, and sometimes arrest those who are neither users nor dealers (Cooper, 1990).

Table 12.2	**Substance Abuse Among U.S. Adolescents**						
Percent Using Drugs and Alcohol, by Gender and Age (12-13, 14-15, 16-17), 1994-1996							
		GROUP					
		Males			**Females**		
	Total	**12-13**	**14-15**	**16-17**	**12-13**	**14-15**	**16-17**
Cigarettes - past month	**19.1**	9.2	18.6	31.0	9.1	20.5	27.1
Alcohol - past month	**20.5**	8.3	20.0	36.2	6.8	21.4	30.9
Any illicit drug - past month	**9.4**	3.9	9.2	17.1	3.1	9.8	13.4
Marijuana - past month	**7.1**	2.1	7.2	14.6	1.5	7.1	10.7
Other illicit - past month[1]	**4.4**	2.3	4.3	7.1	2.5	4.2	6.3
Alcohol - binge drinking[2]	**7.6**	1.5	7.9	18.0	1.4	5.7	11.5
Alcohol - drunk past year[3]	**10.8**	1.8	10.4	23.6	2.0	8.4	18.9
Cigarettes almost daily[4]	**5.3**	0.4	3.6	11.3	0.7	4.7	11.7
Dependence on alcohol or illicit drugs	**6.0**	1.1	5.5	11.0	1.9	6.4	10.4
Alcohol dependence	**4.4**	0.8	3.8	7.5	1.4	4.8	8.6
Illicit drug dependence	**3.2**	0.4	3.4	6.1	1.0	3.2	5.5
Need treatment for illicit drug	**4.5**	1.4	4.4	7.6	1.9	4.5	7.3

[1] Respondent reported using an illicit drug other than marijuana in the past month.

[2] Respondent reported consuming five or more drinks on at least one occasion during the past month.

[3] Respondent reported being drunk three or more times during the past year.

[4] Respondent reported smoking on at least 25 days during the past month.

Source: Office of Applied Studies, SAMHSA, National Household Survey on Drug Abuse, 1994-1996.

As Oetting and Beauvais (1990) point out, many important differences among some categories of adolescents—for example, Latina girls in the western states—may get lost in large national surveys. Results from these national studies may also show less agreement because of the use of different sampling methods. Another problem is that such surveys are carried out through schools, so they miss all the youths who have left school, many of whom may have more extensive drug problems than those who remain in school.

ALCOHOL AND ILLEGAL DRUGS

It has been difficult to say with precision how widespread substance abuse is. Until recently, studies have differed from year to year, region to region, and from one another, even when year and region were the same. In the last few years, methodologies have improved. A survey of more than 3,000 high school students in Massachusetts indicates that alcohol use declined somewhat in the period from 1990 to 1993. Whereas 38 percent reported heavy use (drinking five or more drinks in a row) in 1990, that percentage dropped to 28 percent in 1993. Sixty percent of 9th- through 12th-graders reported alcohol use in the past 30 days in 1990, with 48 percent reporting alcohol use in the same period in 1993 (Wong & Hart, 1994). Data collected in the Midwest (Benson, 1993) revealed that 3 percent of middle school students and 11 percent of high school students have used illicit drugs six or more times in the past year, and 13 percent of middle school and 31 percent of high school students were frequent users of alcohol or binge drinkers. In a review of literature on the status of substance use and abuse among children and adolescents, Newcomb and Bentler (1989) point out that although a number of recent studies have found a decline in adolescents' use of most drugs, teens in the United States *still* have the highest rates of drug use and abuse in the industrialized world.

In one national survey (Johnston & others, 1991), 3.7 percent of high school seniors reported that they used alcohol daily, 57.1 percent in the past month, and 80.6 percent in the past year. Male high school students drink more alcohol and begin to drink earlier than high school females. Among college students, 41 percent had consumed five or more drinks in a row at least once in the past two-week period. Another study (University of Michigan Institute for Social Research, 1994) reveals that almost 15 percent of 8th-graders, 24 percent of 10th-graders, and 28 percent of 12th-graders reported having used 5 or more drinks in a row during the past 2 weeks. Teenagers who drink to become intoxicated, to cope with problems, or in reaction to feelings of anger and frustration are likely to become problem drinkers (Milgram, 1993).

Alcohol use on college campuses contributes to a number of problems, ranging from fights, vandalism, drops in academic performance, and blackouts to accidents and deaths (Milgram, 1993). College presidents continue to rank alcohol abuse as the number one problem on college campuses (Wechsler, 1999). According to a Harvard School of Public Health survey of 720 freshmen at 13 colleges where binge drinking was prevalent, students who binge frequently reported missing classes (55 percent), forgetting where they were or what they did (57 percent), doing something they regretted (60 percent), having unplanned sex (38 percent), and having a hangover (82 percent). In comparison, among students who never binge, only 6 percent reported missing a class, 3 percent were behind in schoolwork, and 5 percent had engaged in unplanned sex. Nonbingers reported additionally that their studying and sleep had been interrupted (67 percent), that they had to take care of a drunken student (57 percent), and that they had experienced an unwanted sexual advance (29 percent) (Kong & Brelis, 1995). Sixty-eight percent of all freshmen at these schools had binged at least once before the end of the first semester, usually as part of a drinking game. Alcohol use generally levels off in the college years as students gain in maturity and goal directedness. Daily drinking by college students is abnormal and should be taken as a sign of more serious psychological difficulties (Weiner, 1993). For some people, drinking habits that begin in college are the precursors of long-term alcoholism.

One of the newer drugs to hit the adolescent subculture is MDMA (also known as Adam or Ecstasy). It produces a smoother, longer euphoria than cocaine and is one of the so-called designer drugs (Buffum, 1988). Because MDMA comes in pill form it is easily concealed and can be used in public places. Though use of MDMA is definitely on the rise, it is a dangerous drug that can cause severe dehydration, brain damage and overdose. MDMA and other similar drugs are often associated with all night parties or "raves."

It has become quite common for adolescents to abuse both illegal drugs and alcohol. A study of 627 adolescents by Wirtz and Harrell (1990) showed that abuse of illegal drugs was the most likely of 10 problem areas to accompany destructive drinking. The combination of alcohol and cocaine is especially likely to occur. This is because alcohol is often used to moderate the crash sometimes experienced after the high produced by cocaine has passed. Grant and Hartford (1990), reporting on a national survey, found that this pattern of dual use was likely to increase with age. There were fewer dual users among 12- to 17-year-olds than in the 18- to 34-year-old group. Of course, this may be the result of cohort rather than age differences (see Chapter 1). That is, it may be that the older group simply has learned practices different from the younger group, and that the age difference is irrelevant.

ANABOLIC STEROIDS

> How serious is the problem of alcohol use among teens today? To what extent is it characterized by use or abuse?

anabolic steroids
An artificial form of the male hormone, testosterone, taken to look more muscular and to perform better athletically.

Adolescents' use of **anabolic steroids** to look more muscular and to perform better athletically is also a matter of increasing concern. Most users are male, but some females in competitive sports also use steroids, hoping to increase their strength and endurance. So far, little is known about the long-term effects of using these drugs, which are an artificial form of the male hormone, testosterone (Schwerin & Corcoran, 1990; Yesalis & others, 1989). Yesalis and associates conducted a nationwide survey of high school seniors. They found that 7 percent of the males had used steroids. Thirty-five percent of these users did not play a school sport. At least one-quarter of the steroid users in this study said that they

WHAT'S YOUR VIEW?

Drinking Behavior of First-Year College Students

What follows is a quotation from the summary of a recent study (Wechsler & Isaacs, 1991) of the drinking behavior of college freshmen in the northeastern section of the country.

❑ The results of a survey of 1,669 freshmen at fourteen colleges in Massachusetts indicate that drinking is a popular activity among these students:

- Nearly all freshmen drink alcoholic beverages.
- A third of the men and a quarter of the women drink more than once a week.

❑ A sizable proportion of the students are heavily involved with alcohol.

- Half of the men and a fifth of the women usually have five or more drinks in a row on a single occasion.
- Half of the men and a third of the women were drunk at least once in the past month.
- A third of the freshmen drink to get drunk.

❑ The legal drinking age has some relationship to the nature of drinking among the underage freshmen.

- Most drinking occurs in private settings: friends' residences, dormitories and fraternities, sororities or clubs.
- However, some drinking occurs in violation of minimum drinking age laws—a quarter of the freshmen drink in bars.

❑ Students manifest many problems related to drinking.

- One-third of the freshmen report having hangovers and doing something they regret.
- One out of six report engaging in unplanned sexual activity after drinking.

❑ In contrast, almost no students viewed themselves as problem drinkers.

- Only 37 of the 1,669 freshmen indicated that they had a drinking problem.
- Only 6 percent of the men and 2 percent of the women rated themselves as heavy or problem drinkers.

❑ Illicit drug use and smoking were less common than drinking among the freshmen.

- Half of the men and two-fifths of the women had ever used marijuana.
- One out of three freshmen had used marijuana in the past year. Most used it infrequently.
- Three percent of the freshmen used cocaine in the past year.
- One of five women and one of eight men were current smokers.

❑ Compared to usage patterns in 1977 [according to a previous study by Wechsler] at many of the same schools, current heavy use of alcohol remained constant while only half as many freshmen used marijuana or cocaine or smoked cigarettes.

- During both 1977 and 1990 the same proportion of freshmen were frequent heavy drinkers—31 percent of the men and 13 percent of the women.
- The only change in drinking patterns was an increase in the proportion of abstainers (from 3 percent to 12 percent) and the virtual disappearance of frequent-light drinkers (from 15 percent to 1 percent).
- In contrast, half as many freshmen use marijuana now compared to 1977 (35 percent versus 64 percent among men and 28 percent versus 55 percent among women).

continued

WHAT'S YOUR VIEW?

- Cocaine use is also less widespread than it was in 1977 (4 percent versus 14 percent among men and 2 percent versus 8 percent among women).

- Smoking is also half as frequent (12 percent versus 23 percent among men and 19 percent versus 33 percent among women).

❑ Automobile safety is a major problem among college freshmen.

- Many drive after one or two drinks (55 percent of men and 34 percent of women) and some after five or more drinks (19 percent of men and 4 percent of women).

- A third of the freshmen have ridden in a car with a driver who was high or drunk.

- Nearly all students have driven much faster than the speed limit and a third of the men and a sixth of the women have gotten a speeding ticket in the past year. One-sixth of the men have drag raced.

- Half have gone through a stop sign and two in five have gone through a red light.

- Three in 10 men and two in 10 women had an automobile accident in the past year.

❑ Drinking does not start in college. Heavy drinking patterns begin in high school.

- Two of five men and one in seven women usually drank five or more drinks in a row when in high school.

- One in three men and one in six women drank alcohol at least once a week in high school.

❑ Drinking, however, increases in college.

- Half of the freshmen increased their drinking in college, while only one in six decreased it.

❑ College drinking is a highly social activity.

- Most students report that their friends drink.

- Most students think it is appropriate to drink enough to get high or drunk at a party or with friends.

- Most students think it is not appropriate to drink any alcohol—even one or two drinks—alone at home.

❑ "Binge" drinking is the typical form of college drinking today.

- Over half of the men and one-third of the women are "binge" drinkers—as measured by having had five drinks in a row on one or more occasions in the past two weeks.

- Ninety percent of men and 80 percent of women who drink more than once a week are "binge" drinkers.

Although this quotation summarizes a great deal of information about freshmen drug use, you should study it closely in order to see if you can detect any major changes in attitudes among college students (the group from whom most of our leaders will be chosen). What's your opinion as to the general picture these data present? Are you alarmed by any of these statistics? Do you see any trends from 1977 to 1990 that might be of concern? If so, what should college officials or groups of college students do?

From D. Wechsler and N. Isaacs, "Binge Drinking at Massachusetts Colleges," in *Journal of the American Medical Association*, 267 (21):2929–2931, 1991. Copyright © 1992 American Medical Asssociation.

were not willing to stop, regardless of health risks. These researchers suggest that steroid use may follow the pattern of an addiction. Schwerin and Corcoran offer a model describing how a person may become psychologically addicted to steroid use. The teenage boy who lacks self-confidence may begin to use steroids in an attempt to improve his masculine appearance. The drug-induced physical changes may bring favorable social reactions and at first increase his self-confidence. However, he may then begin to wonder, "Is it me or the steroids?" Not wishing to lose his new status, he continues to take the drug, and may become psychologically dependent on it. In some cases, overzealous coaches have been implicated in this process.

AN APPLIED VIEW

Symptoms of Chronic Steroid Abuse

Below are lists of the two main types of symptoms of steroid abuse among teenagers, problems of the cardiovascular system and of the reproductive system. Explain this list to a teen who is contemplating taking steroids.

Cardiovascular System

Heart disease
Changes in cholesterol level
Anaphylactic shock
Septic shock
High blood pressure

Reproductive System

Genital swelling
Genital atrophy
Sexual dysfunction
Impotence
Prostate enlargement
Menstrual irregularities
Damage to fetus

There are many other less serious symptoms, such as acne, hives, and diarrhea.

TOBACCO

In spite of the well-publicized health risks posed by smoking, large numbers of adolescents continue to smoke cigarettes. The use of tobacco among college students has risen an alarming 28 percent since 1993 (Wechsler, 1998). Cigar smoking has also become fashionable with older adolescents and young adults. A recent study on tobacco use (CDC, 2000) surveyed middle and high school students in order to determine the prevalence of cigarette, smokeless tobacco, cigar, pipe, bidi, and kretek use. The data indicate that 12.8 percent of the students in grades 6-8 use some form of tobacco. Cigarettes (9.2 percent) were the most common type in use by both males and females and among racial and ethnic groups. Some adolescents, especially girls, believe that smoking will help them either lose weight or keep them from gaining weight. The highly publicized results of a University of Michigan study (Walker, 1995) indicate that 18.6 percent of 8th graders smoked tobacco in the past month, which is a 30 percent increase since 1991. Among high school seniors, 31.2 percent had smoked in the past month, representing a 12 percent increase since 1991. Declines shown from

1988 to 1991 have been reversed with cigarette smoking again on the rise. One study of 276 11th-graders in a middle-class suburb found that being in a particular social crowd influenced teenage smoking behavior. The highest incidence of smoking was found in the least-respected crowd, the "burnouts." The major influence on smoking behavior appeared to be encouragement by the adolescent's best friend, who was also likely to be a member of the same social group (Urberg, 1990).

The use of smokeless tobacco products (chewing tobacco, snuff) is rapidly increasing among adolescent and even preadolescent males. A study done in Oklahoma found that 12 percent of third-grade boys surveyed used smokeless tobacco (Newcomb & Bentler, 1989). The dangers to health from this form of tobacco, although not widely known, are extremely serious. In addition to nicotine addiction, users may suffer from high blood pressure, destruction of mouth tissues, and cancer of the mouth. This type of cancer is often fatal (Sussman & others, 1989).

As you can see, the abuse of both legal and illegal drugs among teens is a serious problem in this country. However, people who use drugs do not become abusers overnight.

Combating Substance Abuse

PREDICTION

The first step toward reducing the effects of drugs and alcohol on our youth must be improving our ability to determine, in the early stages, who is at risk. There is good news on this front. One study (Christiansen & others, 1989) found that a number of factors could predict drinking behavior of seventh- and eighth-graders one year later. These factors include attitudes toward drinking, parental ethnic background, religious affiliation, parents' occupations, and parental drinking attitudes.

Teens who look forward to drinking are likely to have a future problem with drinking (Christiansen & others, 1989; Miller & Smith, 1990). Miller and Smith identified two types of risk factors. The first type involved a person's social anxiety and the tendency to act on impulse. The other risk factor was the person's expectation of the results of alcohol's use. A combination of personal traits, genetic inheritance, and alcohol use history determines the risk of becoming a problem drinker. For example, if a young teenager who was often uncomfortable in social situations had parents who used alcohol regularly, then he or she would be much more likely to begin drinking than a peer who did not have these influences.

PREVENTION

Probably the best way to deal with a problem is to keep it from happening. The best way to keep drug and alcohol abuse from happening is to teach children and teenagers about the dangers involved. In order to educate young people, adults

A SOCIOCULTURAL VIEW

No to Drugs, Yes to Helping Others

In spite of their difficult environment, many inner-city teenagers find ways not only to avoid drug use and the drug culture, but to actively help others. A recent article in *The Boston Globe* described several of these high school students who were honored at a special dinner for their exceptional contributions as volunteers to neighborhood agencies. Some suburban teens also received honors for giving their time to assist at inner-city agencies. Among the organizations served were planning action committees, an Indian Council, a community comprehensive health center, and a neighborhood community center.

This year, the highest award for volunteer service was given to Julio Martinez, the son of a shoemaker. In the summer of 1991, Julio worked as a volunteer every day at the Tobin Community Center on Boston's Mission Hill. He gave so much time at the center that one of the directors thought he was a salaried employee. Here is an excerpt from his interview with the reporter:

❑   Asked why he volunteers, Julio shrugs.
"My mother encouraged me."

"Lots of mothers encourage sons who don't volunteer."
He shrugs again.
"I just don't like hanging around. I like doing stuff. Some kids do a lot of bad stuff, but I'm not into that."
On troubled streets, he manages to avoid trouble.
"It's not hard. If you put your heart in it, if you say you're not going to do it. I don't do drugs, and I've been offered lots of 'em. But that don't make you a man or nothing. A lot of kids who get into trouble say friends put them up to it, but nobody can make you do anything you don't want to do. That's what I say."
What convinced Julio that the dinner was a major event was the price of a ticket, $35.
"Then they told me there are some special people—I don't know who—that will pay $100 a ticket, and I'm like . . ." He rolled his eyes to the ceiling.
"I did do a lot of work, it's true, but there's a lot of other people who did a lot of work, too, and maybe they deserve it more, and so, like, I'm saying, you know, why me?"

must be willing to talk about drugs and alcohol. Many programs to educate youngsters about drug and alcohol abuse have been tried, but few have been successful (Newcomb & Bentler, 1989). Much has been learned from prevention programs that do not work, and some encouraging findings also suggest what does work.

Prevention programs usually follow one of four models: information only model, alternatives model, affective education/social competency model, and social environmental model. Norman and Turner (1993) describe each model and what we know about their effectiveness. During the 1960s and 1970s, drug education programs sought to teach young adolescents about the dangers of drugs. It was hoped that increased knowledge would change behavior. The programs did not achieve their hoped-for goals because many young people did not believe the exaggerated negative descriptions of drug effects or became more curious based on what they heard. A second prevention approach is aimed at promoting desirable alternatives to the use of drugs. The idea is that youths who are productive, busy, and satisfied will be less likely to turn to drug use. Involvement in responsible activity should also relieve boredom, increase self-esteem, and increase commitment to community values. Cohen (1977) described three types of substitutes for drug use: (1) those that provide a deep feeling of relationship to another person or to humanity; (2) those that contribute to self-knowledge or self-reliance; and (3) those that offer a satisfying experience, either physical, mental, or emotional. Although involvement in any or all of these activities is psychologically healthy for teens, there is little published evidence indicating that they actually deter drug use. Affective education and social competency models are directed toward enhancing self-esteem, clarifying values, and teaching stress management, problem solving, decision making, and communication skills. Like the alternatives model, the affective education/social competency approach targets broad mental health goals rather than drug abuse specifically. The direct effects on reducing substance abuse have not been proven. To date, social environmental approaches have proven most effective in reducing cigarette smoking, alcohol use, and substance abuse. Based on findings that peer, parent, and media influences contribute to substance abuse, these programs teach specific strategies for resisting negative influences (Dielman, 1994). Some recent approaches have combined elements of the information and affective education/social competency models with the social environmental model and have obtained promising results.

Although most programs focus on the individual adolescent, it has become increasingly clear that the climate in the family, school, and community are critically important. Leventhal and Keeshan (1993) suggest that many prevention programs have failed because they have not paid enough attention to the environment in which the adolescent lives. We need to consider what drug use means to the teen in the context of the peer group and family. Drug use may contribute to close-knit, intimate bonding among a group of teens. Any attempt to deter drug use must take that strong bonding process into consideration and must find a way to motivate teens to actually use the resistance skills they have learned.

A number of prevention programs seek to strengthen family functioning through parent training (Norman & Turner, 1993). Courses such as Parent Effectiveness Training and the five-week Family Prevention Strategy include sessions on basic knowledge of alcohol and other drugs, good communications within the family, and problem-solving techniques. The addition of a parent component, including parent-child homework exercises, parenting classes, parent participation in a community-wide drug abuse prevention meeting, and establishment of a school-based parent advisory committee, have been shown to add to the effectiveness of school-based social influence curricula (Rohrbach & others, 1994). Awareness of how the media glamorizes the use of alcohol and discussion within the family may also help the teenager to make responsible decisions about drinking. Wodarski (1990) noted that teens do not learn what their parents *tell* them about drinking but rather what they observe their parents doing about drinking. Thus, modeling appropriate behavior is essential. Adolescents need models of responsible drinking, as well as coping skills to be able to resist peer pressure to join in substance abuse. Wodarski concludes:

> ❏ The solution to the problem of substance abuse requires an all-out effort by societal forces that are capable of affecting change. Families, schools, peers, communities, businesses, and the media all have power to eradicate this social problem. Combined, cooperative efforts are essential (p. 684).

What do you recommend to prevent abuse?

No prevention program is likely to be effective if the climate in the school is conducive to drug abuse. Schools where students feel involved, respected, and heard by the school contribute to student commitment. Kumpfer and Turner (1990) maintain that such commitment or bonding to the school deters substance abuse and antisocial behaviors. According to Norman and Turner (1993), effective prevention strategies need to change community, school, and adolescent norms, values, and expectations about substance abuse. Tessler (1980) argues that to find out what the climate is in any particular school, it is necessary to administer a survey such as the one in the Applied View box, titled "The School Climate Survey." This survey is easy to administer to students, teachers, and administrators alike. It only takes a few minutes to complete. Tessler claims that:

> ❏ The results will help everyone get in touch with some of the silent agents responsible for a poor school environment which would make the success of any effective drug education impossible (p. 114).

You may wish to answer the survey questions in the Applied View box for your high school. That would be a good way to remind yourself of the factors that can lead to serious substance abuse.

Questions 1 to 4 relate to situations that can cause antisocial behavior. School and police records can also give information as to whether this kind of problem exists in the school. Questions 5 to 9 concern school services and special programs. Negative responses here can indicate a lack of school unity.

AN APPLIED VIEW

The School Climate Survey

Does your school:

1. Have racial or ethnic problems?   Yes ___ No ___
2. Have a high truancy rate?   Yes ___ No ___
3. Have cases of vandalism? Violence? Serious fights? Gangs? Drug problems?   Yes ___ No ___
4. Have many cases of student arrests?
   Yes ___ No ___
   (If you don't know the answer to all the above, local law enforcement agencies may be able to give you some information.)
5. Plan events which encourage school unity?
   Yes ___ No ___
6. Have good recreational and extracurricular activities which are well supported by the student body and staff?   Yes ___ No ___
7. Provide good counseling and health services?   Yes ___ No ___
8. Give everyone the opportunity to respect their own heritage and those of others?   Yes ___ No ___
9. Involve parents in important school decisions and events?   Yes ___ No ___
10. Seek a constructive bond with the community through work with urban or neighborhood improvement projects, law enforcement, the handicapped, the elderly, etc.?   Yes ___ No ___

11. Use community agencies or the expertise of residents to help with school programs?
    Yes ___ No ___
12. Direct students and parents to community resources which will improve their lives?
    Yes ___ No ___
13. Provide students with real leadership opportunities (not simply token positions)?   Yes ___ No ___
14. Have an "emotional climate" in which students, faculty, and administrators feel free to express their thoughts and feelings?   Yes ___ No ___
15. Encourage students to be creative and curious?   Yes ___ No ___
16. Help students explore and appreciate their own special talents?   Yes ___ No ___
17. Allow students to clarify values?   Yes ___ No ___
18. Help students to effectively deal with inner and outer conflicts?   Yes ___ No ___
19. Provide problem-solving and decision-making experiences for students?   Yes ___ No ___
20. Help students develop goals for the future?
    Yes ___ No ___

From Drugs, *Kids and Schools* by Diane Jane Tessler. A GoodYearBook. Copyright © 1980 by Diane Jane Tessler. Reprinted by permission of Scott, Foresman and Company.

Questions 10 to 12 reveal how members of the school community feel the school fits into the life of the larger community it serves. It also indicates whether neighborhood resources are used for the benefit of those attending the school. Questions 13 and 14 deal with whether the three main segments of the school community—students, faculty, and administrators—are able to communicate effectively with each other. Questions 15 to 20 give an indication of whether students are given opportunities to develop self-respect and self-awareness.

Responses to each of the questions should be tabulated both in terms of total responses and in terms of the students, faculty, and administrators. If these data indicate a problem in one of the five areas—antisocial responses, services and programs, community relations, communications, and affective areas—then committees should be established to improve the situation in the problem area. Tessler says that her experience with this approach indicates that such changes can help steer students away from substance abuse.

The Search Institute in Minneapolis recently examined how communities impact youths in positive and negative ways (Blyth & Roehlkepartain, 1993). They surveyed more than 33,000 adolescents in grades 9 to 12 from 112 communities. Although most of the communities were in the midwestern United States and were relatively small in size, they differed greatly in the percentage of youths who were involved in at-risk behaviors, including use of alcohol and illicit drugs, sexual activity, depression, suicide, antisocial behavior, and school problems. In some communities, frequent alcohol use (defined as having used alcohol 6 or more times in the past 30 days) by high school students was as low as 2 percent, and in other communities was as high as 30 percent. Binge drinking (5 or more drinks in a row, once or more in the past 2 weeks) ranged from 13 percent to 50 percent, and frequent use of illicit drugs ranged from 1 percent to 22 percent. The communities in which drug and alcohol use were lowest were described by teens as providing many opportunities for youths to participate in community life through religious, sports, and other structured activities, and to develop relationships with caring nonparental adults. Moreover, those youths who had fewer internal and external assets (positive values, educational commitment, well-developed social skills, perceived support by parents and teachers, parental supervision and discipline) were less likely to engage in at-risk behaviors, including drug and alcohol use, when they lived in healthy communities that offered opportunities for youth involvement and connection with caring adults. It seems that community level institutions can play a significant role in providing constructive opportunities for youths, which, in turn, decrease the incidence of drug and alcohol abuse.

> How can communities be strengthened to reduce substance abuse?

TREATMENT

Efforts to combat substance abuse should first consider the type of abuse under consideration. Experimental and social drug use rarely require treatment. Teens either stop using substances on their own or use them infrequently enough that they do not interfere with daily functions. Medicinal and addictive drug use almost always interferes with normal development and requires professional treatment. It is estimated that as many as 15 percent of all American teens need treatment for compulsive drug and/or alcohol use (Falco, 1988). Drug treatment approaches range from outpatient care to residential programs depending on the severity of the abuse problem. Treatment approaches are often designed to combat the personal, peer, and family factors that have contributed to the current level of abuse. Teens who enter substance abuse programs have typically experienced family conflict, a parental history of substance abuse, an abundance of peers with alcohol and drug problems, and school problems (Brown, 1993). Emotional problems and deviant behaviors, including reckless and illegal behaviors, are common.

Feigelman and Feigelman (1993) describe five types of treatment approaches as well as their strengths and limitations. Outpatient care allows the adolescent to live at home, go to school, and attend counseling and therapy sessions. Outpatient therapy is usually advised when the abuse is judged as not posing serious risks to the teen or to society. One of the first challenges of therapy with many young abusers is getting them to recognize that a problem exists and motivating them to change their lives. The therapist often tries to find some aspect of their lives that they want to change. The therapist begins where the client is and hopes to develop a workable treatment relationship. In most outpatient programs for teenagers, parents are expected to participate. Family therapy is often provided in addition to individual treatment. Although some of the work involving the families of substance-abusing youths shows promise, many youths needing treatment will never be able to take advantage of this type of program (Lewis, Piercy, & others, 1990). One of the problems of providing treatment for those who need it is the difficulty in reaching those youths who have left school and are at even higher risk for abuse. In many of our larger cities, nearly half of all teens fall into this category (Falco, 1988). Another limitation of outpatient treatment is that many teens improve for a short time and then return to some form of drug use.

Detoxification may be a part of treatment for those teens who have developed psychological or physical dependence on drugs or alcohol. Physiological dependence on alcohol usually requires hospital detoxification because of the serious withdrawal symptoms, such as convulsions, that accompany withdrawal from use. Detoxification is mandatory prior to other treatment for heroin and amphetamine addictions.

Inpatient or residential programs are used for people whose drug dependencies or social environments are more dangerous. Adolescents generally require a residential program when their dependency is accompanied by other serious medical or emotional disorders (such as suicidal behavior, anorexia nervosa, or psychosis), when they are physically addicted to alcohol or drugs, when family members do not support outpatient treatment, and when outpatient treatment has not been successful. Teens entering treatment typically use multiple substances on a regular basis. Residential approaches prohibit contact between abusers and their drug-abusing peers, which is important for those teens who lack internal controls. Contact with families is regulated and monitored, which is important for both the teen and parents when a high degree of family conflict exists. Peer confrontation, drug education, academic and career counseling, behavioral therapy, and other traditional therapies are usually part of the comprehensive treatment program. Inpatient treatment may last from several days to two years. Skills such as stress management, resisting social pressures, impulse control, and verbal problem solving are taught so that the adolescent will not resume drug use as a way of solving problems once the teen leaves treatment. For those who have been in treatment for any period of time, readjustment to life outside the facility is often difficult. Daycare programs provide supervised treatment at

Antidrug media campaigns have encouraged drug addicts to seek help through drug rehabilitation programs. A number of these programs have a high success rate in rehabilitating these addicts.

AN APPLIED VIEW

Talking to Teenagers About Drinking

The U.S. Department of Transportation has made a number of suggestions for talking to teenagers about their drinking behavior:

- Honestly explore your own drinking behavior before you talk with teenagers.
- Be honest in expressing your feelings and in stating your own values and preferences. Encourage the same from them.
- Be calm, firm, and consistent.
- Remember that you're sharing ideas and information about drinking.
- Don't put teenagers on the witness stand or demand a confession.

- Recognize that adolescents are not always able to control the situation they find themselves in.
- Tell teenagers you want to hear what they have to say and to learn what they know about drinking and driving.
- Be a good listener, even when you may not agree.
- Keep to the point. No matter where the discussion leads, and no matter what kind of reaction you may get or may feel, keep forcefully in mind that this discussion concerns only the problem of drinking.

One final point about working with teenagers: professionals caution against using street slang. It changes too often, and its use does not build rapport with teens.

the facility during the day, with patients returning to their homes on evenings and weekends. The treatment is similar to that provided in residential facilities, but relies heavily on parents to monitor progress in the home environment. Because it relies so heavily on parent supervision, it is not a common treatment form among teenagers.

Fortunately, a growing body of research suggests that treatment programs are gaining in their ability to assist abusing teens. Effectiveness is usually measured by a reduction in drug use, reduced criminal behavior, and increased productivity at school or on the job. One study (Hubbard & others, 1983) of teens following residential drug treatment found that alcohol use and criminal activity declined, although marijuana use continued. Other research (Brown, Mott, & Myers, 1990; Brown, 1993) found that approximately one-third of teens completing inpatient treatment abstained from drug and alcohol use for the first six months following treatment. Another 24 percent of the teens had improved, with only minor relapses of drug or alcohol use. The remaining 43 percent returned to more frequent use. Although early relapse is generally a bad sign, some teens who have a bad initial outcome actually do better over time. The greatest risk for initial relapse is in the first month after treatment. Of those teens who are able to maintain abstinence for the first six months, more than 75 percent will continue their abstinence over the next year. Adults' relapses following treatment most often follow stressful emotional and interpersonal experiences. For teens, however, relapses are most often a response to social pressures among peers. Teens who abstain report that developing a new group of friends and finding new activities is highly stressful. Surprisingly, family conflict is high in families where teens abstain. Conflict may be long-term consequences of teen abuse and may accompany the adjustment to a nonabusing family member. The teen might also express anxieties in the home environment that trigger family conflicts. Teens who maintain abstinence for the first year or more are likely to have stayed in school, to have improved their grades, to have less exposure to drugs and alcohol in their environment, to stay away from drug-abusing friends, to develop new friends with lower rates of drug and alcohol use, and to attend self-help groups such as Alcoholics Anonymous (AA), Narcotics Anonymous, (NA) or Cocaine Anonymous (CA) (Brown, 1993).

When it comes to treatment, Narcotics Anonymous and Alcoholics Anonymous are excellent treatment modes for the teen who is already addicted. Many AA groups conduct special meetings for beginners and for young people. Attendance at these meetings is free, although a collection basket is passed for those who wish to make a voluntary contribution. In metropolitan areas, there is usually at least one meeting every day that is within a reasonable distance. Some teens may resist self-help groups because they can't identify with the stories shared by the adult AA members. Participation of teens in 12-step programs has been growing, however. Many adolescent drug treatment programs now include the 12-step model and either hold meetings at the facility or recommend that their teenage clients attend those meetings in their home communities. Counselors who refer their teenage clients to 12-step programs provide a strong message that the counselors are not all-powerful and are willing to accept extra help. It is hoped that this message will allow the teen to also be willing to accept help.

CONCLUSIONS

Drugs and alcohol are "equal opportunity destroyers." They attack the wealthy and the poor, black and white, male and female, old and young. No one is immune. Under the most ideal conditions, adolescents today still have to deal with a myriad number of stressors—clashes between peer and family expectations, school demands, changing bodies, and a bombardment from the media as to what is the "right" way to be. Add to these the additional and all-too-common realities of unstable home situations, pressures to be prematurely sexual within the context of the AIDS reality, poor self-esteem, and pressures to be prematurely adult, and we have adolescents at risk for substance abuse.

There is a false image associated with drug use that pictures the typical drug abuser as an urban youth of color. The high-media profile of the urban drug dealer, a youth drawn into crime for the incentive of the large money associated with dealing, is true to a degree, but has been largely exaggerated. In fact, some studies show that a higher percentage of white adolescents abuse substances.

Vast amounts of national dollars have been spent in an attempt to stem the tide of drug use in the United States. Much more is known today about the nature and effect of drugs and alcohol than was previously known. Drug education programs now exist in virtually every school system in the United States. The war against drugs has escalated right alongside the escalation in drug use itself.

Research has given us valuable information on all levels and about all factors related to alcohol and drug use: AIDS, crack, nicotine, steroids, cocaine, sexual behavior, economics of drug dealing, development of physiological and psychological dependence, and the role of family, peers, and self-concept. There is not a corner that has not been scrutinized in an attempt to understand the nature of this particular beast.

Clearly, however, since drug and alcohol abuse stubbornly persists, the answer does not lie in the simple dispensation of factual information about substance abuse. More effort is being directed toward programs that focus on what motivates drug use. Some such programs look at the psychological needs that seem to be superficially met via drug use, and explore ways to meet these needs in healthier ways.

Many of the same kinds of problems exist for scientific research on youthful criminal behavior, which is also a paramount problem for today's society. In the next chapter, we enlighten you on this sad, yet illuminating, side of life.

C H A P T E R   H I G H L I G H T S

Some Definitions

- It is important to understand basic terminology related to substance abuse: drug abuse, drug addiction, drug dependence, drug tolerance, abuse of inhalants, drug overdose, victimless crime, controlled drugs, possession, dealing, and trafficking.

Stages and Types of Drug Use

- Experimental drug use, social drug use, medicinal drug use, and addictive drug use represent four very different and increasingly serious reasons for using drugs.

Personal, Peer, and Family Influences

- Personal characteristics, such as impulsivity, rebelliousness, poor school performance, depression, alienation, and pleasure seeking, have been associated with movement from less to more serious types of drug use.
- When peers are involved in heavy substance abuse, teens are also likely to become involved.
- Neglect, indifference, parenting style, and modeling of drug and alcohol abuse are some of the parental factors associated with adolescent substance abuse. Adolescents who live in unhappy family environments are more susceptible to substance abuse.
- The quality of relationship with parents seems to be a more important factor in patterns of drug use than family structure (divorce, etc.).

Crime and Abuse

- Not surprisingly, drug use and crime are highly correlated.
- Use of drugs is an effective predictor of violent delinquency.

Sex and Abuse

- The common conception that drug and/or alcohol use enhances sexual enjoyment may be true in small doses, but only if the user believes that it is.
- Generally, it is found that drugs actually suppress sexual performance.

Ethnicity and Abuse

- Drug arrests of youths of color outnumber white arrests, but may be more related to legal system bias than actual drug use.
- African American teens become involved in alcohol use at an older age than white teens.
- Identification with drug-involved peers provides the most likely prediction of drug use for minority youth.

Prevalence of Use

- Drug and alcohol abuse is still prevalent, although some important changes have been noted; for example, tobacco use is on the rise among teenagers, while marijuana use is on the decline.

Combating Substance Abuse

- Prevention programs to combat substance abuse follow one of several models, including information only model, alternatives model, affective education/social competency model, and social environmental model. Recent approaches have combined elements of several models.
- Some factors that can be predictive of drinking behavior are attitudes toward drinking, parental ethnic background, lack of religious affiliation, parents' occupations, parental drinking attitudes, anticipation of drinking, and expectations of the results of alcohol use.
- Fortunately, recent research suggests that treatment programs are effective in helping teens to fight abuse. Treatment options include outpatient therapy, detoxification, inpatient programs, daycare programs, and self-help groups.

K E Y   T E R M S

abuse of inhalants 398
addictive drug use 401
anabolic steroids 413
controlled drugs 398
control theory 404
dealing 401

drug abuse 398
drug addiction 398
drug dependence 398
drug overdose 398
drug tolerance 398
experimental use 401

medicinal use 401
possession 400
problem behavior theory 402
social use 401
trafficking 401
victimless crime 398

W H A T   D O   Y O U   T H I N K ?

1. Suppose you had a friend who had a drug abuse problem and that person was unwilling or unable to admit it. What would you do to help her or him?
2. How would you know if you were developing a serious addiction? What would you do about it if you did know?
3. What are some things that school authorities could do to combat substance abuse?

4. What are some things that the business community could do to combat substance abuse?
5. If you know someone who has become seriously addicted, could you discern clear-cut stages that person went through in the course of becoming addicted?

S U G G E S T E D   R E A D I N G S

Abel, E. L. (1981). *Marijuana: The first twelve thousand years*. New York: Plenum. Gives an excellent description of marijuana use throughout history.

Cohen, S., & Cohen, D. (1986). *A six-pack and a fake I.D.: Teens look at the drinking question*. New York: M. Evans. According to the authors of this book, the decision to drink or not to drink is personal rather than moral. They recognize the tragedy that alcohol can bring into people's lives, but they still "do not see moderate drinking as a problem; indeed, it is often a positive pleasure." They do, however, feel that before coming to conclusions about the use of alcohol, you should have reliable and believable information to help you make the best and most informed decision.

Harris, J. (1987). *Drugged athletes: The crisis in American sports*. New York: Four Winds Press. Athletes take drugs to increase speed, strength, and accuracy; to mask pain; to relax muscles; to relieve stress; to improve performance; and to gain pleasure. Harris provides an overview and discusses specific problems of drugs in sports at all levels.

Kennedy, W. (1983). *Ironweed*. New York: Penquin. Although about adult alcoholics, this book offers an engrossing look at the problem that applics to all ages.

G U I D E D   R E V I E W

1. Some people feel that drug abusers are the only ones who suffer from the use of drugs and call drug abuse a(n)
   a. victimless crime.
   b. societal problem that has been criminalized.
   c. affliction of the weak.
   d. legitimate $1 billion industry.

2. Joe bought a bag of marijuana from a friend, but was arrested two hours later by an eyewitness who called the police. Joe is charged with
   a. dealing.
   b. trafficking.
   c. possession.
   d. none of the above

3. What reduces the risk for adolescent substance abuse?
   a. attachment to family
   b. being religious
   c. authoritative parents
   d. all of the above

4. To relieve tension and anxiety is the primary motivation for which type of drug use?
   a. experimental
   b. medicinal
   c. addictive
   d. social

5. According to Newcomb and Bentler, which of the following is true about hard drug use in adolescence?
   a. It generally has few negative consequences by young adulthood.
   b. It contributes to decreased college involvement.
   c. It contributes to decreased religious commitment.
   d. It prevents adolescents from accomplishing the preparatory tasks for young adult adjustment.

6. Adolescent drug users have conflicts with their parents in all the following areas, EXCEPT
   a. they do not believe that they should be limited to only those friends approved by their parents.
   b. they want more affection from their parents.
   c. they want to be allowed to make decisions without parental advice.
   d. they do not wish to imitate their parents.

7. According to Gold, all of the following are common warning signs for substance abuse, EXCEPT
   a. difficulty sleeping.
   b. frequent outbursts of anger.
   c. drop in school performance.
   d. substantial changes in mood.

8. There is no withdrawal syndrome for
   a. marijuana.
   b. alcohol.
   c. cocaine.
   d. LSD.

9. According to findings of the Search Institute, communities in which drug and alcohol use are low are characterized by
   a. low unemployment.
   b. few working parents.
   c. low rates of divorce.
   d. opportunities for youth to participate in community life.

10. All of the following are types of substance abuse prevention programs, EXCEPT
    a. affective education.
    b. detoxification.
    c. alternative model.
    d. information only.

11. Drug _____ (tolerance/dependence) occurs when there is a physical or psychological need, or both, resulting from continuous drug use.

12. _____ (Physical/Psychological) need occurs when the person feels anxious, depressed, or irritable when unable to have the drug.

13. Daily drinking by college students is _____ (normal/abnormal) and should be taken as a sign of _____ (normal maturation/more serious psychological difficulties).

14. According to _____ (problem behavior/control) theory, adolescents with strong bonds to society are less likely to deviate from conventional behavior.

15. (Anabolic steroids are/MDMA is) _____ an artificial form of the male hormone testosterone.

16. Longer, more intense "trip" episodes, psychosis, and possible death are the effects of _____ (heroin/LSD) overdose.

17. Steroid abuse can cause _____ (high blood pressure/decreased sexual functioning).

18. Because _____ (alcohol/tobacco) products are more easily obtained than any other abusable substance, experimentation with them often begins in preadolescence.

19. Outpatient therapy is a possible _____ (prevention/treatment) approach for substance abuse and other problems.

20. Substance-abusing teens who are successful in maintaining abstinence following inpatient treatment often participate in _____ (social competency/self-help groups such as AA, NA, or CA) following discharge from treatment.

Answers

1. A, 2. C, 3. D, 4. B, 5. D, 6. B, 7. A, 8. D, 9. D, 10. B, 11. dependence, 12. Psychological, 13. abnormal, more serious psychological difficulties, 14. control, 15. Anabolic steroids are, 16. LSD, 17. high blood pressure, 18. tobacco, 19. treatment, 20. self-help groups such as AA, NA, or CA.

ANTISOCIAL BEHAVIOR

As early as the early 1960s, Leonard Bernstein's *West Side Story* signaled society's concern that glib sociological excuses would only worsen the handling of behavior disorders. Public opinion and public institutions are turning back to a punishment model. The pendulum has swung away from the idealization of Rousseau's "noble savage," who would be good only if society did not overcontrol and frustrate him, toward today's bleak cynicism of the essential untreatability of the *Lord of the Flies*.

antisocial behaviors
Acts that inflict physical or mental harm, property loss, or other damage on others.

Antisocial behaviors are acts that "inflict physical or mental harm or property loss or other damage on others" (Tolan & Loeber, 1993). Some of these behaviors may be against the law and others may not. Antisocial behavior can thus range from being uncaring and obnoxious to being extremely harmful and violent. The mild forms of antisocial behavior characterize most adolescents at one time or another, whereas very few adolescents engage in the most serious antisocial behaviors.

Juvenile delinquency is defined as "any illegal act by a minor." Delinquency emphasizes the criminal and legal aspects of the behavior, and generally applies to the more serious antisocial behaviors. However, not all illegal activities are equally serious. As we think about antisocial behavior and juvenile delinquency, it is important to remember that adolescents differ greatly in terms of the frequency, seriousness, and chronicity of their involvement in antisocial behaviors. Most teens are not problem delinquents. According to the National Youth Survey (Elliot & others, 1989), almost 90 percent of adolescents had engaged in one delinquent act during their teenage years. For most of these teens, delinquent activity was limited to a few minor offenses, such as dime store theft, using a fake ID, or having taken an alcoholic drink while under age, and lasted for just a short time. In a survey of teens in the Midwest, 28 percent of 9th- through 12th-grade students admitted to committing two or more antisocial acts such as vandalism, theft, group fighting, weapon use, or getting into trouble with the police within the last year (Benson, 1993). Less than 20 percent of all crime in the United States is committed by teens (FBI, 1992), and only 8 percent of teens are involved in serious and repetitive offenses (Tolan & Loeber, 1993).

This chapter concerns those adolescents who display the more serious and persistent forms of antisocial behavior. Because of the increasing national concern over teenage violence, we will devote considerable attention to the causes and prevention of violence. Three types of offenders will be considered. The **juvenile delinquent** often acts alone, committing theft or destruction to property or persons. The **aggressive gang** engages in a variety of illegal group activities, some of which result in interpersonal violence. **Nonaggressive offenders** (runaways, prostitutes) commit crimes that are mainly harmful to themselves. They may become involved in other illegal activities in their efforts to survive on the streets, or may become the victims of the dangers that surround them.

juvenile delinquent
A juvenile who commits theft or destruction to property.

aggressive gang
A group or gang that engages in a variety of illegal group activities.

nonaggressive offenders
Persons who commit crimes that are mainly harmful to themselves, such as runaways or prostitutes.

Delinquency is a complex issue, reflecting the complexity of problems in 1990s America including, but not limited to, increased drug traffic, a sagging economy, rising unemployment, and an increasing number of homeless individuals and families. It is our hope to better understand the dynamics of delinquency so that the problems of delinquent youth may be better addressed and, ultimately, prevented.

After reading this chapter, you should be able to:

- Distinguish between three major forms of delinquent behavior.
- Extract relevant information from a table about the demographics related to delinquent behavior and interpret the meaning of such information.
- Identify individual and family factors associated with the development of aggressive and antisocial behavior.
- Describe the connection between school performance and delinquent behavior.
- Describe how the media and the availability of guns are blamed for contributing to a culture of violence.
- Explain the role of ethnicity and gender in the experience of violence.
- Describe primary, secondary, and tertiary programs for the prevention of violence.
- State the relationship and relative importance of the connection between drug use, mental health problems, and delinquent behavior.
- List and describe at least three reasons why youths join gangs.
- Create a profile portrait, including personality characteristics, of the typical aggressive gang joiner.
- Explain at least one rehabilitative technique being employed with gang members and comment on its effectiveness.
- Identify at least three common causes associated with running away and understand why knowing these reasons is useful.
- Describe at least two strategies used to help runaways.
- Explain the connection between running away and prostitution and describe the reality of life for a teenage female or male prostitute.
- Discuss these issues from an applied, a sociocultural, and your own point of view.

Delinquency and Violence

When a delinquent act is defined as the violation of legally established codes of conduct, delinquency includes a wide range of illegal behavior—from misdemeanors to major crimes against persons and property. Although we may be dismayed at the variety of ways youths get into trouble, it is important to remember that most of these violations are not serious. Only a small minority of American teenagers have committed major crimes, been arrested, or live a consistently delinquent way of life. However, as table 13.1 indicates, the number of adolescents who have been arrested has increased between 1989 and 1998.

Among all the crimes listed in Table 13.1, those involving interpersonal violence have captured the most public attention, concern, and fear. Youth and violence are constantly linked in newspaper headlines and television news. When asked in a *Boston Globe* telephone survey what was their greatest worry about the world in which their children are growing up, parents indicated that crime and violence was their top concern (Matchan, 1995). Fortunately, only a small percentage of teens experience violence, but statistics on youth violence are nevertheless alarming. The incidence of violent crimes, including murder,

Table 13.1	**Total Arrest Trends, 1989 - 1998**
	(7,259 agencies; 1998 estimated population 161,341,000; 1989 estimated population 149,265,000)

	Number of persons arrested								
Offense Charged	**Total All Ages**			**Under 18 years of age**			**18 years of age and over**		
	1989	**1998**	**% change**	**1989**	**1998**	**% change**	**1989**	**1998**	**% change**
TOTAL[1]	8,395,012	8,975,084	+6.9	1,298,436	1,608,518	+23.9	7,096,576	7,366,566	+3.8
Murder and nonnegligent manslaughter	14,063	11,208	-20.3	1,759	1,354	-23.0	12,304	9,854	-19.9
Forcible rape	23,211	19,328	-16.7	3,443	3,349	-2.7	19,768	15,979	-19.2
Robbery	89,705	78,457	-12.5	19,579	21,330	+8.9	70,126	51,127	-18.5
Aggravated assault	282,086	318,157	+12.8	37,080	45,020	+21.4	245,006	273,137	+11.5
Burglary	287,534	203,024	-29.4	91,957	71,602	-22.1	195,577	131,422	-32.8
Larceny-theft	970,828	825,325	-15.0	275,014	264,287	-3.9	695,814	561,038	-19.4
Motor vehicle theft	139,548	96,462	-30.9	57,250	34,722	-39.4	82,298	61,740	-25.0
Arson	11,284	10,587	-6.2	5,003	5,485	+9.6	6,281	5,102	-18.8
Violent crime[2]	409,065	427,150	+4.4	61,861	71,053	+14.9	347,204	356,097	+2.6
Property crime[3]	1,409,194	1,135,398	-19.4	429,224	376,096	-12.4	979,970	759,302	-22.5
Crime Index total[4]	1,818,259	1,562,548	-14.1	491,085	447,149	-8.9	1,327,174	1,115,399	-16.0
Other assaults	589,190	829,497	+40.8	86,273	145,369	+68.5	502,917	684,128	+36.0
Forgery and counterfeiting	60,362	70,678	+17.1	4,567	4,463	-2.3	55,795	66,215	+18.7
Fraud	220,731	220,262	-.2	4,604	6,638	+44.2	216,127	213,624	-1.2
Embezzlement	9,649	10,585	+9.7	1,009	958	-5.1	8,640	9,627	+11.4
Stolen property, buying, receiving, possessing	109,895	85,485	-22.2	28,307	20,805	-26.5	81,588	64,680	-20.7
Vandalism	184,158	185,517	+.7	72,324	78,620	+8.7	111,834	106,897	-4.4
Weapons, carrying, possessing, etc.	138,961	120,613	-13.2	24,831	28,585	+15.1	114,130	92,028	-19.4
Prostitution and commercialized vice	65,072	63,090	-3.0	996	935	-6.1	64,076	62,155	-3.0
Sex offenses (except forcible rape and prostitution)	65,524	58,553	-10.6	10,043	9,922	-1.2	55,481	48,631	-12.3
Drug abuse violations	829,181	992,048	+19.6	70,029	129,996	+85.6	759,152	862,052	+13.6
Gambling	9,939	7,680	-22.7	543	1,008	+85.6	9,396	6,672	-29.0
Offenses against family and children	42,341	75,813	+79.1	1,626	4,331	+166.4	40,715	71,482	+75.6
Driving under the influence	1,033,660	810,972	-21.5	11,777	12,164	+3.3	1,021,883	798,808	-21.8
Liquor laws	343,736	383,093	+11.4	79,172	95,309	+20.4	264,564	287,784	+8.8
Drunkenness	583,991	459,478	-21.3	15,457	15,700	+1.6	568,534	443,778	-21.9
Disorderly conduct	439,090	421,566	-4.0	67,245	108,282	+61.0	371,845	313,284	-15.7
Vagrancy	17,972	19,040	+5.9	1,728	1,709	-1.1	16,244	17,331	+6.7
All other offenses (except traffic)	1,686,602	2,377,487	+41.0	180,121	275,496	+53.0	1,506,481	2,101,991	+39.5
Suspicion	7,483	2,897	-61.3	1,765	785	-55.5	5,718	2,112	-63.1
Curfew and loitering law violations	44,529	123,878	+178.2	44,529	123,878	+178.2	—	—	—
Runaways	102,170	97,201	-4.9	102,170	97,201	-4.9	—	—	—

[1] Does not include suspicion.

[2] Violent crimes are offenses of murder, forcible rape, robbery, and aggravated assault.

[3] Property crimes are offenses of burglary, larceny-theft, motor vehicle theft, and arson.

[4] Includes arson.

SOURCE: Federal Bureau of Investigation, 1998.

AN APPLIED VIEW

Interpreting Important Tables

To many of us, long complex tables like the one in Table 13.1 look like hieroglyphics. They seem to be just row after row, column after column, of numbers.

In truth, however, such tables can be the bearers of very important information, information you need to know to deal well with adolescents. To get used to interpreting this information, as well as to get a clearer picture of the changes in patterns of delinquency over the past decade or so, we suggest that you try to answer the following set of questions.

1. Among those age groups listed, which group accounts for the most crimes?
2. What type of crime is most prevalent among teenagers? Second most? Third most?
3. What, in your mind, is the most serious type of crime on the list? What percentage of total crimes does this type of crime account for?
4. What crime is most typically committed by sixteen-year-olds?
5. What are the three types of crimes that increase the most across the teenage years?
6. What are the three types of crimes that decrease the most across the teenage years?
7. Can you think of any crimes that have been omitted from this list?
8. Can you figure out what the crime index total means?
9. What do you suppose the crime "suspicion" means (third from the bottom of the list on p. 433)?

rape, aggravated assault, and robbery, increases steadily throughout adolescence, peaking between the ages of 17 and 19 (Earls & others, 1993). Survey findings of 246 inner-city youths between the ages of 14 and 23 revealed that 22 percent had seen someone killed, 42 percent had seen someone shot or knifed, 44 percent could obtain a gun within a day, and 18 percent had carried a gun within the past 3 months (Schubiner & others, 1993).

❏ A 17-year-old African American girl from Boston reported that she had attended funerals of 16 friends and classmates who have died by violence. One death was the consequence of an argument about who was next in line for a haircut (reported by Eron & Slaby, 1994).

Accounts like this lead many people to believe that the problem of violence among youths has reached epidemic proportions for some teens (Eron & Slaby, 1994). Perhaps an understanding of the factors that contribute to violent behavior among teens can be used to prevent further violence and to help those who are already behaving in violent ways. Identifying the causes of violence would be a much easier task if there was one single cause. Violent behavior seems to result, however, *only* when a number of factors occur together. For those of you interested in preventing violence, an awareness of the multiple factors contributing to violence suggests many points—the individual, family, culture—at which efforts to halt the progression of violence may be directed. Let us now turn to a consideration of those factors.

INDIVIDUAL FACTORS

❏ There ain't nothing more exciting than sneaking into somebody's house at night to steal their stuff. God, you can hear with your skin!
—*Eddie, an ex-juvenile delinquent*

As the above quotation suggests, alienation and thrill seeking are often characteristic of delinquent juveniles. Some research suggests that individual differences in thrill seeking are evident early in life. Some children are more fearless and less inhibited than others. They are more difficult to teach and to discipline, not responding as quickly to positive reinforcement and punishment. In grade school and adolescence, they may be drawn to friends and activities that provide excitement, adventure, and daring (Pepler & Slaby, 1994). Some aggressive and violent teens were described as having many temper tantrums at a younger age. Most children who have frequent temper tantrums or are attracted to dangerous activities will not become violent or aggressive as teenagers, but these characteristics of temperament may be risk factors. That is, when individuals who are fearless, uninhibited, and have bad tempers are exposed to other family, social, and cultural risks, they are more likely to become aggressive and violent.

Studies suggest that aggressive and violent adolescents and adults were often more aggressive than their peers as early as the preschool years (Eron & Slaby, 1994). Although biological factors like temperament may be partly responsible for early differences in aggressive behavior, the tendency to behave in aggressive ways is also a learned behavior. We will later consider how the family, the peer group, and the culture sometimes teach and support violent and aggressive behavior. The tendency to behave aggressively has also been linked to differences in social-cognition (see Chapters 5 and 6). From preschool through adulthood, aggressive and violent persons show lower levels of social problem solving skills and express beliefs that support the use of violence (Eron & Slaby). For example, aggressive persons are more likely to interpret the behavior of another person, such as a friend or teacher, as posing a threat to them. They are likely to consider few ways of responding to that threat other than acting in violent or aggressive ways. In addition, they do not consider the consequences of their behavior.

Drugs and alcohol have also been associated with acts of violence. Violent and aggressive youths are much more likely than other youths to use and abuse a number of substances, legal and illegal. Watts and Wright (1990) found that the frequency of alcohol, tobacco, marijuana, and other illegal drugs are highly predictive of both minor and violent delinquency in adolescent males of white, African American, and Mexican American origins. They suggest several reasons for this link. Both drug use and delinquency may serve the youths by maintaining a "tough guy image," getting parental attention, or defying parental authority. Delinquency and drug use may also be a way to escape from the painful realities of their environment, or to relieve chronic boredom. Eron and Slaby (1994) suggest, additionally, that many drug users become involved in robberies involving violence in order to support their drug habit. Because drugs are illegal and valuable, violent crime often surrounds the marketing and sales of drugs. Many people show less self-control after they have used alcohol, and thus may be more likely to react violently. According to Eron and Slaby, however, low self-control is not the cause of violence. Teens as well as adults who have good social problem solving skills and who have learned ways of resolving conflicts and handling anger are not likely to become violent, even when their self-control is diminished by alcohol use. When teens learn that aggression is a good way to get what they want, however, they are more likely to become aggressive and possibly violent when self-control.is low.

FAMILY FACTORS

As mentioned earlier, the family is one context in which aggressive behavior may be learned. Patterson and associates (1992) have studied the families of aggressive children and adolescents to identify patterns of family interaction that predict antisocial behavior. They have identified a **coercive cycle** of interaction that if left unchecked is likely to lead to chronic delinquency and crime in adolescence and adulthood.

coercive cycle
A pattern of family interaction through which the child learns to control the parents by behaving negatively and sometimes aggressively.

The first stage begins with family members directly training the child to be antisocial by reinforcing coercive, belligerent behaviors in the child. The child learns to control other family members through gradually escalating coercive tactics. For example, when the parent asks a child to do something, the child begins to whine or have a tantrum. The child keeps this up until the parent backs down and forgets the original request. In the process, the child's whining or tantrum has been reinforced (the child doesn't have to follow parental requests) and the child has learned to control the parent through negative, and sometimes aggressive, behavior. In addition, few positive social skills are taught.

Patterson and associates (1992) also list "disruptors" that have a negative effect on parenting. These family disruptors make it more difficult for parents to follow through with consistent discipline and increase the likelihood that children's belligerence and aggression will be reinforced. Disruptors include:

- An intergenerational history of antisocial behavior in the family.
- Disadvantaged socioeconomic status.
- Family stressors such as unemployment, marital conflict, family violence, or divorce.

McLoyd (1990) believes that Patterson's model is useful in understanding the effects of poverty on poor African American families and youths. The psychological distress caused by poverty is believed to interfere with consistent and supportive parenting and to increase coercion both by the parents and the children. Coercive parenting, in turn, increases aggressive behavior in children and teens.

Social rejection by peers and school failure are the predictable reactions to antisocial children from this sort of antisocial environment. These failures increase the risk that the antisocial children will join deviant peer groups which, in turn, act as positive feedback for further antisocial behavior.

Patterson and associates (1992) hypothesize that training in antisocial behavior begins in the family as early as preschool and elementary school. This may deprive these children of positive socialization with peers, and hence, heavily tilt the balance toward chronic antisocial behavior and, eventually, adult criminality.

Other parenting practices have also been identified as contributing to the development of aggression. Lack of parental supervision, parental rejection, low levels of parental involvement, and the use of harsh physical punishment as a means of discipline have all been identified as predictors of later aggressive behavior problems and delinquency (Earls & others, 1993; Pepler & Slaby, 1994). Straus (1991) argues that parents should not use physical punishment as a means of discipline. When parents use physical punishment, he argues, children learn that physical force and aggression are acceptable and effective ways to solve problems. Physical punishment may increase compliance in the short run, but it increases the probability of delinquency and eventual adult violent crime over time. Despite Straus's strong argument that physical punishment should never be used, some researchers (Slaby & Roedell, 1982) argue that the line between caring discipline and harmful physical punishment and abuse may be an important, although difficult, distinction. Witnessing marital violence and being a victim of child abuse represent other means for learning aggression within the family. Widom (1990) warns, however, that caution should be taken in assuming that abused children will themselves become child abusers. Fewer than 20 percent of adults who were abused abuse their own children, although they are more likely than other parents to become abusers.

When, if at all, is physical punishment justified? Necessary?

THE SCHOOL CONTEXT

As mentioned in our discussion of family factors, children who have learned aggressive behavior at home are not well prepared for social and academic achievement at school. They are more likely to get in trouble at school, to be out of their seats, to disrupt other children, and to experience academic problems (Pepler & Slaby, 1994). Having fallen behind in the early grades, they are more likely to be suspended and quit school in the teenage years (Kupersmidt & Coie, 1990).

The school may, in conjunction with the neighborhood and family, provide conditions under which antisocial behavior is learned. In addition, as we have seen recently, school may also become the focus of violent acts for troubled

A SOCIOCULTURAL VIEW

Impoverished Inner-City Schools and Delinquency

Some investigators believe that delinquent behavior is caused by characteristics of impoverished inner-city schools themselves. This includes both academic and extracurricular aspects. On the academic side, they cite the passivity required of students in many inner-city classrooms, excessively large classes, poor academic quality, and an unstable student population, indicated by high rates of newly-admitted and transfer students. Perhaps even more important is "the exclusion of marginal youths from 'sponsored' school activities such as clubs, artistic groups, athletics, and student govern-

ment. Such exclusion may be a byproduct of a tracking system that rewards the academically successful and ignores or denigrates the academically unsuccessful" (Henggeler, 1989). In addition, being accepted in these after-school activities often depends on being able to pay for them.

As with anything else in life, when we are excluded from something, we often wish we were included. Anger that results from unfair exclusion based on characteristics over which the adolescent has no control (poverty, race, gender) quite frequently leads to delinquent behavior.

youths. An inner-city high school is a source of stress for many youths, especially when they are receiving poor grades, failing a subject, or being suspended (Mosley & Lex, 1990). Numerous studies have found that "chronic underachievement and a poor school record are . . . predictive of rule breaking and antisocial behavior" (Feldman & others, 1984). In fact, much evidence indicates that there is a causal link between problems in achieving academic competence and delinquency (Cullinan & Epstein, 1979; Jerse & Fakouri, 1978; Kauffman, 1981; Siegel & Senna, 1981; Whelan, 1982).

In spite of the fact that many severely underfunded inner-city high schools are seriously underserving their students, there are dedicated teachers who daily give courage and hope to the adolescents under their care. The love and respect that these men and women show for their students is often the most important motivation for students who might otherwise despair of ever escaping their poor

A SOCIOCULTURAL VIEW

Teachers Can Make the Difference

In New York City, Seward Park High School serves a wide variety of underprivileged students. These adolescents range from those recently arrived from such far-flung places as the peasant villages of China and the Dominican Republic, to those who live in the city's welfare hotels. The school is old and overcrowded, and has been ranked among the worst 10 percent of high schools in the state. It stands in a neighborhood of violent, drug-ridden streets. Nevertheless, 92 percent of its graduates go on to further education.

Dedicated teachers like Jessica Siegel, whose story is told in Samuel Freedman's 1990 book Small Victories, can make all the difference to students whose talents might otherwise remain buried under the burdens of their often bleak lives. Ms. Siegel taught literature and journalism, and agreed to be faculty advisor of the school newspaper. Here is an excerpt from an article written for that paper by one of her students, Lun Cheung:

❏   The road to the ultimate sense of achievement is too often a long and hard one. For many this never happens and dropping out is the solution. No one or at least not many drop out because they want to. Sometimes they just can't help it. For example, many students have serious family problems. The next day is hard to face. This is not an excuse because I realize the students are mature enough to handle it.

Or you're a student who has a nine-period day working after school. Straight after school you go to work until 11:00 PM. You come home and realize

there is a lot of homework to do. You just don't have enough strength left and you fall asleep. Before you realize it you're falling behind. The teachers get on your case. You want to explain to them but you are just too afraid because you think they feel you are just making excuses. And so the inevitable words come out, 'Who cares, forget about it.' How annoying it is when people tell you 'Why don't you just quit your job.' If you quit where will the money come from, who will pay the phone bill, sometimes, who will pay the rent.

Do not take this as an excuse, instead as a better way to understand a student. A student feels lousy when a teacher embarrasses a student by saying, 'Hey look who showed up' or 'No, I don't believe you.' Instead of facing up to the problem students take the easy way out.

There are some teachers who really care though, those teachers who believe in you and have complete faith. They are the ones who want to see you go to college and will help you to achieve those means. But there is that little feeling inside of you that you can't help but feel and your problems pile up and up. This is when the dilemma pops up because you just don't want to disappoint those teachers who have been pushing for you (Freedman, 1990, pp. 393–394).

Not surprisingly, research indicates that students who eventually drop out often miss school, have little respect for teachers, and give little effort to schoolwork.

background. When the students realize how much such a teacher respects them and their abilities, their self-esteem is improved. They are therefore encouraged to make a much greater effort to do their very best work. Many are surprised to discover talents they did not know they possessed, and are inspired to persevere in their education. Many not only graduate from high school, but also continue their education and are enabled to break the cycle of poverty.

Fagan and Pabon (1990) studied 243 male and 133 female dropouts from six inner-city high schools. African American, Latino, and white students were all represented in this group. Among the many reasons given for quitting school, the largest number of former students mentioned loss of interest in school and the need to get a job. These two categories showed no differences by race. Other reasons included too much homework, inability to get along with teachers or fellow students, trouble at home, pregnancy, and drug and alcohol abuse or other health problems. Many left school permanently after they had been suspended. Up to 20 percent had been expelled.

Although only a few of the dropouts cited alcohol and drug problems as reasons for leaving school, the researchers caution that this should not be taken as evidence that many others were not substance abusers. It is quite likely that although many of the former students reported "no problems" with drugs or alcohol, the use of these substances was, in fact, influential in creating a loss of interest in school.

When the dropouts were compared with age-mates who remained students, both groups reported similar levels of drug use, weapons possession, and violence in the schools. Male dropouts reporting on their own behavior while at school admitted they abused drugs, attended school while under the influence of drugs or alcohol, and committed crimes more often while in school than did male students. They also missed school more often, had less respect for teachers, and did not "try hard" in school. Female dropouts more closely resembled female students, and said they were far less involved in drug use or school crime than the males. Fagan and Pabon (1990) conclude that although substance abuse and delinquency may directly influence school dropout for some, these problem behaviors may also be symptoms of other problems such as a poor home life and poverty. A relationship between large family size, suggesting possible parental neglect, and increased delinquency has been demonstrated (Tygart, 1991). Gray-Ray and Ray (1990) suggest that a youth's *perception* of parental neglect is more influential than its actual extent.

Youths who leave school are at a great disadvantage in looking for work, and many remain unemployed for long periods of time. Enforced idleness, boredom, and hopelessness may turn them to delinquency.

Perlmutter (1987) looked specifically at the relationship between learning disabilities and delinquency. He found that learning-disabled adolescents are "more likely to develop severe delinquent behaviors than are their non-disabled peers" (p. 89). To counteract the link between learning problems and delinquency, Rosenberg and Anspach (1973) recommend "educational therapy." This approach includes direct and continuous measurement of student learning activity, individualized instruction, a variety of classroom-wide procedures, and intensive, continuing self-study by administrators and faculty.

AMERICAN CULTURE AND VIOLENCE

culture of violence
The belief that support for violence is prevalent in our society, typified by, for example, violence in the media, limited gun control laws, and respect for family privacy.

Is the United States a culture of violence? What effect does it have on teens?

According to a 1992 study of violence in industrialized nations (Fingerhut & others, 1992), the United States leads the world in homicides and interpersonal violence. Some claim as a result that "violence is as American as apple pie." They blame **the culture of violence** for the high prevalence of violence among youths and teens. The reluctance of the law enforcement community to become involved in domestic violence (between spouses or parents and children) reflects a strong cultural support for family privacy and lessened concern for stopping violent behavior. Our reluctance to pass gun control laws and the widespread availability of alcohol and illicit drugs have been cited as characteristics of a culture that supports violence (Earls & others, 1993). Easy access to guns attaches a high and lethal risk to aggressive and violent behavior among adolescents (Berkowitz, 1994). Among adolescents 15 to 19 years of age, the use of firearms accounts for 71 percent of all homicides (Fingerhut & Kleinman, 1990). According to the Uniform Crime Reports issued by the Federal Bureau of Investigation (1992), the arrest rate for violent offenses of juveniles between the ages of 10 and 17 increased 27 percent from 1980 to 1990. This rise in violence occurred across all social classes and races. Callahan and Rivara (1992) surveyed 11th-grade high school students in Seattle, Washington. Six percent of the students indicated that they owned handguns. Only 25 percent of those who owned handguns stated that they were used for hunting or target shooting. Fifty percent of those who owned guns said that they carried their guns to school. In the midst of considerable national debate about the pros and cons of handgun regulation, both the proponents and opponents of gun control generally agree that we need to find a way of keeping guns out of the hands of those youths and teens who are prone to aggression and violence (Berkowitz, 1994).

Television, film and other media sources have also been blamed for incidents of youth violence. A large body of research demonstrates that aggression can be learned by viewing television and film violence (Eron & Slaby, 1994). Observational learning (see social learning theory in Chapter 6) and desensitization through exposure to media violence are two ways in which viewing violence in film may contribute to violent behavior (Donnerstein & others, 1994). The glorification of cowboys and the Wild West represent traditional cultural symbols that sanction violence. These traditional symbols appear quite mild in comparison with the heroes of recent television and films (e.g., *Die Hard 2, Robocop, Total Recall, The Terminator, Natural Born Killers*). VCR access has increased the likelihood that children and teenagers will be exposed to more movies, including those with violent content, than was typical of the past. Of particular concern is the exposure of teens to X-rated movies and slick movie videos that glorify sex and violence. Cable TV has increased the amount of violence aired through television (Donnerstein & others). By the time a typical youngster reaches the late teens, he or she will have witnessed more than 200,000 violent acts on television (Huston & others, 1992). A recent study by Dr. Sara Stein of Stanford University suggests that even watching TV news can be harmful for young adolescents. Stein and her research team interviewed 959 children ages 9 to 14 concerning the news reports of the kidnapping and murder of a 12-year-old girl in California. Eighty percent of the children interviewed could not stop thinking or dreaming about the crime or were trying desperately to stop thinking about it two months after witnessing the news reports (Bass, 1995).

VIOLENCE AND MINORITY YOUTHS

Teens from ethnic minority backgrounds, gay and lesbian adolescents, and youths with disabilities are more likely than other teens to become victims and perpetrators of violence. African American teens are more likely to die as a result of violence than by any other cause (Eron & Slaby, 1994). Although the rate of death by violence is lower for African American females than it is for males, violence is also the leading cause of death for African American females. Unfortunately, death by violence is becoming increasingly common among young Latinos, especially in cities such as Los Angeles, Chicago, New York, and Miami (Soriano, 1994). Some of the violence directed toward ethnic and racial minorities, lesbian and gay youths, and persons with disabilities is a form of hate violence, stemming from learned prejudice against certain social groups. Ethnic and racial minorities are also more likely to experience violence when they are poor.

Researchers have been interested in learning just what it is about ethnic and racial minority status that is associated with violence. In the process, it is important to try to disentangle the effects of poverty from those of race and ethnicity. One part of the answer has been found by looking at social risk factors, such as unemployment, poverty, drug abuse, and high population density (large numbers of people living in a small geographical area), that contribute to violence. All these risk factors are rampant among the inner-city

A SOCIOCULTURAL VIEW

Some Myths About African American Male Teens

In her review of the literature, Jewelle Gibbs (1991) has discovered a number of myths about African American male teens, and has summarized the evidence against these myths. They are as follows:

Myth 1. Most young African American males are frequently involved in criminal delinquent behavior.

Some reports indicate that nearly one out of every four African American males ages 20 to 29 is in jail, on probation, or on parole. Although these figures have been questioned by some experts, it should be pointed out that even if the figures are correct, more than three out of every four young African American males have no criminal complaints against them.

Myth 2. Most crimes committed by young African American males are directed against whites.

FBI statistics indicate that the great majority of crimes committed by African American males are against other African Americans, not whites.

Myth 3. If homicide rates were not so high among African American males, they would live just as long as white males.

It is a tragic fact that African American males in Harlem have a shorter life expectancy than those in rural areas of Bangladesh, an extremely poor country. It is also true that some of this death rate is due to homicide, but much of it is due to the number of

illnesses African Americans contract that could have been treated if they had been detected early enough.

Myth 4. Young African American males are fully expendable and disposable. If they don't shape up, they can be jailed, isolated in inner-city ghettos, or restricted to an urban underclass.

This is probably the most dangerous myth of all, not only because it is extremely inhumane to African American youths and because it is much more expensive to jail people than it is to alleviate their poverty, but also because it would create a dangerous subculture in our society that ultimately will cause far more problems than it would take to try to repair the circumstances that caused the problem in the first place.

Myth 5. The problems of young African American males are unique to the United States and simply reflect their inability to assimilate like other immigrant groups.

First it should be recognized that African American persons in Great Britain and other Western European countries also have high rates of poverty and have problems that prevent them from being integrated into those societies, too. In addition, even middle-class African Americans find themselves the victims of persistent racial prejudice and cannot be said to be fully assimilated.

neighborhoods where many ethnic and racial minorities reside (Hill & others, 1994). Poverty, in terms of low income, is not by itself the cause of violence. Perceptions of inequity, discrimination, and lack of opportunity contribute to anger, frustration, and feelings of futility. As noted by Prothrow-Stith (1991), "Having a future gives a teenager reasons for trying and reasons for valuing his life" (p. 57). Unfortunately, some teens do not look forward to a better future. According to Erikson, adolescents need to develop an identity and project a life for themselves in the future. This may be difficult for the teen growing up in the midst of social and economic inequality. Homicide rates are highest in those countries where the gap between the wealth enjoyed by the rich and the poor is greatest (Hawkins, 1993).

Early theorists (e.g., Rose, 1978) suggested that a "subculture of violence" exists among the poor. That is, lower classes were assumed to hold deviant values and child-rearing practices that were supportive of violence. Hill and associates (1994) maintain that the concept of a subculture of violence was based on an ignorance of African American, Latino, Asian, and Native American cultures. Many ethnic cultural traditions oppose rather than promote community violence. Respecting elders, valuing family and interpersonal relationships, and maintaining peace are Latino cultural values that are antithetical to violence. African American cultural values of spirituality, interdependence, harmony, and reciprocity ("what goes around comes around") also stand in opposition to violence. Hill and associates believe that the enhancement of a positive cultural identity and biculturalism (see Chapter 6) among youth can help protect ethnic minority youths against violence. When youths have a positive sense of their cultural identity and adhere to the traditions that oppose violence, they may retain a sense of direction and hope despite social and economic inequities. Teens who have a positive sense of ethnic identity may also be less likely to respond aggressively to racial and ethnic slurs. Instead, they may be more able to recognize this as stereotypical racist behavior and not respond to the comments as personal attacks or insults.

Sparks (1994) argues that the daily experience of violence is confusing to inner-city youths developing moral beliefs about what is right and wrong. Sparks interviewed nine inner-city teens and found that these teens believed that shooting was morally unjustified. They did, however, understand that some people become involved in selling drugs as a means of economic survival and recognized that shooting is often an inevitable part of the drug business. Such contradictions make it hard for these teens to resolve moral issues. Sparks believes that inner-city youths need to be helped to understand the social and political circumstances that make it hard to find employment other than drug sales in their community. Youths need, furthermore, to be helped to think of political and social solutions to these problems. They need to be shown ways of accomplishing change rather than be overwhelmed by feelings of hopelessness.

VIOLENCE AND GENDER

Aggression, delinquency, and violence are behaviors associated more frequently with adolescent boys than with girls. Males are 10 times more likely to be arrested for violence than are females (Earls & others, 1993). Among both inner-city and upper middle class teens between the ages of 11 and 24, males more often than females reported having been victims of violence, knowing victims, and witnessing violent acts (with the exception of sexually related crimes, where the statistics for females are higher) (Gladstein & others, 1992).

Unfortunately, we do not know a lot about the development and outcome of aggression among girls because most research has studied boys and has not focused on indirect and interpersonal violence. Differences between boys and girls are most evident in terms of physical aggression (Pepler & Slaby, 1994). Boys, for example, are more likely to get involved in physical fights than are girls. Aggression tends to be accepted more in boys; girls are viewed more negatively, by peers and adults, for any act of physical aggression. As discussed in Chapter 11, girls are hypothesized to internalize their anger (contributing to depression), whereas boys are more likely to externalize anger, resulting in aggressive, violent, and delinquent behavior. It is incorrect, however, to conclude that girls are not aggressive. Verbal, indirect, and interpersonal aggression are common among girls. Indirect aggression may be shown by leaving someone out of the group or talking about them behind their backs. Women are more likely to express physical aggression within intimate and family relationships, but are unlikely to commit violent crimes in the community (Magnusson, 1988).

Although gender differences in aggression were often assumed to be biological (Macoby & Jacklin, 1974), there is increasing evidence that biology is not the primary reason that boys are more physically aggressive (Adams, 1992). Socialization practices in the family, school, and peer group continue to favor passive and nurturing behavior among women and girls and independent, competitive, and aggressive behavior among men and boys (Pepler & Slaby, 1994). Adolescent males are also more likely than girls to hang out in places where violence occurs and to socialize with other males who are offenders (Lauritsen & others, 1992). Societal attitudes and socialization practices, in conjunction with the larger physical size and strength of males in midadolescence and beyond, also contribute to the risks of girls and women to become victims of violence, including homicide, physical abuse, childhood sexual abuse, and dating violence (Sorenson & Bowie, 1994).

> Why do you think boys are more likely to commit violent and aggressive acts?

VIOLENCE PREVENTION

Before examining the factors that contribute to violent and aggressive behavior, we expressed the hope that knowledge of these factors would provide suggestions for the prevention of violence. In recent years, a number of researchers and youth service providers have been using the growing body of knowledge about youth violence to develop prevention programs. Programs have been developed to address individual factors as well as the contexts (i.e., school and family) in which aggression is learned. Because aggressive behavior can be

WHAT'S YOUR VIEW?

School Violence

Sadly, over the last few years we have heard numerous reports about cases of school violence. These high profile cases often involve shootings in which there are multiple victims (for example, the tragedy at Columbine High School). What explanations for violence (family, peers, school achievement, culture of violence, availability of handguns...) discussed in this and other chapters do you believe are contributing to this violence?

learned at a young age, some prevention programs, called primary prevention, are directed at children before any signs of aggressive behavior are apparent. Secondary prevention is directed at those children who are showing early signs of aggression. The hope here is to reduce aggressive behavior before the behaviors become serious. Tertiary or remedial programs are directed at those youths who are already showing serious behavior problems. Because the probability for future violence is greatest among those first arrested at an early age (Wodarski & Hedrick, 1989), primary and secondary prevention appear most promising in reducing increases in violent crimes.

Many preventive interventions on an individual level focus on the development of social skills, social problem solving, and self-control. Spivak and Shure (1974; Shure, 1992) have taught skills in solving interpersonal problems, thinking of alternative solutions, and considering consequences of behavior to children as young as kindergarten age. Spivak and associates (1989) developed a social problem solving skills program taught as part of the high school curriculum. Students learn that anger is a normal emotion that needs to be expressed in healthy ways. Safe and constructive ways to express anger and alternatives to violence are taught through role playing and class discussion. Fighting is considered as one way to handle conflict, but the dangers associated with fighting are also carefully analyzed. This primary prevention program is supported by a mass media campaign that uses T-shirts, posters, and public service announcements to increase awareness of the dangers of violence and the need to develop alternatives. Secondary and tertiary components of Spivak's program involve the identification and referral of adolescents by mental health clinics and hospital emergency rooms. Selman and associates (1992) have modified the pair therapy approach described in Chapter 6 for use as a primary prevention technique for both withdrawn and aggressive children. The program, which is delivered through the public schools, not only provides factual information about the risks associated with aggressive behavior and teaches interpersonal negotiation, but also considers the personal meaning of the dangerous behaviors to the adolescent and his or her relationships with friends, family, and community. According to Selman, the personal meaning of the behavior makes it difficult to change. As long as a teen believes that "I have to fight to survive" or "I have to fight to be respected by my friends," it is difficult to reduce fighting behavior.

Hypothesizing that criminal behavior is linked to dropping out of school, and that both can be the direct result of a lack of self-control, Dacey and associates (1993) designed a program of 14 45-minute lessons in self-control, which were administered in two Boston middle schools to 12 classes of 151 eighth-graders in 1990. Nine of the classes were selected because the students in them were known to have serious problems controlling their behavior, and three were randomly chosen from among the regular classrooms. The lessons incorporated several methods known (on the basis of previous research) to promote greater self-control, including awareness of internal states and behavior modification, visualization and goal orientation, somatic control, stress management, self-image and impulse control, and role play. Two years after the completion of these lessons, the dropout rate of participants was compared to the rate of 23 percent for Boston's 10th-graders. The rate for the experimental group was 14 percent, a decrease in the dropout rate of 40 percent. Whether there was also a decrease in criminal behavior was not studied, but because it is known that those who stay in school through graduation perform significantly fewer criminal acts on the average, it seems likely that this type of instruction reduced criminality, too.

Given that much aggressive behavior is originally learned in the family, parent training and family therapy approaches are also important in the prevention and treatment of antisocial behavior. Goldstein (1991) has considered the ways in which schools need to change so that they will be safe places for all children and teenagers. Strategies include the training of teachers in self-defense strategies and the installation of better lighting and metal detectors. Community prevention efforts range from placing more police on the streets to creating better educational and job opportunities for those youths who perceive little hope in their futures. Because many of these programs are fairly new, data concerning their long-term effectiveness is just beginning to accumulate. Many of these efforts appear quite promising. Because violence and antisocial behavior are complex problems, however, it is likely that individual, family, social, and community interventions are all necessary to make a large and lasting impact (Guerra & others, 1994).

> What approach would you choose to reduce violence?

MULTIPLE-PROBLEM YOUTHS

❑ In his eyes is the fixed stare of the blasted spirit.
Poet Ned O'Gorman

In contrast to the traditional focus of research on single aspects of adolescent problems, a recent study based on the National Youth Survey (NYS) by Elliott and associates (1989) sought to identify and understand the connections between delinquency, drug and alcohol abuse, and mental health problems. The National Youth Survey is a continuing longitudinal study of a nationwide probability sample of adolescents, and is representative of all American youths. It consists of 1,725 adolescents, with members of each race, sex, and age group in the same proportions as are found in the U.S. population of adolescents as a whole. These researchers are also interested in finding out whether delinquency and drug use peak in the middle to late teens and begin to lessen as the youth matures, as some studies have suggested, a process that some observers have called "maturing out" of delinquent behavior. There is some evidence that many youths do grow out of delinquent behavior as they enter their twenties.

Involvement in delinquent behaviors, drug use, and mental health problems were reported by the teenagers themselves in confidential interviews conducted in their homes. These self-reports included all of the 40 different offenses listed in the FBI's "Total Arrests, Distribution by Age" (see Table 13.1). The interviews were also designed to bring out "hidden" delinquency and other problems—those that are not reflected in official sources. Elliott and associates

(1989) found that *delinquency and substance abuse are more related to each other than either is to mental health problems.* They also noted that drug use tends to delay or prevent "maturing out" of delinquency. Another interesting finding was that alcohol was used just before 80 percent of the sexual assaults that were committed, and other drugs were used just before half of the car thefts.

Although individual crimes by a few adolescents may be serious, the criminal activities of gangs have come to be a more critical concern. In the next section, we investigate trends in gang behaviors in the early 1990s.

The Aggressive Gang

Gangs often offer youths the fulfillment of basic needs (Popenoe, 1996; Sanyika, 1996; Vigil, 1996). Some of their functions clearly coincide with those of the larger society. Gangs typically provide protection and recognition of the desire to feel wanted. In addition, the gang becomes a family to its members (Dishion & others, 1995). Furthermore, they offer activities that mark achievement, status, and acceptance, such as the initiation rite of a potential gang member. Gang activities may offer the lure of excitement or may be followed as part of a family or community tradition. Since gang members are often isolated both from legitimate opportunities to earn money and from daily interaction with people outside the gang, it becomes a source of social status and often also of money gained through illegal activities, especially drug dealing (Fagan, 1989).

In a study conducted in a middle-sized midwestern city, both gang members and young people who thought they might join a gang gave "to have more friends" as the primary reason for joining. Other important reasons for gang membership were "because I have nothing else to do" and "so that people will look up to me" (Takata & Zevitz, 1990).

According to a study commissioned by the New York City Youth Board (1989), urban gangs possess the following characteristics:

- Their behavior is normal for urban youths; they have a high degree of cohesion; roles are clearly defined.
- They possess a consistent set of norms and expectations, understood by all members.
- They have clearly defined leaders.
- They have a coherent organization for gang warfare.

The gang provides many adolescents with a structured life they never had at home. What makes the gang particularly cohesive is its function as a family substitute for some adolescents whose strong dependency needs are displaced onto the peer group. The gang becomes a family to its members, and they view the street as their home (Dishion & others, 1995; Short, 1990). Gangs members are sometimes recruited through activities resembling fraternity and sorority rushes, and sometimes through threats and intimidation.

Friedman and associates (1976) studied the victimization of youth by urban street gangs and found that

❏ Rituals of street gang warfare and the practices of victimizing both gang members and nonmembers by having them commit serious crimes and violent offenses may serve to maintain the continuity of the group, to give it structure, and to symbolize the gang's power of life and death over others (p. 527).

Curry and Spergel (1988) make a similar point in their study of communities where both delinquency and gang homicides occur:

❏ For disadvantaged youths, uncertain in the face of the unstable urban social world, the gang is responsive and provides quasi-stable, efficient, meaningful social, and perhaps economic, structures. In gang membership, there is the opportunity to obtain the psychic rewards of personal identity and minimal standards of acceptable status and sometimes the material benefits of criminal income (p. 401).

Thus, the gang becomes a vehicle for tearing its members away from the main social structures and authorities, in particular the family and school.

In this country, the formation of juvenile gangs has typically followed a sudden increase of new ethnic groups due to immigration. The children of new immigrants have a difficult time breaking through cultural barriers, such as a new language and racism. Perceiving their prospects of succeeding in the new society as bleak, some of these children form gangs, which provide the structure and security discussed, but also serve as an outlet to attack the society that seemingly will not accept them. In times past, these gangs were composed of Jewish, Irish, and Italian Americans. Then came the African American gangs. Today's new gangs are frequently formed by Latino and Asian Americans (Burke, 1990; Vigil, 1988). According to Spergel and others (1989), more than half of gang members are African American and another third are U.S. Latino. These estimates may be biased, however. They reflect those youths who have been arrested for gang activity and not actual gang membership (Goldstein & Soriano, 1994). Because different gangs often follow different and specific cultural practices, those who work with gang members must be sensitive to those cultural traditions. The reasons for joining a gang, and the meaning of status, honor, and reputation in that gang, are shaped by cultural traditions (Goldstein & Soriano).

Youth gangs are not delinquent by definition, although both professional literature and the popular media encourage the view that they always exist primarily as law-breaking groups (Hagedorn & Macon, 1988). Some gangs are formed in reaction to violence and the fear of violence on the streets (Walker, 1991). Other gangs are actually informal social groups of adolescents whose main activity is hanging out or doing nothing, just being friends. Others are organized around an activity (e.g., hip-hop music) and provide an outlet for creative and artistic strivings. The use of marijuana and alcohol and various acts of minor delinquency may be condoned by some of these groups, but serious drug use and true criminality are not. These groups of friends appear to support one another in avoiding such pitfalls. Groups that begin in this way usually remain social in character, although in urban settings there is the possibility that conflict with other similar groups may lead them to evolve into delinquent gangs (Baron, 1989; Hagedorn & Macon, 1988).

Although aggressive and delinquent behavior are a small part of gang activity, those behaviors attract the most media attention and public concern. In the United States, gangs are growing rapidly in public concern, as well as in numbers, breadth of location (urban, suburban and rural), increased involvement with drugs, and amount and seriousness of violent behavior (Goldstein & Soriano, 1994). Estimates suggest that there are more than 2,000 different gangs in the United States, with more than 200,000 members (Goldstein & Soriano). Whereas the typical age of gang members used to be between 12 and 21, gang members today may be as young as 9 and as old as 30. Older members are staying in gangs, it seems, because of limited employment opportunities outside of gang activities (Goldstein, 1991). Although male gang members still outnumber female gang members by 15 to 1, the number of female gangs and gang members have been increasing. Females tend to join gangs at an older age than boys and leave the gang at a younger age (Goldstein).

Gangs and Violence

Though there is a long history of gangs in this country, today's gangs have some disturbing characteristics that differentiate them from those of years past (Sheley & others, 1996). Today's gangs are much more heavily armed and seemingly willing to use their weapons. Earlier films about delinquent teenagers—like *Blackboard Jungle* (1955) and *West Side Story* (1961)—showed gang members carrying knives, chains, and pipes. Gang members in recent films like *Colors* (1988) and *Boyz 'N the Hood* (1991), both of which feature actual gang members as extras, are armed with assault rifles and Uzi submachine guns. They do not hesitate to use these highly destructive weapons, not only on rivals and suspected rivals, but also on innocent bystanders and police officers. The showing of these films in itself often causes outbreaks of heightened violence. Snyder (1991) points out that viewing such films can reinforce the violent and antisocial values already held by gang members. Other delinquents may also identify with the characters in the film and become further confirmed in their beliefs and attitudes.

drive-by shootings
When innocent bystanders are injured or killed by violent gangs usually shooting from moving automobiles.

Law enforcement officials have noted that attacks by gangs on police officers are now quite common and continue to increase (Gates & Jackson, 1990; Sessions, 1990). Many innocent bystanders are also injured or killed by violent gangs, often in **drive-by shootings**. Statistics from the Los Angeles Police Department indicate that 50 percent of the victims of gang violence have no connection at all with any gang activity (Gates & Jackson, 1990). In addition to killing members of rival gangs, gang members also frequently kill one another, even within their own "set" or subdivision. This is especially true of large gangs like the Crips of Los Angeles (Bing, 1991; Ewing, 1990).

Gang violence has increased dramatically in the past decade. Twenty-five percent of all juvenile crimes are committed by urban gangs. Los Angeles, with more than 450 street gangs involving more than 36,000 members, has perhaps suffered most from this upsurge in violence (Gates & Jackson). That city saw gang-related homicides increase from an already staggering 150 in 1985 to an unfathomable 387 in 1987. In 1988, nearly 10 percent of all homicides committed in Chicago were gang-related (Ewing, 1990). Many of those killed are themselves teenagers. Youths in smaller cities also suffer from a large number of gang-related deaths. In Boston, for example, 107 young people have been killed by gangs since 1985 (Wall, 1991).

What do you believe is the key factor in gang violence?

Many observers have cited the lack of responsible parents and an unstable family life as major contributors to the tragic growth in the number of "kids killing kids" (Dolan, 1991; Kelly, 1991; Prothrow-Stith & Spivack, 1991; Reid, 1991; Yancey, 1991). Curry and Spergel (1988) suggest that poverty and social disorganization are key factors in violent gang behaviors.

Bing (1991) was allowed to observe a class required of Los Angeles gang members in a juvenile detention camp. When the teacher asked the boys to suggest "a real good reason to kill somebody," they came up with 37 reasons for which they would be willing to kill. Many of these reasons seem shockingly trivial, but most are related to personal pride or protection of the gang's territory. Some of the reasons given were: "Cause he asked me where I was from," "Cause he wearin' the wrong color," "Cause he give me no respect," "For the way he walk," "Cause I don't like his attitude," and "Cause they ugly" (pp. 122–123). Small incidents may serve as triggers that set off gang violence.

AN APPLIED VIEW

An American Nightmare

It is astonishing how the easy purchase of drugs creates an environment in which crime breeds viciously. Take, for example, this description of life in northern Manhattan, New York City:

❏ As Broadway cuts up through the Upper West Side of Manhattan and into Washington Heights, it gradually turns into a giant Caribbean bazaar. . . . As the ever-present crowds make their way up and down the street, the Heights seem a living embodiment of the American Dream—a vibrant, energetic urban melting pot.

Wander off Broadway, though, and the neighborhoods quickly seem like an American nightmare. On side streets in the 150's and 160's, clusters of tough teenagers wearing beepers, four-finger gold rings, and $95 Nikes offer $3 vials of crack, the high-octane, smokable derivative of cocaine. On every block there are four or five different 'crews,' or gangs, each touting its own brand of the drug, known to aficionados as 'Scotty' (as in "Beam me up"). Some blocks are 'hotter' than others, depending on the availability of the crack. On the hottest blocks Scotty

is available '24/7'—24 hours a day, seven days a week. So much business is transacted on these streets that Washington Heights has gained a reputation as the crack capital of America.

The experience of the Heights has been repeated in large cities throughout the country. And now, in smaller communities, too, crack is striking with swift fury. From rural woodlands to shady suburbs, prairie townships to Southern hamlets, no community seems immune. The roster of the infected reads like a roll call of Middle America itself: Roanoke, Va.; Seaford, Del.; Sioux Falls, S.D.; Cheyenne, Wyo.; Sacramento, Calif.; Portland, Ore. Fort Wayne, Ind., once known as the 'City of Churches,' is now home to an estimated 70 crack houses, causing law-enforcement personnel to christen it 'the crack capital of Indiana' " (Massing, 1990, p. 39).

These are the situations that many human service workers are facing, not in large urban ghettos, but in "Hometown, U.S.A." We are going to have to devise better plans for dealing with this insidious danger. What ideas do you have as to what we might do?

Such triggers include fights over girls, rumors, ethnic tensions, and parties (Goldstein & Soriano, 1994).

In addition to pride and territory, money and drugs are increasingly motivating the violence. Police in Los Angeles cite the influence of crack (see Chapter 12) as a major influence in gang violence. Organized urban gangs have now become profit-making enterprises. Taylor (1990) states, "The fact is that drugs have taken street gangs and given them the capability and power to become social institutions" (p. 114). Drive-by shootings are committed by heavily armed youths in expensive cars protecting their "market." The availability of large amounts of money through the drug trade is the main reason that gangs who deal drugs are seeking expanded markets and are no longer limited to the large urban areas.

cultural gangs
Those gangs that are centered in a neighborhood, loosely organized, and not necessarily involved in any criminal activities.

instrumental gangs
Gangs formed for the purpose of carrying out criminal activities.

William Sessions of the FBI observes that gangs may be divided into two categories: **cultural gangs** or **instrumental gangs**. Cultural gangs are those centered in a neighborhood. They are loosely organized and are not necessarily involved in any criminal activities. These gangs are usually made up of a single cultural or ethnic group, especially if that group is newly arrived in this country, as was mentioned earlier. In contrast, instrumental gangs are formed for the purpose of carrying out criminal activities. The lure of big money has drawn many into drug dealing. Some formerly cultural gangs have evolved into instrumental gangs through becoming participants in the cocaine trade (Sessions, 1990).

Smaller cities and towns in the United States have recently seen an increase in the formation of juvenile gangs. These gangs are often related to other, well-established gangs from the larger cities. In effect, gangs such as the Crips and the Bloods of Los Angeles can set up "franchises" in cities like Seattle, Racine, and Shreveport. Cocaine distribution by these two gangs is known to exist in 26 cities across the country. Police in Los Angeles have received questions from 48 states about the Crips's and the Bloods's activities (Sessions, 1990). Residents of smaller cities resist accepting the fact that gang activity has spread to their home area (Pierce & Ramsay, 1990; Takata & Zevitz, 1990).

home invader gangs
Gangs that enter homes and terrorize the occupants while stealing anything of value.

A newly developing form of drug-related gang violence is the **home invader gang**. These gangs enter homes and terrorize the occupants while stealing anything of value. According to Burke (1990), these gangs enjoy their ability to intimidate and control their victims, forcing them to hand over all their valuables, and threatening to kill them if they don't cooperate. They actually prefer to attack while the victims are at home. Although some of these gangs are made up of adults, juvenile home invaders are not unusual. Particularly in vacation areas like Florida, victims are confronted and robbed in motels. Many of these gangs are made up of Asian males.

Gang activity is not restricted to urban settings. Suburban gangs also exist (Muehlbauer & Dodder, 1983). These suburban gangs usually are less organized and formal than those in the cities. They typically get their thrills from the malicious destruction of property. It has been predicted that violence directed by youths toward one another will invade the white suburbs within the next few years (Radin, 1991). Reasons advanced for this prediction include:

- Abdication by many parents, teachers, and community workers of the responsibility to teach children lessons in civility and self-restraint.
- Making lethal violence trivial, not just among children, but in society as a whole.
- Denial of basic necessities—food, shelter, supervision—to an ever-broadening circle of youths.
- Swelling rage among youths at the unfairness and apparent hopelessness of their situation (Radin, 1991, p. 1).

CHARACTERISTICS OF GANG JOINERS

At the beginning of our discussion of gangs, we described some of the reasons why adolescents may be attracted to gang membership. Fortunately, the majority of adolescents find more constructive ways to satisfy the needs fulfilled by gangs. We are just beginning to learn why some teens join a gang while others go to great extremes to avoid doing so (Hochhaus & Sousa, 1988). According to Jankowski (1991), who has made an extensive study of urban gangs, most of the youths who join them do not have psychological problems but are "intelligent, self-motivated and goal-oriented—all traits that American culture values" (p. 15). The youths believe that joining the gang will improve their lives, especially in the face of the extremely limited resources available to them and the heavy competition for what little does exist. Jankowski believes that it is living in this type of environment that creates aggressive and violent behavior.

Gang members are much more likely to have divorced parents or parents with a criminal history. They are more likely to do poorly in school. Friedman and others (1976) showed that what most differentiates the street gang member from the nonmember is the enjoyment of violence. Gang members also have more unrealistic expectations of success than nonmembers (Burton, 1978). Gangs, in effect, promise a more equal opportunity for members to succeed in life than does society. In general, gang members are found to have more drug-abuse problems, more mental disturbance, and are more angry and violent than the average youth.

Female gang members, whose numbers also have increased recently, face many of the same societal barriers that cause males to join gangs, with the additional problem of sexism. As in the larger society, within the context of the gang itself, females are accorded lower status than males (Campbell, 1990).

ATTEMPTS TO ELIMINATE GANG BEHAVIOR

recidivism
Returning to committing delinquent and criminal activities after having been incarcerated.

Recidivism—returning to delinquent and criminal activities after being incarcerated—is very high among adolescent offenders. Some investigators believe that because adolescence is a period of high risk for rejection of societal standards of behavior, these teenagers are even more likely than adult criminals to return to illegal activities. Adolescents being freed from detention are likely to continue to suffer from social problems and to continue to respond in antisocial ways. Their lack of job skills and consequent inability to obtain money in legitimate ways may lead them back into drug-dealing and robbery. Their inability to resolve conflicts peacefully is likely to result in more violence. Their inability to trust or to form healthy relationships often adds to the problem.

Although many gang members grow out of their attachment to the group, a few continue their unlawful conduct into adulthood. In one study, which followed 95 former juvenile delinquents who had spent time in confinement for their offenses, 89 also had an adult criminal record (Lewis, Lovely, & others, 1990). Note that this study investigated only youths who had been in jail or detention, so it does not tell us anything about the many who have never been confined. Research is especially needed to discover why some violent juveniles give up this behavior as they mature, and do *not* become violent adults.

Mark Umbreit (1991) describes efforts to teach juvenile offenders how to settle conflicts without resorting to violence by means of a technique called **conflict mediation**. In this program, a face-to-face meeting between the offender and the victim is arranged and attended by a third person, called the mediator.

conflict mediation
A technique to teach juvenile offenders how to settle conflicts without resorting to violence by means of a face-to-face meeting between the offender and the victim, attended by a third person, called the mediator.

This person has been trained in ways to avoid or resolve conflicts. In talking together under the guidance of the mediator, victim and offender learn how and why the incident occurred, and what it meant for each of them. The young offender is held responsible for his or her behavior and is given a chance to express remorse and to make amends to the person who was harmed. Surprisingly, the researcher found that these meetings are rarely emotionally violent or marred by further conflict.

Some success has been enjoyed within this program, but there are several important questions about the mediation process that must be considered. The two most important ones are imbalance of power and coercion to participate. The mediator must be able to establish the same amount of power for the victim as for the offender. Maintaining this balance throughout the process is also the responsibility of the mediator. It is most desirable for the offender to agree voluntarily to participate in mediation. An unwilling participant is not likely to gain anything from the process and may cause further psychological harm to the victim.

Youths who get into trouble with the law are more likely to lack interpersonal skills, such as the ability to communicate clearly, than youths who do not get into trouble with the law.

Several studies of incarcerated youths have demonstrated that lack of social and interpersonal skills and delinquency are often related (Simonian & others, 1990). The delinquents' beliefs about how unchangeable the causes of their failures are have also been shown to be related to the likelihood of their responding violently to many kinds of situations (Guerra & Slaby, 1990). Some efforts have been made to remedy this situation by teaching delinquent youths cognitive mediation and conflict mediation skills (Umbreit, 1991).

Guerra and Slaby (1990) ran a 12-session program with 120 male and female adolescents who were confined to a state juvenile correctional facility because of violent crimes that included assault and battery, robbery, rape, attempted murder, and murder. The participants were randomly assigned to one of three groups: cognitive mediation training, attention control, and a no-treatment control group. There were 20 males and 20 females in each group. The youths in the treatment groups were divided into small discussion groups that met once a week for one hour. The instructor guided them in learning such skills as considering more than one interpretation of a situation in which they might react violently and responding in thoughtful, responsible ways instead of reacting with passivity or aggression.

Although the youths did learn to change some of their aggressive reactions, this change was, unfortunately, neither large nor lasting. Two years after the participants were released from confinement, no differences could be found between the treatment groups and the control group.

Researchers have had some success in developing tests that will indicate whether a juvenile offender who has served some time in jail is likely to return to criminal activities if paroled. These could provide more concrete guidelines for those making the decision about whether it is safe to return a confined adolescent to the streets (Ashford & LeCroy, 1990).

The Nonaggressive Offender

Persons in this category are technically lawbreakers, but they usually do not inflict physical harm to another person's body or property. Females constitute the majority of teenagers in this category, although the number of males is growing. About 25 percent of juvenile court cases are of this type (Haskell & Yablonsky, 1982).

THE RUNAWAY

According to the Department of Health and Human Services, the number of children leaving home without permission of their parents has been increasing in recent years. It is estimated that one out of ten 12- to 17-year-olds has run away from home at least once. Any youth between the ages of 10 and 17 who has left home overnight without parental permission can be defined as a runaway. Many parents do not report their children as runaways because they expect the children to return in a few days. The amount of time youths remain away from home, the situations they are seeking to escape from or are running to, and the consequences of their runaway efforts are varied. There are also cases of adolescents who have been reported as runaways even though they have been thrown out of the house by parents. By reporting these teens as runaways, parents hope to avoid charges of abuse and neglect.

The number of female runaways is considered to be slightly higher than male runaways, although this may only reflect the number of runaways who are arrested. Because teenage female runaways are more visible on the streets, they are more likely to be arrested than males and thus counted as runaways (Burgess, 1986). The numbers of adolescent males and females served in runaway shelters are about equal (Rotheram-Borus & others, 1991).

Since the passage of the Runaway, Homeless and Youth Act of 1974, running away has not been considered a crime. Today, running away is seldom viewed as the act of a disturbed delinquent or pleasure-seeking teen. Running away is being increasingly understood by mental health professionals as an adolescent's effort to cope with extreme levels of family disorganization, conflict, and abuse.

CAUSES OF RUNNING AWAY

Many parents mistakenly do not report their children as runaways because they expect the children will return in a few days.

Why do children run away from home? When asked why they ran away, runaways most frequently named an unhappy life, verbal abuse, and physical abuse (Burgess, 1986). Almost 70 percent of the runaways interviewed have low achievement and little or no involvement with school. Clashes with family members and the commission of petty crimes are considered to be other major causes. Although research findings on runaways point to parent-child conflict and school difficulties as primary causes for running away, Jones (1988) carefully analyzed the existing research and found that the reasons for running away are often complex. According to Jones, a variety of family and individual factors are important in understanding runaways.

Family Factors

Adolescents who run away are frequently running from a troubled family environment. These adolescents leave home voluntarily because they choose to get away from the problems at home. According to White (1989), almost 75 percent of runaways leave an unhappy situation with the hope that they will find acceptance and happiness elsewhere. Jones identified a variety of family situations that teens may seek to flee. Sometimes runaway adolescents are trying to escape family problems such as an alcoholic parent, incest, or parental violence. Conflict between parent and adolescent over such issues as curfews, dating partners, dress, school grades, and church attendance can also lead to running away. The adolescent generally views the parent as uncaring, not understanding, and too strict, and the parent views the adolescent as disobedient. Improving communication between parents and the teen is critical in this type of family. A family crisis, such as divorce, separation, or financial loss can also trigger running away. Other teens run away as a cry for help—to call attention to themselves or their families so that help will be offered. Finally, some teens run away because they are keeping a secret from their parents, such as pregnancy, school failure, or homosexuality, and are afraid of their parents' reaction.

A second group of adolescents also have troubled family environments but leave home because they have been abandoned, neglected, or abused by their parents. Adolescents in this group feel rejected by their parents and parents often do not oppose their child's leaving. This group of runaways is commonly referred to as **pushouts**, **castaways**, and **throwaways**. Some estimates suggest that only 5 to 10 percent of runaways fit this category, while other research indicates that 40 to 50 percent of youth classified as runaways are actually pushouts, castaways, or throwaways (Adams & Gullotta, 1983; Langway, 1982). This disparity (5 to 50 percent) is probably caused by the difficulty in proving exactly why a teen has run away from home.

Financial difficulties can sometimes cause parents to push out adolescents, especially when there are still younger children in the home who need financial support. Some adolescents are pushed out when parents feel that there is nothing they can do to control their teen's behavior. When repeated attempts to control drug abuse and sexual promiscuity have not worked, some parents feel desperate and believe the best thing is for the teen to leave home. These parents clearly need help in developing more effective ways to control their teen's behavior.

Another group of runaways are referred to as **system kids** (Rotheram-Borus & others, 1991). These are adolescents who were removed from their parents' homes by social welfare agencies because of parental neglect or abuse. After being placed in foster care or a group home, these adolescents decide to run.

One note of caution: the data in this section may suggest that parents are always to blame for runaways. In fairness, it must be stated that although some parents do their best, their child still runs away, for reasons that are not always clear.

According to Robertson (1989), half of runaway youths come from foster homes, group homes, and delinquent detention facilities. Accompanying the recent increase in the number of homeless adults in this country are an increasing number of homeless teens. A growing number of adolescents are entering shelters for runaways because their parents are homeless and unable to care for them (Rotheram-Borus & others, 1991).

Miller and associates (1990) used a case study approach and a survey of nine adolescent runaways to explore several levels of severity. They found three levels:

- *First degree runners*—generally run *from* something as a solution to their problems.
- *Second degree runners*—run both *from* and *to* something and see it as both a solution and a problem.
- *Third degree runners*—run *to* a street culture lifestyle that they believe will prove to be a solution to their problems.

---

pushouts, castaways, throwaways
Adolescents who leave home because they have been abandoned, neglected, or abused by parents.

system kids
Adolescents who, after being placed in foster care or a group home, decide to run.

Running is an impulsive reaction for the first two groups and is planned by the latter group. The researchers also note that youths run away from home because of family conflicts and from residential care because of peer influences. They believe running becomes more severe when youths are removed from their families; thus, residential treatment should not be prescribed when running away is the problem.

Simons and Whitbeck (1991) state that one of the primary reasons that adolescents leave home is to escape *from* parental abuse, primarily physical abuse for boys and sexual abuse for girls. Once on the street, runaways tend to engage in deviant behavior such as prostitution to support themselves. The researchers investigated whether early sexual abuse has an indirect or direct effect on prostitution with a sample of 40 primarily chronic (60 percent) adolescent runaways and 95 homeless women who also had been chronic runaways in their youth (42 percent). More than 40 percent of the runaways and about a quarter of the homeless women reported being sexually abused. More than a third of the runaways gave sexual abuse as their reason for leaving home.

An understanding of the reasons for running away helps to predict both the likelihood that an adolescent will return home and the adolescent's response to treatment. When feelings of caring and concern continue to exist among family members, the adolescent is more likely to want to return home and will be more motivated to work out problems in therapy. Teens who run away because of a crisis generally do not stay away for long and often respond well to short-term treatment. When family problems are not resolved, a cycle of running away is more likely to develop.

Teens who run away repeatedly are more likely to stay away longer, to have more difficulties in school, to be in trouble with the law, and to come from the most troubled families. When on the street, runaways sometimes hope that life has improved at home or they may forget how unhappy they were while at home. Some runaways even feel guilty, believing they may have been responsible for conflict and abuse at home and decide to return home. If they return to an abusive family situation, they are likely to leave again (Burgess, 1986).

Adolescents who feel abandoned by their families are the least likely to return home. Generally, by the time they run away, pushout children have experienced many years of failure at home and at school and the chance that they will be able to adequately resolve family conflicts is low. They feel they have no home to return to and are likely to stay on the streets for long periods of time.

Individual Factors

Jones (1988) also identified a number of individual personality types among adolescent runaways. A small percentage of runaways are youths who suffer from serious emotional disturbances. Generally they call attention to themselves in the street because of their unusual and sometimes bizarre behavior and are quickly brought in for psychiatric treatment. Teens who feel lost or without direction sometimes run away to "find themselves." Some feel so close to their families that they want to get away to figure out things on their own. These youths generally return home quickly because of the problems they find in the street. Teens who are experiencing problems in school and find their friendships more satisfying sometimes run away to avoid school and to spend more time with peers.

Another group of runaways find home life boring and, in the absence of strong family relationships to keep them at home, leave to find excitement on the street. Some youths—Nye (1980) reports them to be as much as 20 percent of the runaway population—are believed to be relatively well adjusted, but leave home to find adventure, fun, and pleasure. (Cutbacks in afterschool and out-of-school activities may account for some of this.)

These runaways resemble the fictional character of Huckleberry Finn and the real-life American hero Benjamin Franklin, who left his hometown of Boston as an adolescent to experience the adventure of the sea and to explore the cities of New York and Philadelphia. This last group of runaways idealizes the world away from home and the adventure it offers. In contemporary America, however, runaways are unlikely to find the kind of adventure they seek. Although they may not have psychological difficulties before leaving home, the circumstances they encounter on the street can quickly lead to psychological distress.

STRESS, COPING, AND RUNNING AWAY

As you will recall from Chapter 11, stress often contributes to the development of physical and psychological difficulties, especially when the individual lacks sufficient coping strategies to deal with the stress he or she is experiencing. While acknowledging the importance of family and school factors, Roberts (1982) also suggested that adolescents who run away are likely to have experienced many stressful life events and few effective coping strategies for responding to that stress. According to this view, running away results when adolescents experience high levels of stress and, in the absence of more effective coping strategies, view running away as the only solution to their problems. Roberts set out to test his ideas by interviewing runaway adolescents and high school students. Not surprisingly, Roberts found that, in comparison with the high school students, the runaways reported more stressful life events, many of which were related to family conflicts. Some stressful life events mentioned only by the runaways were being physically beaten by a parent; being thrown out of the house; being placed in a youth shelter; being caught dealing drugs; the death of a parent; a parent's lover moving into the house. Roberts gives an example of the kind of stressors reported by one runaway:

> ❏ Johnny, age 15, had been living with his father for two years. His parents were divorced. His father was an alcoholic. Johnny recalled the night he had been thrown to the ground and stomped on by his father during one of his drunken rages. To escape from further beatings, Johnny fled as soon as his father went to sleep (p. 21).

In contrast, the stressful life events mentioned more often by the high school students included the death of a grandparent and a youth's broken romance.

A study by Rotheram-Borus and associates (1991) also indicates that runaways have experienced high levels of life stress. In their study, runaways reported four times more stressful life events than the average adolescent. About 40 percent of the runaways reported that their parents had problems with drugs. Many of their parents were also frequently absent from the home. During the past three months, one-fifth of the youths had been physically assaulted and another fifth had been raped or sexually assaulted.

Runaways also describe ineffective ways of coping with stress. The overwhelming majority (83 percent) of the runaways interviewed by Roberts (1982) reported that they solve personal problems by taking drugs and alcohol, leaving the house temporarily, crying, attempting suicide, trying to forget it, and running away. In contrast, the methods used by the nonrunaways—thinking it through, talking to mother, a friend, or an older brother, writing about it—are generally more effective. Robert's study helps us understand runaway behavior as a combination of excessive life stress and poor coping strategies. It seems unlikely, however, that many adolescents have the coping strategies needed to handle such high levels of family stress.

CONSEQUENCES OF RUNNING AWAY

Running away may be viewed as the only solution for some runaways, but it generally leads to even more serious difficulties. Whereas some runaways return home, obtain the adult guidance they needed, settle back into school, and gain a new appreciation of their families, the runaways who remain on the street often become the victims of pimps, drug pushers, and other criminals. Most runaways have no job skills, no money, no place to stay, and no plan about how to survive. They are easily exploited sexually and often become involved in crime. Pimps learn of the places where runaways arrive and gather in the cities and greet them

AN APPLIED VIEW

Warning Signs of Running Away

Teens who are thinking about running away often display behaviors that can serve as warning signs. Teens who display several of these signs need help from mental health professionals. These include:

- Sudden changes in friends, mood, behavior, or habits
- A sudden lack of interest in school or truancy from school
- Increased rule breaking, rebellion, or outbursts of temper
- Accumulating money or other possessions
- Talking about running away or about friends running away
- Withdrawal from family and friends

- Taking drugs or drinking as a way to solve problems or relieve boredom
- High levels of depression, anxiety, or fear

As you have seen in the section about runaways, running away can lead to very serious consequences. Therefore, if you suspect a teen of preparing to run away, you must notify a competent professional, even though you may feel like you are betraying an adolescent's confidence. Such professionals include the teen's school principal, guidance counselor, clergyperson, and/or psychotherapist.

with offers of money and a place to stay. Half or more of the runaways who remain on the streets for more than a month become involved in prostitution. The unfortunate result is that teens who run away from home to escape physical and sexual abuse often become repeated victims of abuse.

Related to their involvement in sexual activity is the concern that runaways are at risk for AIDS. One survey of runaways (Rotheram-Borus & Koopman, 1991), revealed that 65 percent of the youths had been sexually active within the past three months. The median number of sexual partners during that period was 2.7 for runaway males and 1.3 for runaway females. Only 18 percent of these youths (those who were also most knowledgeable about AIDS prevention) indicated that they used condoms consistently. A screening completed at a large shelter for runaways in New York City found 7.4 percent of the adolescent boys and 5.4 percent of the adolescent girls to be HIV positive (Stricof & others, 1988).

Runaway youths report high levels of physical and emotional difficulties. According to Robertson (1989), 57 percent of runaways report problems in getting enough food to eat and 19 percent describe serious medical problems for which they have not been able to get treatment. Seventy percent of the runaways studied by Robertson said that they were depressed or had behavior problems and 49 percent reported that they had attempted suicide within the past year. Rotheram-Borus (1993) interviewed 576 consecutive admissions to publicly funded runaway programs in New York City. Thirty-seven percent of the youths had previously made a suicide attempt. Forty-four percent of those attempts had occurred within the past month. Although the females were more likely than the males to be depressed and attempt suicide, the rate of current suicidal thinking and planning was as high for males as females. For some teens, suicide may be a response to the difficulties encountered on the streets. For others, it may be another ineffective strategy for escaping the stressors in their lives at home.

street kids
Runaways who join with other runaways and become skillful at fending for themselves through involvement in illegal activities.

Although most runaways face a grim existence on the streets, some join with other runaways and manage to survive. These teens, known as **street kids**, become skillful at fending for themselves through involvement in illegal activities. A small number of runaways maintain a hopeful attitude about the future and continue their education on a part-time basis (White, 1989). They become resourceful, self-reliant, and self-confident, and are examples of the resilient youths discussed in Chapter 11.

TREATMENT APPROACHES

As society's understanding of runaways has changed, so has its treatment approaches. When running away was thought of as the behavior of delinquent and emotionally disturbed adolescents, runaways were most often sent to youth detention facilities or to psychiatric hospitals. Now, runaway behavior is increasingly viewed as an effort to escape abusive and conflicted family situations. Consequently, treatment efforts are being directed toward helping runaway adolescents develop the coping strategies, problem-solving skills, and support systems needed to resolve stressful family situations.

runaway shelters
Temporary shelters for youths who have left home.

The **runaway shelter** was developed in the 1960s to provide temporary shelter for youths who left home as part of the youth culture and the social protest and hippie movements. Runaway shelters continue to be an important part of the treatment system for runaways. They exist as a place where runaways are provided with individual counseling and, later, family therapy as efforts are made to work out family conflicts. Runaway counselors try to help family members improve communication and problem-solving methods, reestablish feelings of love, and help the adolescent become independent and responsible while remaining in the family (Palmer & Patterson, 1981). Given the high rate of depression and suicide among runaways, shelters need to screen for these serious problems (Rotheram-Borus, 1993).

Runaway shelters are intended to provide treatment for a short period of time, usually 14 to 30 days, before the runaway returns home. Sometimes, after 30 days of intensive counseling, it is evident that the conflicts between the parents and the adolescent are too severe for the adolescent to return home. A series of alternative living arrangements, including youth homes, foster homes, and supervised apartments, have been developed to serve these youths. Within these programs, runaways are taught skills such as budgeting, job interviewing, and personal and health care that will help them live on their own. Although these programs are helpful to many runaways, too few exist to serve the growing numbers of runaway youths. Other runaways are too fearful or distrustful to make use of the shelter system.

A number of additional services to runaways and their parents have been instituted in recent years. Among these experiments are toll-free phone numbers, which are open on a twenty-four-hour basis to counsel runaways; school programs that explain the causes and problems of running away to teenagers; and conferences, training sessions, and literature made available to the parents of runaways (see Appendix).

For those teens who do return home, family therapy may be indicated to improve family communication and reduce those problems that led the teen to leave home. Running away behavior may sometimes be a teen's effort to obtain a safe time-out from family problems. Family therapists can work with shelters and other community agencies to begin family therapy before the teen returns home (Gavazzi & Blumenkrantz, 1991).

Burgess (1986) believes that more needs to be done to reduce the negative consequences of running away. She has recommended a number of efforts.

- Runaways need safe, alternative places to live.
- They need immediate relief from the stress they are experiencing. The relief provided needs to counter the relief offered by drugs and alcohol.
- Runaway youths need ways to earn money so they are not forced to become involved in prostitution and drug sales.
- They need alternative education services so that they can become more employable.

PROSTITUTION

❑ "It's not so bad, honey," Sherry said. "Flatbackin' ain't the worst thing can happen to ya."

Although there are no reliable statistics concerning the number of teenage prostitutes, many professionals believe that teenage female prostitution has been growing in recent years and that male teenage prostitution has been growing at an even faster rate. There is a definite relationship between the number of runaways and the number of teenagers who turn to prostitution to support themselves. It has been difficult to determine the seriousness of this problem because of the variety of legal definitions of prostitution. Statistics may be distorted by the fact that most police officers are extremely reluctant to charge a female teenager with prostitution but are much less hesitant to do so with a male.

Psychologists disagree as to the reasons that bring a teenager to prostitution. The psychoanalytic school believes that prostitutes have a highly negative self-image, usually because of rejection by the father. Prostitution is a symbolic way to degrade oneself and a way to defend against the need for love. Prostitution also serves as the defense mechanism of compensation (see Chapter 2). It can make the teenager feel free of the internal conflicts and anxieties of being unloved. Erikson's theory explains prostitution as a negative identity. It represents a rejection of society's values in general, as well as a rejection of self.

The old idea that most prostitutes are nymphomaniacs who simply cannot get enough sex seems to be pretty well discounted among the prostitutes themselves (Kurz, 1977). Kurz interviewed many teenage prostitutes and found that for almost all of them sex is decidedly unpleasant and something that they do because they see it as the only way they can survive on their own.

While there does not seem to be any one condition that leads to adolescent prostitution, Schaffer and DeBlassie (1984) identified four background characteristics commonly found among adolescent prostitutes.

> For many years, female runaways have turned to prostitution to support themselves. Also of great concern is the increasing number of teen male prostitutes.

- *Alienation.* Gibson-Ainyette and others (1988) found that adolescent female prostitutes expressed less moral concern and more cynicism than other adolescents, including delinquents. They tend to be alienated from their families, to reject conventional values, and to be distrustful of most people. Williams and Kornblum (1985) believe that many teenage prostitutes have lacked adult role models, such as aunts and mothers, from whom they could learn traditional moral values. The examples set for them suggested instead that traditional values such as marriage, education, and employment were unimportant. The following excerpt sums up one adolescent prostitute's feelings of alienation.

❑ I hate what society considers normal. So, I find other ways of living with this world, without letting it bother me. If it bothers others, that's their problem. Every man for himself. When it comes to money you will find very few are going to help you make it. And if you're the type that helps others, you'll find yourself taken for a sucker. So you resign to helping yourself (Willams & Kornblum, 1985, p. 65).

- *Physical and sexual abuse.* Parental neglect and abuse is considered to be common among girls who become prostitutes. Some prostitutes were sexually abused as children or as adolescents by adults outside their family. In addition, many prostitutes described a series of negative sexual experiences, including rapes and repeated "one-night stands" (Gibson-Ainyette & others, 1988). As a result of their negative sexual experiences, many female adolescent prostitutes hold very negative attitudes about themselves and their bodies and about men.
- *Education and employment.* Many teenage prostitutes have a history of academic failure, school absenteeism, and dropping out. Among the adolescent prostitutes studied by Gibson-Ainyette and others (1988), 65 percent were in special education classes at some point in their schooling. Failure in school contributes to feelings of alienation from traditional values and institutions. Following school failure, adolescent prostitutes sometimes find a sense of personal worth among peers on the streets. With poor academic credentials, school dropouts have few ways to support themselves. Much of the work available to them is boring and pays poorly. Prostitution offers high income, without taxes or time cards, and may bring some feeling of success, power, and importance. In some poor communities, prostitutes, pimps, and drug dealers are the only ones who earn much money. Many runaways become involved in prostitution as a means of economic survival. Teenagers who get involved in using drugs may come in contact with pimps and prostitutes in the process of buying drugs. They may then find that prostitution is an easy way or the only way to finance their expensive drug habit. Other teenagers enter into prisons and detention centers for other crimes and are introduced to prostitution while in jail. After leaving jail, this may be the most sure means of financial support. Although prostitution may bring in much money at first, many prostitutes learn that they lose much of their earnings to their pimp.
- *Family.* In addition to physical and sexual abuse, the family backgrounds of prostitutes are marked by other difficulties. They often describe their parents as unaffectionate and feel closer to peers. The peer groups to which they attach themselves are often a negative influence, however. Parental absence and a lack of adult supervision in early adolescence are also common. High

AN APPLIED VIEW

Prostitution and the Family

Family problems are common in the prostitute's life. Freudenberger (1973) cites the case of an 11-year-old girl, Maria, who was introduced to prostitution by Dolores, an older girl. A member of a large family, Maria received little love or attention from her parents, but she was expected to be a useful provider for the family. She turned to Dolores, who enjoyed taking care of her. Maria became familiar with the prostitute's life and admired Dolores for it. When Dolores died from an overdose of drugs, Maria's sense of personal isolation and loss drove her to become a prostitute.

WHAT'S YOUR VIEW?

Should Nonaggressive Offenders Be Treated as Criminals?

Some people have suggested that because their crimes are victimless, nonaggressive offenders should not be prosecuted. They argue that because these youths mainly hurt themselves, they are really the victims of their "crimes." They need help, not punishment by incarceration. In fact, the parents who drive their children from their homes, and the "johns" who solicit the sexual favor of teenagers are the ones who should be punished by the law. When nonaggressive offenders are placed in the juvenile justice system, they are exposed to prostitution, homosexuality, rape, beatings, and other conditions, which add to their problems and decrease the likelihood of rehabilitation.

Parents of runaways or the wives of husbands who catch a venereal disease from a prostitute would likely disagree. The parents of the runaway, faced with an incorrigible child, typically feel helpless. They feel that without police involvement, they have no place to turn to get their child back. Furthermore, it is argued that prostitutes should be treated criminally because they violate the societal rule that the family must be respected—anyone who tempts people to disregard their marriage vows must be punished for it.

You be the judge. Which side is right, or is there a third possible solution?

rates of parental separation, divorce, and foster home placements are found among adolescent prostitutes. Some teenage girls try to obtain the attention and affection they feel is lacking in their families through sexual gratification. For some, the pimp is the only person who has ever cared. Although he may also be abusive, he offers security, manages the money, and may provide a place to live.

These life circumstances seem to contribute to much personal unhappiness and psychological distress. Gibson-Ainyette and others (1988) found that adolescent prostitutes report poor self-concepts and high levels of depression, anxiety, and other psychological symptoms. The levels of psychopathology reported by prostitutes were much higher than those of delinquents, suggesting that prostitution should be viewed more as a mental health problem than as a criminal problem.

chickens
Young male prostitutes.

chicken hawks
Older male homosexuals who pay young boys for prostitution.

Young male prostitutes are called **chickens** and those who specialize in buying young boys for prostitution are called **chicken hawks**. Although many male prostitutes describe themselves as heterosexual, they are engaged as prostitutes in homosexual activity. Little research has been done with adolescent male prostitutes, but they are, like female adolescent prostitutes, likely to have been abused or neglected in childhood. As prostitutes, males have a lonely existence on the street, whereas female prostitutes often develop their own support networks (Schaffer & DeBlassie, 1984).

C O N C L U S I O N S

As you have learned, there are many different actions that could cause a person to be labeled a juvenile delinquent. The definitions, causes, and consequences of these acts are extremely intricate, and may differ widely from one part of the country to another. The good news is that the number of criminal acts committed by youths has been declining in recent years, but this fact seems to be offset by the growing number of violent crimes being committed by youth gangs. You can see that continuing research is necessary in all areas that relate to troubled young people, especially those teens who see illegal and destructive behavior as their only option.

What can society do to alleviate these problems?

Unfortunately, as Schneider (1990) points out, changes in social institutions that would provide greater opportunities for young people are often seen as too indirect and expensive. The juvenile justice system is relying more and more on severe punishment or the threat of such punishment to discourage criminal behavior.

Perhaps the answer, or at least part of it, lies in helping adolescents prepare themselves better for their role as adults. Industrialized countries do not seem to be doing this very well. In the next chapter, we examine the ways by which societies induct youth into adulthood, and consider some suggestions about how this might be improved.

C H A P T E R   H I G H L I G H T S

Introduction
- The three main types of antisocial behavior are exemplified by the juvenile delinquent, the aggressive gang member, and the nonaggressive offender.

Delinquency and Violence
- Although only a small number of teens are involved in acts of violence, statistics on youth violence are nevertheless alarming.
- Alienated, thrill seeking, uninhibited, fearless, aggressive, and lacking problem-solving skills are some of the individual characteristics associated with youths who are at-risk for violent and antisocial behavior.
- Coercive patterns of family interaction are often a precursor to delinquency and crime in adolescence.
- Chronic academic underachievement and school failure are highly correlated with delinquent behavior.
- American culture has been criticized for supporting the expression of violence.
- Teens from ethnic minority backgrounds, gay and lesbian teens, youths with disabilities, and females are more likely than other teens to become victims of violence.
- Violence prevention efforts are increasing because of increased understanding of the factors that contribute to violent and aggressive behavior.
- Drug use, mental health problems, and delinquency are related, but the relationship is complicated.

- Many of the myths surrounding the association between African American male youths and delinquent behavior are inaccurate.

The Aggressive Gang
- Gangs typically have a high degree of cohesion and organization, a consistent set of norms, clearly defined leaders, and coherent organization for warfare.
- Gangs often spring up after ethnic group immigration to the United States, in response to feeling alienated from the dominant culture in which they find themselves.
- Gang joiners typically have certain characteristics, such as intelligence, self-motivation, and goal-directedness, as well as a greater enjoyment of violence and need for acceptance by others.
- The three types of gangs are cultural, instrumental, and home invader gangs.
- Gangs have become much more violent in the past decade, probably in response to increased drug trafficking.

The Nonaggressive Offender
- Some of the common causes for running away are a stressful home environment (verbal and/or physical abuse), being ordered to leave by parents, getting "lost in the system" of government intervention, and serious personality or emotional problems.
- Runaways typically have poor coping skills, which ill prepare them for dealing with the generally higher levels of stress that they are exposed to.

- Treatment of runaways includes providing them with coping strategies, problem-solving skills, and more effective support systems.

- Causes of prostitution include alienation from society, physical and/or sexual abuse, undereducation, lack of legitimate employment, and family difficulties.

K E Y   T E R M S

aggressive gang 431
antisocial behaviors 431
castaways 456
chicken hawks 464
chickens 464
coercive cycle 436
conflict mediation 453

cultural gangs 451
culture of violence 440
drive-by shootings 449
home invader gangs 451
instrumental gangs 451
juvenile delinquent 431
nonaggressive offenders 431

pushouts 456
recidivism 452
runaway shelters 461
street kids 460
system kids 460
throwaways 456

W H A T   D O   Y O U   T H I N K ?

1. Try to remember every illegal act you ever committed (we hope there won't be too many of them). Can you imagine how any one of those acts could have led you into becoming a juvenile delinquent?

2. In what situation do you think it is proper to try a juvenile delinquent as an adult? What are your reasons?

3. If you were the chief of police in a medium sized town, what specific actions would you take to combat adolescent crime?

4. If you were a mayor of a large city, what specific actions would you take to combat adolescent crime?

5. What role should schools play in combating delinquent behavior?

6. The consequences of running away from home can be dire. What do you think should be done with a runaway who has been caught by the police?

7. Some people say that because prostitutes only hurt themselves, they should not be treated as criminals but rather as emotionally troubled. What's your position?

S U G G E S T E D   R E A D I N G S

Garbarino, J., Dubrow, N., Kostelny, K. & Padro, C. (1992) *Children in Danger: Coping with the consequences of community violence.* San Francisco: Jossey-Bass. A psychological understanding of the consequences of violence on the development of children and adolescents.

Hinton, S.E. (1967). *The Outsiders.* New York: Dell. This remarkable book describes gang rituals, class warfare and the coming of age in this truly intense environment. Another popular book by this author is *That was Then, This is Now* (1971), a sequel to *The Outsiders.*

Rodriguez, L. (1993). *Always Running.* Willimnatic, CT: Curbstone Press. Describes in vivid detail life among the "Crips" and the "Bloods" two rival gangs that started in the barrios of Los Angeles and have since branched out. Another excellent book on the same subject: Bing, L. (1991). *Do or Die.* New York: Harper Collins.

Shulman, I. (1981). *West Side Story.* New York: Pocket Books. This novelization of the classic musical is a primer on how it feels to be conflicted over loyalties to ethnic group, gang and love. The story is adapted from Shakespeare's famous play *Romeo and Juliet.*

GUIDED REVIEW

1. The tendency to behave aggressively has been linked to individual differences in all of the following areas, EXCEPT
   a. temperament
   b. social cognition
   c. high need for social approval
   d. attraction to dangerous activities

2. Drug and alcohol use may be associated with acts of violence for a number of reasons, EXCEPT
   a. both are ways to relieve boredom
   b both are psychotic processes
   c. both serve a tough guy image
   d. many drug users become involved in robberies

3. McLoyd suggests that poverty may contribute to learned aggression, because it interferes with consistent and supportive parenting. Patterson labeled factors that have a negative effect on parenting as
   a. disruptors
   b. breakers
   c. weights
   d. disabilities

4. In the inner city, poverty contributes to violence through
   a. concern for lack of food
   b. perceptions of inequity and lack of opportunity
   c. lack of money
   d. teenage pregnancy

5. When both gang members and young people who thought they might join a gang were asked the reasons for joining a gang, the number one reason was
   a. "because I have nothing else to do."
   b. "so people will look up to me."
   c. "for personal protection."
   d. "to have more friends."

6. Most of the youths who join gangs are
   a. irrational and conceited
   b. suspicious and paranoid

   c. bored, lacking skills and possessing below-average intelligence
   d. intelligent, self motivated and goal oriented

7. Efforts are being made to teach juvenile offenders how to settle conflicts without resorting to violence using a technique called
   a. conflict mediation
   b. confrontation analysis
   c. dispute arrangement
   d. third-party intervention

8. Runaway behavior is increasingly viewed as a way for adolescents to
   a. seek better economic and financial opportunities
   b. avoid the responsibility placed on them by adults
   c. free themselves to engage in criminal behavior
   d. escape abusive and conflicted family situations

9. Sarah was removed from her parents' home by the department of social services because of parental neglect and abuse and was placed in a series of foster homes. Unhappy with her life in foster homes, Sarah ran away. Runaways like Sarah are commonly referred to as
   a. system kids
   b. pushouts
   c. street kids
   d. throwaways

10. Young male prostitutes are called
    a. chicken hawks
    b. johns
    c. chickens
    d. tricks

11. According to Patterson's research _____ (demanding/coercive) family interactions contribute to aggressive behavior.

12. According to Hill and associates many ethnic cultural traditions _____ (promote/oppose) community violence.

13. Among both inner-city and suburban teens _____ ((males/females) between the ages of 11 and 24 report more often than _____ (males/females) of the same age being the victim of violence, knowing victims of violence and witnessing violent acts, with the exception of sexual violence.

14. Aggression can be learned by viewing film and television through _____ (repression/observational learning).

15. What makes a gang particularly cohesive is its function as a _____ (police/family) for some adolescents whose strong dependency needs are displaced on the peer group.

16. _____ (Cultural/Instrumental) gangs are loosely organized, are centered in a neighborhood, and are usually made up of a single cultural or ethnic group, newly arrived in this country.

17. Females constitute the majority of teenagers who meet the definition of _____ (juvenile delinquent/nonaggressive offender).

18. What most distinguishes the street gang member from the nonmember is the _____ (ethnicity/enjoyment of violence).

19. A group of runaways who have troubled family environments and leave home because they have been abandoned, neglected, or abused by their parents and whose parents often do not oppose their child's leaving are commonly referred to as _____ (street kids/pushouts).

20. _____ (Freud's/Erikson's) theory explain prostitution as a negative identity.

Answers

1. C, 2. B, 3. A, 4. B, 5. D, 6. D, 7. A, 8. D, 9. A, 10. C, 11. coercive, 12. oppose, 13. males, females, 14. observational learning, 15. family, 16. Cultural, 17. nonaggressive offender, 18. enjoyment of violence, 19. pushouts, 20. Erikson's.

The End of Adolescence

INITIATION INTO ADULTHOOD

Life is difficult. This is a great truth, one of the greatest truths. It is a great truth because once we truly see this truth, we transcend it. Once we truly know that life is difficult—once we truly understand and accept it—then life is no longer difficult. Because once it is accepted, the fact that life is difficult no longer matters.

M. S. Peck, *The Road Less Travelled*

The young men standing in the living room of the old fraternity house had solemn faces. As the fraternity president began intoning the sacred words that would lead to their induction, Dave and Bill glanced cautiously at each other. The two were in the front row, waiting along with nine other sheepish freshmen "pledges." Each knew the other was remembering the same thing—the long ordeal of their pledge period, which had begun at the start of the semester.

For weeks they had had to wear ridiculous beanies on their heads, and act as virtual slaves to the fraternity brothers. Despite their best efforts to obey the complex rules, each had numerous violations, which the brothers had noted in the young men's pledge books.

Last night, for the initiation opener, they had endured one blow across their buttocks from a thick magazine (rolled up and taped for the purpose) for each of their rules violations. Bent over and holding their ankles, they had managed to get through the beatings without crying out, but not without becoming very black-and-blue. Then they were taken together and dropped off in groups of three in the middle of a dark woods, with instructions to get back to the frat house by 10 a.m. if they hoped to be initiated.

After numerous mishaps, all made it to the main road and hitched into town. On their return to the house, they thought their initiation was finished, but the ordeal was far from over. Next came a series of lesser trials called a **hazing**. It included:

hazing
The often dangerous practices used by some fraternities to initiate new members.

- Being made to lie on their backs while tablespoons of baking soda and then vinegar were poured into their open mouths. They were ordered not to swallow, to close their mouths and keep them that way no matter what. The brothers laughed uproariously when, inevitably, the mixture exploded, shooting a geyser from their tightly-pressed lips high into the air.

- Being blindfolded and made to eat warm "dog manure." Actually they had eaten doughnuts soaked in warm water, but the bag of manure held under their noses made them believe it was the real thing.

- Having a mixture of Liederkranz cheese and rotten eggs smeared on their upper lips, then being made to run around while inhaling the dreadful odor.

Now, as the torture was over, and the final ceremony underway, both Dave and Bill had the same thought in their minds: I've done it! I've survived! I'm in!

initiation rite
A cultural and sometimes ceremonial task that signals progress to some new developmental stage.

What you have just read is a description of an **initiation rite** that actually took place some years ago. In recent years, the horrors of the fraternity initiation have been softened by legal restrictions and by more humane attitudes. For example, Baier and Williams (1983) surveyed 440 active members and 420 alumni members of the fraternity system of a large state university. They compared attitudes of these men toward 22 hazing practices (such as those just described) known to be used by the frats. They found the active members were considerably more opposed to the practices than the alumni.

Nevertheless, most of us have heard of recent cases of maimings and even deaths of young men who have been put through hazing not only by fraternities but in the military as well. A recent report of hazing aboard a Coast Guard cutter (Brelis, 1998) describes humiliating and degrading acts of verbal and physical abuse to new recruits. Although many military groups have anti-hazing policies, hazing activities are reported to be common.

Initiation Rites

Have you ever participated in an initiation rite? What effect did it have on you?

In this chapter, you learn about the role of initiation rites in fostering the journey from adolescence to adulthood. You see how such rites function in preindustrial societies and examine the effects of their presence and absence in our culture. Keep in mind as you are reading that initiation rites were primarily designed for a different transition than we experience today, namely from childhood to adulthood. Urban-industrial life has made possible the addition of adolescence, a developmental phase that is constantly being studied and discussed. The chapter ends with the description of several activities that could serve as initiation rites for our youths.

After reading this chapter, you should be able to:

- Discuss the purpose of initiation rites and the effects on adolescents of the absence of a formal rite of passage into adulthood in our culture.
- Describe the components of some initiation rites in preindustrial societies.
- List activities in our own society that may be parallel to these rites.
- Compare and contrast the initiation rites of American adolescents in the industrial era with those of today.
- Discuss the concept of the adolescent moratorium and its strengths and weaknesses.
- Specify some high school activities that could serve some of the purposes of an initiation rite.
- Form a personal list of characteristics you believe necessary for a person of either gender to exhibit to be considered a mature adult.
- Discuss these issues from an applied, sociocultural, and your own point of view.

INITIATION RITES IN OTHER CULTURES

Why do some people, and the groups they wish to be associated with, place so much value on holding initiation rites? And why are so many adolescents, many of them otherwise highly intelligent and reasonable, willing and eager to endure such pain? Is it simply because they want to join the group, to feel that they belong? There seems to be more to it than that. Throughout the world, adolescents readily engage in such activities because they seem to want to be tested, to prove to themselves that they have achieved the adult virtues of courage, independence, and self-control. And the adults seem to agree that adolescents should prove they have attained these traits before being admitted to the "club of maturity." Compare the activities that Dave and Bill were put through to those of two members of a primitive African tribe, Yudia and Mateya, described in the Sociocultural View box.

A SOCIOCULTURAL VIEW

Yudia and Mateya Come of Age in Africa

Yudia cannot believe how rapidly her feelings keep changing. One moment she is curious and excited, the next, nervous and afraid. Tonight begins her *igubi*, the rite that celebrates her induction into adulthood. Yudia has longed for this day most of her 11 years, but now she wonders if she really wants the responsibilities of a grown-up.

Though it seems much longer, only a week has passed since the excruciating beginning of her initiation. The memory of it is already dimming: the bright fire, her women relatives pinning her down on the table, her grandmother using a thin sharp stone to cut out her clitoris, the searing jolts of pain.

The women had held and consoled her, empathizing fully with her feelings. Each had been through the same agony. For them, too, it occurred shortly after their first menstruation. They had explained to her that this was just the beginning of the suffering she must learn to endure as an adult woman. All during the past week, they had been teaching her about the pain her husband would sometimes cause her, about the difficulties of pregnancy and childbirth, about the many hardships she must bear stoically. For she is Kaguru, and all Kaguru women accept their lot in life without complaint.

It has been a hard week, but tonight the pleasure of the igubi will help her forget her wound. There will be singing, dancing, and strong beer to drink. The ceremony, with its movingly symbolic songs, will go on for two days and nights. Only the women of this Tanzanian village will participate, intoning the time-honored phrases that will remind Yudia all her life of her adult duties.

In a large hut less than a mile from the village, Yudia's male cousin Mateya and seven other 13-year-old Kaguru boys huddle together, even though the temperature in the closely thatched enclosure is a stifling 110 degrees. Rivulets of sweat flow from their bodies and flies dot their arms, backs, and legs. They no longer pay attention to the flies, nor to the vivid slashes of white, brown, and black clay adorning all their faces. Their thoughts are dominated by a single fear: will they cry out when the elder's sharpened stone begins to separate the tender foreskin from their penises? Each dreams of impressing his father, who will be watching, by smiling throughout the horrible ordeal.

Three months of instruction and testing have brought the young men to this point. They have learned many things together: how to spear their own food, how to tend their tiny gardens, how to inseminate their future wives, and most importantly, how to rely on themselves when in danger. The last three months have been exhausting. They have been through many trials. In some, they had to prove they could work together; in others, their skill in self-preservation was tested. For most of them, being out of contact with their mothers was the hardest part. They have not seen any of the female members of their families since they started their training. Unlike Yudia's initiation, which is designed to draw her closer to the adult women of the tribe, Mateya's initiation is designed to remove him forever from the influence of the females, and to align him with the adult men.

Now it is evening. Mateya is the third to be led out to the circle of firelight. Wide-eyed, he witnesses an eerie scene. His male relatives are dancing in a circle around him, chanting the unchanging songs. The grim-faced elder holds the carved ceremonial knife. Asked if he wishes to go on, the boy nods yes. Abruptly the ritual begins: the hands of the men hold him tight; the cold knife tip touches his penis; a shockingly sharp pain sears his loins; he is surprised to hear a piercing scream; then, filled with shame, he realizes it comes from him.

Thus far, Yudia's and Mateya's initiations have been different. Mateya's has been longer and harder than Yudia's. She is being brought even closer to the women who have raised them both, but Mateya must now align himself with the men. (Freud stated that, unlike females, all males must give up their identification with their mother; this makes adult males subconsciously doubtful of their sexual identity.)

The initiations are similar, though, in that both youths have experienced severe physical pain. In both cases, the operations were meant to sensitize them to the vastly greater role that sex will now play in their lives. Furthermore, their mutilations made them recognizable to all adult members of the Kaguru tribe. At this "coming out" ceremony, males and females also receive new names, usually those of close ancestors. This illustrates the continuity of the society. The beliefs of the tribe are preserved in the continuous flow from infant to child to adult to elder to deceased and to newborn baby again. When all is done, Yudia and Mateya can have no doubt that they have passed from childhood to adulthood.

A SOCIOCULTURAL VIEW

Initiation Rites in Europe

The problem of initiation rites is not limited to the United States. For example, in his study of French juvenile delinquents, Garapon (1983) saw a "symbolic, sacrificial dimension" to their crimes. He notes that cars and other stolen goods were often either dumped in the canals or burned, which he feels is similar to the sacrifices of prized goods that preindustrial tribes carry out with water or fire. He points out other links to tribal initiations: most crimes are committed at night, and offenders wind up in courtrooms surrounded by symbolic costumes such as the judge's robes.

In Germany, Zoja (1984) has suggested similar parallels in the path to drug addiction. He states, "Drug addiction can be an active choice, allowing the user to acquire a solid identity and social role, that of the negative hero, as well as access to an esoteric glimpse of an 'other world' " (p. 115). Obviously, in countries throughout the world, drug use serves as another type of initiation rite.

Analysis of an Initiation Rite

Before discussing the implications of American initiation rites (or the lack of them), we'll provide a more detailed description of the purposes and components of such rites.

The first analysis of initiation ceremonies in preindustrial societies such as the Kaguru was completed by anthropologist Arnold Van Gennep in 1909 (Van Gennep, 1909/1960). His explanation is still regarded highly, as can be seen in more recent studies (e.g., Anderson & Noesjirwan, 1980; Brain & others, 1977; Hill, 1987; Kitahara, 1983; Lidz & Lidz, 1984; Morinis, 1985; Ramsey, 1982). Van Gennep argued that the purpose of the initiation rites, as with all rites of passage (marriage, promotion, retirement, and so on), is to cushion the emotional disruption caused by a change from one life status to another. Hans

Sebald (1992) agreed with this interpretation, further delineating three functions that rites of passage serve:

> *Informing*—the individual is informed of his or her new rights and duties. A special training period may be involved, with the indoctrination led by select elders of the community.
> *Announcing*—the new status of the individual is announced to the community. The group is now expected to interact with the initiate in accordance to this new role.
> *Emotionally anchoring*—A profound sense of belonging, both to the community and to the new status, will illicit loyalty and commitment. The public ritual provides the individual with important validation and reinforcement. In the future, the initiate "does not play the new role, he or she *becomes* the role" (p. 115).

Although the ritual or celebration may differ between cultures, all rites of passage are designed to move the individual from one stage of human experience to the next.

For males, this transition also involves the end of dependence on their mothers and other older women and the beginning of their inclusion in the world of men. This ceremony is often scheduled to coincide with the peak in adolescent physiological maturation, and therefore has often been called a **puberty rite**. Van Gennep argues that this is inappropriate, because initiation may be held by one tribe when the children are 8 years old, and in another at 16. Children of 8 have not yet started puberty; those of 16 are halfway through it. Also, the age at which puberty starts differs from individual to individual, and now occurs approximately three years earlier than it did 100 years ago (see Chapter 3). Nevertheless, the initiation rite is usually held at one age within each tribe, regardless of the physical maturity of the individual initiates.

puberty rite
An initiation ceremony often scheduled to coincide with the peak in adolescent physiological maturation.

Serving as an introduction to sexuality and separation from mother is one purpose of the initiation rite. Several other purposes have been suggested. In his classic text *Totem and Taboo* (1914/1955), Sigmund Freud offered the psychoanalytic explanation. In his view, such ceremonies are necessitated by the conflict between fathers and sons over who will dominate the women of the tribe. Adolescent males are seen as challenging the father's authority and right to control the women.

To make clear their supremacy in the tribe and to ensure the allegiance of the young males, the adults set a series of trials for the youths at which the adults are clearly superior. The ultimate threat held over the young is castration, the loss of sexual power. Most rites include trials of strength, endurance, prowess, and courage. These usually involve forced ingestion of tobacco and other drugs, fumigations, flagellations, beatings with heavy sticks (running the gauntlet), tattooing, cutting of the ears, lips, and gums, and that most Freudian of inflictions, the circumcision of the foreskin of the penis.

The message is clear: "We, the adult males, are in charge. Join us and be loyal, or else!" Psychologist Bruno Bettleheim (1969) agreed with Freud that there is a fear of castration among the males, but argues that the main role of the initiation rite is to ease the stress of becoming an adult, not to exaggerate it.

The Passage to Adulthood

In this section, we examine the transition to adulthood in the United States. We also consider the effects of the absence of a definitive initiation rite.

THE TRANSITION TO ADULTHOOD IN THE UNITED STATES

In the industrial past of the United States, it was fairly clear when one became an adult. In their late teens, boys and girls usually got married and assumed an adult role. Males were accepted as partners in the family farm or business; females became housewives. This has changed in many ways. What Black (1974) has suggested is still true:

> ❏ Today, in modern society, initiation of the boy and girl into adult life is far more complicated. Society is fast, heterogeneous, a network of interdependent groups with many different backgrounds, traditions, and outlooks, the products of religious, racial, national, and class differences. In our age of technology, the young have to learn to deal with cars and trains and planes, machines and other electronic equipment, typewriters, television sets, computers, and mass production assembly lines. They face high concentrations of population, high mobility, and relationships on regional, national, and global levels. All this they have to know and understand at a time when customs, laws and institutions are undergoing drastic and rapid change in the midst of a high degree of human differences and human conflict (p. 25).

There are no specific rituals comparable to those in preindustrial societies to help Western youths through this difficult period. For example, religious ceremonies like confirmation and bar and bas mitzvah no longer seem to play the role they had in earlier times. Kilpatrick (1975) argues that

> ❏ At some point we grew too sophisticated, at some point the rituals lost their vitality and became mere ornaments. We may still keep their observance, but they are like old family retainers, kept on in vague remembrance of their past service (p. 145).

In his anthology of adolescent literature, Thomas Gregory (1978) points out that many modern writers find the decline of the initiation rite in the United States a noteworthy theme. He describes their thinking as follows:

> ❏ Today's adolescents are faced with not knowing when they have reached maturity. . . . the absence of a formal rite of passage ceremony necessitates a larger and more uncertain transition, with much groping, as adolescents not only try to establish their new adulthood, but also their identity (p. 336).

WHAT'S YOUR VIEW?

Did You Go Through Initiation Rites Yourself?

Do we have initiation rites in United States society? In the spaces below, write down all the initiation rites that you have participated in that you can remember. Did these rites help you become an adult? Were they sufficient? Were they formal or informal? Taken together, do they indicate an American definition of maturity? Some activities that may be considered initiation rites are suggested in the text below.

1. _____

2. _____

3. _____

4. _____

5. _____

6. _____

7. _____

8. _____

9. _____

10. _____

TYPES OF INITIATION ACTIVITIES IN THE UNITED STATES

This is not to say that Americans have no activities that signal the passage to maturity. We have numerous types of activities, which usually happen at various stages and ages of adolescence. Here is a list of the types and some examples of each, some of which will be discussed later:

- *Religious*
    Bar mitzvah or bat mitzvah
    Confirmation or baptism
    Participating in a ceremony such as reading from the Bible

- *Sexual*
    Menarche (first menstruation)
    Nocturnal emissions (male wet dreams)
    Losing one's virginity

- *Social*
    Sweet sixteen or debutante parties
    Going to the senior prom
    Joining a gang, fraternity, or sorority
    Beginning to shave
    Being chosen as a member of a sports team

Graduation from high school is one of the few clear-cut initiation rites we have in our society today.

Moving away from one's family and relatives
Joining the armed forces
Getting married
Becoming a parent
Voting for the first time

- *Educational*
  Getting a driver's license
  Graduating from high school
  Going away to college

- *Economic*
  Getting a checking or credit card account
  Buying one's first car
  Getting one's first job

- *Physical*
  Body piercing (ears, nose, etc.)
  Cosmetic surgery (e.g. breast implants, tummy tucks)
  Tattoos

THE ADOLESCENT MORATORIUM

adolescent moratorium
A time-out period during which the adolescent experiments with a variety of identities without having to assume the responsibility for the consequences.

In the late 20th century, the attitude that youths need a time-out period to explore possibilities and to continue education has become widespread. This phase of life is known as the **adolescent moratorium** (see Chapter 6).

Perhaps it is natural that we have discarded rites of passage into adulthood for the more leisurely moratorium. In preindustrial societies, children must take over the responsibilities of adulthood quickly. The survival of the tribe depends on getting as much help from all individuals as possible. In industrial societies, and even more so in our information-processing society, the abundance of goods makes it less necessary that everyone contribute to the society. There is also a need for more extensive schooling in preparation for technical types of work. The moratorium, then, comes about because of our advanced economic system. For these reasons, the initiation ritual has declined considerably since the 19th century. Is our society better or worse for this change?

IMPLICATIONS OF THE LACK OF AN INITIATION CEREMONY

Today we are having doubts that moratoriums are effective. In fact, it appears that crime is one of the ways that some youth are *initiating themselves* into adulthood.

Michael Ventura (1989) suggests a cause for the ineffectiveness of the moratorium and, consequently, this dangerous self-initiating behavior. Emerging at puberty within each person is a "craving for extremes," which represents all the inner intensity associated with adolescence. These extremes manifest themselves

The Influence of Media and Advertisements on Teen Values

The leisure time created for identity development in this country has made teens a lucrative target audience for advertising. Not only do they have the time, but many possess the resources as well. In a school setting, they are a captive audience. The presence, and some might say the intrusion, of advertising and the media into the classroom has brought the issue to public attention.

Although advertising has long adorned vending machines, athletic scoreboards, and school yearbooks, the addition of television news programs such as *Channel One* has alarmed some social scientists. (*Channel One* is a news program oriented to teens, with ten minutes of news and two minutes of advertising. Subscribing schools receive equipment and agree to let 90 percent of their students watch the program daily.) Studying the effects of such a program on 10th-grade students in a Michigan school, Greenberg and Brand (1993) found that students viewing Channel One were more aware of the news but also that they reported more materialistic attitudes than nonviewers.

In his book, Warning: Nonsense is Destroying America, Vincent Ruggiero (1994) quotes an advertising vice president as admitting, "You've got to reach kids throughout the day—in school, as they're shopping in the mall . . . or at the movies. You've got to become part of the fabric of their lives" (p. 122).

In your opinion, what role should advertising and media play in education?

in the radical behaviors common in adolescence (i.e., heavy metal music, jargon language code, unusual fashions and hairdos, and so on). These cravings for extremes produce an inherent need for social survival skills, which teenagers cannot provide for themselves. Thus, there is a need for initiation. According to Ventura, initiation rites satisfy the craving while providing the necessary survival skills for life in this world.

Although obviously not appropriate for our society, ancient tribal rituals such as those of Yudia and Mateya appear to satisfy the craving for extremes. This satisfaction is accomplished through the extreme nature of the ceremony (involving danger and pain) while the survival skills are given, at least psychologically, along with the new status of adult. Having proven themselves worthy by passing the rite, the youth are now considered to be among the mature.

By contrast, Ventura says, American culture "denies the craving [and] can't possibly meet the need" (p. 47). As a result, adolescence is prolonged and the eventual transition to adulthood is almost impossible to pinpoint. Ventura claims that adolescents are forced to

❏ generate forms—music, fashions, behaviors—that prolong the initiatory moment, i.e., that cherish and elongate adolescence—as though hoping to be somehow initiated by chance somewhere along the way (p. 47).

Do you accept the idea that gang warfare is really a form of self-initiation into manhood?

Thus, the tendency toward self-initiating behavior arises. In the most frustrated of adolescents, this self-initiation takes on criminal characteristics. One need only glance at the headlines of any urban newspaper (e.g., the Los Angeles riots) for examples of this kind of criminal behavior.

Adolescent males (and, increasingly, females) seem to need to do something dangerous and difficult. Males raised without fathers or father substitutes are especially vulnerable to the attractions of criminality. As they are leaving adolescence, many of them seem to feel that they must prove their adulthood by first proving their manhood in risk-taking behavior.

For much of their childhood, males in the United States are also highly dependent on female attention. This is often true among males whose fathers are absent. Such a youth sometimes compensates by:

> ❐ tribe-like gangs and undergoing harsh initiation rites, all in the service of proving his manhood. Much of the trouble that these youth get into serves the same function as primitive rituals. To compensate for the dominant role of mother in his childhood, the boy needs a dramatic event or a series of them to establish male identity (Kilpatrick, 1975, p. 155).

In the 1960s and early 1970s, many American youths sought to establish their identities by imitating the very rituals of the preindustrial tribes described earlier in this chapter. Known as hippies and flower children, they yearned for a return to a simpler life. Many of them moved to the wilderness, living on farms and communes away from the large cities in which they were brought up. Many totally rejected the cultural values of their parents. The most famous symbol of their counterculture was the Woodstock musical marathon in 1969. With its loud, throbbing music, nudity, and widespread use of drugs, it was similar to many primitive tribal rites. On the other hand, many of these teens also worked hard to promote such universal tribal values as cultural diversity and concern for the downtrodden. These self-designed initiation rites seem to have been unsuccessful as passages to maturity. Most of the communes and other organizations of the youth movement of the 1960s have since failed. Most American youths have decided "you can't go back again."

Organized sport is another example of our efforts to include initiation rites in American life. The emphasis on athletic ability has much in common with the arduous tasks given to preindustrial youths. In particular, we can see a parallel in the efforts by fathers to encourage their sons to excel in Little League baseball and Pop Warner football. Fathers (and often mothers) are seen exhorting the players to try harder, to fight bravely, and when hurt, to "act like a man" and not cry.

Thus, in delinquency, the counterculture, and in sports, we see evidence that members of several age groups today need the establishment of some sort of initiation rite. Adolescents and adults alike seem to realize that something more is needed to provide assistance in this difficult transitional period. But what?

Traditional initiation rites are inappropriate for American youths. In preindustrial societies, individual status is ascribed by the tribe to which the person belongs. Social scientists call this an **ascribed identity**. The successes or failures of each tribe determine the prestige of its members. Family background and individual effort usually make little difference. In earlier times in the United States, the family was the prime source of status, which was rather stable. For

ascribed identity
In preindustrial societies, individual status that is ascribed by the tribe to which the person belongs.

The American emphasis on athletic ability during the teenage years has much in common with the arduous tasks given as part of the initiation rites to youths in preindustrial cultures.

achieved identity
In industrial societies, individual status that is brought about by personal effort and early commitment to a career path.

Do you know anyone who has an ascribed identity? Anyone who has an achieved identity?

example, almost no children of the poor became merchants, doctors, or lawyers. Today, personal effort and early commitment to a career path play a far greater role in the individual's economic and social success. This is called an **achieved identity**. For this reason (and others), preindustrial customs are not compatible with Western youths today.

Nevertheless, the problem of knowing when to treat people as children and when to treat them as adults remains. This is a problem not only for adults but for the adolescents themselves. What can we do to resolve this serious problem? Some argue that we should lend more credence toward institutions already in place, namely, milestones celebrated within the school and family, and regard them as legitimate rites of passage.

Methods That Facilitate the Transition to Adulthood

THE HIGH SCHOOL AS A SETTING FOR THE RITE OF PASSAGE

Fasick (1988) suggests that secondary education is the best setting for the transition to adulthood. He asserts that secondary education "is necessary for responsible citizenship in modern society" (p. 467). Fasick presses his point with reference to guidelines established by Van Gennep (1909/1960; see previous discussion). Van Gennep proposed three universal steps in his rites of passage:

> ❑ separation, where individuals are given a special, secluded, and separate status from their community; a transitional period in which initiates are given instruction in basic cultural assumptions; and a reincorporation of each person into a socially recognized new status (Roy, 1990, p. 63).

By traversing these steps, one emerges as an accepted and mature member of one's "tribe." Fasick holds that in modern society, the high school most closely fits Van Gennep's guidelines. This threshold into adulthood culminates in the high school graduation ceremony.

More specifically, Larson (1988) argues that within the high school experience, the junior theme (a research project done during the 11th grade) could serve as a meaningful rite of passage itself. Larson believes that an adult is defined by his or her "capacity to influence the world and exercise control over his or her life" (p. 267). This capacity is facilitated by certain skills that adolescents must obtain: the ability to act thoughtfully toward a self-determined end; the ability to direct attention and energy toward this end; and the ability to generate a unique and personal product.

These abilities are best acquired through tests that demand their use, says Larson, and the most common in Western society is the high school junior theme project. Larson's research on 154 students in a suburban Chicago high school led him to conclude that those who complete the junior theme project have undergone a meaningful initiation rite toward adulthood. He writes:

> ❑ The young initiate who can endure the notorious personal trials of these projects, who can control his or her attention and come up with a final product, has made a significant step toward the autonomous status of adulthood in our society (p. 268).

These projects serve initiatory purposes because Western society has defined adulthood in terms of personal autonomy and productivity.

Inherent in an initiation rite is a confrontation with the fearful uncertainties of adult life. This confrontation requires conquering these fearful uncertainties and often involves pain, stress, and the opposite extreme, boredom. Conquering assumes that hard work and patience will go hand in hand; this is called endurance. The initiated is the one who endures the initiation test.

This, purports Larson, is the precise nature of the junior theme project. It requires endurance. Throughout as many as seven stages, which may require several months to complete, the young initiate will experience a variety of challenges as well as a wide range of accompanying emotions, from agony to ecstasy. In the end, those initiates who turn in their papers, and are subsequently graded as having passed, have survived the rite of passage.

By the end of the time mandated by the junior theme project, the initiates not only will have learned the abilities required to perform the task and produce the product, they will have learned to master emotional excesses as well. These, together with intentional productivity and self-control, mark the Western adult. Thus, writes Larson, "the Junior Theme . . . becomes an enactment of the identity quest and, in some cases, an identity crisis" (p. 281). The identity sought in this case is that of a responsible adult.

Another proposal, the walkabout, goes quite a bit further than Larson's suggestion.

THE WALKABOUT APPROACH

walkabout
Originally an aborigine initiation rite, the American version attempts to focus the activities of secondary school by demonstrating to the student the relationship between education and action.

In the remote regions of Australia, the aborigines have a rite of passage for all 16-year-old males. It is known as the **walkabout**. In the walkabout, the youth, having received training in survival skills throughout most of his life, must leave the village and live for six months on his own. He is expected not only to stay alive, but to sustain himself with patience, confidence, and courage. During this six-month estrangement from home and family, he learns to strengthen his faith in himself. He returns to the tribe with the pride and certainty that he is now accepted as an adult member. According to educator Maurice Gibbons (1974),

❐ The young native faces an extreme but appropriate trial, one in which he must demonstrate the knowledge and skills necessary to be a contributor to the tribe rather than a drain on its meager resources. By contrast, the young North-American is faced with written examinations that test skills very far removed from actual experience he will have in real life (p. 597).

As a result of Gibbons's article, a group was set up by Phi Delta Kappa (PDK), the national education fraternity, to see what could be done about promoting walkabouts for boys and girls in this country. They have produced a booklet that makes specific suggestions. In it, the PDK Task Force suggests that

❐ The American walkabout has to focus the activities of secondary school. It does so by demonstrating the relationship between education and action. It infuses the learning process with an intensity that is lacking in contemporary secondary schools. The walkabout provides youth with the opportunity to learn what they can do. It constitutes a profound maturing experience through interaction with both older adults and children. The walkabout enriches the relationship between youth and community (PDK Task Force, 1976, p. 3).

The PDK Task Force also puts out a monthly magazine that prints stories about the various types of walkabouts devised by students in participating schools.

The process has three phases: prewalkabout, walkabout, and postwalkabout. Each of these phases calls for learning specific skills. In the prewalkabout, adolescents study personal, consumer, citizenship, career, and lifelong learning skills. In the walkabout itself, the categories of skills to be mastered are logical inquiry, creativity, volunteer service, adventure, practicality, world of work, and cognitive development. The task force suggests numerous activities that foster learning in each of these skills. Most involve at least six months of supervised study and activity outside the school, such as working in a halfway house for mental patients.

The postwalkabout is a recognition that the student has engaged in a major rite of passage on his or her way to adulthood. It is not enough to recognize this experience in a ceremony where members are confirmed en masse, such as the typical graduation. Instead, an individual ceremony involving the graduate's family and friends is held for each walkabout that the student undergoes.

> ❏ The celebration of transition could take a variety of forms. The ceremonies are varied according to family tastes and imagination, but in each celebration the graduate is the center of the occasion. Parents and guests respond to the graduate's presentation. Teachers drop by to add their comments and congratulations. The graduate talks about his or her achievements, sharing some of the joys and admitting the frustrations (PDK Task Force, 1976, p. 34).

THE FAMILY AS A SETTING FOR THE RITE OF PASSAGE

> ❏ John and Debra grew up just four miles apart. They never knew each other until they met one day, years later, and eventually married. John was very self-confident and possessed a strong sense of purpose for his life, which was reflected in his career choice and his subsequent success within it. Debra, while herself an obviously bright and gifted woman, possessed little confidence in her abilities. This often paralyzed her capacity for making decisions, especially in regard to her career. Consequently, five years into their marriage, Debra found that she resented John's satisfaction with his life and choices and was frustrated with her own lack of direction.
>
> By comparing their families, we can see some of the reasons why John and Debra differ. John's family celebrated significant milestones in his life with recognition, parties, "bragging" within earshot of John, and by granting increased responsibilities with every passing accomplishment. Debra, by contrast, comes from a family where birthdays were frequently forgotten and thus interpreted as unimportant. When celebrations were attempted, they often resulted in arguments over missed events in the past. As a result, Debra grew to view herself as undeserving of recognition, while John emerged as an adult assured of his talents and cognizant of his limitations.

In what ways could the American family be encouraged to provide a more successful introduction into adulthood for its teen members?

Initiation signals not only the arrival of adulthood, it also signifies one's acceptance into a society. This socialization most effectively occurs within a microcosm of society, namely, the family. Family structures typically resemble those of the society at large. The rules and norms expected of mature participants in a society are most effectively communicated and enforced within the family. Families that fail to convey these rules and norms often have children who have trouble adjusting to life within the society.

Within many non-Western cultures, these societal norms are explicit and universally agreed upon. In Western cultures, however, most previously explicit norms have given way to vague substitutes. As a result, societal expectations are difficult for adolescents to grasp. It has been proposed that this shift has thwarted efforts by the family to serve as a socializing agent for teenagers. Closely related to this is a diminishing of family rites of passage that may have formerly assisted in the socializing process.

Quinn and associates (1985) assert that while children fully expect, even demand, the right to adulthood, our society is no longer in agreement as to what this process should look like:

> ❏ In our view, adolescents live in families with few clearly marked, uniform, or inevitable transition points. . . . The absence of developmental markers has created a false sense of security and an invalid confirmation of a progression toward future stages of development (p. 103).

This results in an "identity struggle," leaving the adolescent alone to define the transitions in growing up. As previously mentioned, these definitions regularly take the form of self-initiating behaviors.

What is needed, according to these researchers, are rites of passage within the family itself. However, these rites must not simply be "gifts for surviving the years" (p. 105). Rather, they should recognize significant accomplishments by giving the adolescent increased autonomy, responsibility, and privilege. These rites of passage place the onus of growing up on the adolescent, with a gradual expansion of his or her boundaries.

Two examples of family initiation rites are celebrations of significant accomplishments (e.g., learning to drive the family car) and the increased sharing of family responsibilities (using the car to do the shopping). In most families, obvious events (such as birthdays and graduations) are celebrated. However, there may be other milestones that, while possibly insignificant to the adult observer, may be very meaningful to the adolescent. In the research of Quinn and associates, teenagers were given greater responsibilities in the family and then granted higher status within the family after having proven themselves competent. This rite of passage took the form of an "official" induction as an "influential factor in the family functioning, thereby gaining increased recognition of his worth and more autonomy" (p. 109).

Obviously the family system benefits as well as the adolescent. Rites of passage assist in the often traumatic developmental dilemmas that naturally come with growing up. They ease the transitions of life for all involved by recognizing and affirming change as normal and essential. At the same time, they allow the family to be involved in powerful ways. Selvini-Palazzoli and associates (1978) found that rites of passage served to unify the family by creating a sense of "we-ness" that transcended the individual differences that potentially bring about family stress.

AN APPLIED VIEW

Using Family Rituals as Rites of Passage

The use of initiation rites offers valuable implications for therapists and other professionals. Roy (1990) claims that:

❑ Clergy and therapists can increase their effectiveness by promoting change through rituals. Rites of passage are the most obvious arena for change, individual as well as family. . . . family members [can

resolve] family disputes around a rite of passage and facilitate the entry or exit of family members from nuclear family units (p. 63).

With the apparent conflict that surrounds families with adolescents, creating and employing meaningful rites of passage for the sake of healthy growth and family harmony seems a natural option.

Roy (1990) suggests that the regular inclusion of family rituals—of which rites of passage are one type—promotes communication and healing within the family system. Through the marking of change in an individual, the family also realizes that it is changing. By traveling the developmental road together, it seems that adolescents and their families are able to adjust to the inherent difficulties of transition more readily.

A number of prominent thinkers have suggested that human development could be enhanced if the transitions from adolescent to adult were used to learn to deal better with fears. But how to do it? Here is a suggestion that has been gaining in popularity.

OUTWARD BOUND

Outward Bound
A program where people increase their sense of self-worth and self-reliance by participating in a series of increasingly threatening experiences.

The **Outward Bound** program (Outward Bound, 1988) was founded during World War II to help merchant seamen in England survive when their ships were torpedoed. Early in the war, it was learned that many sailors died because they became paralyzed with fear when their ship was hit. Outward Bound was designed to prepare these men to handle their fears in dangerous situations. The program was so successful that after the war it was redesigned for much broader training in conquering fear.

Its basic premise is that when people learn to deal with their fears by participating in a series of increasingly threatening experiences, their sense of self-worth increases and they feel better able to rely on themselves. The program uses such potentially threatening experiences as mountain climbing and rappelling, moving about in high, shaky rope riggings, and living alone on an island for several days. Some of the experiences in the program also involve cooperation of small groups to meet a challenge, such as living in an open rowboat on the ocean for days at a time.

Outward Bound has grown rapidly in recent years, and has installations throughout the country. Each program emphasizes the use of its particular surroundings. For example, the Colorado school emphasizes rock climbing,

The Outward Bound program has served as a basic rite of passage by offering a chance to prove one's self-worth and to have this feeling validated by others.

What are some other social and psychological effects you might expect an initiation program such as Outward Bound to have on its participants?

recidivism rate
The percentage of convicted persons who commit another crime once they are released from prison.

rappeling, and mountaineering. The Hurricane Island school in Maine uses sailing in open boats on the ocean as its major challenge. The school in Minnesota emphasizes reflection and development of appropriate spiritual needs. The Vision Quest program in Pine River, Michigan (Wiland, 1986) is a related type of challenge. This approach uses Native American ceremonies to prepare participants for a four-day solo experience designed to "demarcate the questor's entrance into society as a fully responsible and mature adult" (p. 30).

Outward Bound has proven its special worth for teenagers. It originally started with males, but most of its sessions now include equal numbers of males and females. The program operates as a basic rite of passage by offering a chance to prove one's self-worth and to have this feeling validated by others. The philosophy of the program is that participants cannot be told what they are capable of, but must discover it for themselves.

Since the Outward Bound experience effectively reduces the **recidivism rate** (the percentage of convicted persons who commit another crime once they are released from prison), it has widespread implications in treating juvenile delinquents. This program is not punitive like reform schools are, and is considerably less expensive than prison. However, the effectiveness of the program seems to decrease significantly after the enrollees have spent some time back in their neighborhoods. Perhaps this just means that they must be brought back for "booster" sessions from time to time.

Although there is a lack of extensive experimental evidence on the effects of Outward Bound, there is no scarcity of testimony from the participants themselves. As one short teenager put it, "Size really doesn't matter up there. What really counts is determination and self-confidence that you can do it!"

Many graduates say that they find life less stressful and feel more confident about their everyday activities as a result of participating in Outward Bound.

AN APPLIED VIEW

Should Teachers Be Restricted to Only Teaching Cognitive Goals?

A number of prominent educators believe that schools should concern themselves only with intellectual matters, not with personal and social development. Do you agree? These educators are opposed to ideas like Outward Bound and the walkabout. How do you feel about these two ideas? If you like these ideas, how can you use them as a springboard to other methods by which we adults might help youths in their "passage to maturity"?

AN APPLIED VIEW

The Components of Maturity

Think of one woman and one man who are the most mature persons you know—people with whom you are personally familiar, or people who are famous. Then ask yourself, "Why do I think these people are so much more mature than others?" In the spaces below, for both the male and the female, create a list of the characteristics that seem to distinguish them in terms of their maturity.

How do the lists differ? Is male maturity significantly different from female maturity? Which of the two people is older? Which of the two do you admire more? Which of the two are you more likely to want to imitate? Were you able to think of many candidates for this title of "most mature adult," or was it difficult to think of anyone? Are either or both of the people you picked professionals? Are either or both of these persons popular with their own peer group? What is the significance of your answers to you?

Female

_____

_____

_____

Male

_____

_____

_____

One of the most positive aspects of the program is its effect on women. Many say that they are surprised to discover how much more self-reliant they have become. Probably the most important result is that most graduates say they feel more responsible and grown-up after having been through the experience.

The Outward Bound and walkabout procedures are becoming better known, and they will almost certainly help alleviate the need for initiation to adulthood, but they are clearly not sufficient in themselves. The movement from high school to college encompasses each of the three stages of passage—separation, transition, and reincorporation—that Van Gennep (1909/1960) described. In the absence of sanctioned rites of passage, students have designed their own tests of worthiness in order to belong: hazing, rape and other sexual exploits, violent crime, and alcohol abuse. The practice of binge-drinking, consuming more than five drinks in one sitting, was once considered a harmless rite of passage on college campuses but has now been cited as the number one substance abuse problem in American college life. It is estimated that one in three college students now drinks primarily to get drunk (Commission on Substance Abuse, 1994).

In designing new student orientation programs for colleges and universities, safer but nonetheless challenging alternatives to meet the rites of passage needs should be considered. For example, a New England college strongly encourages all their incoming freshmen to participate in a short-term wilderness program. Some high schools are now offering chemical-free events, such as graduation parties or proms, that demonstrate to the students they can enjoy themselves without the use of alcohol. This is one specific way of changing cultural expectations and therefore the rites of passage that accompany them. The complexity of American adulthood requires a variety of such approaches if we are to develop in our youths the kind of mature women and men we want.

AN APPLIED VIEW

The Solo Experience

Let us give you two examples of modern initiation ceremonies. One involves suburban children experiencing life in the deep woods, and the other highlights an inner-city African American youth.

For thirteen years, one of the authors of this book (Dr. Dacey) directed a weekend camp experience for young adolescents. One of the activities at that camp was preparing to go on a solo, which meant taking care of yourself in an isolated part of the woods for 24 hours (cook your own meals, make your own shelter, and so on). While on their solos, these campers were continuously watched through binoculars but had no idea they were being observed.

The great majority of campers completed the solo and were extremely proud of themselves. Even those who quit sometime before the end of the exercise were helped to see that they had accomplished a major feat. Upon returning to the group, each camper shared a portion of their personal journal that recorded their thoughts and feelings while on solo. There was no doubt that graduates of this camping program believed themselves to have made a major step toward maturity as a result of this solo.

Trotter (1991) describes the inner-city initiation rite: "Two youths clad in African costumes pound out a thudding rhythm on ceremonial drums as a column of boys advances toward the front of the crowded auditorium. A matching column of girls moves down another side. The boys wear colorful vests and hats called ìkufisî; the girls wear black patterned wrap skirts and white headcloths . . . . The audience claps to the drumbeat.

"The dancers gather on the stage, which is adorned with African masks and sculptures that evoke cultures a thousand miles away from the junior high school in Washington, D.C. Seated nearby is a group of the schoolìs ìeldersî—teachers and administrators, also in African garb. Some of the elders address the assembly in solemn tones about these students' achievement: They have completed a training program for adulthood modeled after African coming-of-age rituals.

"Then, one by one, the boys and girls step across a symbolic threshold, accept a colorful kinte cloth around their shoulders, and speak into a microphone their new African names, a symbol of achieving a new state. 'My name is Lezia,' says one girl, 'and the rites of passage taught me hope' " (p. 48).

Programs such as this are rapidly emerging in urban areas to provide the African American community with a focus on cultural values. The curriculum consists of topics such as household budgeting, human sexuality, African art and history, music, and spirituality. Whether it is conducted in-school, after-school, or elsewhere in the community, the overall goal of the program is to instill a strong, positive sense of self and achievement.

C O N C L U S I O N S

We began this chapter by comparing how youth are inducted into adulthood in preindustrial tribes and in modern America. We concluded that our situation is much more complex than that of the tribes, and that although there have been several good suggestions for initiation activities in postindustrial societies, the absence of clear initiation rites still causes problems for us.

Before we can specify initiation activities that would be useful in inducting youth into adulthood, perhaps we need a clearer idea of the successful adult. To put this another way: "What is a mature person?" This question has intrigued thinkers throughout recorded history. It has been variously described as a search for peace, for the knowledge of God, for satori, for nirvana, for self-actualization, or for wisdom.

In many ways, this whole book has been about the "passage to maturity." We hope that in reading it, you have gained an improved sense of what we adults need to do to help adolescents with their passage. Perhaps even more importantly, we hope that in studying the many theories and research reports detailed in this book as well as in carrying out some of the suggested activities, you have made significant progress in your own quest.

C H A P T E R   H I G H L I G H T S

Initiation Rites

- Initiation rites in other cultures offer a formal ceremony marking the transition from child to adult.
- Our own culture provides no such universal and formal way for taking leave of adolescence and being accepted into the adult community.
- Fraternities and other organizations often require aspiring members to pass through a series of trials to gain acceptance into the group.
- Some researchers have suggested parallels to this process in the crimes juveniles are often required to commit in order to become a member of a gang, and in the sequence of experiences that so often lead to drug addiction.

Analysis of an Initiation Rite

- The purpose of initiation rites in preindustrial societies is to cushion the emotional disruption arising from the transition from one life status to another.
- For the male, these rites also formally end dependence on his mother and the other women in the community, and bring him into the group of male adults.
- Initiation rites also serve as an introduction for both genders to the sexual life of an adult.

The Passage to Adulthood

- In our industrial past, the transition to adulthood was fairly clear.
- Entry into adulthood is far more complex for adolescents today, largely due to the increase of sophisticated technologies and the need for many more years of formal education.

- No specific rituals exist in present Western societies to aid youths through this difficult change.
- There are five types of activities that signal the passage to maturity in America today: religious, social, sexual, educational, or economic.
- The adolescent moratorium, or time-out, allows teens to explore possibilities and extend education. It comes about because of our advanced economic system.
- Some researchers have suggested that the moratorium is ineffective and causes self-initiating behavior, which may be dangerous.
- American adolescence is often prolonged, and the transition to adulthood almost impossible to pinpoint.
- Organized sports, the adolescent counterculture, and delinquency all provide some sort of trials that must be passed for acceptance into adulthood.

Methods That Facilitate the Transition to Adulthood

- Some high school projects that require self-determination, focus, and sustained work to achieve a final product may serve as an initiation rite.
- The Australian aborigines require all 16-year-old males to spend six months alone surviving in the wilderness. This ritual, called the "walkabout," confers adulthood on the youth who completes it. An American form of the walkabout, incorporating varied work in real-world settings, has been suggested.
- The Outward Bound program, with its emphasis on various forms of wilderness survival, serves as a rite of passage for many youths.

K E Y   T E R M S

achieved identity 481
adolescent moratorium 478
ascribed identity 480

hazing 471
initiation rite 471
Outward Bound 486

puberty rite 475
recidivism rate 487
walkabout 483

W H A T   D O   Y O U   T H I N K ?

1. Do you remember any experiences from your own youth that were particularly helpful in your transition to adulthood? In what ways were these experiences helpful?
2. The adolescent moratorium seems to be getting longer and longer. Do you think this is a good thing? Why or why not?
3. Some theorists have suggested that because of our capitalistic financial system, we need to delay adolescent entrance into the adult world for longer and longer periods. Can you think why they would say so? Do you agree?
4. Can you think of a third plan for initiation along the lines of the walkabout and Outward Bound?
5. How would you define maturity?

S U G G E S T E D   R E A D I N G S

Bly, R. (1990). *Iron John*. Reading, MA: Addison-Wesley. This book is a result of Bly's ten years of work with men to discover the truths of masculinity. He goes beyond the stereotypes of popular culture in this discussion of male initiation and the role of the mentor.

Bronowski, J. (1973). *The ascent of man*. Boston: Little, Brown. This classic work makes exciting reading when one is beginning to study the psychology of human development.

Erikson, E. (Ed.) (1978). *Adulthood*. New York: Norton. A collection of essays on what it means to become an adult. Written by experts from a wide variety of fields.

G U I D E D   R E V I E W

1. As long ago as 1909, Arnold Van Gennep argued that the purpose of initiation rites, as with all rites of passage, is
   a. related to the competitive nature of all human beings.
   b. to pass on information necessary for people to survive and flourish.
   c. to allow all individuals to feel that they belong to some larger group.
   d. to cushion the emotional disruption caused by a change from one life status to another.
2. Van Gennep proposed which three universal steps in his rites of passage?
   a. a separation, a transitional period, and a recognition period
   b. a separation, an instrumental period, and a resolution period
   c. a separation, a transitional period, and a reincorporation period
   d. a separation, an instructional period, and a reincorporation period
3. Because rites of passage ceremonies often coincide with the peak in adolescent physiological maturation, they have often been called
   a. manhood rites.
   b. castration rites.
   c. passage rites.
   d. puberty rites.
4. The simplest explanation for why adolescents go through painful initiation rites is because they want to
   a. feel that they belong.
   b. do what is expected of them.
   c. prove something to themselves.
   d. join an exclusive group.

5. After living at home for his first year of college, Pete is going to move into the dorms. This is an example of what type of initiation rite?
   a. social
   b. religious
   c. economic
   d. educational

6. Shelly has just graduated from college but is not sure what career she wants to pursue. She decides to take some time and travel around the United States for a few months to explore possibilities. This time-out period is also known as a(n)
   a. adolescent moratorium.
   b. premature foreclosure.
   c. nurturing period.
   d. adolescent sabbatical.

7. In the 1960s and 1970s, many American youths sought to establish their identity by
   a. using drugs.
   b. rejecting traditional values.
   c. living on cooperative farms.
   d. imitating primitive tribal rites.

8. What is the basic premise of Outward Bound?
   a. Teamwork in the face of adversity can make friends out of strangers.
   b. Being forced to depend on others builds trust.
   c. People will have a profound maturing experience through interaction with both older adults and children.
   d. People who face their fears by participating in threatening experiences increase their sense of self-worth and self-reliance.

9. In preindustrial societies, individual status was determined by the tribe to which the person belonged rather than by individual or family efforts. Social scientists call this a(n)
   a. linked identity.
   b. ascribed identity.
   c. corrective identity.
   d. achieved identity.

10. Researchers analyzing the transition to adulthood in the United States generally conclude that there are
    a. still many formal initiation rites, but they are on the decline.
    b. few formal initiation rites.
    c. more formal initiation rites than ever before.
    d. a few formal initiation rites, but the number of such rites is increasing.

11. In contemporary American society, the transition from adolescence to adulthood is _____ (clear/obscure).

12. Freud thought that the ultimate threat held over the young during the initiation rite is _____ (castration/parental rejection).

13. Becoming a parent is an example of a _____ (social/sexual) type of initiation activity.

14. Organized sport is an example of our efforts to include _____ (hazing/initiation rites) in American life.

15. The adolescent moratorium comes about because of our _____ (changing family types/advanced economic system).

16. The most famous symbol of the adolescent _____ (moratorium/counterculture) was the Woodstock music festival of 1969.

17. David's interest in news reporting during high school led to working on the school newspaper, freelance writing, and an internship during the summer for the local city paper. David's early commitment to reporting paid off, because he got a job right out of college. David has an _____ (achieved identity/ascribed identity).

18. Moe is a 16-year-old Australian aborigine male. He will soon leave his village and live for six months on his own, using his survival skills as he goes through the rite of passage called _____ (walkabout/lone retreat).

19. Within many _____ (non-Western cultures/Western cultures), societal norms are explicit and universally agreed upon.

20. An example of a _____ (puberty/family initiation) rite is a celebration of significant accomplishments.

Answers

1. d, 2. c, 3. d, 4. a, 5. a, 6. a, 7. b, 8. d, 9. b, 10. b, 11. obscure, 12. castration, 13. social, 14. initiation rites, 15. advanced economic system, 16. counterculture, 17. achieved identity, 18. walkabout, 19. non-Western cultures, 20. family initiation

APPENDIX

SERVICES AVAILABLE TO YOUTHS IN MOST MEDIUM-TO-LARGE U.S. CITIES, AND THEIR 800 NUMBERS

CHARITABLE ORGANIZATIONS
Catholic Relief Services World Headquarters 235–2772

Child Works International 533–3315

Children Incorporated 538–5381

United Community Charities Incorporated 833–6650

CLINICS—MENTAL HEALTH
All Women's Health & Medical Services 223–3909

New Life Treatment Centers Inc. 227–5433

DRUG ABUSE & ADDICTION
Adolescent Chemical Dependency Unit North Mississippi Med. 442–2238

Alcohol & Drug Abuse Center 352–7873

Chemical Emergency Preparedness Program Hotline 535–0202

Cocaine Abuse Triple A Abuse Alternatives 452–9300

"800 Cocaine" Information 262–2463

Narconon International 468–6933

EATING DISORDERS
Anorexia & Bulimia Resource & 24-hr. Helplinc 772–3390

H O P E Eating Disorders Program 635–1022

The Rader Institute 255–1818

EDUCATIONAL ALTERNATIVES
American Association of Overseas Studies 338–2748

Audubon Girl Scout Council 852–8421

Student Camp & Trip Advisors 522–5883

United States Sports Camps 468–7007

ENVIRONMENTAL, CONSERVATION & ECOLOGICAL ORGANIZATIONS
Global Perspectives 221–8897

National Wildlife Federation 432–6564

GENERAL SERVICES
Child Abuse Prevention—Kids Peace 257–3223

Child Care Resource & Referral 343–3470

Child Support Hotline—Department of Social Services 831–4573

Children's Hospice International 242–4453

Foster Care Coalition 367–8373

National Center for Missing & Exploited Children 843–5678

National Day-Care Referral 554–5437

Teen Help 637–0701

Vegetarian Awareness Network 872–8343

HANDICAPPED SERVICES
Family Support & Information Network 852–0042

National Information System For Health-Related Services 922–1107

Project Reach 537–3224

HEALTH SERVICES
Comprehensive Health Education Foundation 323–2433

Human Health Systems 448–8378

National AIDS Hotline Centers for Disease Control 342–2437

National Children Cancer Society 532–6459

National Health & Fitness Alliance 348–9297

National Sexually Transmitted Disease Hotline 227–8922

Rape Hotline—serviced on a local basis

PREGNANCY COUNSELING

Adoption Hotline 444–4844

Catholic Adoption Services 833–5878

Crisis Pregnancy Center 683–1023, 344–4272

Dimensions Medical Center-Abortion Services 553–3939

Femcare 228–1560

Florence Crittenton Services 448–0024

Foster Care Coalition 367–8373

Gynecological Surgical Services 932–3312

Healthy Mother Healthy Baby Line 422–2968

Life Saver Ministry 648–4357

Lovejoy Sugicenter 752–6189

Mercy Ministries 922–9130

Pregnancy Center Archdiocese 492–5530

Volunteers of America 426–5934, 222–3196

RUNAWAYS

Missing Children Help Center USA-kids

Missing Children Safety Council Nat'l Child Watch 222–1464

National Runaway Switchboard 621–4000

Operation Lookout—National Center for Missing Youth 782–7335

Vanished Children Alliance 826–4743

SUICIDE

Suicide Prevention Center 352–7873

GLOSSARY

A

abstract thinking Thinking in the formal operations stage, which allows reality to be represented by symbols that can be manipulated mentally. (p. 105)

abuse of inhalants Sniffing substances such as airplane glue, paint thinners, and gasoline to get high. The most popular new inhalant is amyl nitrite, popularly known as poppers. These substances are especially dangerous because they can permanently damage the nervous system. (p. 398)

accommodation Modifying our existing schemata because we cannot make our perception of the environment fit. What we are seeing is so new and different that we must change in order to adapt to it; that is, we learn. (p. 103)

achieved cultural identity When individuals find a way of resolving the differences between their cultural background and the dominant culture as well as the problem of the lower status of their cultural group in society. (p. 193)

achieved identity In industrial societies, individual status that is brought about by personal effort and early commitment to a career path. (p. 481)

adaptation Piaget states that a tendency in all human beings is to adapt to the environment. Adaptation consists of two complementary processes: assimilation and accommodation. (p. 103)

addictive drug use Use of drugs as a necessary means for feeling good physically and emotionally. (p. 401)

adolescent life change event scale A psychological measure developed to evaluate the amount of stress in an adolescent's life based on the number of stressful life events experienced in the last year. (p. 360)

adolescent moratorium A time-out period during which the adolescent experiments with a variety of identities without having to assume the responsibility for the consequences. (p. 478)

age cohort A group of people born about the same time. (p. 28)

aggressive gang A type of adolescent delinquent behavior distinguished by a group or gang that engages in a variety of illegal group activities. (p. 431)

AIDS (acquired immune deficiency syndrome) A virus that attacks certain cells of the body's immune system, leaving the person vulnerable to any number of fatal complications, such as cancer and pneumonia. (p. 306)

amenorrhea The absence or suppression of menstruation, often seen in anorexics and bulimics. (p. 76)

anabolic steroids An artificial form of the male hormone, testosterone, taken to look more muscular and to perform better athletically. Most users are male, but some females in competitive sports also use steroids, hoping to increase their strength and endurance. It appears that long-term effects of using these drugs are dangerous. (p. 413)

ancestral worldview The belief, common among ethnic minority cultures, that the well-being of the larger group, such as the family or the community, is more important than the success of a single individual. (p. 222)

androgynous Refers to those persons who have higher than average male and female elements in their personalities. (p. 197)

anorexia nervosa A syndrome of self-starvation that mainly affects adolescent and young adult females. It is characterized by intense fear of gaining weight, disturbed body image, significant weight loss, body weight below minimal normal, and amenorrhea, none of which can be accounted for by a known physical problem. (p. 368)

antisocial behaviors Acts that inflict physical or mental harm, property loss, or other damage on others. (p. 431)

artistic type One of Holland's six vocational personality types. The artistic type is creative, expressive, and nonconforming and may choose occupations such as musician or interior decorator. (p. 339)

asceticism A defense mechanism against the sexual, "sinful" drives of youth that often causes the teenager to become extremely religious. (p. 44)

ascribed identity In preindustrial societies, individual status that is ascribed by the tribe to which the person belongs. (p. 480)

assimilation Perceiving the environment in a way that fits our existing schemata. That is, we make reality fit what already exists in our minds. (p. 103)

attachment theory A theory proposing that caretakers who are consistently responsive and sensitive contribute to the development of secure attachment, which in turn helps children develop greater self confidence and a willingness to be more independent. (p. 215)

authoritarian parenting style Baumrind's term to describe parents who are demanding, give their children little autonomy, control their children by rules with little explanation, and fail to show warmth to their children. (p. 218)

authoritative parenting style Baumrind's term to describe the most common and most successful style of parenting, in which parents hold high expectations, provide explanations for rules, and create an environment of warmth and caring. (p. 218)

automatic processing Information-processing procedures that occur in situations that are consistent and provide a lot of opportunity for practice, such as driving a car. They are generally quick and effortless and become faster with practice. (p. 117)

autosexuality The love of oneself; the stage at which the child becomes aware of himself or herself as a source of sexual pleasure, and consciously experiments with masturbation. (p. 283)

B

beer goggle sex Sexual activity resulting from drinking too much alcohol. (p. 302)

behavioral model of depression The view that depression results from low levels of positive reinforcement or from an inability to escape from punishment. (p. 375)

behavioristic model A model that sees adolescent subcultures starting out as a result of a series of trial-and-error behaviors, which are reinforced if they work. The behavioristic model sees peer group members behaving the way they do because they have no other choice. (p. 269)

being needs Needs that increase as they are attended to (thus they are also termed growth needs). An example of a being need is the appreciation of music; the more we come to like music, the more we desire the joys it can provide. (p. 49)

biculturalism Strong identification with both the majority culture and a person's own minority culture. (p. 192)

bidirectional research Research that takes into account the two-way nature of person-environment interactions. The two are always affecting each other. (p. 57)

biological theory of homosexuality A theory that postulates that the reaction of a fetus's brain to sex hormones during the second through sixth month of gestation may create a genetic tendency toward homosexuality. This theory states further that persons born with this tendency (called a predisposition) can be influenced by the environment to either select or avoid homosexuality. (p. 287)

biopsychosocial The idea that development proceeds by the interaction of biological, psychological, and social forces. (p. 36)

body image How people believe they look to others. (p. 89)

bulimia nervosa A disorder characterized by episodic binge eating and depressed mood, with low self-worth thoughts following the eating binges. Bulimics are aware that their eating pattern is abnormal, but feel that they cannot control their eating voluntarily. (p. 369)

C

case studies In-depth looks at individuals. (p. 25)

castaways Adolescents who leave home because they have been abandoned, neglected, or abused by their parents. They feel rejected by their parents, who often do not oppose their child's leaving. (p. 456)

chicken hawks Older male homosexuals who specialize in paying young boys for prostitution. (p. 464)

chickens Young male prostitutes. (p. 464)

chlamydia This is now the most common STD, with about five to seven million new cases each year. There often are no symptoms. It is diagnosed only when complications develop. (p. 307)

cliques Relatively small and intimate groups of close friends of similar ages, backgrounds, and interests. (p. 260)

close friendships Friends viewed as being supportive and available to provide support when needed. This basic building block of the larger, peer group structure is positively associated with the social, psychological, and academic adjustment of adolescents. (p. 260)

codification The tendency of adolescents to establish detailed rules. (p. 134)

coercive cycle A pattern of family interaction through which the child learns to control the parents by behaving negatively and sometimes aggressively. (p. 436)

cognitive model of depression The view that depression results from negative cognitions or thoughts. Beck discusses three kinds of negative cognitions: negative view of the self, negative interpretation of one's experiences, and negative views of the future. (p. 374)

cognitive structures Specific quantitative and qualitative mental abilities that develop in stages as a child's intelligence develops. (p. 101)

commitment The third of Perry's intellectual and ethical stages in which "because of the available evidence and my own understanding of my values, I have come to new beliefs." (p. 152)

community service learning
A method of educational reform that involves students in service activities in community settings in conjunction with related academic study in the classroom. (p. 333)

compensation The cognitive ability to recognize an inequality of quantity and then add to the lesser amount to create an equality. (p. 112)

compulsory education Refers to laws that require children to be in school between the ages of 6 and 16. It was instituted in part to make sure that children would be freed from the terrible conditions of the sweatshops. (p. 17)

concrete operational stage
Piaget's stage of mental development in which children become concerned with why things happen. The intuitive thinking style of the preoperational stage is replaced by elementary logic. (p. 104)

conflict mediation A technique to teach juvenile offenders how to settle conflicts without resorting to violence by means of a face-to-face meeting between the offender and the victim, attended by a third person, called the mediator. The young offender is held responsible for his or her behavior and given a chance to express remorse and to make amends to the person who was harmed. (p. 453)

consolidation A quantitative cognitive change from childhood to adulthood, whereby improved problem-solving techniques are employed in a wider variety of situations and with greater skill. (p. 112)

construct An idea about some aspect of the human being. The ego and intelligence are examples. (p. 37)

contextualism Richard Lerner's theory that to understand an adolescent, we must carefully describe the ever-present interaction between the person's internal state and the environment in which that person is operating. The internal state and the environment are embedded in each other. (p. 55)

controlled drugs Those drugs that have been limited in their distribution and manufacture under the Controlled Substances Act of 1970. This act empowers the Attorney General of the United States and designates to punish those who use or sell drugs illegally. There are five levels of punishment under this law. (p. 398)

controlled processing An information-processing procedure in which new information lacking consistency in rules and sequence is manipulated, therefore making specific attention to each step a necessity because there is no established pattern. Controlled processing takes place when the information to be processed does not provide the opportunity for practice that leads to speed and ease. (p. 117)

control theory The theory that adolescents with strong bonds to society are less likely to deviate from conventional behavior. (p. 404)

conventional level In Kohlberg's theory of moral reasoning, the level at which the person wants to fulfill society's expectations and be fair to all. (p. 136)

conventional type Among Holland's six vocational personality types, the conventional types enjoy working systematically with numbers, filing clerical records, and copying printed information. (p. 339)

convergent thinking In problem solving, when we converge or close in on the one correct answer. (p. 119)

coping strategies Methods people use to handle stress. (p. 362)

core self Hart describes a self that reflects our deepest values and is consistent across social relationships. (p. 184)

correlational studies Studies that analyze the co-relationship between two variables. (p. 26)

critical thinking Involves the ability to think logically, to apply this logical thinking to the assessment of situations, and to make good judgments and decisions. (p. 119)

cross-sectional studies Studies that compare groups of individuals of several ages at the same time in order to investigate the effects of aging. (p. 29)

crowds Peer groups with a reputation for certain values, attitudes, or activities. Common crowd labels among high school students today include jocks, brains, druggies, populars, nerds, burnouts, and delinquents. (p. 260)

cultural gangs Those gangs that are centered in a neighborhood. They are loosely organized and are not necessarily involved in any criminal activities. These gangs are usually made up of a single cultural or ethnic group, especially if that group is newly arrived in this country. (p. 451)

cultural identity That part of a person's self-concept that comes from the knowledge and feelings about belonging to a particular cultural group. Cultural identity includes self-identification, a sense of belonging, an attitude toward one's cultural group, and involvement. (p. 191)

cultural identity search A process of exploring one's cultural identity. (p. 192)

culture of violence The belief that support for violence is prevalent in our society, typified by, for example, violence in the media, limited gun control laws, and respect for family privacy. (p. 440)

culture transmission model A model explaining the rise of adolescent subcultures that posits that a new subculture arises as an imitation of the subculture of the previous generation. This takes place through a learning process by which younger teenagers model themselves after adults in their twenties. (p. 268)

D

dealing The sale of illegal drugs on a small scale, usually carried out with a friend of the seller. (p. 401)

defense mechanisms Unconscious attempts to prevent awareness of unpleasant or unacceptable ideas. (p. 41)

deficiency needs Those needs that decrease as they are attended to; they can be fully satisfied. When some physical or psychological deficit occurs (thirst, loneliness) and there is an action that can be taken to eliminate it, the need is a deficiency need. (p. 49)

delayed puberty The stages of pubertal change do not begin until a significant time after the normal onset. (p. 67)

depression A condition marked by a sorrowful state, fatigue, and a general lack of enthusiasm about life. It can be of long or short duration and of low or high intensity, and can occur in a wide variety of conditions at any stage of development. (p. 372)

depressive equivalents The expression of depression in adolescence through symptoms that are different from those displayed by adults. Concentration difficulty, running away, sexual acting out, boredom and restlessness, and aggressive behavior and delinquency are examples of depressive equivalents. (p. 377)

descriptive studies Information is gathered on subjects without manipulating them in any way. (p. 25)

developmental diversity The concept espoused by contextual theorists that we ought to pay particular attention to the rich and beneficial differences in the ways people of various ethnic groups develop throughout life. (p. 59)

developmental tasks Skills, knowledge, and attitudes that are needed by an individual in order to succeed in life at each stage of life; these tasks lie midway between the needs of the individual and the goals of Western society. (p. 48)

disengaged family A term used by family systems' theorists to describe families in which there is too little involvement or caring among family members. (p. 217)

divergent thinking The type of thinking used when the problem to be solved has many possible answers; especially important in creative thinking. (p. 119)

diversity According to Toffler, stress is increased by the percentage of our lives that is in a state of change at any one time. (p. 359)

double standard Engaging in premarital sexual relations is acceptable for males but not for females. (p. 291)

drive-by shootings When innocent bystanders are injured or killed by violent gangs usually shooting from moving automobiles. Many of the victims of gang violence have no connection at all with any gang activity. (p. 449)

drug abuse Use of a drug in such a way that the individual's physical, mental, or emotional well-being is impaired. (p. 398)

drug addiction Historically, this term has referred to dependence on narcotics. Today it has so many

meanings that experts are now beginning to use the term "drug dependence" instead. (p. 398)

drug dependence When a physical or psychological need, or both, result from continuous drug use. Psychological need occurs when the person feels anxious, depressed, or irritable when he or she doesn't have the drug. Physical dependence, on the other hand, occurs only when negative physical symptoms result from drug withdrawal, such as vomiting, sweating, muscle tremors, joint pain, delusions and hallucinations, and almost always a strong sense of anxiety. (p. 398)

drug overdose Taking so much of a drug that it causes an acute reaction, usually extreme anxiety, which is sometimes followed by stupor, low breathing rate, and, in rare cases, coma. (p. 398)

drug tolerance A condition that develops from continuous use of a drug when a larger and larger amount is needed to produce the same effect. (p. 398)

dualism The first of Perry's three intellectual and ethical stages, in which "things are either absolutely right or absolutely wrong." (p. 152)

dysmenorrhea Menstrual cramps. (p. 82)

E

early formal operations stage The first stage of formal operations (from 11 to about 14), in which abstract thought, logic, metacognition, and hypothetical reasoning occur. (p. 105)

eating disorders Disorders such as anorexia nervosa and bulimia nervosa characterized by drastically reduced food intake or episodic binge eating. (p. 75)

ecological theory A view of the growing child and adolescent as an active agent in a series of interacting systems. This theory was first presented by Urie Bronfenbrenner. (p. 209)

ego The central part of our personality, according to Freud. It is the (usually) rational part that does all the planning. It keeps us in touch with reality. It begins to develop from the moment of birth. (p. 42)

egocentric thinking Adolescent's style of thought in which they think more about themselves, and watch themselves as though from above. This trait is composed of two specific factors: the imaginary audience and the personal fable. (p. 114)

ego psychology Psychoanalytic view that emphasizes the ego more and the id and superego less than Freud did. (p. 45)

emotional attachment During adolescence, an individual may give up the dependence of a younger child while developing interdependence with his or her parents. The need for emotional attachment remains; emotional distance from parents is usually not healthy. (p. 248)

emotional separation Identity is formed at adolescence through detaching emotionally from the family and shifting affection to peers. (p. 248)

empirico-inductive method A method of problem solving used by young children, in which they look at available facts and try to induce some generalization from them. (p. 110)

endocrine system The ductless glands such as the pituitary, thyroid, and sex glands that secrete directly into the bloodstream. (p. 68)

enmeshed family A family systems term used to describe families in which members are very close and limit the freedom necessary for individual growth. (p. 218)

enterprising type Among Holland's six vocational personality types, enterprising types often enjoy business activities that are directed towards making money or achieving organizational goals. (p. 339)

environmental model of depression A theoretical model that describes depression as a consequence of environmental circumstances, such as negative life events, or high levels of life stress and few sources of support. (p. 376)

escape Perry's term for refusing responsibility for making commitments. Since everyone's opinion is "equally right," the person believes that no commitments need be made, and thus escapes from the dilemma. (p. 153)

essential schools Model of education based on the philosophy of Theodore Sizer, which maintains that every student should master a number of essential skills. (p. 336)

experimental use Limited use of drugs out of curiosity or to have a new experience. (p. 401)

exploration stage According to Super, most teens are involved in this stage of career development. (p. 340)

extended family A family structure in which grandparents, aunts and uncles, and other family members beyond the nuclear family (parents and siblings) play an important role in providing emotional, physical, and financial support. (p. 222)

externalizing Expressing depressive feelings outwardly, such as through aggressive behavior, delinquency, running away, or rebellion. (p. 376)

F

family system Families develop rules and boundaries that help to maintain the status quo or keep things the way they have always been. A family system that is too

resistant to change can create difficulties for adolescents who are leaving home and making career decisions. (p. 342)

family therapy A form of psychological treatment in which all family members work together with a therapist to help the family as a whole to function better and to meet the needs of all family members. (p.214)

fixated Refers to being stuck at a developmental stage and, therefore, unable to become a fully mature person. (p. 43)

formal operations stage Piaget's fourth stage of mental development, in which children are able to perform abstract operations entirely in their minds. (p. 105)

full service schools A model of educational reform in which a variety of educational, health, and psychological services are provided to children and families in a school setting. (p. 331)

future shock The illness, according to Toffler, that results from having to deal with too much change in too short a time. (p. 359)

G

gays A name referring to male homosexuals. (p. 286)

gender Our conceptions of what it means to be male or to be female. (p. 195)

gender aschematic Persons who do not use gender as a standard for evaluating or classifying behavior. Bem maintains that being gender aschematic is more healthy psychologically than restricting one's behavior based upon gender-role expectations. (p. 200)

gender identity Self-identification as a male or female based on biological characteristics. (p. 196)

gender intensification hypothesis According to Hill and Lynch, pubertal changes in early adolescence contribute to increased concern about conformity to gender stereotypes. (p. 195)

gender role A pattern of behavior that results partly from genetic makeup and partly from the specific traits that are in fashion at any one time and in any one culture. For example, women are able to express their emotions through crying more easily than men, although there is no known physical cause for this difference. (p. 195)

gender-role adaptation Defined by whether the individual's behavior may be seen as in accordance with her or his gender. (p. 196)

gender-role behavior The extent to which a person's behaviors, occupations, and interests are considered masculine and feminine according to cultural norms. (p. 196)

gender-role system The set of attitudes and beliefs about the ways in which the abilities and personalities of men and women differ and are suited to different career choices. (p. 342)

gender schema A cognitive process through which behaviors are labeled as masculine or feminine. (p. 200)

genital herpes This is an incurable sexually transmitted disease with about 500,000 new cases every year. With no cure, there are now estimated to be about 30 million people in this country who experience the recurring pain of this infection. (p. 309)

gonadotropins The hormones secreted by the pituitary gland. (p. 73)

gonorrhea This well-known venereal disease accounts for between one-and-a-half and two million cases per year. One quarter of those were reported among adolescents. The most common

symptoms are painful urination and a discharge from the penis or the vagina. (p. 308)

goodness of fit The term contextual theorists use to describe the quality of the interaction between people and their environments. If this interaction produces desirable results, there is "goodness of fit." (p. 57)

gynecomastia Inappropriate physical development marked by male breast growth. (p. 84)

H

hazing The often dangerous practices used by some fraternities to initiate new members. (p. 471)

hepatitis B This viral disease is transmitted through sexual contact, and also through the sharing of infected needles. The symptoms include high fever and aches; liver damage may result. (p. 309)

heterosexuality Preference for intimate, interpersonal relationships and sexual interaction with members of the opposite sex. (p. 290)

hierarchy of needs Needs that Maslow believes overlap in stages. The basic needs are present at birth and higher-order needs show up as the person grows older. Furthermore, satisfaction of these needs is sequential; the basic needs must be met before later, more complex needs can be successfully fulfilled. (p. 49)

Holland's theory A theory describing vocational choice as the matching of an individual's personality type to the type of environment in which the work takes place. (p. 338)

home invader gangs A newly developing form of drug-related gang violence. These gangs enter homes and terrorize the occupants while stealing anything of value. These gangs enjoy their ability to

intimidate and control their victims, forcing them to hand over all their valuables, and threatening to kill them if they don't cooperate. (p. 451)

homosexual A person who prefers sexual interaction and intimate, interpersonal relationships with members of the same sex. (p. 285)

hormones Chemical agents produced by the endocrine system that trigger physical change such as puberty. (p. 68)

hypothetical reasoning Forming conclusions based on hypothetical possibilities, a type of formal stage thinking. (p. 105)

hypothetico-deductive method A method of problem solving used by adolescents, in which they hypothesize about the situation and deduce from it what the facts should be if the hypothesis were true. (p. 110)

I

id The simplest of Freud's mind structures, it operates only in the pursuit of pleasure. (p. 42)

ideal self The self that one would like to be. The development of the ideal self is made possible by hypothetical thinking. (p. 169)

identity crisis Erikson's term for the situation, usually in adolescence, that causes us to make major decisions about our identity. (p. 53, 185)

identity status Refers to Marcia's four types of identity formation. (p. 188)

imaginary audience One aspect of adolescent egocentric thinking in which the adolescent thinks that everyone is noticing everything he or she does and every aspect of his or her appearance in fine detail. The adolescent feels "on stage" most of the time, in front of an imaginary audience. (p. 114)

individuation The first separation process, which takes place in early childhood when the young child comes to realize that she or he is separate and different from her or his parents. (p. 170)

information processing The study of how children (and adults) perceive, comprehend, and retain information. (p. 117)

information-processing strategy The process of understanding information and acting on that understanding. (p. 112)

initiation rite A cultural and sometimes ceremonial task that signals progress to some new developmental stage. (p. 471)

instrumental gangs Gangs formed for the purpose of carrying out criminal activities. The lure of big money has drawn many into drug dealing. (p. 451)

intellectualization A defense mechanism, discovered by Anna Freud, in which the adolescent defends against emotional feelings of all kinds by becoming extremely logical about life. (p. 44)

interdependence When people depend upon one another for help, and define themselves as members of a group, such as a family, a church, or a community. In contrast with the belief that self is defined by independence, Gilligan, Josselson, and others suggest that connection with other people helps, rather than interferes with, the development of the self, particularly for women and non-Western cultures. (p. 171)

interdependence theory A new model of parent-adolescent relationships in which adolescents gain independence not through rebellion, but through gradual increases in freedom and responsibility. (p. 215)

intergroup competition Coleman suggests applying the process and the rules of sports to the classroom

so that students experience the joy of being a member of a winning team. (p. 267)

internalizing Expressing depressive feelings by worrying, becoming anxious, and keeping one's feelings inside. (p. 376)

interpersonal competition According to Coleman, the traditional classroom model in which one classmate's success may well mean lower grades for others. (p. 267)

interpropositional thinking The ability to think of the ramifications of combinations of propositions. (p. 111)

intrapropositional thinking The ability to think of a number of possible outcomes that could result from a single choice. (p. 111)

introspection Inward-looking, or thinking about what is going on in one's mind. An example is the question, "Who am I, really?" (p. 169)

inversion The cognitive ability to recognize an inequality of quantity and then subtract from the greater amount to create an equality. (p. 112)

investigative type The Holland vocational personality type who prefers working with ideas and is generally analytical, methodical, and precise. (p. 339)

J

job ceiling According to Ogbu, many minority youths observe unemployment and underemployment among relatives, friends, and neighbors and develop the belief that no matter how hard they work, they are limited by their race in how much they will be allowed to achieve. (p. 345)

juvenile delinquent A juvenile who commits theft or destruction to property. (p. 431)

juvenile justice Involves special hearings, the confidentiality of records, and a separate jailing and punishment of youngsters. (p. 17)

L

later formal operations stage The second phase of formal operations (age 15 to 19) in which the abilities of dealing with systems of symbols, propositional logic, individual thinking patterns, and scientific reasoning have developed. (p. 107)

learned helplessness model of depression A behavioral model of depression proposing that people become depressed because they cannot escape from a bad situation. When their efforts to improve their conditions do not work, they learn that trying is not worthwhile and give up their efforts. (p. 375)

learning theory of homosexuality Homosexuality is thought to be the result of learned experiences from significant others, according to this view. (p. 287)

lesbians A name referring to female homosexuals. (p. 286)

logic Thinking in the formal stage, which is much more orderly and systematic. (p. 105)

longitudinal studies The researcher makes several observations of the same individuals at two or more times in their lives. (p. 28)

looking-glass self Cooley's theory, which suggests that our self-concept is formed through the reflection of the attitudes of others back onto ourselves. (p. 176)

loss of reinforcement model of depression A behavioral model of depression proposing that people become depressed because events or situations no longer provide positive reinforcement. (p. 375)

M

manipulative experiments The experimenter attempts to keep all variables (all the factors that can affect a particular outcome) constant except one, which is carefully manipulated (see "treatment"). The goal is to learn whether this treatment is effective. (p. 26)

marginalism Weak identification with both the majority culture and your own minority culture. A person who feels isolated and alienated from both cultures. (p. 192)

medicinal use Use of drugs to reduce tension and anxieties. (p. 401)

menarche The onset of menstruation. (p. 23, 79)

mental structures The blueprints in our minds that equip us to affect our environment. They are the tools of adaptation. (p. 102)

metacognition "Thinking about thinking," or being able to analyze one's own thoughts. This type of thinking characterizes the formal stage. (p. 105)

midlife transition Feelings of anxiety which sometimes occur at midlife as individuals examine what they have done with their life and think about what they would like to do in the remaining years. (p. 213)

moral components In James Rest's theory, the four steps that are used in the process of solving moral dilemmas. (p. 146)

moratorium of youth A time-out period during which the adolescent experiments with a variety of identities, without having to assume the responsibility for the consequences of any particular one. (p. 187)

mutual collaboration In Selman's model, the ability to understand and respect one's own needs and those of other people. It requires the capacity for intimacy, or sharing of experiences, and the capacity of autonomy, or ability to define one's interests and negotiate them with another person. (p. 185)

N

naturalistic experiments In these experiments, the researcher acts solely as an observer and does as little as possible to disturb the environment. "Nature" performs the experiment, and the researcher acts as a recorder of the results. (p. 27)

nonaggressive offenders Persons who commit crimes that are mainly harmful to themselves, such as runaways or prostitutes. (p. 431)

normal range of development The stages of pubertal change occurring at times that are within the normal range. (p. 85)

novelty Toffler's term for the dissimilarity of new situations in our lives, which contribute to stress. (p. 359)

nurturant parenting style A style of parenting identified by Dacey in which parents are very interested in their children, but provide few firm rules to govern it. Instead, by modeling and family discussions, they espouse a set of well-defined values and expect their children to make personal decisions based upon those values. (p. 221)

O

observational learning Influence of modeling on personality development as stressed in Bandura's social learning theory. (p. 46)

observational studies Studies that describe people simply by counting the number and types of their behaviors. (p. 25)

one-time, one-group studies These are studies that are carried out only once on one group of subjects. (p. 28)

operations Mental events that take the place of actual behavior. (p. 104)

organization Our innate tendency to organize causes us to combine our schemata more efficiently. The schemata of the infant are continuously reorganized to produce a coordinated system of higher-order structures. (p. 103)

outcome expectations Beliefs learned from teachers, parents, friends, and past experience about one's abilities and expectations for success and failure in areas such as school and career. (p. 343)

Outward Bound A program where people increase their sense of self-worth and self-reliance by participating in a series of increasingly threatening experiences. (p. 486)

P

pair therapy A therapeutic technique developed by Selman in which two adolescents work together with a therapist with the goal of developing more advanced interpersonal skills. The pair therapist plays the role of providing a third-person perspective and helps to work through conflicts in the social interaction. (p. 185)

Parsons's theory The first theory of vocational choice, suggesting that people can make better choices if they match their skills and interests to the requirements of the world of work. (p. 337)

pelvic inflammatory disease (PID) This disease often results from chlamydia or gonorrhea, and frequently causes prolonged problems, including infertility. Symptoms include lower abdominal pain and a fever. (p. 308)

permissive-indifferent parenting style Parents who have little control over their children and have few methods of discipline. Lack of discipline results because parents are not interested enough or have too many other stressors in their lives to adequately supervise and discipline their children. (p. 219)

permissive-indulgent parenting style Parents who are warm, but not demanding. They set few rules for their children because they believe that children should have a lot of freedom. (p. 219)

permissive parenting style Describes parents who give their children lots of freedom and hope the children will do what is best. (p. 218)

personal fable An aspect of adolescent egocentric thinking in which adolescents make stories about themselves. Most of these fables have two aspects: a sense of being unique and all-powerful. (p. 115)

personal identity The set of beliefs about the self concerning how one changes over time yet remains the same individual, how one is different from others, and how one is able to act independently. The development of a clear, realistic, and integrated self-concept lays the basis for identity development. (p. 165)

perspective taking The ability to see the environment from someone else's point of view. A child's personal theory about what other people are like, which affects the way she or he relates to other children. This view of what other people are like changes over time. (p. 115, 150)

pituitary gland The main endocrine gland that secretes gonadotropins into the bloodstream, which, upon reaching the brain, stimulate the production of more hormones in other glands. (p. 73)

possession Illegal possession refers to obtaining drugs from someone not legally sanctioned to distribute them. (p. 400)

postconventional level In Kohlberg's theory, people at the postconventional level are

concerned with moral principles that they have thought carefully about and chosen as their own. (p. 137)

precociousness The ability to do what others are able to do, but at a younger age. (p. 126)

preconventional level In Kohlberg's theory, people at the preconventional level are concerned with avoiding punishment and gaining satisfaction. (p. 136)

premature foreclosure A situation where a teenager chooses an identity too early, usually due to external pressure. (p. 188)

premenstrual syndrome (PMS) A series of behavioral, emotional, and physical symptoms that occur to many women around a week before menstruation. (p. 82)

preoperational stage Piaget's second stage, during which the ability to represent objects symbolically in the mind begins. (p. 104)

primary prevention Programs directed at eliminating problems before they begin. (p. 366)

principle of caring In Gilligan's moral theory, the traits of care and sensitivity to the needs of others based on relationships, which so often typifies women's moral judgments. (p. 141)

principle of justice In Kohlberg's theory, refers to our inherited potential to recognize when we are being fair or unfair with each other. (p. 140)

probabilistic The term used by contextual theorists to describe their predictions about behavior. We can never say how people will behave in a certain situation, but only how they will probably behave. (p. 56)

problem behavior theory The theory that early problem behaviors contribute to substance abuse and that persons who have one deviant behavior are likely to have others as well. (p. 402)

prodigiousness The ability to do what is qualitatively better than the rest of us are able to do. Such a person is referred to as a prodigy. (p. 126)

protective factors Characteristics of resilient individuals that protect them from stress. Supportive family environments, support from the social network outside of the family, positive self-concept, an easy-going disposition, and good social skills are some of the known protective factors. (p. 365)

psychoanalysis Freud's explanation of the psychic development of humans. Also his method of psychological therapy. (p. 41)

psychoanalytic model of depression According to this model, the causes of depression are best explained by a history of loss. Depressed adolescents, for example, may have a history of parental separation or death, death of a pet, loss of friends, or a series of moves. (p. 374)

psychoanalytic theory of homosexuality Freud's theory suggested that if the child's first sexual feelings about the parent of the opposite sex are strongly punished, the child may identify with the same-sex parent and develop a permanent homosexual orientation. (p. 287)

psychogenic model One of three models explaining the rise of adolescent subcultures, it assumes that teens are psychologically disturbed by the world they are living in and therefore form a subculture to escape or avoid reality. (p. 268)

puberty A relatively abrupt and qualitatively different set of physical changes that normally occur at the beginning of the teen years. (p. 67)

puberty rite An initiation ceremony often scheduled to coincide with the peak in adolescent physiological maturation. (p. 475)

pushouts Adolescents who leave home because they have been abandoned, neglected, or abused by their parents. They feel rejected by their parents, who often do not oppose their child's leaving. (p. 456)

R

realistic type The Holland vocational personality type who prefers working with tools, machinery, animals, and the outdoors. (p. 339)

real versus the possible The ability of the adolescent to imagine possible and even impossible situations. (p. 109)

recidivism Returning to committing delinquent and criminal activities after having been incarcerated. (p. 452)

recidivism rate The percentage of convicted persons who commit another crime once they are released from prison. (p. 487)

reciprocal determinism According to Bandura, the process through which beliefs, behavior, and environment affect one another as an individual's view of himself or herself is learned. (p. 174)

relativism The second of Perry's three intellectual and ethical stages in which "anything can be right or wrong depending on the situation; all views are equally acceptable." (p. 152)

repudiation As described by Erikson, by choosing one identity we repudiate (turn down) all other choices. (p. 54, 185)

resilience Individuals who deal well with stress and who have few psychological, behavioral, or learning problems as a result of it are said to have resilience. (p. 365)

retreat According to Perry's theory of ethical development, when someone retreats to an earlier ethical position. (p. 153)

retrospective accounts Involves an individual's backward look at their earlier lives. (p. 18)

risk factors Stressors that place an individual at risk for academic or psychological difficulties. Risk factors include poverty, chronic illness, parental mental illness and drug abuse, physical and sexual abuse, exposure to violence, parental divorce, and teenage motherhood. (p. 365)

runaway shelters Temporary shelters for youth who have left home. Runaway shelters continue to be an important part of the treatment system for runaways. (p. 461)

S

schema Mental representations of our roles, of other people and their roles, and of the situations in which they might interact. These schemas function like files into which we sort and classify new information. (p. 149)

schemata Patterns of behavior that infants use to interact with the environment. (p. 102)

school-to-work transition A method of educational reform that seeks to improve the linkages between high school learning and job entry. (p. 335)

secondary prevention Intervention during the early stages of a problem to reduce its severity or the length of time it will last. (p. 366)

second individuation According to Blos, in adolescence another individuation occurs when the individual grows from a dependent child to an independent adult. (p. 170)

secular trend The decreasing age of the onset of puberty in females in Western countries, particularly the average age of menarche. (p. 75)

self-concept Those beliefs, attitudes, and thoughts about the self that are descriptions about one's physical, social, and psychological qualities (e.g., what I look like, what I am good at and how I feel). (p. 165)

self-efficacy Our self-expectations or beliefs about what we can accomplish as a result of our efforts, which influence our willingness to attempt the task and the level of success we achieve. (p. 174, 343)

self-esteem Evaluating one's success in meeting set goals. (p. 165)

self-in-relation A theory developed by the Stone Center, which states that a clear understanding of our relationships with others is most important in the definition of the self, especially for women. (p. 173)

self-report studies Studies that ask people their opinions about themselves or other people by use of interviews or questionnaires. (p. 25)

sensorimotor stage Piaget's first stage, during which mental operations are not yet possible. (p. 102)

sequential (longitudinal/cross-sectional) studies Cross-sectional studies done at several times with the same groups of individuals. (p. 30)

sexual identity Development of a cohesive sense of self as a sexual being in relation to culturally determined categories. (p. 195)

sexual orientation Choice of a sexual partner of the same sex or opposite sex. (p . 196)

sexual revolution The extraordinary change in human sexual behavior that occurred in the 1960s and 1970s. (p. 277)

sexually transmitted disease (STD) A class of diseases, such as AIDS, gonorrhea, herpes, chlamydia, etc. that are transmitted through sexual behavior. (p. 306)

social capital The knowledge, values, and beliefs passed on from one generation to the next through time spent together. (p. 226)

social cognition The ability to think critically about interpersonal issues, which develops through age and experience. Used to make sense of other people and to decide how to interact with them. (p. 114, 148)

social construction The idea that self-esteem represents, in large part, the attitudes that significant others hold. (p. 176)

social isolates Persons who are avoided by others, and typically avoid each other because they have low prestige. (p. 259)

social learning According to this theory, children develop their attitudes toward sexuality by modeling and conditioning. For instance, parents in American culture generally offer their children the model that sex is an embarrassing topic. (p. 296)

social norms The behavior that society says is appropriate. (p. 146)

social readjustment rating scale A scale developed by Holmes and Rahe to measure life events that require considerable adjustment and to determine the relationship between those events and the physical health. (p. 360)

social type Among Holland's vocational personality types, the social type enjoys working with and helping other people. (p. 339)

social use Infrequent use of drugs that is limited to parties, dances, and other special occasions. (p. 401)

sociocultural diversity Many cultures coexist within a country such as the United States or Canada. The term "sociocultural" emphasizes social rather than biological or psychological differences. (p. 6)

sociogram Graphic picture resulting from sociogram technique (see Figure 8.2). (p. 271)

sociogram technique Asking each member of a group to write down the names of the three people in the group whom they like the most. To score, each person who was named

first is given a 3, second a 2, and third a 1. Then the scores for each person in the group are added, thereby indicating a range from most popular to social isolate. (p. 259)

solidification Cognitive growth in which the thinker is more certain and confident in the use of the newly gained mental skills and is more likely to use them in new situations. (p. 113)

state of identity Erikson not only describes identity as the general picture one has of oneself, he also refers to it as a state toward which one strives. If you were in a state of identity, the various aspects of your self-concept would be in agreement with each other; they would be identical. (p. 185)

stereotype Generalizations about the characteristics—either positive or negative—of a group that are supposedly shared by its members. (p. 3)

storm and stress Hall's description of emotional upheavals between the ages of 12 and 25 that also typify human history for the past 2,000 years. (p. 38)

street kids Runaways who join with other runaways and become skillful at fending for themselves through involvement in illegal activities. (p. 460)

stress Emotional tension arising from life events and feelings of threat to one's safety or self-esteem. (p. 355)

subculture Any group that has its own customs (ways of dressing, for example) but is also part of a larger cultural group. (p. 265)

superego One of Freud's three structures of the psyche. It is comparable to the conscience. (p. 42)

Super's theory An influential theory of career development presenting a series of developmental stages or tasks essential for the development of career identity and maturity. (p. 339)

syphilis A sexually transmitted disease which presents a great danger in that in its early stage, there are no symptoms. Its first sign is a chancre ("shan-ker"), a painless open sore that usually shows up on the tip of the penis and around or in the vagina. After a while, this disappears, and if there is no treatment, the disease enters its third stage, which can lead to death. (p. 309)

system kids A group of runaway adolescents who were removed from their parents' homes by social welfare agencies because of parental neglect or abuse. After being placed in foster care or a group home, these adolescents decide to run. (p. 460)

systems theory A view of the family and other parts of the social environment, such as the school and peer group, in which the experiences of one individual are believed to affect other individuals and the system as a whole. (p. 216)

T

temporizing An aspect of Perry's theory of ethical development, in which some people remain in one position for a year or more, exploring its implications, but hesitating to make any further progress. (p. 153)

testosterone The hormone produced by the testes that triggers the physical changes of puberty in boys. (p. 73)

throwaways Adolescents who leave home because they have been abandoned, neglected, or abused by their parents. They feel rejected by their parents, who often do not oppose their child's leaving. (p. 456)

trafficking The sale of large amounts of illegal drugs. (p. 401)

transience Toffler's term for the lack of permanence of things in our lives that leads to increased stress. (p. 359)

treatment Action taken with an experimental group, but not with a control group, in order to measure its effects. Examples of treatments are instruction, medication, and therapy. (p. 26)

U

unexamined cultural identity Early adolescents and others who may have given little thought to their identity. They may have absorbed their parents' attitudes without question, or they may prefer views of the dominant culture. (p. 192)

V

varicocele A condition occurring in male puberty in which one testicle is noticeably larger than the other. (p. 84)

victimless crime Some people feel that drug abusers are the only ones who suffer from the use of drugs, so they should not be fined or imprisoned for their actions. However, drug abusers tend to be poor financial risks, dangerous drivers, and often resort to theft to support their habit; they are usually dependent on their families and often on society for support, so their acts are seldom victimless. (p. 398)

W

walkabout Originally an aborigine initiation rite, the American version attempts to focus the activities of secondary school by demonstrating to the student the relationship between education and action. (p. 483)

Z

zone of proximal development In Vygotsky's theory, the area at the upper edge of our present abilities in any task in which cognitive growth takes place. (p. 114)

REFERENCES

Aalsma. M & Lapsley, D. (1999). Religiosity and adolescnt narcissim: Implications for value counseling. *Counseling and Values, 44 (1), 17-29.*

AAUW, (1999). *Gender Gaps.* NY: Marlowe.

Abramson, L. Y., Seligman, M. E. P., & Teasdale, J. (1978). Learned helplessness in humans: Critique and reformulation. *Journal of Abnormal Psychology*, 87, 49–74.

Ackerman, J. W. (1958). *The psychodynamics of family life.* New York: Basic Books.

Ackerman, P. T., Anhalt, J. M., Holcomb, P. J., & Dykman, R. A. (1986). Presumably innate and acquired automatic processes in children with attention and/or reading disorders. *Journal of Child Psychology and Psychiatry and Allied Disciplines*, 27 (4), 513–529.

Adams, D. (1992). Biology does not make men more aggressive than women. In K. Bjorkqvist & P. Niemela (Eds.), *Of mice and men: Aspects of female aggression* (pp. 17–26). San Diego: Academic Press.

Adams, G. R., & Gullotta, T. S. (1983). *Adolescent life experiences.* Monterey, CA: Brooks-Cole.

Adams, J. (1986). *Conceptual blockbusting* (3rd ed.). Reading, MA: Addison-Wesley.

Adler, T. (1991, March). Little is known of teens' pregnancy motivations. *APA Monitor*, pp. 14–15.

Ainsworth, M. D. S., Blehar, M. C., Walters, E., & Wall, S. (1978). *Patterns of attachment: A psychological study of the strange situation.* Hillsdale, NJ: Erlbaum.

The Alan P. Guttmacher Institute. (1994). *Sex and America's teenagers.* New York: The Alan P. Guttmacher Institute.

Alexander, A., & Kempe, R. S. (1984). The role of the lay therapist in long-term treatment. *Child Abuse and Neglect*, 6 (30), 329–334.

Alexander, P. A. (1985). Gifted and nongifted students' perceptions of intelligence. *Gifted Child Quarterly*, 29 (3), 137–143.

Alfieri, T., Ruble, D. & Higgins, E. (1996). Gender stereotypes during adoelscence: Developmental changes and the transition to junior high school. *Developmental Psychology,* 32 (6), 1129-1137.

Alishio, K., & Maitland-Shilling, K. (1984). Sex differences in intellectual and ego development in late adolescence. *Journal of Youth and Adolescence*, 13, 213–224.

Allen, J. P., Hauser, S. T., Bell, K. L., & O'Connor, T. G. (1994). Longitudinal assessment of autonomy and relatedness in adolescent-family interactions as predictors of adolescent ego development and self-esteem. *Child Development*, 65, 179–194.

Allen, L., & Majidi-Ahi, S. (1989). Black American children. In J. T. Gibbs, L. N. Huang, & Assoc., *Children of color: Psychological interventions with minority youth.* San Francisco: Jossey-Bass.

Allen, S.F. & Stoltenberg, C.D. (1995). Psychological separation of older adolescents and young adults from their parents: An investigation of gender differences. *Journal of Counseling and development,* 73 (5), 542-546.

Allgood-Merten, B., Lewinsohn, P., & Hops, H. (1990). Sex differences in adolescent depression. *Journal of Abnormal Psychology*, 99 (1), 55–63.

Alliance for Service Learning in Education Reform. (1993). Standards of quality for school-based service learning. *Equity & Excellence in Education*, 26,71–73.

Allison, K. W., & Takei, Y. (1993). Diversity: The cultural contexts of adolescents and their families. In R. M. Lerner (Ed.), *Early adolescence: Perspectives on research, policy, & intervention.* Hillsdale, NJ: Erlbaum.

Almeid, D.M., Wethington, E. & Cxhandler, A. (1999). Daily transmissions of tensions between marital dyads and child dyad. *Journal of Marriage and the Family, 61 (1), 49-61.*

Amabile, T. M., Hennessey, B. A., & Grossman, B. S. (1986). Social influences on creativity: The effects of contracted-for reward. *Journal of Personality and Social Psychology*, 50, 14.

American Council on Education. (1995). *Annual Status Report.* Washington, DC: American Council on Education, Office of Minority Concerns.

American Psychiatric Association. (1985). *Diagnostic and statistical manual of mental disorders (DSM-III-R)* (3rd, rev. ed.) Washington, DC: American Psychiatric Press.

Andersen, A. E., & Hays, A. (1985). Racial and socioeconomic influences in anorexia nervosa and bulimia. *International Journal of Eating Disorders*, 4, 479–488.

Anderson, A. M., & Noesjirwan, J. A. (1980). Agricultural college initiations and the affirmation of rural ideology. *Mankind*, 12 (4), 341–347.

Anderson, E.R., Greene, S.M. Hetherington, E.M. & Clingempeel, W.G. (1999). The dynamics of parental remarriage: Adolescent, parent, and sibling influences. In Hetherington, E.M. (Ed.) *Coping with divorce*.

Anderson, S., Russell, C., & Schumm, W. (1983). Perceived marital quality and family life-cycle categories: A further analysis. *Journal of Marriage and the Family*, 45, 127–139.

Anderson, V., Kinsley, C., Negroni, P., & Price, C. (1991). Community service learning and social improvement in Springfield, Massachusetts. *Phi Delta Kappan*, 72, 761–764.

Andre, T., Frevert, R., & Schuchmann, D. (1989). From whom have college students learned what about sex? *Youth and Society*, 20 (3), 241–268.

Angle, C. (1983). Adolescent self-poisoning: A nine-year follow-up. *Developmental and Behavior Pediatrics*, 4 (2), 83–87.

Apter, T. (1990). Mothers on a see-saw: Friends and peers. *Altered loves: Mothers and daughters during adolescence*. New York: St. Martin's Press.

Archer, S. L. (1992). A feminist's approach to identity research. In G. R. Adams, T. P. Gullotta, & R. Montemayor (Eds.), *Adolescent identity formation: Advances in adolescent development* (pp. 25–49). Newbury Park, CA: Sage.

Archer, S. L. (1993). Identity in relational contexts: A methodological proposal. In J. Kroger (Ed.), *Discussion on ego identity* (pp. 75–99). Hillsdale, NJ: Erlbaum.

Aries, P. (1962). *Centuries of childhood. New York*: Vintage Books.

Arlin, P. K. (1975). Cognitive development in adulthood: A fifth stage? *Developmental Psychology*, 11 (5), 602–606.

Armistead, L., Wierson, M., & Forehand, R. (1990). Adolescents and maternal employment: Is it harmful for a young adolescent to have an employed mother? *Journal of Early Adolescence*, 10 (3), 260–278.

Armsden, G., McCauley, E., Greenberg, M., Burke, P., & Mitchell, J. (1990). Parent and peer attachment in early adolescent depression. *Journal of Abnormal Child Psychology*, 18 (6), 683–697.

Armsden, G. C., & Greenberg, M. T. (1987). The inventory of parent and peer attachment: Individual differences and their relationship to psychological well-being in adolescence. *Journal of Youth and Adolescence*, 16 (5), 427–454.

Armstrong, J., & Roth, D. M. (1989). Attachment and separation difficulties in eating disorders: A preliminary investigation. *International Journal of Eating Disorders*, 8, 141–155.

Arnett, J.J. (1999). Adolescent storm and stress, revisited. *American Psychologist, 54 (5), 317-326.*

Ary, D.V., Duncan, T.E., Duncan, S.C. & Hops, H. (1999). Adolescent problem behavior: The influence of parents and peers. *Behaviour Research and Therapy,* 37 (3), 217-230.

Ashford, J., & LeCroy, C. (1990). Juvenile recidivism: A comparison of three prediction instruments. *Adolescence*, 25 (98), 441–450.

Ashton, J., & Donnan, S. (1981). Suicide by burning as an epidemic phenomenon: An analysis of 82 death and inquests in England and Wales in 1978–79. *Psychological Medicine*, 11 (4), 735–739.

Asian-American Drug Abuse Program, Inc., Report. (1978). Los Angeles, CA.

Associated Press. (1995, April 28). Reading test shows drop in students' proficiency. *The Boston Globe*, p. 3.

Association for Supervision and Curriculum Development (ASCD). (1991). Plan for "African-American immersion schools" raises questions. *Update*, 33, 6–8.

Atkinson, D., Froman, T., Romo, J., & Mayton, D. (1977). The role of the counselor as a social activist: Who supports it? *School Counselor*, 25, 85–91.

Atkinson, R. (1989). Respectful, dutiful teenagers. *Psychology Today*, 22, 22–26.

Attie, I., & Brooks-Gunn, J. (1989). Development of eating problems in adolescent girls: A longitudinal study. *Developmental Psychology*, 25 (1), 70–79.

Attie, I., Brooks-Gunn, J., & Petersen, A. C. (1990). A developmental perspective on eating disorders and eating problems. In M. Lewis & S. Miller (Eds.), *Handbook of developmental psychopathology*. New York: Plenum.

Bachman, J. G., Johnston, L. D., & O'Malley, P. M. (1987). *Monitoring the future: Questionnaire responses from the nation's high school seniors*, 1986. Ann Arbor: Survey Research Center, Institute for Social Research.

Bachman, J. G., O'Malley, P. M., & Johnston, P. (1978). *Adolescence to adulthood: Change and stability in the lives of young men*. Ann Arbor, MI: Institute for Social Research.

Baer, J., Barr, H., Bookstein, F., Sampson, P. & Streissguth, A. (1998). Prenatal alcohol exposure and family history of alcoholism in etiology of adolescent problems. *Journal of Studies on Alcohol*, September, 533-543.

Baier, J. L., & Williams, P. S. (1983). Fraternity hazing revisited: Current alumni and active member attitudes toward hazing. *Journal of College Student Personnel*, 24 (4), 300–305.

Bailey, S. (1992). *Shortchanging girls, shortchanging America.* American Association of University Women Educational Foundation.

Baker, W. (1989, Winter). The global teenager. *Whole Earth Review*, 65, 2–35.

Bakken, L., & Romig, C. (1992). Interpersonal needs in middle adolescents: Companionship, leadership and intimacy. *Journal of Adolescence*, 15, 301–316.

Balk, D. E. (1995). *Adolescent development. Early through late adolescence.* Pacific Grove, CA: Brooks/Cole.

Bancroft, L. (1979). The reasons people give for taking overdoses: A further inquiry. *British Journal of Medical Psychology*, 52, 353–365.

Bandura, A. (1978). The self system in reciprocal determinism. *American Psychologist*, 344–358.

Bandura, A., Ross, D., & Ross, S. (1963). Imitation of film-mediated aggressive models. *Journal of Abnormal and Social Psychology*, 66, 3–11.

Bandura, A., & Walters, R. (1959). *Adolescent aggression.* New York: Ronald Press.

Barnes, G., Farrell, M., & Banerjee, S. (1994). Family influence on alcohol abuse and other problem behaviors among black and white adolescents in a general population sample. *Journal of Research on Adolescence*, 4, 183–202.

Baron, P., & Peixoto, N. (1991). Depressive symptoms in adolescents as a function of personality characteristics. *Journal of Youth and Adolescence*, 20, 493–500.

Baron, S. (1989). Resistance and its consequences: The street culture of punks. *Youth and Society*, 21 (2), 207–237.

Barrera, M., Jr. (1981). Social support in the adjustment of pregnant adolescents: Assessment issues. In B. H. Gottlieb (Ed.), *Social networks and social support* (pp. 69–96). Beverly Hills, CA: Sage.

Baruch, G. K., Biener, L., & Barnett, R. C. (1987). Women and gender in research on work and family stress. *American Psychologist*, 42, 130–136.

Bass, A. (1992, January 6). Teen sex. *The Boston Globe*, pp. 31, 34, 35.

Bass, A. (1995, June 3). Measuring TV news' toll on children. *The Boston Globe*, pp. 1, 9.

Baumrind, D. (1968). Authoritarian vs. authoritative parental control. *Adolescence*, 3, 255–272.

Baumrind, D. (1971). Current patterns of parental authority. *Developmental Psychology Monographs*, 4, 1–103.

Baumrind, D. (1989). Parenting styles and adolescent development. In J. Brooks-Gunn, R. Lerner, & A. C. Petersen (Eds.), *The encyclopedia of adolescence.* New York: Garland.

Baumrind, D. (1991a). Effective parenting during the early adolescent transition. In P. A. Cowan & M. Hetherington (Eds.), *Family transitions* (pp. 111–163). Hillsdale, NJ: Erlbaum.

Baumrind, D. (1991b). The influence of parenting style on adolescent competence and substance abuse. *Journal of Early Adolescence*, 11 (1), 56–95.

Baumrind, D. (1996). The discipline controversy revisited. *Family Relations: Journal of Applied Family and Child Studies*, 45 (4), 405-414.

Beaty, L.A. (1999). Identity development of homosexual youth and parental and familial influences on the coming out process. *Adolescence*, 34 (135), 597-601.

Beardslee, W. R., & Podorefsky, D. (1988). Resilient adolescents whose parents have serious affective and other psychiatric disorders: Importance of self-understanding and relationships. *American Journal of Psychiatry*, 145, 63–69.

Beck, A. (1976). *Cognitive therapy and the emotional disorders.* New York: International Universities Press.

Beck, A. T. (1967). *Depression: Clinical, experimental and theoretical aspects.* New York: Hoeber.

Becker, H. (1987). *Addressing the needs of different groups of early adolescents: Effects of various school and organizational practices on students from different social backgrounds and abilities.* Report no. 16. Baltimore: Johns Hopkins University, Center for Research on Elementary and Middle Schools.

Belenky, M., Clinchy, B., Goldberger, N., & Tarule, J. (1986). *Women's ways of knowing.* New York: Basic Books.

Bell, C., & Ward, G. R. (1980). An investigation of the relationship between dimensions of self-concept and achievement in mathematics. *Adolescence*, 15, 895–901.

Bellah, R., Madsen, R., Sullivan, W., Swidler, A., & Tipton, S. (1985). *Habits of the heart: Individualism and commitment in American life.* Berkeley: University of California Press.

Bem, A. (1987). Youth suicide. *Adolescence*, 22 (86), 271–290.

Bem, D. (1996). Exotic becomes erotic: A developmental theory of sexual orientation. *Psychological review, 103 (2), 320-335.*

Bem, S. L. (1975). Sex role adaptability: One consequence of psychological androgyny. *Journal of Personality and Social Psychology*, 31, 634–643.

Bem, S. L. (1981). Gender schema theory: A cognitive account of sex typing. *Psychological Review*, 88 (4), 354–364.

Benedict, R. (1950). *Patterns of culture*. New York: New American Library.

Benedict, R. (1954). Continuities and discontinuities in cultural conditioning. In W. E. Martin & C. B. Stendler (Eds.), *Readings in child development*. New York: Harcourt, Brace.

Benson, J., Eiserman, P., & Wardlaw, G. (1989). Relationship between nutrient intake, body mass index, menstrual function, and ballet injury. *Journal of the American Dietetic Association*, 89 (1), 58–63.

Benson, P. (1993). *The troubled journey: A portrait of 6th–12th grade youth*. Minneapolis, MN: Search Institute.

Berkowitz, L. (1994). Guns and youth. In L. D. Eron, J. Gentry, & P. Schlegel (Eds.), *Reason to hope: A psychosocial perspective on violence and youth* (pp. 251–279). Washington, DC: American Psychological Association.

Berlin, I. N. (1987). Suicide among American Indian adolescents: An overview. *Suicide and Life Threatening Behavior*, 17, 218–232.

Berman, A., & Jobes, D. A. (1991). *Adolescent suicide: Assessment and intervention*. Washington, DC: American Psychological Association.

Berman, A. L., & Schwartz, R. (1990). Suicide attempts among adolescent drug users. *American Journal of Diseases of Children*, 144, 310–314.

Berman, M. (1975, March 30). Review of Life, *history and historical moment* by E. Erikson. *New York Times Magazine*, 2.

Berman, P. (1990). *The search for meaning: Americans talk about what they believe and why*. New York: Ballantine/Random House.

Berndt, T. J. (1989). Friendships in childhood and adolescence. In W. Damon (Ed.), *Child development today and tomorrow*. San Francisco: Jossey-Bass.

Berndt, T. J., & Mekos, D. (1995). Adolescents' perceptions of the stressful and desirable aspects of the transition to junior high school. *Journal of Research on Adolescence*, 5, 123–142.

Bersoff, D. M., & Miller, J. G. (1993). Culture, contexts, and the development of moral accountability judgments. *Developmental Psychology*, 294, 664–676.

Berzonsky, M. & Kuk, L. (2000). Identity status, identity processing style and transition to university. *Journal of Adolescent Research*. 15 (1), 81-98.

Besharov, D. & Gardiner, K. (1997). Trends in teen sexual behavior. *Children and Youth Services Review, 19 (5-6), 341-367*.

Bettleheim, B. (1969). *The children of the dream*. New York: Macmillan.

Biederman, J & Spencer, T. (1999). Depressive disorders in childhood and adolescence: A clinical perpsective. *Journal of Child and Adolescent Psychopharmacology, 9 (4), 233-237*.

Billy, J., Landale, N., Grady, W., & Zimmerle, D. (1988). Effects of sexual activity on adolescent social and psychological development. *Social Psychology Quarterly*, 51 (3), 190–212.

Bing, L. (1991). *Do or die*. New York: HarperCollins.

Birtchnell, J., & Alarcon, J. (1971). The motivation and emotional state of 91 cases of attempted suicide. *British Journal of Medical Psychology*, 44, 45–52.

Bjarnason, 1998). Parents, religion and perceived social coherence: A Durkheimian framework of adolescent anomie. *Journal for Scientific Study of religion, 37 (4), 742-754*.

Black, A. (1974). *Without burnt offerings*. New York: Viking.

Blau, G. M., & Gullotta, T. P. (1993). Promoting sexual responsibility in adolescence. In T. P. Gullotta, G. R. Adams, & R. Montemayor (Eds.), *Adolescent sexuality, advances in adolescent development: An annual book series, Vol. 5* (pp. 181–203). Newbury Park, CA: Sage.

Block, J., & Robins, R. W. (1993). A longitudinal study of consistency and change in self-esteem from early adolescence to early adulthood. *Child Development*, 64, 909–923.

Bloom, B. (1964). *Stability and change in human characteristics*. New York: Wiley.

Bloom, J. K. (1991, October 1). Fewer black men attend college. *The Boston Globe*, pp. 1, 8.

Blos, P. (1979). *The adolescent passage*. New York: International Universities Press.

Blumberg, F. (1999). Developmental differences at play: Children's selective attention and performance in video games. *Journal of Applied Developmental Psychology*, 19 (4), 615-624.

Blustein, D., Devenis, L. E., & Kidney, B. A. (1989). Relationship between identity formation process and career development. *Journal of Counseling Psychology*, 36, 196–202.

Blyth, D., Simmons, R., & Zakin, D. (1985). Satisfaction with body image for early adolescent females: The impact of pubertal timing within different school environments. *Journal of Youth and Adolescence*, 14 (3), 207–225.

Blyth, D. A., & Roehlkepartain, E. C. (1992). Working together. *Search Institute Source*, 8 (2), 1–3.

Blyth, D. A., & Roehlkepartain, E. C. (1993). *Healthy communities, healthy youth: How communities contribute to positive youth development.* Minneapolis, MN: Search Institute.

Boas, F. (1911). Growth. In H. Kiddle (Ed.), *A cyclopedia of education*. New York: Steiger.

Boehm, K., Schondel, C., Marlowe, A., & Manke-Mitchell, L. (1999). Teens' concerns: A national evaluation. *Adolescence, 34* (135),

Bogenschneider, K. , Wu, M., Raffaelli, M., & Tsay, J. (1998). "Other teens drink, but not my kid": Does parental awareness of adolescent alcohol use protect adolescents from risky consequences? *Journal of Marriage and the Family*, 60, 356-373.

Bogenschneider, K., Wu, Ming Yeh, Raffaelli, M. & Tsay, J.C. (1998). Parent influences on adolescent peer oriemtation and substance use: The interface of parenting practices and values. *Child Develoipment, 69* (6), 1672-1688.

Boldizar, J. P., Wilson, K. L., & Deemer, D. K. (1989). Gender, life experiences, and moral judgment development: A process-oriented approach. *Journal of Personality and Social Psychology*, 57, 229–238.

Bollen, K., & Phillips, D. (1982). Imitative suicides: A national study of the effects of television news stories. *American Sociological Review*, 47, 802–809.

Bookman, M. (1999). Phantoms slain: Reading Gilligan as a revolutionary text. *Mind Culture and Activity*, 6 (3), 237-252.

Borchard, D. C., Kelly, J. J., & Weaver, N. P. K. (1988). *Your career: Choices, chances and changes*. Dubuque, IA: Kendall/Hunt.

Boston Women's Health Book Collective. (1984). *Our bodies, ourselves*. Boston: Touchstone.

Botta, R.A. (1999). Television images and adolescent girtls' body image disturbance. *Journal of Communication*, 49 (2), 22-41.

Bourne, R. (1913). *Youth and life*. Boston: Little, Brown.

Bovee, T. (1991, September 29). Wide gap found in earnings by white, black college grads. *The Boston Globe*, pp. 1, 7.

Bower, B. (1989). Teenagers reap broad benefits from authoritative parents. *Science News*, 136, 117–118.

Bowlby, J. (1969). *Attachment and loss: Vol. I: Attachment*. New York: Basic Books.

Bowlby, J. (1973). *Separation: Anxiety and anger*. New York: Basic Books.

Bowlby, J. (1988). Developmental psychiatry comes of age. *American Journal of Psychiatry*, 145, 1–10.

Boxer, A. M., Cohler, B. J., Herdt, G., & Irvin, F. (1993). Gay and lesbian youth. In P. H. Tolan & B. J. Cohler (Eds.), *Handbook of clinical research and practice with adolescents* (pp. 249–280). New York: John Wiley.

Boxer, A.M., Cook, J.A. & Herdt, G. (1999). Experience of coming out among gay and lesbian youth: Adolescents alone? In Blustein, J. & Levine, C. (Eds.), *The adolescent alone: Decision making in health care in the United State* (pp. 121-138). NY: Cambridge University Press.

Boyer, E. (1983). *High school: A report on secondary education in America*. New York: Harper & Row.

Boykin, A. W., & Toms, F. (1985). Black child socialization. In H. McAdoo & J. McAdoo (Eds.), *Black children: Social, educational, & parental environments* (pp. 33–51). Newbury Park, CA: Sage.

Brabeck, M. (1984). Longitudinal studies of intellectual development during adulthood. *Journal of Research and Development in Education*, 17 (3), 12–25.

Brabeck. M. (1989). *Who cares? Theory, research, and educational implications of the ethic of care*. New York: Praeger.

Brabeck, M., & Weisgerber, K. (1988). Responses to the Challenger tragedy: Subtle and significant gender differences. *Sex Roles*, 19, 639–650.

Brack, C., Orr, D., & Ingersoll, G. (1988). Pubertal maturation and adolescent self-esteem. *Journal of Adolescent Health Care*, 9, 280–285.

Brain, J. L., Blake, C. F., Bluebond-Langner, M., Chilungu, S. W., Coelho, V. P., Domotor, T., Gorer, G., LaFontaine, J. S., Levy, S. B., Loukotos, D., Natarajan, N., Raphael, D., Schlegel, A., Stein, H. G., & Wilder, W. D. (1977). Sex, incest, and death: Initiation rites reconsidered. *Current Anthropology*, 18 (2), 191–198.

Braisted, J., Mellin, L., Gong, E., & Irwin, C. (1985). The adolescent ballet dancer: Nutritional practices and characteristics associated with anorexia nervosa. *Journal of Adolescent Health Care*, 6, 365–371.

Bratcher, W. E. (1982). The influence of the family on career selection: A family systems perspective. *Personnel and Guidance Journal*, 87–91.

Brelis, M. (1998). Lewd hazing on cutter detailed. Boston: *Boston Globe*, p. B1,4.

Brengden, M., Vitaro, F. & Bukowski, W. (2000). Deviant friends and early adolescents' emotional and behavioral adjustment. *Journal of Research on Adolescence, 10 (2)*, 173-189.

Brent, D.A., Baugher, M., Bridge, J., Chen, T. & Chiapetta, L. (1999). Age and sex related risk factors for adolescent suicide. *Journal of the American Academy of Child and adolescent Psychiatry*, 38 (12), 1497-1505.

Brent, D. A., Perper, J. A., Goldstein, C. E., Kolko, D. J., Allan, M. J., Allman, C. J., & Zeleank, J. P. (1988). Risk factors for adolescent suicide. *Archives for General Psychiatry*, 45, 581–588.

Bronfenbrenner, U. (1979). *The ecology of human development*. Cambridge, MA: Harvard University Press.

Brooks, A. (1989). *Children of fast-track parents*. New York: Viking.

Brooks-Gunn, J. (1987). Pubertal processes: Their relevance for developmental research. In V. Van Hasselt & M. Hersen (Eds.), *Handbook of adolescent psychology* (pp. 11–129). Elmsford, NY: Pergamon.

Brooks-Gunn, J. (1988). Antecedents and consequences of variations in girls' maturational timing. *Journal of Adolescent Health Care*, 9, 365–373.

Brooks-Gunn, J. (1991). How stressful is the transition to adolescence for girls? In M. E. Colten & S. Gore (Eds.), *Adolescent stress: Causes and consequences* (pp. 131–149). Hawthorne, NY: Aldine de Gruyter.

Brooks-Gunn, J. & Chase-Lansdale, P.L. (1995). *Adolescent parenthood*. In Handbook of parenting (3): Status and social conditions of parenting (pp. 113-149). NJ: Lawrence Elbaum Associates.

Brooks-Gunn, J., & Furstenberg, F. (1986). The children of adolescent mothers: Physical, academic and social outcomes. *Developmental Review*, 6 (3), 224–251.

Brooks-Gunn, J., & Furstenberg, F. F., Jr. (1989). Adolescent sexual behavior. *American Psychologist*, 44 (2), 249–257.

Brooks-Gunn, J., & Graber, J. (1999). What's sex got to do with it? The development of sexual identities during adolescence. In Contrada, R. & Ashmore, R. (Eds.), *Self, social identity, and physical health: Interdisciplinary explorations. 155-182.* NY: Oxford University Press.

Brooks-Gunn, J., & Paikoff, R. L. (1993). Sex is a gamble, kissing is a game: Adolescent sexuality and health promotion. In S. G. Millstein, A. C. Petersen, & E. O. Nightingale (Eds.), *Promoting the health of adolescents: New directions for the twenty-first century* (pp. 180–208). New York: Oxford University Press.

Brooks-Gunn, J., & Petersen, A. (1984). Problems in studying and defining pubertal events. *Journal of Youth and Adolescence*, 13 (3), 181–195.

Brooks-Gunn, J., & Petersen, A. C. (1991). Studying the emergence of depression and depressive symptoms during adolescence. *Journal of Youth and Adolescence*, 20 (2), 115–119.

Brooks-Gunn, J., Petersen, A., & Eichorn, D. (1985). The study of maturational timing effects in adolescence. *Journal of Youth and Adolescence* 14 (3), 149–161.

Brooks-Gunn, J., & Reiter, E. (1990). The role of pubertal processes. In S. Feldman & G. Elliott, (Eds.), *At the threshold: The developing adolescent*. Cambridge, MA: Harvard University Press.

Brooks-Gunn, J., & Warren, M. (1988). The psychological significance of secondary sexual characteristics in nine- to eleven-year-old girls. *Child Development*, 59, 1061–1069.

Brooks-Gunn, J., & Warren, M. (1989). Biological and social contributions to negative effect in young adolescent girls. *Child Development*, 60, 40–55.

Brophy, J. (1985). Interactions of male and female students with male and female teachers. In L. C. Wilkinson & C. B. Marrett (Eds.), *Gender influences in classroom interaction*, pp. 115–142. New York: Academic Press.

Brown, B., Clasen, D., & Eicher, S. (1986). Perceptions of peer pressure, peer conformity dispositions, and self-reported behavior among adolescents. *Developmental Psychology*, 22, 521–530.

Brown, B. B. (1989). The role of peer groups in adolescents' adjustment to secondary schools. In T. Berndt & G. Ladd (Eds.), *Peer relationships in child development* (pp. 188–215). New York: John Wiley.

Brown, B. B. (1990). Peer groups and peer cultures. In S. Feldman and G. Elliot (Eds.), *At the threshold: The developing adolescent*. Cambridge, MA: Harvard University Press.

Brown, B. B., Eicher, S. A., & Petrie, S. (1986). The importance of peer group ("crowd") affiliation in adolescence. *Journal of Adolescence*, 9, 73–96.

Brown, G. W., Bhrolchain, M. N., & Harris, R. (1975). Social class and psychiatric disturbance among women in an urban population. *Sociology*, 9, 225–54.

Brown, J., Childers, K., & Waszak, C. (1990). Television and adolescent sexuality. *Journal of Adolescent Health Care*, 11, 62–70.

Brown, L. J. P., Powell, J., & Earls, F. (1989). Stressful life events and psychiatric symptoms in black adolescent females. *Journal of Adolescent Research*, 4, 140–151.

Brown, L. M., & Gilligan, C. (1992). *Meeting at the crossroads: Women's psychology and girls' development*. Cambridge, MA: Harvard University Press.

Brown, S. A. (1993). Recovery patterns in adolescent substance abuse. In S. Barr, G. A. Marlott, & R. J. McMahon (Eds.), *Addictive behaviors across the lifespan* (pp. 161–183). Newbury Park, CA: Sage.

Brown, S. A., Mott, M. A., & Myers, M. A. (1990). Adolescent alcohol and drug treatment outcome. In R. R. Watson (Ed.), *Drug and alcohol abuse prevention* (pp. 373–403). Totowa, NJ: Human.

Brown, S. V. (1983). The commitment and concerns of black adolescent parents. *Social Work Research and Abstracts*, 19(4), 27–34.

Browne, A., & Finkelhor, D. (1986). Impact of child sexual abuse: A review of the research. *Psychological Bulletin*, 99, 66–77.

Bruch, H. (1981). *Eating disorders*. Canada: Basic Books.

Buchanan, C., Maccoby, E. E., & Dornbusch, S. M. (1991). Caught between parents: Adolescents' experience in divorced homes. *Child Development*, 62, 1008–1029.

Buchanan, C. M., Maccoby, E. E., & Dornbusch, S. N. (1992). Adolescents and their families after divorce: Three residential arrangements compared. *Journal of Research on Adolescence*, 2 261–291.

Buffum, J. (1988). Substance abuse and high-risk sexual behavior: Drugs and sex, the dark side. *Journal of Psychoactive Drugs*, 20 (2), 165–168.

Buffum, J., & Moser, C. (1986). MDMA and human sexual function. The MDMA Conference, Oakland, California. *Journal of Psychoactive Drugs*, 18 (4), 355–359.

Bukowski, W., Sippola, L. & Hoza, B. (1999). Same and other: Interdependency bewteen particip[ation in same and other sex friendships. *Journal of Youth and Adolescence, 28 (4), 439-459.*

Burgess, A. W. (1986). *Youth at risk: Understanding runaway and exploited youth*. Washington, DC: National Center for Missing and Exploited Children.

Burgos-Debray, E. (1984). *I, Rigoberta Menchu*. New York: Alpine Press.

Burke, T. (1990). Home invaders: Gangs of the future. *The Police Chief*, 57, 23.

Burnham, W. (1911). Hygiene and adolescence. In H. Kiddle (Ed.), *A cyclopedia of education*. New York: Steiger.

Busch-Rossnagel, N. A., & Vance, A. K. (1982). The impact of schools on social and emotional development. In B. Wolman (Ed.), *Handbook of developmental psychology*. Englewood Cliffs, NJ: Prentice-Hall.

Butler, K. (1991). Spirituality and therapy: Toward a partnership. *Utne Reader*, 43, January/February, 75–83.

Byrne, D. & Mazanov, J. (1999). Sources of adolescent stress, smoking and the use of other drugs. *Stress and Medicine, 15 (4), 215-227.*

Calderone, M. (1985). Adolescent sexuality: Elements and genesis. *Journal of the American Academy of Pediatrics*. (supplement), 699–703.

Callahan, C. M., & Rivara, F. P. (1992). Urban high school youth and handguns. *Journal of the American Medical Association*, 267, 3038–3042.

Campbell, A. (1990). Female participation in gangs. In C. R. Huff (Ed.), *Gangs in America*. Newbury Park, CA: Sage.

Campbell, J. D., & Lavallee, L. F. (1993). Who am I? The role of self-concept in understanding the behavior of people with low self-esteem. In R. F. Baumeister (Ed.), *Self-esteem: The puzzle of low self-regard* (pp. 3–20). New York: Plenum Press.

Canterbury, R. J., Clavet, G., McGarvey, E. & Koopman, C. (1998). HIV risk related attitudes of incarcerated adolescents: Implications for public school students. *High School Journal, 82 (1), 1-10.*

Canterbury, R. J., Grossman, S. J., & Lloyd, E. (1993). Drinking behaviors and lifetime incidents of date rape among high school graduates upon entering college. *College Student Journal, 27, 75-84.*

Cantwell, D. P., & Baker, L. (1991). Manifestations of depressive affect in adolescence. *Journal of Youth and Adolescence*, 20 (2), 121–133.

Cantwell, D. P., & Carlson, G. A. (Eds.). (1983). *Affective disorders in childhood and adolescence*. New York: Spectrum.

Carlson, G., & Cantwell, D. (1982). Suicidal behavior and depression in children and adolescents. *Journal of the American Academy of Child Psychiatry*, 21, 361–368.

Carlson, G. A. (1983). Depression and suicidal behavior in children and adolescents. In D. P. Cantwell & G. A. Carlson (Eds.), *Affective disorders in childhood and adolescence.* (pp. 335–351). New York: Spectrum.

Carlson, N. R. (1994). *Physiology of behavior.* Boston, MA: Allyn & Bacon.

Carnahan, S. (1994). Preventing school failure and dropout. In J. Simeonsson (Ed.), *Risk, resilience and prevention: Promoting the well-being of all children* (pp. 103–124). Baltimore: Paul H. Brooks.

Carnegie Corporation. (1990). Adolescence: Path to a productive life or a diminished future? *Carnegie Quarterly,* 25 1, 2.

Carnegie Council on Adolescent Development. (1995). *Great transitions: Preparing adolescents for a new century.* New York: Carnegie Corporation.

Carnegie Council on Adolescent Development. Task Force on Youth Development and Community Programs. (1992). A matter of time: *Risk and opportunity in the nonschool hours.* New York: Carnegie Corporation.

Carnegie Council on Adolescent Development's Task Force on Education of Young Adolescents Staff. (1989). *Turning points: Preparing American youth for the 21st century.* New York: Carnegie Council on Adolescent Development.

Carpenter, L. (1998). From girls into women: Scripts for sexuality and romance in Seventeen magazine, 1974-1994. *Journal of Sex Research, 35 (2), 158-168.*

Carroll, J. (1988). Freshman retention and attrition factors at a predominantly black urban community college. Journal of College Student Development, 29 (1), 52–59.

Carruth, B., & Goldberg, D. (1990). Nutritional issues of adolescents: Athletics and the body image mania. Journal of Early Adolescence, 10 (2), 122–140.

Cary, L. (1991). Black ice. New York: Knopf.

Case, R. (1987). The structure and process of intellectual development. International Journal of Psychology, 22, 571–607.

Catania, J. A., Turner, H., Kegeles, S. M., Stall, R., Pollack, L., & Coates, T. J. (1989). Older Americans and AIDS: Transmission risks and primary prevention research needs. The Gerontologist, 29, 373–381.

Cate, R., & Sugawara, A. I. (1986). Sex-role orientation and dimensions of self-esteem among middle adolescents. Sex Roles, 15, 145–157.

Caughey, J. (1988). Masking the self. Adolescent Psychiatry, 15, 319–332.

Celotta, B., Jacobs, G., & Keys, S. G. (1987). Searching for suicidal precursors in the elementary school child. Special Issue: Identifying children and adolescents in need of mental health services. American Mental Health Counselors Association Journal, 9 (1), 38–50.

Centers for Disease Control (CDC). (1986). Youth suicide in the United States, 1970–1980. Atlanta, GA.

Centers for Disease Control (CDC). (1989a). AIDS and human immunodeficiency virus infection in the United States: 1988 update. Morbidity and Mortality Weekly Report, 38, 1–35.

Centers for Disease Control (CDC). (1989b). First 100,000 cases of acquired immunodeficiency syndrome—United States. Morbidity and Mortality Weekly Report, 38, 561–562.

Cernkovich, S. A., & Giordano, P. C. (1987). Family relationships and delinquency. Criminology, 25, 295–321.

Chan, J.P. & Chamg, A.M. (1999). The effects of an educational programme on adolescents with premenstrual syndrome. Health Education Research, 14(6), 817-830.

Chandler, M. (1975). Relativism and the problem of epistemological loneliness. Human Development, 18, 171–180.

Chao, R. K. (1994). Beyond parental control and authoritarian parenting style: Understanding Chinese parenting through the cultural notion of training. Child Development, 65, 1111–1119.

Chebator, P. (1993). The bar exam. Chestnut Hill, MA: Boston College Press.

Children's Defense Fund. (1987). Preventing children having children. Washington, DC: Children's Defense Fund.

Chilman, C. (1985). Feminist issues in teenage parenting. Child Welfare, 64 (3), 225–234.

Chilman, C. S. (1983). Adolescent sexuality in a changing American society: Social and psychological perspectives for the human services professions (2nd ed.). New York: Wiley.

Chiu, M. L., Feldman, S. S., & Rosenthal, D. A. (1992). The influence of immigration on parental behavior and adolescent distress in Chinese families residing in two Western nations. Journal of Research on Adolescence, 2, 205–239.

Chodorow, N. (1978). The reproduction of mothering. Berkeley: University of California Press.

Christiansen, B., Roehling, P., Smith, G., & Goldman, M. (1989). Using alcohol expectancies to predict adolescent drinking behavior after one year. Journal of Consulting and Clinical Psychology, 57, 93–99.

Cicchetti, D., & Schneider-Rosen, K. (1984). Toward a transactional model of childhood depression. In D. Cicchetti & K. Schneider-Rosen (Eds.), Childhood Depression (pp. 5–28). San Francisco: Jossey-Bass.

Clarizio, H. F. (1994). Assessment and treatment of depression in children and adolescents. Brandon, VT: Clinical Psychology.

Clark, K., & Clark, M. (1947). Racial identification and preference in Negro children. In T. Newcomb & E. Hartley (Eds.), Readings in social psychology, (pp. 551–560). New York: Holt.

Cohen, J. (1983). Commentary: The relationship between friendship selection and peer influence. In J. L. Epstein and N. Karweit (Eds.), Friends in school. Orlando, FL: Academic Press.

Cohen, S. (1977). Alternatives to adolescent drug abuse. Journal of the American Medical Association, 238 (14), 1561–1562.

Cohn, S. J., Carlson, J. S., & Jensen, A. R. (1985). Speed of information processing in academically gifted youth. Personality and Individual Differences, 6 (5), 621–629.

Çok, F. (1990). Body image satisfaction in Turkish adolescents. Adolescence, 25 (98), 409–413.

Coleman, J. C. (1978). Current contradictions in adolescent theory. Journal of Youth and Adolescence, 7 (1), 1–11.

Coleman, J. S. (1961). The adolescent society. New York: Free Press.

Coleman, J. S. (1987). Families and schools. Educational Researcher, 16, 32–38.

Coleman, J. S., Campbell, E. Q., Hobson, C. J., McPartland, J., Mood, A. M., Weinfeld, F. D., & York, R. L. (1966). Equality of educational opportunity. Washington, DC: Government Printing Office.

Colletta, N. D. (1981). Social support and the risk of maternal rejection by adolescent mothers. The Journal of Psychology, 109, 191–197.

Colligan, J. (1975). Achievement and personality characteristics as predictors of observed tutor behavior. Dissertation Abstracts International, 35, 4293–4294.

Commission on Skills of the American Workforce. (1990). America's choice: High skills or low wages! Rochester, NY: National Center on Education and the Economy.

Commission on Substance Abuse. (1994). Rethinking rites of passage: Substance abuse on America's campuses. New York: Columbia University.

Compas, B. (1987). Coping with stress during childhood and adolescence. Psychological Bulletin, 101, 393–403.

Compas, B., & Wagner, B. M. (1991). Psychosocial stress during adolescence: Intrapersonal and interpersonal processes. In M. E. Colten & S. Gore (Eds.), *Adolescent stress:Causes and consequences.* New York: Aldine de Grnyter.

Compas, B. E., Davis, G. E., Forsythe, C. J., & Wagner, B. M. (1987). Assessment of major and daily stressful events during adolescence: The adolescent perceived events scale. *Journal of Consulting and Clinical Psychology, 55,* 534—541.

Compas, B. E., & Hammen, C. L. (1994). Child and adolescent depression: Covariation and comorbidity in development. In R. J. Haggerty, L. R. Sherrod, N. Garmezy, & M. Rutter, *Stress, risk and resilience in children and adolescents: Processes, mechanisms, and interventions.* New York: Cambridge University Press.

Connell, J. P., Halpern-Feshler, B. L., Clifford, E., Crichlow, W., & Usinger, P. (1995). Hanging in there: Behavioral, psychological, and contextual factors affecting whether African-American adolescents stay in school. *Journal of Adolescent Research, 10,* 41-63.

Connolly, J., Goldberg, A. & Pepler, D. (1999). Conceptions of cross-sex friendships and romantic relationships in early adolescence. *Journal of Youth and Adolescence, 28* (4), 481-494.

Cooley, C. H. (1902). *Human nature and the social order.* New York: Charles Scribner & Sons.

Coombs, R., Paulson, M., & Richardson, M. (1991). Peer vs. parental influence in substance use among Hispanic and Anglo children and adolescents. *Journal of Youth and Adolescence, 20* (1), 73-88.

Cooper, C. (1990). Caught in the crossfire. *Drugs, Society, and Behavior, 90/91.* Guilford, CT: The Dushkin Publishing Group.

Corcoran, J., Frankline, C. & Bell, H. (1997). Pregnancy prevention from the teen perspective. *Child and Adolescent Social Work Journal, 14,* 365-382.

Corder, B. G., Shor, W., & Corder R. E. (1974). A study of social and psychological characteristics of adolescent suicide attempts in an urban, disadvantaged area. *Adolescence, 9* (33), 1—6.

Cosse, W. J. (1992). Who's who and what's what? The effects of gender on development in adolescence. In B. R. Wainrib (Ed.), *Gender issues across the life cycle* (pp. 5—21). New York: Springer.

Council on Families in America (1995, March). *Marriage in America: A report to the nation.* New York: Institute for Values.

Cournoyer, R. J., & Mahalik, J. (1995). Cross-sectional study of gender role conflict examining college-aged and middle-aged men. *Journal of Counseling Psychology, 42,* 11—19.

Crawford, M (1997). Agreeing to differ: Feminist epistemologies and Women's Ways of Knowing. In Gergen M. & Davis, S. (Eds.) *Toward a new psychology of gender.* NY: Routledge.

Cropper, D., Meck, D., & Ash, M. (1977). The relationship between formal operations and a possible fifth stage of cognitive development. *Developmental Psychology, 13,* 517—518.

Crowe, P., Philbin, J., Richards, M.H. & Crawford, I. (1998). Adolescent alcohol involvement and the experience of social environments. Journal of Research on Adolescence, 8 (4), 403-422.

Crumley, F. (1982). The adolescent suicide attempt: A cardinal symptom of a serious psychiatric disorder. *American Journal of Psychotherapy, 36* (2), 158—165.

Cullinan, D., & Epstein, M. H. (1979). *Special education for adolescents: Issues and perspectives.* Columbus, OH: Merrill.

Cummings, M. (1983). Family planning among the urban poor: Sexual politics and social policy. *Family Relations: Journal of Applied Family and Child Studies, 32* (1), 47—58.

Curran, D. (1984). Peer attitudes toward attempted suicide in mid-adolescents. *Dissertation Abstracts International, 44* (12), 3927B.

Curry, G. D., & Spergel, I. (1988). Gang homicide, delinquency, and community. *Criminology, 26* (3), 381—405.

Cyranowski, J.M., Frank, E., Young, E. & Shear, K. (2000). Adolescent onset of the gender differences in lifetime rates of major depression. Archives of General Psychiatry, 57 (1), 21-27.

Dacey, J. (1976). *New ways to learn.* Stamford, CT: Greylock.

Dacey, J. (1986). Adolescent attitudes toward their bodies. *Adolescents Today* (3rd ed.). Glenview, IL: Scott, Foresman.

Dacey, J., Amara, D., & Seavey, G. (1993, Winter). Reducing dropout rate in inner city middle school children through instruction in self-control. *Journal of Research on Middle Level Education, 17* (10), 109—116.

Dacey, J., & Travers, J. (1991). *Human development across the life span.* Dubuque, IA: Win. C. Brown.

Dacey, J., & Travers, 1. (1996). *Human development across the lifespan* (3rd ed.). Madison, WI: Brown & Benchmark.

Dacey, J. (1989). Discriminating characteristics of the families of highly creative adolescents. *The Journal of Creative Behavior, 24* (4), 263—271.

Dacey, J. & Lennon, K. (1998). *Understanding creativity: The interplay of biological, psychological, and social factors.*CA: Jossey-Bass

Dacey, J. S., & Packer, A. J. (1992). *The nurturing parent.* New York: Fireside.

Dale, K.S. & Landers, D.M. (1999). Weight control in wrestling: Eating disorders or disordered eating? Medicine and Science in Sports and Exercise, 31 (10), 1382-1389.

DAmico, R., & Maxwell, N. L. (1994). The impact of post-school joblessness on male black-white wage differentials. *Industrial Relations, 33,*

Damon, W., & Hart, D. (1988). *Self-understanding in childhood and adolescence.* New York: Cambridge University Press.

Daniels, D., & Moos, R. (1990).Assessing life stressors and social resources among adolescents. *Journal of Adolescent Research, 5,* 268—289.

D'Augelli, A. R. (1993). Gay men in college: Identity processes and adaptation. In R. A. Pierce and M. A. Black (Eds.), *Life-span development: A diversity reader* (pp. 190—208). Dubuque, IA: Kendall/Hunt.

D'Augelli, A. R., & Rose, M. L. (1990). Homophobia in a university community: Attitudes and experiences of heterosexual freshmen. *Journal of College Student Development,* 31,484-491.

D'Augelli, D. A. (1988). The adolescent closet: Promoting the development of the lesbian or gay male teenager. *The School Psychologist, 42,* 2—3.

D'Augelli, D. A., Collins, C., & Hart, M. M. (1987). Social support patterns of lesbian women in a rural helping network. *Journal of Rural Community Psychology, 8,*12—22.

D'Aurora, D. L., & Fimian, M. J. (1988). Dimensions of life and school stress experienced by young people. *Psychology in the Schools, 25,* 44-53.

Davies, E., & Furnham, A. (1986).The dieting and body shape concerns of adolescent females.*Journal of Child Psychology and Psychiatry, and Allied Disciplines* 27, 417-428.

Davies, M., & Kandel, D. B. (1981). Parental and peer influences on adolescents' educational plans: Some further evidence. *American Journal of Sociology,87,* 363—387.

Davis, S. S., & Davis, D. A. (1989). *Adolescence in a Moroccan town.* New Brunswick, NJ: Rutgers University Press.

Davis, T. C., Peck, G. Q., & Storment, J. M. (1993). Acquaintance rape and the high school student. *Journal of Adolescent Health,* 14, 220-224.

Dawson, D. A. (1990). AIDS knowledge and attitudes for January—March 1990. *Advance data,* 193. National Center for Health Statistics, Washington, DC.

Dekoric, M., Noom, M.J. & Meeus, W. (1997). Expectations regarding development during adolescence: Parental and adolescent perceptions. *Journal of Youth and Adolescence, 26 (3),* 253-272.

De Las Fuentes, C. & Vasquez, M. (1999). Immigrant adolescent girls of color: Facing American challenges. In Johnson, N., Roberts, M. & Worell, J. (Eds.) Beyond appearance: A new look at adolescent girls. Washington, DC: APA.

Dembo, R. (1981). Examining a causal model of early drug involvement among inner-city junior high school youths. *Human Relations, 34* (3), 169—193.

Diamond, M. (1982). Sexual identity, monozygotic twins reared in discordant sex roles, and a BBC follow-up. *Archives of Sexual Behavior, 11,* 181—185.

Diamond, M., & Diamond, G. (1986). Adolescent sexuality: Biosocial aspects and intervention strategies. *Journal of Social Work and Human Sexuality, 5* (1), 3—13.

DiClemente, R. J., Boyer, C. B., & Mills, S. J. (1987). Prevention of AIDS among adolescents: Strategies for the development of comprehensive risk-reduction health education programs. *Health Education Research, 2* (3), 287—291.

Dielman, T. B. (1994). School-based research on the prevention of adolescent alcohol use and misuse: Methodological issues and advances. *Journal of Research on Adolescence, 4,* 271-294.

Dill, ,K.E. & Dill, J.C. (1999). Video game violence: A review of the empirical literature. Aggression and Violent Behavior, 3 (4), 407-428.

Dishion, T., Patterson, G., Stoolmiller, M., & Skinner, M. (1991). Family, school, and behavioral antecedents to early adolescent involvement with antisocial peers. *Developmental Psychology, 27(1),* 172—180.

Dittus, P., Jaccard, J. & Gordon, V. (1999). Direct and nondirect communication of maternal beliefs to adolscents: Adolescent motivations for premarital sexual activity. *Journal of Applied Social psychology, 29 (9), 1927-1963.*

Dixon, R. A., & Lemer, R. M. (1992). A history of systems in developmental psychology. In M. H. Borustein & M. E. Lamb (Eds.), *Developmental psychology: An advanced textbook.* Hillsdale, NJ: Erlbaum.

Dodge, K. A., & Murphy, P. R. (1983). The assessment of social competence in adolescents. In P. Karoly & J. J. Steffans (Eds.), *Adolescent behavior disorders: Foundations and contemporary concerns,* (pp. 61—96). Lexington, MA: Lexington Books.

Doherty, K. (1998). A mind of her own: Effects of need for closure and gender reactions to noncomformity. *Sex Roles, 38 (9-10), 801-819.*

Doherty, W., & Needles, R. (1991). Psychological adjustment and substance use among adolescents before and after a parental divorce. *Child Development, 62,* 328-337.

Dolan, J. (1991, April 29). Kids killing kids: What can be done? "You did not care." *The Boston Globe.*

Donnerstein, B., Slaby, R. G., & Eron, L. D. (1994). The mass media and youth aggression. In L. D. Eron, J. Gentry, & P. Schlegel (Eds.), *Reason to hope: A psychosocial perspective on violence and youth* (pp. 219—250). Washington, DC: American Psychological Association.

Dom, L., Susman, B., Nottelmann, B., Inoff-Germain, G., & Chrousos, G. (1990). Perceptions of puberty: Adolescent, parent and health care personnel. *Developmental Psychology, 26* (2), 322—329.

Dori, G. & Overholser, J. (1999). Depression, hopelessness, and self-esteem: Accounting for suicidality in adolescent psychiatric inpatients. *Suicide and Life Threatening Behavior, 29 (4), 309-318.*

Dorn, L.D., Nottlemann, E.D., Susman, E.J., Inoff, G.g. & Cutler, G.B. Jr. (1999). Variability in hormonal concentrations and self-reported menstrual histories in young adolescents: Menarche as an integral part of a developmental process. Journal of Youith and Adolescence, 28 (3), 283-304.

Douvan, E., & Adelson, J. (1966). *The adolescent experience.* New York: Wiley.

Dranoff, S. M. (1974). Masturbation and the male adolescent. *Adolescence, 9 (34),* 166—176.

Dryfoos, J. (1994). Full service schools: *A revolution in health and social services for children, youth, and families.* San Francisco: Jossey-Bass.

Dryfoos, J. G. (1995). Full service schools: Revolution or fad? *Journal of Research on Adolescence, 5,* 147—172.

Dumont, M & Provost, M. (1999). Resilience in adolescents: Protective role of social support, coping strategies, self-esteem, and social activities on experience of stress and depression. *Journal of Youth and Adolescence, 28 (3), 343-363.*

Duncan, S.C., Duncan, T.E., Biglan, A. & Ary, D. (1998). Cointributions of the social context to the development of adolescent substance use: a multivariate altent growth modeling approach. Drug and Alcohol Dependence, 50 (1), 57-71.

Duncan, P., Ritter, P., Dornbusch, S., Gross, R., & Carlsmith, M. (1985). The effects of pubertal timing on body image, school behavior, and deviance. *Journal of Youth and Adolescence, 14 (3),* 227—235.

Dunphy, D. (1963). The social structure of urban adolescent peer groups. *Sociometry, 26,* 230-246.

Durant, R., Jay, S., & Seymore, C. (1990). Contraceptive and sexual behavior of black female adolescents. *Journal of Adolescent Health Care, 11,* 326-334.

Durbin, D. L., Darling, N., Steinberg, L., & Brown, B. B. (1993). Parenting style and peer group membership among European-American adolescents. *Journal of Research on Adolescence, 3,* 87-100.

Earls, F., Cairns, R. B., & Mercy, J. A. (1993). The control of violence and the promotion of nonviolence in adolescents. In S. G. Millstein, A. C. Petersen, & E. 0. Nightingale (Eds.), *Promoting the health of adolescents: New directions for the twenty-first century* (pp. 285—304). New York: Oxford University Press.

Ebata, A., & Moos, R. H. (1994). Personal, situational, and contextual correlates of coping in adolescence. *Journal of Research on Adolescence, 4,* 99—126.

Ebbert, S. (2000). Police say clubs awash in designer drugs. *Boston Globe, March 18, 2000.*

Eccles, J. (1984). Sex differences in mathematics participation. In M. W. Steinkamp & M. L. Maehr (Eds.), *Advances in motivation and achievement. Vol. 2. Women in science* (pp. 283—331). Greenwich, CN: JAI Press.

Eccles, J., Barber, B., Jozefowicz, D., Malenchuck, O. & Vida, M. (1999). Self evaluations of competence, task values and self esteem. In Johnson, N., Roberts, M. & Worell, J. (Eds.) Beyond appearance: A new look at adolescent girls. Washington, DC: APA.

Eccles, J., Midgley, C., Wigfield, A., Buchanan, C. M., Reuman, D., Flanagan, C., & Maclver, D. (1993). Development during adolescence: The impact of stage-environment fit on young adolescents' experiences in schools and in families. *American Psychologist, 48* (2), 90-101.

Eccles, J. S., & Harold, R. D. (1993). Parent-school involvement during the early adolescent years. *Teachers College Record, 94* (3), 568—587.

Ekstrom, R., Goertz, M. E., Pollack, J. M., & Rock, D. A. (1986). Who drops out of high school and why? Findings from a national study. *Teachers College Record, 87,* 356-373.

Elders, J. & Albert, A. (1998). Adoelscent pregnancy and sexual abuse. *Journal of the American Medical Association, 280 (7),* 648-649.

Elkind, D. (1967). Egocentrism in adolescence. *Child Development, 38,* 1025—1034.

Elkind, D. (1976). *Child development and education: A Piagetian perspective.* New York: Oxford University Press.

Elkind, D. (1978). Understanding the young adolescent. *Adolescence, 13,* 127—134.

Elkind, D. (1980). The origins of religion in the child. In J. Tisdale (Ed.), *Growing edges in the psychology of religion.* Chicago: Nelson-Hall.

Elkind, D. (1981). *Children and adolescents: Interpretive essays on Jean Pia get.* New York: Oxford University Press.

Elkind, D. (1984). *All grown up and no place to go: Teenagers in crisis.* Reading, MA: Addison-Wesley.

Elkind, D. (1985). Reply to D.Lapsley and M. Murphy's *Developmental Review* paper. *Developmental Review, 5,* 218—226.

Elkind, D., & Bowen, R. (1979). Imaginary audience behavior in children and adolescents. *Developmental Psychology, 15,* 38—44.

Elliott, D., Huizinga, D., & Menard, S. (1989). *Multiple-problem youth: Delinquency, substance use, and mental health problems.* New York: Springer-Verlag.

Elliott, D., & Morse, B. (1989). Delinquency and drug use as risk factors in teenage sexual activity. *Youth and Society, 2] (1),* 32—60.

Elster, A. B., & Lamb, M. E. (Eds.). (1986). *Adolescent fatherhood.* Hillsdale, NJ: Erlbaum.

England, E. & Petro, K. (1998). Middle schjool students' perception of peer groups: Relative judgments about group chracterisitics. *Journal of Eraly Adolescents, 18 (4),* 349-373.

Enns, C. E. (1991). The "new" relationship models of women's identity: A review and critique for counselors. *Journal of Counseling and Development, 69,* (3), 209—217.

Enright, R. D., Levy, V. M., Harris, D., & Lapsley, D. K. (1987). Do economic conditions influence how theorists view adolescents? *Journal of Youth and Adolescence, 16 (6),* 541—560.

Entwisle, D. (1990). Schools and the adolescent. In S. Feldman & G. R. Elliot (Eds.), *At the threshold: The developing adolescent.* Cambridge, MA: Harvard University Press.

Epps, E. G. (1980). The impact of school desegregation on aspirations, self-concepts and other aspects of personality. In R. L. Jones (Ed.), *Black psychology.* New York: Harper & Row.

Erikson, E. (1958). *Young man Luther: A study in psychoanalysis and history.* New York: W. W. Norton.

Erikson, E. (1959). Identity and the life cycle. *Psychological Issues,1,* 18—164.

Erikson, E. (1963). *Childhood and society* (2nd ed.). New York: W.W. Norton.

Erikson, E. (1968). *Identity: Youth and crisis.* New York: W. W. Norton.

Erikson, E. (1969). *Gandhi's truth: On the origins of militant nonviolence.* New York: W. W. Norton.

Erikson, E. (1975). *Life, history and the historical moment.* New York: W. W. Norton.

Erikson, E. & Coles, R. (2000). *The Erik Erikson reader.* NY: W.W. Norton.

Eron, L. D., & Slaby, R. G. (1994). Introduction. In L. D. Eron, J. Gentry, & P. Schlegel (Eds.), *Reason to hope: A psychosocial perspective on violence and youth* (pp. 1—22). Washington, DC: American Psychological Association.

Estrada, A., Rabow, J., & Watts,R. K. (1982). Alcohol use among Hispanic adolescents: A preliminary report. *Hispanic Journal of Behavioral Sciences, 4*(3), 339—351.

Ewing, C. (1990). *When children kill: The dynamics of juvenile homicide.* Lexington, MA: Lexington Books/D. C. Heath.

Fagan, J. (1989). The social organization of drug use and drug dealing among urban gangs. *Criminology, 27*(4), 633—669.

Fagan, J., & Pabon, E. (1990). Contributions of delinquency and substance abuse to school dropout among inner-city youths. *Youth and Society, 21* (3), 306—354.

Fakouri, M. E. (1976). Some clinical implications of Piaget's theory. *Psychology, 13* (1), 33—36.

Falco, M. (1988). *Preventing abuse of drugs, alcohol, and tobacco by adolescents.* Washington, DC: Camegie Council on Adolescent Development.

Fasick, F. (1988). Patterns of formal education in high school as *rites de passage. Adolescence, 23* (90), 457—467.

Federal Bureau of Investigation (FBI). (1992). *Uniform crime reports, 1991]: Crime in the United States.* Washington, DC: U.S. Government Printing Office.

Federal Bureau of Investigation,U.S. Department of Justice.(1992). *Uniform Crime Reports for the United States.* Washington, DC: U.S.Government Printing Office.

Federal Bureau of Investigation,U.S. Department of Justice.(1998). *Uniform Crime Reports for the United States.* Washington, DC: U.S.Government Printing Office.

Feigelman, B., & Feigelman, W. (1993). Treating the adolescent substance abuser. In S. L. A. Struassner (Ed.), *Clinical work with substance-abusing clients.* New York: Guilford.

Feldman, D. (1979). The mysterious case of extreme giftedness. In A. H. Passow, (Ed.), *The gifted and the talented: Their education and development.* Chicago: University of Chicago Press (NSSE).

Feldman, D., Rosenberg, M. S., & Peer, G. G. (1984). Educational therapy: A behavior change strategy for pre-delinquent and delinquent youth. *Journal of Child and Adolescent Psychotherapy,]* (1), 34—37.

Feldman, S. S., Rosenthal, D. A., Mont-Reynaud, R., Leung, K., & Lau, 5. (1991). Ain't misbehavin': Adolescent values and family environments as correlates of misconduct in Australia, Hong Kong, and the United States. *Journal of Research on Adolescence,* 109—134.

Feldman, S.S. & Rosenthal, D.A. (2000). The effect of communication characteristics on family members' perceptions of parents as sex educators. *Journal of Research on Adolescence, 10 (2), 119-150.*

Feldman, S. S.,, Turner, R.A. & Araujo, K (1999). Interpersonal context as an influence on sexual timetables of youth: Gender and ethnic differences. *Journal of Research on Adolescence, 9 (1), 25-52.*

Feltey, K. M., Ainslie, J. J., & Gleib, A. (1991). Sexual coercion attitudes among high school students. *Youth & Society, 23,* 229—250.

Ferguson, R. (1998). Teachers' perceptions and expectations and the Black-White test score gap. . In Jencks, C & Phillips, M (Eds.) *The Black-White test score gap, Washington DC: Brookings Institution.*

Fimian, M. J., & Cross, A. H. (1987). Stress and burnout among preadolescent and early adolescent gifted students: A preliminary investigation. *Journal of Early Adolescence, 6,*247—267.

Fine, G. A., Mortimer, J. T., & Roberts, D. F. (1990). Leisure, work and the mass media. In S. Feldman & G. Elliott (Eds.), *At the threshold: The developing adolescent* (pp. 225—252). Cambridge, MA: Harvard University Press.

Fine, M. (1986). Why urban adolescents drop into and out of high school: The ideology of school and work. *Teachers College Record, 87,* 393—409.

Fine, M. (1988). Sexuality, schooling, and adolescent females: The missing discourse of desire. *Harvard Educational Review, 58,* 29—53.

Fine, M. (1991). *Framing dropouts: Notes on the politics of an urban school.* Albany, NY: SUNY Press.

Fingerhut, L. A., Ingram, D., & Feldman, J. (1992). Firearm and nonfirearm homocide among persons 15 through 19 years of age. *Journal of the American Medical Association, 267,*3048—3053.

Fingerhut, L. A., & Kleinman, J. C. (1990). International and interstate comparisons of homocide among young males. *Journal of the American Medical Association, 263,* 3292—3925.

Fingerman, K. (1989). Sex and the working mother: Adolescent sexuality, sex role typing and family background. *Adolescence, 24,* (93), 1—17.

Finkelstein, J.W., D'Arcangelo, M.R., Susman, E.J., Chinchilli, V.M., Kunselman, S.J., Schwab, J., Demers, L.M., Liben, L.S. & Kulin, H.E. (1999). Self assessment of physical sexual maturation in boys and girls with delayed puberty. Journal of Adolescent Health, 25 (6), 379-381.

Fisher, M., Trieller, K., & Napolitano, B. (1989). Premenstrual symptoms in adolescents. *Journal of Adolescent Health Care, 10,* 369—375.

Fiske, S., & Taylor, 5. (1984). *Social cognition.* New York, NY: Random House.

Flanigan, B., McLean, A., Hall, C., & Propp, V. (1990). Alcohol use as situational influence on young women's pregnancy risk-taking behvaiors. *Adolescence, 25* (97), 205—214.

Flannery, D. J., Montemayor, R., & Eberly, M. B. (1994). The influence of parent negative emotional expression on adolescents' perception of their relationships with their parents. *Personal Relationships, 1,* 259—274.

Flannery, D.J., Williams, L.L. & Vazsonyi, A.T. (1999). Who are they with and what are they doing? Delinquent behavior, substance abues and early adoeslecnts' after school time. *Americocan Journal of Orthopsychiatry, 69 (2), 247-253.*

Flavell, J. (1977). *Cognitive development.* Englewood Cliffs, NJ: Prentice-Hall.

Flavell, J. (1982). On cognitive development. *Child Development, 53,* 1—10.

Fletcher, A., Elder Jr. G. & Mekos, D. (2000). Parental influences on adolescent involvement in community activities. *Journal of Research on Adolescence, 10 (1), 29-48.*

Floyd, F., Stein, T., Harter, K, Allison, A., & Nye, C. (1999). Gay, lesbian, and bisexual youths: Separation-Individuation, parental attitudes, identity consolidation, and well-being. *Journal of Youth and Adolescence, 28 (6),* 719.

Fogelson, R. D. (1979). Person, self, and identity: Some anthropological retrospects, circumspects, and prospects. In B. Lee (Ed.), *Psychosocial theories of the self* New York: Plenum Press.

Fordham, 5. (1988). Racelessness as a factor in black students' school success: Pragmatic strategy or pyrrhic victory? *Harvard Educational Review, 58,* 54-84.

Fordham, S., & Ogbu, J. U. (1986). Black students' school success: Coping with the burden of "acting white." *Urban Review, 18,* 176—206.

Forstein, M. (1989, April). *Sexuality and AIDS.* Paper presented at the Conference on the Psychiatric Treatment of Adolescents and Young Adults, Harvard Medical School, Boston, MA.

Fowler, J. (1984). *Becoming adult, becoming Christian.* San Francisco:Harper & Row.

Fowler, J. (1986). Stages in selfhood and faith. In *Faith development and pastoral care.* Philadelphia: Fortress Press.

Fowler, J. (1991a). The church and the future: Images of promise and peril. In *Weaving the New Creation. Stages offaith and the public church.* San Francisco: Harper.

Fowler, J. (1991b). Stages in faith consciousness. In *Weaving the New Creation: Stages offaith and the public church.* San Francisco: Harper.

Fowler, J. (1996). *Faithful change: The personal and public challenges of postmodern life.* Nashville: Abington Press

Frankel, K. (1990). Girls' perceptions of peer relationship support and stress. *Journal of Early Adolescence, 10* (1), 69—88.

Frankl, V. (1978). *The unheard cry for meaning.* New York: Touchstone/Simon & Schuster.

Franz, C., & White, K. (1985). Individuation and attachment in personality development: Extending Erikson's theory. *Journal of Personality, 53,* 224-256.

Freedman, R. (1986). *Beauty bound.* Lexington, MA: Lexington Books.

Freedman, 5. (1990). *Small victories.* New York: HarperCollins.

Freeman, E. W. (1982). Self-reports of emotional distress in a sample of urban black high school students. *Psychological Medicine, 12* (4), 809—817.

Freiberg, P., & American Psychological Association (APA). (1991, July). Suicide in family, friends is familiar to too many teens. *Monitor*, pp. 36-37.

French, S. A., Perry, C. L., Leon, G. R., & Fulkerson, J. A. (1994). Food preferences, eating patterns, and physical activity among adolescents: Correlates of eating disorders symptoms. *Journal of Adolescent Health, 15* (4) 286-294.

Freud, A. *(1958)*. Adolescence. In Eissler, R., et al. (Eds.), *The psychoanalytic study of the child.* New York: International Universities Press.

Freud, A. (1968). Adolescence. In A. E. Winder and D. L. Angus (Eds.), *Adolescence: Contemporary studies.* New York: American Book.

Freud, A. (1969a). Adolescence as a developmental disturbance. In G. Kaplan and S. Lebovici (Eds.), *Adolescence: Psychosocial perspectives.* New York: Basic Books.

Freud, A. (1 969b). Adolescence as a psychological disturbance. In G. Kaplan & S. Lebovici (Eds.), *Adolescence: Psychosocial perspectives.* New York: Basic Books.

Freud, 5. (1914/1955). *Totem and taboo.* In J. Strachey (Ed. and Trans.), *The standard edition of the complete psychological works of Sigmund Freud* (Vol. 14, 239—258). London: Hogarth.

Freudenberger, H. (1973). A patient in need of mothering. *Psychoanalytic Review, 60* (1), 1—7

Frey, D., & Rosch M. (1984). Information seeking after decisions: The roles of novelty of information and decision reversibility. *Personality and Social Psychology Bulletin, 10* (1), 91—948.

Friedan, B. (1963). *The feminine mystique.* New York: Norton.

Friedenberg, E. Z. *(1959). The vanishing adolescent.* New York: Random House.

Friedenherg, E. Z. (1967). *Coming of age in America.* New York: Vintage Books.

Friedman, C. J., Mann, F., & Friedman, A. (1976). Juvenile street gangs: The victimization of youth. *Adolescence, 1,* 527—533.

Frisch, R. E. (1988). Fatness and fertility. *Scientific American, 258* (3), 88—95.

Frydenberg, E. & Leweis, R. (1999). Things don't get better just because you are older: A case for facilitating reflection. *British Journal of educational Psychology, 69 (1), 81-94.*

Funk, J. B. (1993). Reevaluating the impact of video games. *Clinical Pediatrics, 32* (2), 86-90.

Furstenberg, F. F. (1990). Coming of age in a changing family system. In S. Feldman and G. Elliot (Eds.), *At the threshold: The developing adolescent.* Cambridge, MA: Harvard University.

Furstenberg, F. F. (1991). As the pendulum swings: Teenage childbearing and social concern. *Family Relations, 40,* 127—138.

Furstenberg, F. F. (1994, August). Fathering in the inner city: Paternal participation and public policy. In W. Marsiglio (Ed.), *Fatherhood: Contemporary scholarship* (pp. 118—147). Newbury Park, CA: Sage.

Furstenberg, F. F. (1994). History and current status of divorce in the Unitede States. *Future of Children, 4 (1), 29-43.* Furstenberg, F. F., Brooks-Gunn, J., & Chase-Lansdale, L. (1989). Teenaged pregnancy and childbearing. *American Psychologist, 44* (2), 313—320.

Furstenberg, F. F., & Crawford, A. G. (1978). Family support: Helping teenage mothers to cope. *Family Planning Perspectives, 10* (6), 322—333.

Furstenberg, F. F., Moore, K. A., & Peterson, J. L. (1986). Sex education and sexual experience among adolescents. *American Journal of Public Health, 75,* 1221—1222.

Furstenberg, F. F., Shea, J., Allison, P., Herceg-Bacon, R., & Webb, D. (1983). Contraceptive continuation among adolescents attending family planning clinics. *Family Planning Perspectives, 15,* 211—217.

Gabriel, A., & McAnarney, E. R. (1983). Parenthood in two subcultures: White, middle-class couples and black, low-income adolescents in Rochester, New York. *Adolescence, 18* (71), 595—608.

Gage, M. G., & Christensen, D. H. (1990). Early adolscents' values about their pets. *The Journal of Psychology,* 124 (4), 417—425.

Galambos, N. L., Sears, H. A., Almeida, D. M., & Kolaric, G. C. (1995). Parents' work overload and problem behavior in young adolescents. *Journal of Research on Adolescence, 5,* 201—224.

Gallup, G. E. (1977). Ninth annual Gallup poll of the public's attitudes. *Phi Delta Kappan, 59* (1), 34—48.

Gamoran, A., & Mare, R. D. (1989). Secondary school tracking and educational inequality: Competition, reinforcement, or neutrality? *American Journal of Sociology, 94,* 1146-1183.

Garapon, A. (1983, August! September). Place de l'initiation dans la d6linquance juvenile. (Inititation role in juvenile delinquency.) *Neuropsychiatrie de lEnfance et de I adolescence, 31* (8—9), 390—403.

Gardner, H. (1982). *Art, mind, and brain: A cognitive approach to creativity.* New York: Basic Books.

Gardner, H., & Winner, E. (1982). Children's conceptions (and misconceptions) of the arts. In Gardner, H., *Art, mind, and brain.* New York: Basic Books.

Garfinkel, P., Garner, D. M., & Goldbloom, D. 5. (1987). Eating disorders: Implications for the 1990s. *Canadian Journal of Psychiatry, 32,* 624—631.

Gargiulo, J., Attie, I., Brooks-Gunn, J., & Warren, M. (1987). Girls' dating behavior as a function of social context and maturation. *Developmental Psychology, 23* (5), 730-737.

Garland, A. F., & Zigler, E. (1993). Adolescent suicide prevention. *American Psychologist, 48* (2), 169—182.

Gamer, D. S., & Garfinkel, P. E. (1980). Sociocultural factors in the development of anorexia nervosa. *Psychological Medicine, 10,* 647—656.

Gamora, A. & Weinstein, M. (1998). Tracking: Differentiation and opportunity in restructured schools. *American Journal of Education, 106 (3), 385-415.*

Garn, S. M., & Petzold, A. (1983). Characteristics of mother and child in teenage pregnancy. *American Journal of Diseases of Children, 137* (4), 365—368.

Garner, D.M., Rosen, L.W. & Barry, D. (1998). Eating disorders among athletes: Research and recommendations. Child and Adolescent Psychiatric Clinics of North America, 7 (4), 839-857.

Gates, D., & Jackson, R. (1990). Gang violence in L.A. *The Police Chief, 57,* 20-21.

Gavazzi, S. M., & Blumenkrante, D. G. (1991). Teenage runaways: Treatment in the context of the family and beyond. *Journal of Family Psychotherapy, 2,* 15—29.

Gavin, L. A., & Furman, W. (1989). Age differences in adolescents' perceptions of their peer groups. *Developmental Psychology, 25,* 827—834.

Gay, P. (1988). *Freud: A life for our time.* New York: Norton.

Geertz, C. (1984). From the native's point of view: On the nature of anthropological understanding. In R. Schweder & R. Levine (Eds.), *Culture theory.* New York: Cambridge University Press.

Gelman, D. (1990). A much riskier passage. *Newsweek Special Issue: The New Teens,* 10-17.

Gelman, R. (1979). Preschool thought. *American Psychologist, 34,* 900-904.

Gelman, R., & Baillargeon, R. (1983). A review of some Piagetian concepts. In P. Mussen (Ed.), *Carmichael's manual of child psychology.* (Vol. 4). New York: Wiley.

Gibbons, J. L., Stiles, D. A., & Adolescents' attitudes towards family and gender roles: An international comparison. *Sex Roles, 25,* 639—657.

Gibbons, M. (1974). Walkabout. *Phi Delta Kappan,* 596—602.

Gibbs, J., Arnold, K., & Burkhart, J. (1984). Sex differences in the expression of moral judgment. *Child Development, 55* (3), 1040-1043.

Gibbs, J. T. (1984). Black adolescents and youth: An endangered species. *American Journal of Orthopsychiatry, 54,* 6-21.

Gibbs, J. T. (1991, August 10). 10 myths about young black males. *The Boston Globe,* 77—78.

Gibson, P. (1989). Gay male and lesbian youth suicide. In ADAMHA, *Report of the secretary's task force on youth suicide* (Vol. 3, pp. 110-142. DHHS Pub. No. [ADM] 89—1623). Washington, DC: U.S. Government Printing Office.

Gibson-Ainyette, I., Tempter, D., & Brown, R. (1988). Adolescent female prostitutes. *Archives of Sexual Behavior, 17,* 431—438.

Gil, S., Stockard, J., Johnson, M., & Williams, 5. (1987). Measuring gender differences: The expressive dimension and critique of androgyny scales. *Sex Roles, 17,* 375—400.

Gilbert, J. (1986). *A cycle of outrage: America's reaction to the juvenile delinquent in the 1950s.* New York: Oxford University Press.

Gilligan, C. (1977). In a different voice: Women's conception of self and morality. *Harvard Educational Review, 47,* 481—517.

Gilligan, C. (1982). *In a different voice. Psychological theory and women's development.* Cambridge, MA: Harvard University Press.

Gilligan, C. (1983). New maps of development: New visions of maturity. *Annual Progress in Child Psychiatry and Child Development, 3,* 98—115.

Gilligan, C. (1988). Adolescent development reconsidered. In C. Gilligan, J. V. Ward, & J. M. Taylor (Eds.), *Mapping the Moral Domain.* Cambridge, MA: Harvard University Press.

Gilligan, C., & Attanucci, J. (1988). Two moral orientations: Gender differences and similarities. *Merrill Palmer Quarterly, 34,* 223—237.

Gilligan, C., Lyons, N. P., & Hanmer, T. J. (1990). *Making connections: The relational worlds of adolescent girls at Emma Willard School.* Cambridge, MA: Harvard University Press.

Gilligan, C., Ward, J., & Taylor, J. (Eds.). (1988). *Mapping the moral domain.* Cambridge, MA: Harvard University Press.

Gilligan, C., & Wiggins, G. (1987). The origins of morality in early childhood relationships. In J. Kagan & S. Lamb (Eds.), *The emergence of morality in young children,* pp. 277—305. Chicago: University of Chicago.

Giordano, P. C., Cemkovich, S. A., & DeMaris, A. (1993). The family and peer relations of black adolescents. *Journal of Marriage and the Family, 55,* 227—287.

Gjerde, P. F., & Block, J. (1991). Preadolescent antecedents of depressive symptomatology at age 18: A prospective study. *Journal of Youth and Adolescence, 20* (2), 217—232.

Gjerde, P. F., Block, J., & Block, J. (1988). Depressive symptoms and personality during late adolescence: Gender differences in the externalization internalization of symptom expression. *Journal of Abnormal Psychology, 97,* 475—486.

Gjerde, P. F., & Westenberg, P. (1998). Dysphoric adolescents as young adults: A prospective study of the psychological sequelae of depressed mood in adolescence. *Journal of Research on Adolescence, 8* (3), 377-402.

Gladstein, J., Rusonis, B. S., & Heald, F. P. (1992). A comparison of inner-city and upper-middle class youths' exposure to violence. *Journal of Adolescent Health, 13,* 275—280.

Glover, R. T., & Marshall, R. (1993). Improving the school-to-work transition of American adolescents. *Teachers College Record, 94* (3), 588—610.

Goethals, G., & Klos, D. (1976). *Experiencing youth.* Boston: Little, Brown.

Gold, M., & Douvan, E. (1969). *Adolescent development: Readings in research and theory.* Boston: Allyn & Bacon.

Goldberger, N. Tarule, J., Clinchy, B. & Belenky, M (Eds.) (1996). *Knowledge Difference and power: Essays inspired by Women's Ways of Knowing.* NY: Basic Books.

Gold, M. D. (1986). *The facts about drugs and alcohol.* New York: Bantam.

Goldman, R. J., & Goldman, J. G. (1982). How children perceive the origin of babies and the roles of mothers and fathers in procreation: A cross-national study. *Child Development, 53,* 491—504.

Goldstein, A. (1991). *Delinquent gangs: A psychological perspective.* Champaign, IL: Research Press. Goldstein, A., & Soriano, F. I. (1994). Juvenile gangs. In L. D. Eron, J. Gentry, & P. Schlegel (Eds.), *Reason to hope: A psychosocial perspective on violence and youth* (pp. 315—333). Washington, DC: American Psychological Association.

Gonsiorek, J. (1988). Mental health issues of gay and lesbian adolescents. *Journal of Adolescent Health Care, 9,* 114-122.

Gonzalez, J., Wagner, F. W., & Brunton, D. (1993). Community service learning at Putnam High School. *Equity & Excellence in Education, 26,* 27—29.

Goodchilds, J., Zellman, G. L., Johnson, P. B., & Giarrusso, R. (1988). Adolescents and their perception of sexual interaction. In A.W. Burgess (Ed.) *Rape and sexual assault, Volume 2* (pp. 245—270). New York: Garland.

Goode, E. (1972). Sex and behavior. *Sexual Behavior, 2* (5), 45—51.

Goodman, E. (1996, February 8). Targeting the men who prey on teen-age girls. *The Boston Globe,* p. 17.

Goodman, P. (1966). A social critic on "moral youth in an immoral society." In *The young Americans* (pp. 18—19). New York: Time Books.

Gore, S., Aseltine, R. H., & Colten, M. E. (1993). Gender, social-relational involvement, and depression. *Journal of Research on Adolescence, 3,* 101—125.

Gottfredson, L. 5. (1978). Providing black youth more access to enterprising work. *Vocational Guidance Quarterly, 27,* 114-123.

Gottfredson, L. 5. (1981). Circumstances and compromise: A developmental theory of occupational aspirations. *Journal of Counseling Psychology, 28,* 545—579.

Gottfredson, M. R., & Hirschi, T. (1990). *A general theory of crime.* Stanford, CA: Stanford University Press.

Gottfredson, M. R., & Hirschi, T. (1994). A general theory of adolescent problem behavior: Problems and prospects. In R. D. Ketterlinus & M. E. Lamb (Eds.), *Adolescent problem behaviors* (pp. 41—56). Hillsdale, NJ: Erlbaum.

Gottlieb, B. H. (1991). Social support in adolescence. In M. E. Colten & S. Gore (Eds.), *Adolescent stress: Causes and consequences.* New York: Aldine de Grnyter. Gottlieb, G. (1992). *Individual development and evolution: The genesis of novel behavior.* New York: Oxford University Press.

Graham, R. (1992, February 19). Susan Bailey: Growing up invisible. *The Boston Globe.*

Grant, B. (2000). Estimates of U.S. children exposed to alcohol abuse and dependence in the family. American Journal of Public Health, 90 (1) 112-115.

Grant, B., & Hartford, T. (1990). Concurrent and simultaneous use of alcohol with cocaine: Results of a national survey. *Drug and Alcohol Dependence, 25,* 97—104.

Gray-Ray, P., & Ray, M. (1990). Juvenile delinquency in the black community. *Youth and Society, 22* (1), 67—84.

Green, A. (1978). Self-destructive behavior in battered children. *American Journal of Psychiatry, 135* (5), 579—583.

Greenberg, B., & Brand, J. (1993). Television news and advertising in schools: The "Channel One" controversy. *Journal of Communication, 43* (1), 143—151.

Greenberger, E., & Steinberg, L. (1981). The workplace as a context for the socialization of youth. *Journal of Youth and Adolescence, 10,* 185—210.

Greenberger, E., & Steinberg, L. (1986). *When teenagers work: The psychological and social costs of adolescent employment.* New York: Basic Books.

Greenberger, E., Steinberg, L., & Vaux, A. (1981). Adolescents who work: Health and behavioral consequences of job stress. *Developmental Psychology, 17,* 691—703.

Greenfield, P. M. (1994). Cognitive effects of video games: Guest editor's introduction. Video games as cultural artifacts. *Journal of Applied Developmental Psychology, 15* (1), 3—12.

Greenfield, P. M., DeWinstanley, P., Kilpatrick, H., & Kaye, D. (1994). Action video games and informal education: Effects on strategies for dividing visual attention. *Journal of Applied Developmental Psychology, 15* (1), 105—123.

Gregory, T. (1978). *Adolescence in literature.* New York: Longman.522 References

Grilo, C., Sanislkow, C., Fehon, D., Lipschitz, D., Martino, S. & McGlashan, (1999). Correlates of suicide risk in adolescent inpatients who report a history of childhood abuse. *Comprehensive-Psychiatry. 40*(6): 422-428.

Grotevant, H. D. (1992). Assigned and chosen identity components: A process perspective on their integration. In G. R. Adams, T. P. Gullotta, & R. Montemayor (Eds.), *Adolescent identity formation: Advances in adolescent development* (pp. 73—90). Newbury Park, CA: Sage.

Grotevant, H. D (1997). Identity processes: Integrating social psychological and developmental approaches. *Journal of Adolescent Research, 12 (3), 354-357.*

Grotevant, H. D., & Cooper, C. R. (1986). Individuation in family relationships. *Human Development, 29,* 82—100.

Guerney, L., & Arthur, J. (1984). Adolescent social relationships. In R. Lerner & N. L. Galambos, *Experiencing adolescents: A sourcebook for parents, teachers and teens* (pp. 87—118). New York: Teachers College Press.

Guerra, N., & Slaby, R. (1990). Cognitive mediators of aggression in adolescent offenders: 2. Intervention. *Developmental Psychology, 26* (2), 269—277.

Guerra, N. G., Tolan, P. H., & Hammond, W. R. (1994). Prevention and treatment of adolescent violence. In L. D. Eron, J. Gentry, & P. Schlegel (Eds.), *Reason to hope: A psychosocial perspective on violence and youth* (pp. 383—403). Washington, DC: American Psychological Association.

Guilford, J. P. (1975). Creativity: A quarter century of progress. InI. A. Taylor & J. W. Getzels (Eds.), *Perspectives in creativity.* Chicago: Aldine. Gullota, I. P., Adams, J. R., & Montemayor, R. (Eds.). (1993). *Adolescent sexuality.* Newbury Park, CA: Sage.

Guttmacher, S., Lieberman, L., Ward, D., Freudenberg, N., Radosh, A. & Des-Jarlais, D. (1997). Condom availability in New York City public schools: Rtelationships to condom use and sexual behavior. *American Journal of Public Health, 87 (9), 1427-1433.*

Guttmann, J. & Lazar, A. (1998). Mother's of father's custody: Dpoes it matter for social adjustment? *Educational Psychology, 18 (2),* 225-234.

Haan, N., Weiss, R., & Johnson, V. (1982). The role of logic in moral reasoning and

Hackett, G., & Betz, N. E. (1981). A self-efficacy approach to the career development of women. *Journal of Vocational Behavior, 18*, 326-339.

Haggerty, R. J., Sherrod, L. R., Garmezy, N., & Rutter, M. (1994). *Stress, risk, and resilience in children and adolescents: Processes, mechanisms, and interventions.* New York: Cambridge University Press.

Hajcak, F., & Garwood, P. (1989). Quick-fix sex. Pseudosexuality in adolescents. *Adolescence, 23* (92), 756—758. Haley, J. (1980). *Leaving home.* New York: McGraw-Hill. Hall, G. 5. (1904). *Adolescence* (2 vols.). New York: Appleton. Hall, G. 5. (1923/1951). *Life and confessions of a psychologist.* New York: Appleton.

Hallinan, M. T. (1991). School differences in tracking structures and tracking assignments. *Journal of Research on Adolescence,], 251—276.

Hallinan, M. T., & Sorenson, A. B. (1987). Ability grouping and sex differences in mathematics achievement. *Sociology of Education, 60,* 63—72.

Hailman, R. (1967). Techniques for creative teaching. *Journal of Creative Behavior, 1* (3), 325—330.

Haliburn, J. (2000). Reasons for adolescent suicide attempts. *Journal of the American Academ,y of Child and Adolescent Psychiatry, 39 (1), 13-14.*

Halmi, K. A., Casper, R. C., Eckert, E. D., Goldberg, S. C., & Davis, J. M. (1979). Unique features associated with the age and onset of anorexia nervosa. *Journal of Psychiatry Research, 1,* 209—215.

Halmi, K. A., Falk, J. R., & Schwartz, E. (1981). Binge-eating and vomiting: A survey of a college population. *Psychological Medicine, 11,*

Hamilton, L. H., Brooks-Gunn, J., & Warren, M. P. (1985). Sociocultural influences on eating disorders in female professional dancers. *International Journal of Eating Disorders, 4,* 465—477.

Hammen, C., & Zupan, B. A. (1984). Seif-schemas, depression and the processing of personal information in children. *Journal of Experimental Child Psychology, 37,* 598—608.

Handelsman, C., Cabral, R., & Weisfeld, G. (1987). Sources of information and adolescent sexual knowledge and behavior. *Journal of Adolescent Research, 2* (4), 455—463.

Hankin, B. & Abramson, L. (1999). Development of gender differences in depression: Description and possible explanations. *Annals of Medicine, 31* (6), 372-379.

Haq, M. N. (1984). Age at menarche and the related issue: A pilot study of urban school girls. *Journal of Youth and Adolescence, 13,* 559—567.

Hare-Mustin, R. (1987). The gender dichotomy and developmental theory: A response to Sayers. *New Ideas in Psychology, 5* (2), 261—267.

Harkavy-Freidman, J. M., Asnis, G. M., Boeck, M., & DiFore, J. (1987). Prevalence of specific suicidal behaviors in a high school sample. *American Journal of Psychiatry, 16,* 313—325.

Harrison, A., Serafica, F., & McAdoo, H. (1984). Ethnic families of color. In R. D. Parke (Ed.), *Review of Child Development Research, No. 7.* Chicago: University of Chicago Press.

Harrison, A., Wilson, M., Pine, C., Cahn, S., & Buriel, R. (1990). Family ecologies of ethnic minority children. *Child Development, 61,* 347—362.

Hart, D. (1988). The adolescent self-concept in social context. In D. K. Lapsley & F. C. Power (Eds.), *Self, ego, and identity: Integrative approaches.* New York: Springer.

Harter, S. (1990a). Self and identity development. In S. Feldman & G. Elliott (Eds.), *At the threshold: The developing adolescent.* Cambridge, MA: Harvard University Press.

Harter, S. (1990b). Processes underlying adolescent self-concept formation. In R. Montemayor, G. R. Adams, & T. P. Gullotta (Eds.), *From childhood to adolescence: A transitional period? Advances in adolescent development, volume 2* (pp. 205—239) Newbury Park, CA: Sage.

Harter, S. (1993). Causes and consequences of low self-esteem in children and adolescents. In R. F. Baumeister (Ed.), *Self esteem: The puzzle of low self regard* (pp. 87—116). New York: Plenum Press.

Harter, S. (1999). The Construction of the Self: A Developmental Perspective. NY: Guilford.

Harter, S., & Marold, D. (1989). *A model of risk factors in adolescent suicide.* Kansas City, MO: Presentation, Society for Research in Child Development.

Harter, S., Whitesell, N. & Junkin, L. (1998). Similarities and differences in domain specific and global self-evaluations of learning disabled and behaviorally disordered and normally achieving adolescents. *American educational research journal, 35* (4), 653-680.

Hartley, P. (1999). Eating disorders and health education. Psychology- Health and Medicine, 3 (1), 133-140.

Hartos, J. & Power, T. (1997). Mothers' awareness of their early adolescents' stressors: Relation between awareness and adolescent adjustment. *Journal of Early Adolescence, 17 (4),* 371-389.

Hartup, W. W. (1982). Peer interaction and behavioral development of the individual child. In W. Damon (Ed.), *Social and personality development: Essays on the growth of the child* (pp. 220-233). New York: Norton.

Hartup, W. W. (1983). Peer relations. In E. M. Hetherington (Ed.), *Handbook of child psychology, Vol. IV: Socialization, personality, and social development* (pp. 103—196). New York: John Wiley.

Hartup, W. W. (1989). Social relationships and their developmental significance. *American Psychologist, 44* (2), 120—126.

Haskell, M. R., & Yablonsky, L. (1982). *Juvenile delinquency.* Chicago: Rand-McNally.

Hauser, S. T., & Bowlds, M. K. (1990). In S. Feldman and G. Elliot(Eds.), *At the threshold:The developing adolescent.* Cambridge, MA: Harvard University Press.

Havighurst, R. 1. (1951). *Developmental tasks and education.* New York: Longmans, Green.

Hawkins, D. F. (1993). Inequality, culture, and interpersonal violence. *Health Affairs, 12,* 80-95.

Hawkins, J. D., Jenson, I. M., Catalano, R. F., & Lsihenr, D. M. (1988). Delinquency and drug abuse: Implications for social services. *Social Service Review, 62,* 258—284.

Hawks, B., & Muha, D. (1991). Facilitating the career development of minorities: Doing it better this time. *The Career Development Quarterly, 39,* 251—260.

Hawton, K. (1982a). Motivational aspects of deliberate self-poisoning in adolescents. *British Journal of Psychiatry, 141,* 286—290.

Hawton, K. (1982b). Adolescents who take overdoses: Their characteristics, problems and contacts with helping agencies. *British Journal of Psychiatry, 140,* 118—123.

Hawton, K. (1982c). Classification of adolescents who take overdoses. *British Journal of Psychiatry, 140,* 124-131.

Heat, D.T. & Orthner, D.K. (1999). Stress and adaptation among male and female single parents. *Journal of Family Issues, 20 (4), 567-587.*

Hechinger, F. M. (1992). *Fatefulchoices: Healthy youth for the 21st century.* NY: Carnegie Corporation of New York, Carnegie Council on Adolescent Development.

Hechinger, F. M. (1993). Schools for teenagers: A historic dilemma. *Teachers College Record, 94 (3),* 522—539.

Heilbrun, A., & Friedberg, L. (1990). Distorted body image in normal college women: Possible implications for the development of anorexia nervosa. *Journal of Clinical Psychology, 46 (4),* 398—401.

Heller, H. C. (1993). The need for a core, interdisciplinary life-sciences curriculum in the middle grades. *Teachers College Record, 94,* 645—652.

Henggeler, 5. (1989). *Delinquency in adolescence.* Newbury Park, CA: Sage Publications.

Henry, T. (1992, August 27). SAT scores see first increase since 1987. *The Boston Globe,* p. 3.

Hemandez, L., Marek, E., & Renner, J. (1984). Relationships among gender, age, and intellectual development. *Journal of Research in Science, 21 (4),* 365—375.

Herrenkohl, E., Herrenkohl, R., Egolf, B. & Russo, M. (1998). The relationship between early maltreatment and teenage parenthood. *Journal of Adolescence, 21 (3),* 291-303.

Hersch, P. (1990). The resounding silence. *Family Therapy Networker, 14* (4), 18—29. Hersch, P. (1991, May). Sexually transmitted diseases are ravaging our children. *American Health,* pp. 42—52.

Hetherington, E. M. (1973). Girls without fathers. *Psychology Today, 6* (9), 47—52.

Hetherington, E. M. (1991). Families, lies and videotapes. *Journal of Research on Adolescence, 1,* 323—348.

Hetherington, E. M., & Clingempeel, W. E. (1988, March). *Coping with remarriage: The first two years.* Symposium presented at the Conference on Human Development, Charleston, SC.

Hetherington, E. M., Cox, M., & Cox, R. (1985). Long-term effects of divorce and remarriage on the adjustment of children. *Journal of the American Academy of Child Psychiatry, 24* (5), 518—530.

Hetherington, E. M., StanleyHagan, M., (1998). *Parent and child deveioipoment in "nontraditional families"* (pp. 137-159). NJ; Lawrence Elbaum Associates.

Hetherington, E. M., StanleyHagan, M., & Anderson, E. R. (1989). Marital transitions: A child's perspective. *American Psychologist, 44* (2), 303—312.

Hewlett, 5. (1991). *When the bough breaks: The cost of neglecting our children.* New York: Basic Books.

Higgins, A. (1989). The just community educational program: The development of moral role-taking as the expression of justice and care. In M. Brabeck (Ed.), *Who cares? Theory, research, and educational implications of the ethic of care.* New York: Praeger.

Higgins, S. T., & Stitzer, M. L. (1986). Acute marijuana effects on social conversation. *Psychopharmacology, 89* (2), 234-238.

Higgins-Trenk, A., & Gaite, A. (1971). Elusiveness of formal operational thought in adolescents. Proceedings of the 79th Annual Convention of the American Psychological Association.

Hightower, E. (1990). Adolescent interpersonal and familial precursors of positive mental health at midlife. *Journal of Youth and Adolescence, 19* (3), 257—275.

Hill, H. M., Soriano, F. I., Chen, S. A., & LaFromboise, T. D. (1994). Sociocultural factors in the etiology and prevention of violence among ethnic minority youth. In L. D. Eron, J. Gentry, & P. Schlegel (Eds.), *Reason to hope: A psychosocial perspective on violence and youth* (pp. 59—97). Washington, DC: American Psychological Association.

Hill, J. P., & Lynch, M. E. (1983). The intensification of gender-related role expectations during early adolescence. In J. Brooks-Gunn & A. C. Petersen (Eds.), *Girls at puberty: Biological, psychological, and social perspectives* (pp. 201—228). New York: Plenum.

Hill, P., Jr. (1987). *Passage to manhood: Rearing the male African-American child.* Paper presented at the Annual Conference of the National Black Child Development Institute, Detroit, MI.

Hirschi, T. (1969). *Causes of delinquency.* Berkeley: University of California Press. Hochhaus, C., & Sousa, F. (1988). Why children belong to gangs: A comparison of expectations and reality. *High School Journal, 71,* 74-77.

Hofferth, S. L., & Hayes, C. D. (Eds.). (1987). *Risking the future: Adolescent sexuality, pregnancy, and childbearing.* (Vol. 2). Washington, DC: National Academy Press. Hoffman, L. W. (1989). Effects of maternal employment in the two-parent family. *American Psychologist, 44* (2), 283—292.

Holcomb, D. R., Holcomb, L. C., Sondag, K. A., & Williams, N. (1991). Attitudes about date rape: Gender differences among college students. *College Student Journal, 25,* 434-439.

Holinger, P. (1979). Violent death among the young: Recent trends in suicide, homicide and accidents. *American Journal of Psychiatry, 139* (9), 1144-1147.

Holinger, P. C., Offer, D., & Ostrov, E. (1987). Suicide and homicide in the United States: An epidemiologic study of violent death, population changes, and the potential for prediction. *American Journal of Psychiatry, 144* (2), 215—219.

Holland, J. L. (1985a). *Making vocational choices: A theory of vocational personalities and work environments* (2nd ed.). Englewood Cliffs, NJ: Prentice-Hall.

Holland, J. L. (1985b). *Professional manual self-directed search.* Odessa, FL.: Psychological Assessment Resources, Inc.

Holmes, J., & Rahe, 5. (1967). A social adjustment scale. *Journal of Psychosomatic Research, 11,* 213—218.

Holtzen, D. W., Kenny, M. E., & Mahalik, J. R. (1995). Contributions of parental attachment to gay or lesbian disclosure to parents and dysfunctional cognitive processes. *Journal of Counseling Psychology, 42,* 350-355.

Horney, K. (1967). The technique of psychoanalytic therapy. *American Journal of Psychoanalysis, 28* (1), 3—12.

Horowitz, A. V., & White, H. R. (1987). Gender role orientations and styles of pathology among adolescents. *Journal of Health and Social Behavior, 28* (2), 158—170.

Hotchkiss, L., & Borow, H. (1990). Sociological perspectives on work and career development. In D. Brown, L. Brooks, et al. (Eds.), *Career choice and development.* San Francisco:Jossey-Bass.

Howard, J., Boyd, G. M., & Zucker, R. A. (1994). Overview of issues. Special Issue: Preventing alcohol abuse among adolescents: Preintervention and intervention research. *Journal of Research on Adolescence, 4* (2), 175—181.

Howard, M. (1985). Postponing sexual involvement among adolescents: An alternative approach to prevention of sexually transmitted disease. *Journal of Adolescent Health Care, 6* (4), 271—277.

Hrabowski, F.A. III, Maton, K.I. & Greif, G.L. (1998). *Beating the odds: Raising academically successful African american males.* Ny: Oxford university Press.

Huang, L. N., & Ying, Y. (1989). Chinese American children and adolescents. In J. T. Gibbs, L. N. Huang, & Assoc., *Children of color: Psychological interventions with minority youth.* San Francisco: Jossey-Bass.

Huba, G., Wingard, J., & Bentler, P. (1979). Beginning adolescents drug use and peer and adult interaction patterns. *Journal of Consulting Clinical Psychology, 47.*

Huba, G., Wingard, J., & Bentler, P. (1980a). A longitudinal analysis of the role of peer support, adult models, and peer subcultures in beginning adolescent substance use. *Multivariate Behavior Research, 15,* 259—280.

Huba, G., Wingard, J., & Bentler, P. (1980b). Framework for an interactive theory of drug use. In D. Lettieri, M. Sayers, & H. Pearson, (Eds.), *Theories on drug abuse.* Rockville, MD: National Institute on Drug Abuse.

Huba, G., Wingard, J., & Bentler, P. (1980c). Applications of a theory of drug use prevention. *Journal of Drug Education, 10,* 25—38.

Hubbard, R., Cavanaugh, E., Graddock, S., & Rachel, J. (1983). *Characteristics, behaviors and outcomes for youth in TOPS study* (Report submitted to the National Institute on Drug Abuse, Contract No. 271-79-3611). Research Triangle Park, NC: Research Triangle Institute.

Huff, C. R. (Ed.). (1990). *Gangs in America.* Newbury Park, CA: Sage.

Humphrey, L. L. (1989). Observing family interactions among subtypes of eating disorders using structural analysis of social behavior. *Journal of Consulting and Clinical Psychology, 57,* 206-213.

Humphrey, L. L., & Stern, S. (1988). Object relations and the family system in bulimia: A theoretical integration. *Journal of Marital and Family Therapy, 14,* 337—350.

Hunter, E. (1998). Adolescent attraction to cults. *Adolescence, 33 (131),* 709-714.

Huston, A. C., Donnerstem, E., Fairchild, H., Feshbach, N. D., Katz, P. A., Murray, J. P., Rubinstein, E. A., Wilcox, B. L., & Zuckerman, D. (1992). *Big world, small screen: The role of television in American society.* Lincoln: University of Nebraska Press.

Idol, L. (1987). A critical thinking map to improve content area comprehension of poor readers. *Remedial and Special Education, 8* (4), 28—40.

Imobekhai, S. Y. (1986). Attainment of puberty and secondary sexual characteristics in some rural and urban Nigerian adolescents. *Nigerian Journal of Guidance and Counseling, 2,* 48—55.

Inclan, J. E., & Herron, D. G. (1989). Puerto Rican adolescents. In J. T. Gibbs, L. N. Huang, & Assoc., *Children of color: Psychological interventions with minority youth.* San Francisco: Jossey-Bass.

Isay, R. A. (1989). *Being homosexual: Gay men and their development.* New York: FarrarStraus-Giroux.

Jacobs, J. (1971). *Adolescent suicide.* New York: Wiley.

Jacobson, K. & Crockett, L. (2000). Parental monitoring and adolescent adjustment: An ecological perspective. *Journal of Research on Adolescence, 10 (1),* 65-97.

Jacobziner, H. (1965). Attempted suicide in adolescence. *Journal of the American Medical Association, 10,* 22—36.

James, W. (1890/1950). *The principles of psychology.* New York: Dover.

James, W. (1892). *Psychology: The briefer course.* New York: Holt, Rinehart, & Winston.

Jankowski, M. 5. (1991). *Islands in the street.* Berkeley: University of California Press.

Janus, C. J., & Janus, 5. (1993). *The Janus Report on Sexual Behavior.* New York: John Wiley.

Jaquish, G., & Ripple, R. E. (1980). Cognitive creative abilities across the adult life span. *Human Development, 143—152.*

Jaquish, G. A., Block, J., & Block, J. H. (1984). *The comprehension and production of metaphor in early adolescence: A longitudinal study of cognitive childhood antecedents.* Unpublished manuscript.

Jarrett, R. L. (1995). Growing up poor: The family experiences of socially mobile youth in low-income African American neighborhoods. *Journal of Adolescent Research, 10,* 111—135.

Jencks, C., Smith, M., Acland, H., Bane, M. J., Cohen, D., Gintis, H., Heyns, B., & Michelson, S. (1972). *Inequality: A reassessment of the effect of family and schooling in America.* New York: Basic.

Jenkins, A. (1982). *The psychology of the Afro-American. A humanistic perspective.* New York: Pergamon.

Jensen, A. R., Cohn, S. J., & Cohn, C. M. (1989). Speed of information processing in academically gifted youth. *Personality and Individual Differences, 6* (5), 621—629.

Jerse, F. W., & Fakouri, M. E. (1978). Juvenile delinquency and academic deficiency. *Contemporary Education, 49,* 108—109.

Jessor, R. (1993). Successful adolescent development among youth in high-risk settings. *American Psychologist,* 48, 117—126.

Jessor, R., & Jessor, 5. (1977). *Problem behavior and psychosocial development.* New York: Academic Press.

Jessor, R., Turbin, M & Costa, F. (1998). Risk and protection in successful outcomes among disadvantaged adolescents. *Appiled Developmental Science, 2 (4),* 194-208.

Johnson, B., Shulman, S., & Collins, W. A. (1991). Systemic patterns of parenting as reported by adolescents: Developmental differences and implications for psychosocial outcomes. *Journal of Adolescent Research, 6* (2), 235—252.

Johnston, L. D., OMalley, P. M., & Bachman, J. G. (1991). *Drug use among American high school seniors, college students, and young adults, 1975—1990, Vol. 1* (DHHS Publication No. (ADM) 91—1835). Washington, DC: Government Printing Office.

Jones, J., & Barlow, D. (1990). Self-reported frequency of sexual urges, fantasies, and masturbatory fantasies in heterosexual males and females. *Archives of Sexual Behavior, 19* (3), 269—279.

Jones, L. P. (1988). A typology of adolescent runaways. *Child and Adolescent Social Work, 5,* 16-29.

Jordan, T. E. (1987). *Victorian childhood.* Albany: State University of New York Press.

Jorgensen, S. R. (1993). Adolescent pregnancy and parenting. In T. P. Gullotta, G. R. Adams, & R. MontemayOr (Eds.), *Adolescent sexuality, advances in adolescent development: An annual book series, Vol. 5* (pp. 103—140). Newbury Park, CA: Sage.

Josselson, R. (1988). The embedded self: I and thou revisited. In D. K. Lapsley & Quintana, S. M. (Eds.), *Self, ego and identity: Integrative approaches* (pp. 91—108). New York: Springer.

Josselson, R. L. (1987). *Finding herself: Pathways of identity development in women.* New York: Jossey-Bass.

Josselson, R. L. (1996). *Revisiting herself: The story of women's identity from college to midlife.* NY: Oxford University Press.

Jurich, A. P., Schumm, W. R., & BoIlman, S. R. (1987). The degree of family orientation perceived by mothers, fathers, and adolescents. *Adolescence, 22* (85), 119—128.

Kaeding, 5. (1985). Some will die: Teenage 5-icide-drinking and dying. Unpublished paper.

Kagan, J., & Moss, H. (1962). *Birth to maturity: A study in psychological development.* New York: Wiley.

Kalakoski, V. & Nurmi, J. (1998). Identity and educational transition: Age differences in adolescent exploration and commitment realted to education, occupation and family. *Journal of Research on Adolescence, 8 (1), 29-47.*

Kandel, D. B. (1980). Developmental stages in adolescent drug involvement. In D. J. Lettieri, M. Sayers, & H. W. Pearson, Theories on drug use: Selected contemporary perspectives. *N.I.D .A -Research*

Kaplan, A. G. (1991). The "self-in-relation": Implications for depression in women. In J. Jordon, A. Kaplan, J. B. Miller, I. Stiver & J. L. Surrey (Eds.), *Women's growth in connection: Writing from the Stone Center.* New York: Guilford.

Kaplan, A. G., & Klein, R. (1985). The relational self in late adolescent women. In *Works in progress* (Rep. No. 17), Wellesley, MA: Wellesley College, Stone Center for Developmental Services and Studies.

Kaplan, E. A. (1990). *MTV and adolescents.* Paper presented at the 98th annual convention of the American Psychological Association, August 10-14, Boston, Massachusetts. APA Audiotape 90—217.

Kauffman, J. M. (1981). *Characteristics of children's behavior disorders* (2nd ed.). Columbus, OH: Merrill.

Kazdin, A. (1993). Adolescent mental health: Prevention and treatment programs. *American Psychologist, 48* (2), 127—141.

Keil, F. (1984). A characteristic to defining shift in the development of word meaning. *Journal of Verbal Learning and Verbal Behavior, 23* (2), 221—236.

Kelliher, M. F. (1993). Community service learning: One school's story. *Equity & Excellence in Education, 26,* 12—14.

Kelly, J. (1991, April 29). Kids killing kids: What can be done? A social meltdown. *The Boston Globe.*

Kelly, J. A., & Hansen, D. J. (1987). Social interactions and adjustment. In V. B. Van Hasselt & M. Hersen (Eds.), *Handbook of adolescent psychology* (pp. 131—146). New York: Pergamon Press. Keniston, K. (1965). *The uncommitted.* New York: Harcourt, Brace, and World.

Keniston, K. (1968). *Young radicals.* New York: Harcourt, Brace, and World.

Kennedy, R. E., & Petersen, A. C. (1989). *Stressfulfamily events and adjustment among young adolescents.* Unpublished manuscript, Pennsylvania State University, College Park.

Kenny, M. E. (1987). The extent and function of parental attachment among first-year college students. *Journal of Youth and Adolescence, 16,* 17—27.

Kenny, M. E., & Donaldson, G. (1991). Contributions of parental attachment and family structure to the social and psychological functioning of first-year college students. *Journal of Counseling Psychology, 38,* 479—486.

Kenny, M. E., & Donaldson, G. A. (1992). The relationship of parental attachment and psychological separation to the adjustment of first-year college women. *Journal of College Student Development, 33,* 431—438.

Kenny, M. E., & Hart, K. (1992). Relationship between parental attachment and eating disorders in an inpatient and a college sample. *Journal of Counseling Psychology, 39* (4), 521—526.

Kenny, M. E., Moilanen, D., Lomax, R., & Brabeck, M. M. (1993). ContributiOns of parental attachments tO view of self and depressive symptoms among early adolescents. *Journal of Early Adolescence, 13,* 408—430.

Ketterlinus, R. D., Lamb, M. E., & Nitz, K. (1994). Adolescent nonsexual and sex-related problem behaviors: Their prevalence consequenecs, and co-occurrence. In R. D. Ketterlinus & M. E. Lamb (Eds.), *Adolescent problem behaviors* (pp. 17-40). Hillsdale, NJ: Erlbaum.

Kienhorst, C. W., Wolters, W. H., Diekstra, R. F., & Otto, E. (1987). A study of the frequency of suicidal behavior in children aged 5 to 14. *Journal of Child Psychology and Psychiatry and Allied Disciplines, 28(1),* 153—165.

Kilpatrick, D.G., Acierno, R., Saunders, B., Resnick, H.S., Best, C.L. & Schnurr, P.P. (2000). Risk factors for adolescent substance abues and dependence: Data from a national sample. Journal of Consulting and Clinical Psychology, 68 (1), 19-30

Kilpatrick, W. (1975). *Identity and intimacy.* New York: Delacorte.

Kim, K. & Smith, P.K. (1998). Childhood stress, behavioural symptoms and mother-daughter pubertal development. Journal of Adolescence, 21 (3), 231-240.

Kinard, E., & Reinherz, H. (1987). School aptitude and achievement in children of adolescent mothers. *Journal of Youth and Adolescence, 16* (1), 69—87.

King, I. (1914). *The high school age.* Indianapolis: Bobbs-Merrill.

Kinney, D. (1999). From "headbangers" to "hippies": Delineating adolescents' active attempts to form an alternative peer culture. In McLellan, J. & Pugh, M.J. (Eds.) *The role of peer groups in adolescent social identity: Exploring the importnace of stability and change.* New directions for child and adolescent developemnt, 84 (pp. 21-35). CA: Jossey-Bass.

Kirkland, M., & Ginther, D. (1988). Acquired immune deficiency syndrome in children: Medical, legal, and school-related issues. *School Psychology Review, 17,* 304-305.

Kirkpatrick, J., Beebe, T., Mortimer, J. & Snyder, M (1998). Volunteerism in adolescence: A process perspective. *Journal of Research on Adolescence, 8 (3), 309-332.*

Kitahara, M. (1983). Female puberty rites: Reconsideration and speculation. *Adolescence, 18* (72), 957—964. Kitahara, M. (1984). Female physiology and female puberty rites. *Ethos, 12* (2), 132—149.

Kitchener, K., & King, P. (1981). Reflective judgment: Concepts of justification and their relationship to age and education. *Journal of Applied Developmental Psychology, 2,* 89—116.

Klassen, A. D., Williams, C. J., & Levitt, E. E. (1989). *Sex and morality in the U.S.: An empirical enquiry under the auspices of the Kinsey Institute.* Middletown, CT: Wesleyan University Press.

Klein, H. (1990, June). Adolescence, youth, and young adulthood. *Youth & Society, 21* (4), 446—471.

Klein, H., & Cordell, A. (1987). The adolescent as mother: Early risk identification. *Journal of Youth and Adolescence, 16(1),* 47—58.

Klemchuk, H., Hutchinson, C., & Frank, R. (1990). Body dissatisfaction and eating-related problems on the college campus: Usefulness of the Eating Disorder Inventory with a nonclinical population. *Journal of Counseling Psychology, 37* (3), 297—305.

Kling, K., Hyde, J., Showers, C & Buswell, B. (1999). Gender differences in self-esteem: A meta-analysis. *Psychological Bulletin, 125 (4), 470-500.*

Knight, G., Dubro, A., & Chao, C. (1985). Information processing and the development of cooperative, competitive, and individualistic social values. *Developmental Psychology, 27* (1), 375.

Kobak, R., & Sceery, A. (1988). Attachment in late adolescence: Working models, affect regulation, and representation of self and others. *Child Development, 59,* 135—146.

Koch, P. (1988). The relationship of first intercourse to later sexual functioning concerns of adolescents. *Journal of Adolescent Research, 3* (3—4), 345—362.

Koch, P. B. (1993). Promoting healthy sexual development during early adolescence. In R. M. Lemer (Ed.), *Early adolescence: Perspectives on research, policy, and intervention* (pp. 293—307). Hillsdale, NJ: Erlbaum.

Koff, E., Rierdan, J., & Stubbs, M. (1990). Gender, body image, and self-concept in early adolescence. *Journal of Early Adolescence, 10* (2), 56—68.

Kogan, N. (1973). Creativity and cognitive style: A life-span perspective. In P. B. Baltes & K. W. Schaie (Eds.), *Lifespan developmental psychology.* New York: Academic Press.

Kogan, N. (1983). Stylistic variation in childhood and adolescence: Creativity, metaphor, cognitive styles. In P. H. Mussen (Ed.), *Handbook of child psychology* (Vol. 3). New York: Wiley.

Kohlberg, L. (1970). Moral development and the education of adolescents. In R. F. Pumell (Ed.), *Adolescents and the American high school.* New York: Holt, Rinehart & Winston.

Kohlberg, L. (1984). *The psychology of moral development* (Vol. 2). San Francisco: Harper & Row.

Kohut, H. (1977). *The restoration of self* New York: International Universities Press.

Kohut, H. (1984). Introspection, empathy and semicircle of mental health. Chicago Psychoanalytic Society and the Chicago Institute for Psychoanalysis Conference: The vital issues (1981), Chicago, Illinois. *Emotions and Behavior Monographs, 3,* 347—375.

Kong, D. (1995, April 28). 60% of US pregnancies unwanted or mistimed, study says. *The Boston Globe,* pp. 1, 14.

Kong, D., & Brelis, M. (1995, April 6). Binge drinking lures many college freshmen, study says. *The Boston Globe,* pp. 1, 17.

Koski, K. J., & Steinberg, L. (1990). Parenting satisfaction of mothers during midlife. *Journal of Youth and Adolescence, 19,* 465—474.

Koss, M. P., Gidycz, C. A., & Wisniewski, N. (1987). The scope of rape: Incidence and prevalence of sexual aggression and victimization in a national sample of higher education students. *Journal of Counseling and Clinical Psychology, 55,*162—170.

Kovacs, M. (1989). Affective disorders in children and adolescents. *American Psychologist, 44,* 209—215.

Koyle, P. F., Jensen, L. C., Olsen, J., & Cundick, B. (1989). Comparison of sexual behaviors among adolescents. *Youth and Society, 20* (4), 461—476.

Krebs, (1967). *Some relations between moral judgment attention and resistance to temptation.* Unpublished Ph.D. dissertation, University of Chicago.

Kreipe, R., & Strauss, J. (1989). Pubertal maturation and parent-adolescent distance: An evolutionary perspective. In G. Adams, R. Montemayor, & T. Gullotta (Eds.), *Biology of adolescent behavior and development* (pp. 71—97). Newbury Park, CA: Sage.

Kreipe, R., Strauss, J., Hodgman, C., & Ryan, R. M. (1989). Menstrual cycle abnormalities and subclinical eating disorders: A preliminary report. *Psychosomatic Medicine, 51(1),* 81—86.

Kubitscchek, W. & Hallinan, M. (1998). Tracking and students' friendships. *Social Psychology Quarterly, 61 (1), 1-15.*

Kuhn, D. (1984). Short-term longitudinal evidence for the sequentiality of Kohlberg's early stages of moral development. *Developmental Psychology, 12,* 162—166.

Kumpfer, K., & Turner, C. (1990). The social ecology model of adolescent substance abuse: Implications for prevention. *The International Journal of the Addictions, 25* (4A), 435—463.

Kupersmidt, J. B., & Coie, J. D. (1990). Preadolescent peer status, aggression, and school adjustment as predictors of externalizing problems in adolescence. *Child Development, 61,* 1350-1362.

Kurdek, L., Fine, M. A., & Sinclair, R. J. (1995). Parenting transitions, family climate, and peer norm effects. *Child Development, 66,* 430—445.

Kurz, 5. (1977, November 23). Teenage prostitutes. *Equal Times, 6.*

Kuttler, A., La Greca, A. & Prinstein, M. (1999). Friendship qualities and social emotional functioning of adolescents with close, cross-sex friendships. *Journal of Research on Adolescence, 9 (3),* 339-366.

Langhinrichsen, R.J., Lewinsohn, P., Rohde, P., Seely, J., Monson, C.M. Meyer, K.A. & Langford, R. (1998). Gender differences in suicide related behaviors of adolescents and young adults. Sex Roles, 39 (11-12), 839-854.

Lambom, S. D., Mounts, N., Seinberg, L., & Dornbusch, S. M. (1991). Patterns of competence and adjustment among adolescents from authoritative, authoritarian, indulgent, and neglectful families. *Child Development, 62,*1049—1065.

Lamborn, S.D., & Steinberg, L. (1993). Emotional autonomy redux: Revisiting Ryan and Lynch. *Child Development, 64,* 483—499.

Lamke, L. K. (1982). The impact of sex-role orientation on self-esteem in early adolescence. *Child Development, 53,* 1530-1535.

Lammers, C., Ireland, M., Resnick, M. & Blum, R. (2000). Influences on adolescents' decision to postpone onset of sexual intercourse: A survival analysis among youths aged 13 to 18 years. *Journal of Adolescent Health, 26 (1), 42-48.*

Lancaster, J. B., & Hamburg, B. A. (1986). *School-age pregnancy and parenthood.* New York: Aldine de Gruyter. Langway, L. (1982, October 18). A nation of runaway kids. *Newsweek,,* 97—98.

Lapsley, D., & Murphy, M. (1985). Another look at the theoretical assumptions of adolescent egocentrism. *Developmental Review, 5, 201—217.*

Lapsley, D. K. (1989). Continuity and discontinuity in adolescent social cognitive development. In R. Montemayor, G. Adams, & T. Gullotta (Eds.), *Advances in adolescence research* (Vol. 2), pp. 183—203. Orlando, FL: Academic Press.

Lapsley, D. K. (1993). Toward an integrated theory of adolescent ego development: The "new look" at adolescent egocentrism. *American Journal of Orthopsychiatry, 63 (4), 562-571.*

Lapsley, D. K., Harwell, M., Olson, L., Flannery, D., & Quintana, S. (1984). Moral judgement, personality and attitude to authority in early and late adolescence. *Journal of Youth and Adolescence, 13* (6), 527—541.

Lapsley, D. K., & Rice, K. (1988). The "New Look" at the imaginary audience and the personal fable: Toward a general model of adolescent ego development. In D. K. Lapsley, & F. C. Power (Eds.), *Self, ego and identity: Integrative approaches* (pp. 109—129). New York: Springer-Verlag.

Larose, S. & Boivin, M. (1998). Attachment to parents, social support ex[ectations and socipemotional adjustment during the high school-college transition. *Journal of Research on Adolescence, 8 (1), 1-27.*

Larsen, R., & Asmussen, L. (1991). Anger, worry and hurt in early adolescence: An enlarging world of negative emotions. In M. E. Colten & S. Gore (Eds.), *Adolescent stress: Causes and consequences.* New York: Aldine de Gruyter.

Larson, R., & Johnson, C. (1981). Anorexia nervosa in the context of daily experience. *Journal of Youth and Adolescence, 10 (6),* 455—471.

Lasch, C. (1979). *The culture of narcissism: American life in an age of diminishing expectations.* New York: Warner.

Lauritsen, J. L., Laub, J. H., & Sampson, R. J. (1992). Conventional and delinquent activities: Implications for the prevention of violent victimization among adolescents. *Victims and Violence, 7,* 91—108.

Laursen, B. (1995). Conflict and social interaction in adolescent relationships. *Journal of Research on Adolescence, 5,* 55—70.

Laursen, B., Coy, K.C. & Collins, W.A. (1999). Reconsidering changes in parent-child conflict across adolescence: A meta-analysis. *Child Development, 69 (3), 817-832.*

Lazarus, R., & Folkman, 5. (1984). *Stress, appraisal and coping.* New York: Springer.

Lazarus, R., & Folkman, 5. (1985). *Ways of coping scale.* Palo Alto, CA: Consulting Psychologists Press.

Leadbeater, B. J., Blatt, S. J., & Quinlan, D. M. (1995). Gender-linked vulnerabilities to depressive symptoms, stress, and problem behaviors in adolescents. *Journal of Research on Adolescence, 5,* 1—29.

LeCroy, C. (1988). Parent-adolescent intimacy: Impact on adolescent functioning. *Adolescence, 23 (89),* 137—147.

Lee, C. (1985). Successful rural black adolescents: A psychosocial profile. *Adolescence, 20,* 129—142.

Lee, V., & Bryk, A. (1986). Effects of single-sex secondary schools on student achievement and attitudes. *Psychological Bulletin, 78,* 381—395.

Leeming, F. C., Dwyer, W. 0., & Oliver, D. P. (1996). *Issues in adolescent sexuality. Readings from the Washington Post writers groups.* Boston: Allyn & Bacon.

Lees, 5. (1986). *Losing out: Sexuality and adolescent girls.* London: Hutchinson. Leon, G. R., Fulkerson, J. A., Perry, C. L., & Dube, A. (1994). Family influences, school behaviors, and risk for the later development of an eating disorder. *Journal of Youth and Adolescence, 23 (5),* 499—515.

Lerner, M. (1990). The fire of "ice." *Drugs, Society, and Behavior, 90/91.* Guilford, CT: The Dushkin Publishing Group.

Lerner, R., & Foch, T. (1987). A life-span perspective for early adolescence. In R. Lerner & T. Foch (Eds.), *Biological-psychosocial interactions in early adolescence* (pp. 9—32). Hillsdale, NJ: Erlbaum.

Lerner, R., Orbs, J., & Knapp, J. (1976). Physical attractiveness, physical effectiveness, and self-concept in late adolescents. *Adolescence, 11,* 313—326.

Lerner, R., & Walls, T. (1999). Revisiting individuals as producers of their development: From dynamic interactionism to developmental systems. In Brandtstaedter, J. & Lerner, R. (Eds.), *Action and self development: Theory and research through the lifespan* (pp. 3-36). Thousand Oaks CA: Sage Publications.

Leslie, L., Huston, T., & Johnson, M. (1986). Parental reactions to dating relationships: Do they make a difference? *Journal of Marriage and the Family, 48 (2),* 57—66.

Levenson, P., Pfefferbaum, B., & Morrow, J. (1987). Disparities in adolescent-physician views of teen health information concerns. *Journal of Adolescent Health Care, 8,* 171—176.

Leventhal, H., & Keeshan, P. (1993). Promoting healthy alternatives to substance abuse. In S. G. Millstein, A. C. Petersen, & E. 0. Nightingale (Eds.), *Promoting the health of adolescents: New directions for the twenty-first century* (pp. 260-284). New York: Oxford University Press.

Levine, A. (1980). *When dreams and heroes died.* San Francisco: Jossey-Bass.

Levine, G., Preddy, D., & Thordike, R. (1987). Speed of information processing and level of cognitive ability. *Personality and Individual Differences, 8 (5),* 599—607.

Levinson, D. (1978). *Seasons of a man's life.* New York: Knopf.

Levinson, D. (1996). *Seasons of a woman's life.* New York: Knopf.

Lewinsohn, P. M., Youngren, M. A., & Grosscup, 5. 5. (1979). Reinforcement and depression. In R. A. Depue (Eds.), *The psychology of depressive disorders.* New York: Academic Press.

Lewis, D. 0., Lovely, R., Yeager, C., & della-Femina, D. (1990). Toward a theory of the genesis of violence: A follow-up study of delinquents. *Journal of the American Academy of Child and Adolescent Psychiatry, 28 (4),* 431—436.

Lewis, J. M., & Looney, J. G. (1983). *The long struggle: Well-functioning working-class black families.* New York: Brunner/Mazel.

Lewis, R., Piercy, F., Sprenkle, D., & Trepper, T. (1990). Family-based interventions for helping drug-abusing adolescents. *Journal of Youth and Adolescent Research, 5 (1),* 82—95.

Lickona, T. (1977). How to encourage moral development. *Learning, 5 (7),* 36—44.

Lidz, T., & Lidz, R. W. (1984). Oedipus in the stone age. *Journal of the American Psychoanalytic Association, 32 (3),* 507—527.

Liebert, R. M. (1984). What develops in moral development? In W. K. Kurtines and J. H. Gewirtz (Eds.), *Morality, moral behavior, and moral development.* New York: Wiley.

Lifson, A., Hessol, N., & Ruther-ford, G. W. (1989, June). *The natural history of HI V infection in a cohort of homosexual and bisexual men. Clinical manifestations, 1978—1989.* Paper presented at the Fifth International Conference on AIDS, Montreal, Canada.

Lightfoot, A. (1978). *Urban education in social perspective.* Chicago: Rand McNally.

Lightfoot, S. L. (1983). *The good high school.* New York: Basic Books.

Lingxin, H. & Bonstead, B.M. (1998). Parent-child differences in educational expectations and academic achievement of immigrant and native students. *Sociology of Education, 71 (3), 175-198.*

Lipsitz, J. (1984). *Successful schools for young adolescents.* New Brunswick, NJ: Transaction.

Lipsitz, J. (1991). Public policy and young adolescents: A 1990s context for researchers. *Journal of Early Adolescence, 11* (1), 20-37.

Lock, J. & Steiner, H. (1999). Gay, lesbian and bisexual youths risks for emotional and social problems: Results from a community based survey. Journal of the American Academy of Child and Adolescent Psychiatry, 38 (3), 297-304.

Longshore, D., & Prager, J. (1985). The impact of school desegregation: A situational analysis. *American Journal of Sociology, II,* 75—91.

Lonky, E., Roodin, P., & Rybash, J. (1988). Moral judgment and sex role orientation as a function of self and other presentation mode. *Journal of Youth and Adolescence, 17(2),* 189—195.

Lopez, F. G., & Andrews, 5. (1987). Career indecision: A family systems perspective. *Journal of Counseling and Development, 56,* 304-307.

Lowenkopf, E. (1982, May/June). Anorexia nervosa: Some nosological consideration. *Comprehensive Psychiatry, 23* (3), 233—239.

Lundholm, J., & Littrell, I. (1986). Desire for thinness among high school cheerleaders: Relationship to disordered eating and weight control behaviors. *Adolescence, 21,* 573—579.

Maccoby, E., & Martin, J. (1983). Socialization in the context of the family: Parent-child interaction. In E. M. Hetherington (Ed.), *Handbook of child psychology: Socialization, personality and social development* (Vol. 4). New York: Wiley.

Maccoby, E. E., & Jacklin, C. N. (1974). *The psychology of sex differences.* Stanford, CA: Stanford University Press.

Mack, J. E., & Hickler, H. (1982). *Vivienne: The life and suicide of an adolescent girl.* New York: New American Library.

Madaus, G. F., Kellaghan, T., Rakow, E. A., & King, D. J. (1979). The sensitivity of measures of school effectiveness. *Harvard Educational Review, 49,* 207—230.

Maggs, J., & Kolaric, G. (1990, August). *After-school supervision and adolescents' peer experiences and self-image.* Paper presented at the annual meeting of the American Psychological Association, Boston, MA.

Magnusson, D. (1988). *Individual development from an interactional perspective: A longitudinal study.* Hillsdale, NJ:Erlbaum.

Mahler, M., Pine, F., & Bergman, A. (1975). *The psychological birth of the human infant.* New York: Basic Books.

Maloney, M., McGuire, J., Daniels, S., & Specker, B. (1989). Dieting behavior and eating attitudes in children. *Pediatrics, 84* (3), 482—489.

Mandler, J. M. (1983). Representation. In P. Mussen (Ed.), *Carmichael's manual of child psychology* (Vol. 4). New York: Wiley.

Manlove, J. (1998). The influence of high school drop out and school disengagement on the risk of school-age pregnancy. *Journal of Research on Adolescence, 8 (2),* 187-220.

Marcia, J. E. (1966). Development and validation of ego identity status. *Journal of Personality and Social Psychology, 3,* 551—558.

Marcia, J. E. (1967). Ego identity status: Relationship to change in self-esteem, general maladjustment and authoritarianism. *Journal of Personality, 35,* 118—133.

Marcia, J. E. (1968). The case history of a construct: Ego identity status. In E. Vinacke (Ed.), *Readings in general psychology.* New York: Van Nostrand-Reinhold.

Marcia, J. E. (1980). Identity in adolescence. In J. Adelson (Ed.), *Handbook of adolescent psychology.* New York: John Wiley.

Marcia, J. E. (1994). The empirical study of ego identity. In Bosma, H. (Ed.), *Identity and development:An interdisciplinary approach.* CA: Sage Publications.

Markstrom, C. (1999). Reliogious involvement and adolescent psychosocial development. *Journal of Adolescence, 22 (2), 205-221.*

Markus, H., & Nurius, P. (1986). Possible selves. *American Psychologist, 41,* 954-969.

Martin, A. D., & Hetrick, E. S. (1988). The stigmitization of the gay and lesbian adolescent. *Journal of Homosexuality, 15,* 163—183.

Marx, F. (1989). *After school programs for low-income young adolescents: Overview and program profiles.* Working Paper No. 194, Wellesley College Center fur Research on Women.

Maslow, A. (1968). *Toward a psychology of being* (2nd ed.). New York: Van Nostrand.

Maslow, A. (1971). *The farther reaches of human nature.* New York: Viking.

Mason, C. A., Cauce, A. M., Gonzales, N., Hiraga, Y., & Grove, K. (1994). An ecological model of externalizing behaviors in African-American adolescents: No family is an island. *Journal of Research on Adolescence, 4,* 639—655.

Mason, E. P. (1967). Comparison of personality characteristics of junior high school students from American Indian, Mexican and Caucasian ethnic backgrounds. *Journal of Social Psychology, 73,* 115—128.

Massachusetts Department of Education. (1994). *Sexually transmitted disease prevention education in Massachusetts schools: Results of focus group research* (No. 1759—10-750-6/94-DOE). Malden, MA: Massachusetts Department of Education.

Massachusetts Department of Public Health. (1991). *Adolescents at risk 1991: Sexually transmitted diseases.* Boston, MA: Author.

Massing, M. (1990). Crack's destructive sprint across America. *Drugs, society, and behavior, 90/91.* Guilford, CT: The Dushkin Publishing Group.

Masten, A. S., Neeman, J., & Andenas, 5. (1994). Life events and adjustment in adolescents: The significance of event independence, desirability, and chronicity. *Journal of Research on Adolescence, 4,* 71—98.

Masters, W., & Johnson, V. (1966). *Human sexual response.* Boston: Little, Brown.

Matchan, L. (1995, June 4). Hectic on the homefront. *The Boston Globe,* pp. 1,30,31.

Maton, K., & Zimmerman, M. (1990). *Psychological predictors of substance abuse among urban black male adolescents.* Paper presented at the 98th Annual Convention of the American Psychological Association, August 10-14, Boston, MA.

Mazza, J.J. & Reynolds, W.M. (1998). A longitudinal investigation of depression, hopelessness, social support, and major and minor life events and their relation to suicidal ideation in adolescents. Threatening Behavior, 28 (4), 358-374.

McCabe, M. P., & Collins, J. K. (1984). Measurement of depth of desired and experienced sexual involvement at different stages of dating. *Journal of Sex Research, 20,* 377—390.

McCarthy, M. (1990). The thin ideal, depression and eating disorders in women. *Behavioral Research & Therapy, 28* (3), 205—215.

McCarthy-Tucker, S.N. (1999). Teaching logic to adolescents to improve thinking skills. *Korean Journal of Thinking and Problem Solving, 8 (1),* 45-66.

McFarlane, A. H., Bellissimo, A., Norman, G. R., & Lange, P. Adolescent depression in a school-based community sample: Preliminary findings on contributing social factors. *Journal of Youth and Adolescence, 23,* 601—620.

McGoldrick, M., & Gersen, R. *(1985). Genograms in family assessment.* New York: W. W. Norton.

McGrath, E., Keita, G. P.,Strickland, B. R., & Russo, N. F. (1990). *Women and depression: Risk factors and treatment issues.* Washington, DC: American Psychological Association.

McIntire, J. (1980). Suicide and self-poisoning in pediatrics. *Resident and Staff Physician,* 72—85.

McKenry, D., Tishler, C., & Kelley, C. (1982). Adolescent suicide: A comparison of attempters and non-attempters in an emergency room population. *Clinical Pediatrics, 21(5),* 911—916.

McLoyd, V. C. (1990). The impact of economic hardship on Black families and children: Psychological distress, parenting, and socioemotional development. *Child Development,* 6], 311—346.

McNelles, L.R. & Connolly, J.A. (1999). Intimacy between adolescent friends: Age and gender differences in intimate affect and intimate behaviors. *Journal of Research on Adolescence, 9 (2), 143-159.*

Mead, M. (1927/1949). *Coming of age in Samoa.* New York: Mentor Books.

Mead, M. (1970). *Culture and commitment.* New York: Doubleday.

Mechanic, D. (1983). Adolescent health and illness behavior: Review of the literature and a new hypothesis for the study of stress. *Journal of Human Stress, 9,* 4—13.

Meeks, J. E., & Cahill, A. J. (1988). In S. C. Feinstein (Ed.), *Adolescent Psychiatry, 15, 475—486.*

Meilman, P. (1979). Cross-sectional age changes in ego identity status during adolescence. *Developmental Psychology, 15* (2), 230—231.

Meston, C. & Gorzalka, B. (1996). Differential effects of sympathetic activation on sexual arousal in sexually dysfunctional and functional women. *Journal of Abnormal Psychology, 105 (4), 582-591.*

Meyer, A. L., Pierce, R., & Burgess, R. L. (1991, April). *Assessment of gender differences in suicidal thoughts, attitudes, and beliefs in a normal adolescent population.* Paper presented at the Biennial Meeting of the Society for Research on Child Development, Seattle, Washington.

Milgram, G. G. (1993). Adolescents, alcohol, and aggression. *Journal of Studies on Alcohol, Suppl. 11,* 53—61.

Miller, B. C., Christopherson, C. R., & King, P. K., (1993). Sexual behavior in adolescence. In T. P. Gullotta, C. R. Adams, & R. Montemayor (Eds.), *Adolescent sexuality, advances in adolescent development: An annual book series, Vol. 5,* (pp. 57—76). Newbury Park, CA: Sage.

Miller, B. C., & Dyk, P. A. (1993). Sexuality. In P. H. Tolan & B. J. Cohler (Eds.), *Handbook of clinical research and practice with adolescents* (pp. 95—123). New York: John Wiley.

Miller, B. C., McCoy, J. K., & Olson, T. D. (1986) Dating age and stage as correlates of adolescent sexual attitudes and behavior. *Journal of Adolescent Research,* 1, 361—371.

Miller, J. B. (1976). *Toward a new psychology of women.* Boston: Beacon Press.

Miller, J. B., Jordan, J.V., Kaplan, A.G. & Stiver, I.R. (1997). *Women's Growth in Diversity.* NY: Guilford.

Miller, J. G., & Bersoff, D. M. (1989). When do American children and adults reason in social conventional terms? *Developmental Psychology, 24,* 366—375.

Miller, J. G., & Bersoff, D. M. (1999). Development in the context of everyday family relationships: Culture, interpersonal morality and adaptation. In Killen M & Hart, D (Eds.) Morality in Everyday life: Developmental perspectives. (259-282). NY: Cambrideg University Press.

Miller, J. G., Bersoff, D. M., & Harwood, R. L. (1990). Perceptions of social responsibilities in India and in the United States: Moral imperatives or personal decisions? *Journal of Personality and Social Psychology, 58* (1), 33—47.

Miller, P., & Smith, G. (1990). *Dispositional and specific learning history risk factors for alcohol abuse.* Paper presented at the 98th Annual Convention of the American Psychological Association, August 10-14, Boston, MA.

Mills. R. & Mills, R. (1996). Adolescents' attitudes toward female gender roles: Implications for education. *Adolescence, 31 (1234), 741-745.*

Mintz, B. I., & Betz, N. E. (1988). Prevalence and correlates of eating-disordered behaviors among undergraduate women. *Journal of Counseling Psychology, 35,* 463—471.

Minuchin, P. P., & Shapiro, E. K. (1983). The school as a context for social development. In P. H. Mussen (Ed.), *Handbook of child psychology, Vol. IV.* (pp. 197—274). New York: John Wiley.

Minuchin, 5. (1974). *Families and family therapy.* Cambridge, MA: Harvard University Press.

Minuchin, S., Rosman, B. L., & Baker, L. (1978). *Psychosomatic families: Anorexia nervosa in context.* Cambridge, MA: Harvard University Press.

Mitchell, I. B., & Eckert, E. D. (1987). Scope and significance of eating disorders. *Journal of Consulting and Clinical Psychology, 55* (5), 628—634.

Mitic, W. (1990). Parental versus-peer influence on adolescents' alcohol consumption. *Psychological Reports, 67,* 1273—1274.

Molidar, C. & Tolman, R.M. (1998). Gender and contextual factors in adolescent dating violence. *Violence Against Women, 4(2), 180-194.*

Money, J., & Ehrhardt, A. (1972). *Man & woman/boy and girl.*New York: New AmericanLibrary.

Montgomery, M. J., & Sorell, G. (1998). Love and dating experience in early and middle adolescence: Grade and gender comparisons. *Journal of Adolescence, 21 (6), 677-689.*

Moore, B. N., & Parker, R. (1986). *Critical thinking.* Palo Alto, CA: Mayfield.

Moore, K. A. (1985). *Facts at a glance.* Unpublished manuscript. Child Trends, Inc., Washington, DC.

Moore, K. A., Peterson, J. L., & Furstenberg, F. F. (1986). Parental attitudes and the occurrence of early sexual activity. *Journal of Marriage and Family, 48,* 777—782.

Morinis, A. (1985). The ritual experience: Pain and the transformation of consciousness in ordeals of initiation. *Ethos 13* (2), 150—174. Morris, J. (1974, July). Conundrum. *Ms. Magazine, 57*—64.

Morris, W. (Ed.). (1971). *The American heritage dictionary of the English language.* Boston: Houghton Mifflin.

Mortimer, J. T., Finch, M. D., Shanahan, M., & Ryu, 5. (1992). Work experience, mental health and behavioral adjustment in adolescence. *Journal of Research on Adolescence, 2,* 25—57

Mosatche, H. (1983). *Searching: Practices and beliefs of the religious cults and human potential groups.* New York: Stravon Educational Press.

Mosley, J., & Lex, A. (1990). Identification of potentially stressful life events experienced by a population of urban minority youth. *Journal of Multicultural Counseling and Development, 18,* 118—125.

Moulton, P., Moulton, M. & Roach, S. (1998). Eating disorders: A means for seeking approval? Eating Disorders: The Journal of Treatment and Prevention, 6 (4), 319-327.

Muehlbauer, G., & Dodder, L. (1983). *The losers: Gang delinquency in an American suburb.* New York: Praeger.

Muir, J. C. (1993, March 31). Homosexuals and the 10 percent fallacy. New York: *The New York Times.*

Mullis, I., Dossy, J., Owen, E., & Phillips, G. W. (1991). *The state of mathematics achievement: NAEP's 1990 assessment of the nation and the trial assessment of the states.* Princeton, NJ: Educational Testing Service.

Mullis, R. L., & McKinley, K. (1989). Gender-role orientation of adolescent females: Effects on self-esteem and locus of control. *Journal of Adolescent Research, 4,* 506—516.

Murphy, J. (1987). Educational influences. In V. B. Van Hasselt & M. Hersen (Eds.), *Handbook of adolescent psychology.* New York: Pergamon Press.

Nagel, K. L., & Jones, K. H. (1992). Sociological factors in the development of eating disorders. *Adolescence,27,* 105, 107—113.

National Center for Education Statistics (1994). *The pocket condition of education 1994.* Washington, DC: U.S. Department of Education, Office of Educational Research and Improvement.

National Center for HealthStatistics. (1990, August 16). Study shows 1 out of 4 babies born to unwed women in 1988. *The Boston Globe,* p. 23.

National Center for Health Statistics. (1994). *Vital statistics of the United States, Annual.* Hyattsville, MD: National Center for Health Statistics.

National Commission on Children. (1991). *Speaking of kids: A national survey of children and parents.* Washington, DC:

National Commission on Children. National Institute of Allergy and Infectious Diseases (1987).

Natriello, G., McDill, E. L., & Pallas, A. M. (1987). *In our lifetime: Schooling and the disadvantaged.* Unpublished manuscript.

Navone, J. (1990). Heroes, saints and leaders: Models for human development. *Studies in Formative Spirituality,]],* 23—34.

Neimeyer, G., & Heesacker, M. (1992). Vocational development: Assessment and intervention in adolescent career choice. In C. E. Walker & M. C. Roberts (Eds.), *Handbook of clinical child psychology,* 2nd ed. (pp. 661—676). New York: John Wiley.

Nelson, W.L., Huighes, H.M., Katz, B. & Searight, H.R. (1999). Anorexic eating attitudes and behaviors of male and female college students. Adolescence, 34 (135), 621-633.

Newcomb, M., & Bentler, P. (1989). Substance use and abuse among children and teenagers. *American Psychologist, 44* (2), 242—248.

Newcomb, M. D., Huba, G. J., & Bentler, P. M. (1981). A multidimensional assessment of stressful life events among adolescents: Derivation and correlates. *Journal of Health and Social Behavior, 22,* 400—415.

New drugs, old danger, (2000). *Boston Globe, April 15, 2000.*

Newman, P. R., & Newman, B. M. (1976). Early adolescence and its conflict: Group identity versus alienation. *Adolescence, 11(42),* 261—273.

New York City Youth Board. (1989, August 18). Study outline traits of urban gangs. *New York Times,* p. 23.

Nichols, J. G., Licht, B. G., & Pearl, R. A. (1982). Some dangers of using personality questionnaires to measure personality. *Psychological Bulletin, 92,* 572—580.

Nightingale, E. 0., & Wolverton, L. (1988). *Adolescent rolelessness in modern society.* Washington, DC: Carnegie Council on Adolescent Development.

Nobles, W. W. (1976). Extended self: Rethinking the so-called Negro self-concept. *Journal of Black Psychology, 2* (2), 15—24.

Nolen-Hoeksema, 5. (1987). Sex differences in unipolar depression: Evidence and theory. *Psychological Bulletin, 101,* 259—282.

Nolen-Hoeksema, S., Girgus, J., & Seligman, M. P. (1991). Sex differences in depression and explanatory style in children. *Journal of Youth and Adolescence, 20* (2), 233—245.

Noles, S., Cash, T., & Winstead, B. (1985). Body image, physical attractiveness, and depression. *Journal of Consulting and Clinical Psychology, 53* (1), 88—94.

Norman, E., & Turner, 5. (1993). Adolescent substance abuse prevention programs: Theories, models, and research in the encouraging 80's. *The Journal of Primary Prevention, 14,* 3—20.

Nottelmann, E., Susman, E., Blue, J., Inoff-Germain, G., Dorn, L., Lonaux, D., Cutler, G., & Chrousos, G. (1987a). In R. Leruer & T. Foch (Eds.), *Biological-psychosocial interactions in early adolescence* (pp. 303—321). Hillsdale, NJ: Erlbaum.

Nottelmann, E., Susman, E., Dorn, L., Inoff-Germain, G., Loriaux, D., Cutler, G., & Chrousos, G. (1987b). Developmental processes in early adolescence: Relations among chronologic age, pubertal stage, height, weight, and serum levels of gonadotropins, sex steroids and adrenal androgens. *Journal of Adolescent Health Care, 8,* 246—260.

Nowak, M. (1998). The weight-conscious adolescent: Body image, food intake and weight related behavior. Journal of Health, 23 (6), 389-398.

Nuttall, R., & Nuttall, E. (1980). *The impact of disaster on coping behaviors of families.* Unpublished manuscript. Chestnut Hill, MA: Boston College.

Nye, F. (1980). A theoretical perspective on running away. *Journal of Family Issues, 1* (2), 274-299.

O'Callaghan, M., Borkowski, J., Whitman, T., Maxwell, S. & Keogh, D. (1999). A model of adolescent parenting: The roles of cognitive readiness to parent. *Journal of Research on Adolescence, 9 (2), 203-225.*

O'Dea, J.A. and Abraham, S. (1999). Onset of Disordered Eating Attitudes and Behaviors in Early Adolescence: Interplay of Pubertal Status, Gender, Weight, and Age. *Adolescence, 34, 671-680.*

Oetting, E., & Beauvais, F. (1987). Common elements in youth drug abuse: Peer clusters and other psychological factors. *Journal of Drug Issues, 17,* 133—151.

Oetting, E., & Beauvais, F. (1990). Adolescent drug use: Findings of national and local surveys. *Journal of Consulting and Psychology, 58* (4), 385—394.

Offer, D. (1969). *The psychological world of the teenager.* New York: Basic Books.

Offer, D., & Offer, J. (1975). *From teenage to young manhood.* New York: Basic Books.

Offer, D., Ostrov, E., & Howard, K. I. (1981). *The adolescent: A psychological self-portrait.* New York: Basic Books.

Ogbu, J. (1983). Schooling the inner city. *Society,* 75—79.

O'Heron, C. A., & Orlofsky, J. L. (1990). Stereotypic and nonstereotypic sex-role trait and behavior orientations, gender identity and psychological adjustment. *Journal of Personality and Social Psychology, 58,* 134-143.

Ohio Cancer Information Service. (1991). *What is it that l have, don't want, didn't ask for, can't give back and how I feel about it.* Columbus, OH: Ohio Cancer Information Service.

Ohye, B. & Henderson Daniels, J. (1999). The "other" adolescent girls: Who are they? In Johnson, N., Roberts, M. & Worell, J. (Eds.) Beyond appearance: A new look at adolescent girls. Washington, DC: APA.

Okagaki, L., & Greenfield, P. M. (1994). Effect of video game playing on measures of spatial performance: Gender effects in late adolescence. *Journal of Applied Developmental Psychology, 15* (1), 33—58.

O'Keefe, M. & Treister, L. (1998). Vicitms of dating violence among high school students: Are the predictors different for males and females? *Violence Against Women, 4 (2), 195-223.*

Okwumabua, J. 0. (1990). Child and adolescent substance abuse: Etiology and prevention. In S. B. Morgan & T. M. Okwumabua (Eds.), *Child and adolescent disorders: Developmental and health psychology perspectives* (pp. 395—427). Hillsdale, NJ: Erlbaum.

OMalley, P., & Bachman, J. (1983). Self-esteem: Changes and stability between ages 13 and 23. *Developmental Psychology, 19,* 257—268.

O'Reilly, B. (1990). Why grade "A" executives get an "F" as parents. *Fortune, 121* (1), 36—46.

Orenstein, P. (1994). SchoolGirls: *Young women, self-esteem, and the confidence gap.* New York: Anchor Books.

Orr, D., Brack, D., & Ingersoll, G. (1988). Pubertal maturation and cognitive maturity in adolescents. *Journal of Adolescent Health Care, 9,* 273—279.

Orthner, D. K. (1990). Parental work and early adolescence: Issues or research and practice. *Journal of Early Adolescence, 10* (3), 246-259.

Otto, V. (1972). Suicidal attempts in childhood and adolescents—today and after ten years: A follow-up study. In A. L. Annell (Ed.), *Depressive states in childhood and adolescence.* New York: Halstead Press.

Outward Bound, U.S.A. (1988). *Outward Bound.* Greenwich, CN: Outward Bound National Office.

Ozorak, E. (1989). Social and cognitive influences on the development of religious belief and commitment in adolescence. *Journal for the Scientific Study of Religion, 28* (4), 448—463.

Paikoff, R., & Brooks-Gunn, J. (1991). Do parent-child relationships change during puberty? *Psychological Bulletin, 110* (1), 47—66.

Palmer, W., & Patterson, B. (1981). If you loved my son you would take out the trash. In J. S. Gordon & M. Beyer (Eds.), *Reaching troubled youth: Runaways and community mental health.* Rockville, MD: U.S. Department of Health and Human Services.

Parham, T. (1989). Cycles of psychological nigrescence. *The Counseling Psychologist, 17,* 187—226.

Parham, T., & Helms, J. (1985). Attitudes of racial identity and self-esteem of black students: An exploratory investigation. *Journal of College Student Personnel, 26,* 143—147.

Parks, 5. (1986). *The critical years: The young adult search for a faith to live by.* New York: Harper and Row. Paroski, P. (1987). Health care delivery and the concerns of gay and lesbian adolescents. *Journal of Adolescent Health Care, 8,* 188—192.

Parsons, F. (1909). *Choosing a vocation.* Boston: Houghton Mifflin.

Patterson, G. R., Reid, J. B., Dishion, T. J. (1992). *Antisocial boys.* Eugene, OR: Castalia.

Patterson, J., & McCubbin, H. I. (1987). Adolescent coping style and behaviors: Conceptualization and measurement. *Journal of Adolescence, 10,* 163—186.

Paul, R. W. (1987). Dialogical thinking. In J. B. Baron, & R. J. Sternberg (Eds.), *Teaching thinking skills.* New York:Freeman.

PDK Task Force. (1976). The walkabout. *Phi Delta Kappan Pamphlet.*

Pepler, D. J., & Slaby, R. G. (1994). Theoretical and developmental perspectives on youth and violence. In L. D. Eron, J. Gentry, & P. Schlegel (Eds.), *Reason to hope: A psychosocial perspective on violence and youth* (pp.

27—58). Washington, DC: American Psychological Association.

Perlmutter, B. F. (1987). Delinquency and learning disabilities: Evidence for compensatory behaviors and adaptation. *Journal of Youth and Adolescence, 16* (2), 89—95.

Perrone, V. (1993). Learning for life: Where do we begin? *Equity & Excellence in Education, 26,* 5—8.

Perry, W. (1968a). *Forms of intellectual and ethical development in the college years.* New York: Holt, Rinehart & Winston.

Perry, W. (1968b). *Patterns of development in thought and values of students in a liberal arts college: A validation of a scheme.* U.S. Department of Health, Education, and Welfare. Office of Education, Bureau of Research, final report.

Perry, W. (1981). Cognitive and ethical growth. In A. Chickering (Ed.), *The modern American college.* San Francisco: JosseyBass.

Petersen, A. (1988). Adolescent development. In M. Rosenzweig & L. Porter (Eds.), *Annual review of psychology.* Palo Alto, CA: Annual Reviews.

Petersen, A. C. (1985). Pubertal development as a cause of disturbance: Myths, realities and unanswered questions—genetic, social, and general. *Psychology Monographs, 111* (2), 205—232.

Petersen, A. C. (1987). The nature of biological-psychosocial interactions. In R. M. Lerner & T. T. Foch (Eds.), *Biological-psychosocial interactions in early adolescence.* Hillsdale, NJ: Erlbaum.

Petersen, A. C., Compas, B., Brooks-Gunn, J., Stemmler, M., Ey, S., & Grant, K. (1993). Depression in adolescence. *American Psychologist, 48* (2), 155—168.

Petersen, A. C., Kennedy, R. E., & Sullivan, P. (1991). Coping with adolescence. In M. E. Colten & S. Gore (Eds.), *Adolescent stress: Causes and consequences* (pp. 93—110). Hawthorne, NY:Aldine de Gruyter.

Petersen, A. C., & Taylor, B. (1980). The biological approach to adolescence. In J. Adelson (Ed.), *Handbook of adolescent psychology.* New York: Wiley.

Peterson, C. (1983). Menarche: Meaning of measure and measuring meaning. In S. Golub (Ed.), *Menarche* (pp. 63—76). New York: Heath.

Peterson, P. L., Hawkins, D., Abbott, R. D., & Catalano, R. (1994). Disentangling the effects of parental drinking, family management, and parental alcohol norms on current drinking by black and white adolescents. *Journal of Research on Adolescence, 4,* 203—228.

Petzel, S. V., & Cline, D. (1978). Adolescent suicide: Epidemiological and biological aspects. *Adolescent Psychiatry, 6,* 249—266.

Philibert, P. (1987). Relation, consensus and commitment as foundations of moral growth. *New Ideas in Psychology, 5* (2), 183—195.

Phillips, D. (1979). Suicide, motor vehicle fatalities, and the mass media: Evidence toward a theory of suggestion. *American Journal of Sociology, 84* (5), 1150—1174.

Phinney, J. 5. (1989). Stages of ethnic identity in minority group adolescents. *Journal of Early Adolescence, 9,* 34—49.

Phinney, J. 5. (1990). Ethnic identity in adolescents and adults. *Psychological Bulletin, 108,* 499—514.

Phinney, J. S., & Alipuria, L. L. (1990). Ethnic identity in college students from four ethnic groups. *Journal of Adolescence, 13,* 171—183.

Phinney, J. S., & Chavira, V. (1995). Parental ethnic socialization and adolescent coping with problems related to ethnicity. *Journal of Research on Adolescence, 5,* 31—54.

Phinney, J. S., & Rosenthal, D. (1992). Ethnic identity in adolescence: Process, context, and outcome. In G. R. Adams, T. P. Gullotta, & R. Montemayor (Eds.), *Adolescent identity formation: Advances in adolescent development* (pp. 145—172). Newbury Park, CA: Sage.

Piaget, J. (1932/1965). *The moral judgment of the child.* New York: Macmillan.

Piaget, J. (1948/1966). *Psychology of intelligence.* New York: Harcourt.

Piaget, J. (1953). *The origins of intelligence in the child.* New York: Harcourt, Brace.

Piaget, J. (1972). Intellectual evolution from adolescence to adulthood. *Human Development, 15,* 1—12.

Picard, C.L. (1999). The Level of Competition as a Factor for the Development of Eating Disorders in Female Collegiate Athletes, *Journal of Youth and Adolescence, 28, 583-594.*

Pierce, D., & Ramsay, T. (1990). Gang violence: Not just a big-city problem. *The Police Chief, 57,* 24-25.

Pierce, K. (1990). A feminist theoretical perspective on the socialization of teenage girls through *Seventeen* magazine. *Sex Roles, 23* (9/10), 491—500.

Pierce, R. A., & Schwartz, R. (1992, March). *A longitudinal study of adolescent-parent relations: Intimacy as a predictor of suicide ideation.* Paper presented at the Biennial Meeting of the Society for Research on Adolescence, Washington, DC.

Pierce, W., Lemke, E., & Smith, R. (1988). Critical thinking and moral development in secondary students. *High School Journal, 71(3),* 120-126.

Piercy, F., Volk, R., Trepper, T., Sprenkle, D., & Lewis, R. (1991). The relationship of family factors to patterns of adolescent substance abuse. *Family Dynamics of Addiction Quarterly, 1* (1), 41—54.

Pipher, M. (1994). *Reviving Ophelia: Saving the souls of adolescent girls.* New York: Ballantine Books. Pleck, J. (1983). The theory of male sex-role identity: Its rise and fall, 1936-present. In M. Lewin (Ed.), *In the shadow of the past: Psychology portrays the sexes.* New York: Columbia University Press.

Pleck, J., Sonenstein, F., & Swain, 5. (1988). Adolescent males' sexual behavior and contraceptive use: Implications for male responsibility. *Journal of Adolescent Research, 3* (3—4), 275—284.

Pleck, J., Sonenstein, F. L., & Ku, L. C. (1994). Problem behaviors and masculine ideology in adolescent males. In R. Ketterlinus & M. E. Lamb (Eds.), *Adolescent problem behaviors.* Hillsdale, NJ: Erlbaum.

Pliner, P., Chaiken, S., & Flett, G. (1990). Gender differences in concern with body weight and physical appearance over the life span. *Personality and Social Psychology Bulletin, 16* (2), 263—273.

Pliszka, S.R., Sherman, J.O., Barrow, M.V. & Irick, S. (2000). Affective disorders in juvenile offenders: A preliminary study. American Journal of Psychiatry, 157 (1), 130-132.

Pollack, W. (1998). *Real Boys.* NY: Henry Holt.

Pombeni, M. L., Kirchler, E., & Palmonari, A. (1990). Identification with peers as a strategy to muddle through the troubles of the adolescent years. *Journal of Adolescence, 13,* 351—369.

Pope, H.G. Jr., Olivardia, R., Gruber, A. & Borowieki, J. (1999). Evolving ideals of male body image as seen through action toys. International Journal of Eating Disorders, 26 (1), 65-72.

Popenoe, D. (1988). *Disturbing the nest: Family change and decline in modern societies,* New York: Aldine De Gruyter. Prothrow-Stith, D. (1991). *Deadly consequences.* New York: HarperCollins.

Prothrow-Stith, D., & Spivack, H. (1991, April 29). Kids killing kids: What can be done? Three ways to ease adolescent violence. *The Boston Globe.*

Pugh, M & Hart, D. (1999). Identity development and peer group participation. In McLellan, J.A. & Pugh, M. (Eds.), The role of peer groups in adolescent social identity: Exploring the importance of stability and change. New Directions for child and adolescent development, 84 (pp. 55-70). San Francisco: Jossey-Bass.

Puka, B. (1989). The liberation of caring: A different voice for Gilligan's 'Different Voice." In M. Brabeck (Ed.), pp. 19—44, *Who cares? Theory, research and educational implications of the ethic of care.* New York, NY: Praeger.

Quinn, W., Newfield, N., & Protinsky, H. (1985). Rites of passage in families with adolescents. *Family Process, 24,* 101—111.

Radin, C. (1991, April 28). Of violence and the young: Adrift in a 'culture of impulse,' children turn on one another. *The Boston Sunday Globe,* 1,14.

Raffaelli, M., Bogenschneider, K. & Flood, M.F. (1998). Praent-teen communication about sexual topics. *Journal of Family Issues, 19 (3), 315-333.*

Ramsey, P. (1982). Do you know where your children are? *Journal of Psychology and Christianity, 1* (4), 7—15.

Raphael, R. (1988). *The men from the boys.* Lincoln, NE: University of Nebraska.

Reasoner, R. W. (1983). Enhancement of self-esteem in children and adolescents. *Family and Community Health,* 6(2), 51—64.

Recklitis, C. & Noam, G. (1999). Clinical and developmental perspectives on adolescent coping. *Child Psychiatry and Human Development, 30 (2), 87-101.*

Rees, J., & Trahms, C. (1989). Nutritional influences on physical growth and behavior in adolescence. In G. Adams, R. Montemayor, & T. Gullotta (Eds.), *Biology of adolescent behavior and development* (pp. 195—222). Newbury Park, CA: Sage.

Reid, B. V. (1990). Weighing up the factors: Moral reasoning and culture change in a Samoan community. *Ethos, 18* (1), 48—70.

Reid, P. (1991, April 29). Kids killing kids: What can be done? Action must come from within the community. *The Boston Globe.*

Reinherz, H., Frost, A., Stewart-Berghauer, G., Pakiz, B., Kennedy, K., & Schille, C. (1990). The many faces of correlates of depressive symptoms in adolescents. *Journal of Early Adolescence,*

Reisman, J. M. (1985). Friendship and its implications for mental health or social competence. *Journal of Early Adolescence, 5,* 383—391.

Remafedi, G. (1987). Male homosexuality: The adolescent's perspective. *Pediatrics, 79,* 326-330.

Remafedi, G. (1988). Homosexual youth. *Journal of the American Medical Association, 258* (2), 222—225.

Remafedi, G., Resnick, M., Blum, R., & Harris, L. (in press). The demography of sexual orientation in adolescents. *Pediatrics.*

Renouf, A. G., & Harter, 5. (1990). Low self-worth and anger as components of the depressive experience in young adolescents. *Development and Psychopathology, 2,* 293—310.

Rest, J. (1983). Morality, pp. 556—629. In P. Mussen (Ed.), *Handbook of child psychology* (Vol. 3). New York: John Wiley.

Rest, J., Narvarez, D., Bebeau, M. & Thoma, S. (1999). A neo-Kohlbergian approach: The DIT and Schema theory. *Educational Psychology, Review, 11 (4),* 294-324.

Rest, J., Narvarez, D., Thoma, S. & Bebeau, M. (1999). DIT2: Devising and testing a revised instrument of moral judgment. *Journal of Educational psychology, 91 (4),* 644-659.

Rhodes, J., & Jason, L. (1990). A social stress model of substance abuse. *Journal of Consulting and Clinical Psychology, 58* (4), 394-401.

Richardson, J. H. (1991, March 31). Sizer on school reform: Let's rethink the basics. *The Boston Sunday Globe,* pp. AI6-17.

Rierdan, J., & Koff, E. (1985). Timing of menarche and initial menstrual experience. *Journal of Youth and Adolescence, 14* (3), 237-244.

Rierdan, J., Koff, E., & Stubbs, M. (1988). A longitudinal analysis of body image as a predictor of the onset and persistence of adolescent girls' depression. *Working Paper No. 188,* Wellesley College Center for Research on Women.

Roazen, P. (1976). *Erik H. Erikson: The power and limits of a vision.* New York: Free Press.

Roberts, A. R. (1982). Stress and coping patterns among adolescent runaways. *Journal of Social Science Research, 5,* 15—27.

Roberts, R.E., Phinney, J., Masse, L., Chen, Y., Roberts, C.R. & Romero, A. (1999). The structure of ethnic identity of young adolescents from diverse ethnocultural groups. *Journal of Ear;ly Adolescence, 19 (3), 301-322.*

Robertson, M. J. (1989). *Homeless youth: An overview of recent literature.* Paper presented at the National Conference on Homeless Children and Youth, Institute for Policy Studies at Johns Hopkins University, Washington, DC.

Rodman, H. (1989). Controlling adolescent fertility. *Society* (1), 35—37.

Rodriguez, R. A. (1988, June). *Significant events in gay identity development: Gay men in Utah.* Paper presented at the 96th Annual Convention of the American Psychological Association, Atlanta. Rogers, C. R. (1961). *On becoming a person.* Boston: Houghton Mifflin.

Rogoff, B. (1990). Apprenticeship in thinking. *Cognitive development in social context.* New York: Oxford University Press.

Rohn, R. (1977). Adolescents who attempt suicide. *The Journal of Pediatrics, 90,* 636-638.

Rohrbach, L., Hodgson, C., Broder, B., Montgomery, S., Flay, B., Hansen, W., & Pentz, M. A. (1994). Parental participation in drug abuse prevention: Results from the Midwestern prevention project. *Journal of Research on Adolescence, 4,* 295—318.

Roper Organization. (1991). *The American Chronicle youth poll.* Storrs, CT: University of Connecticut (Roper). Rosaldo, M. (1984). Towards an anthropology of self and feeling. In R. A. Schweder & R. A. Levine (Eds.), *Cultural theory: Essays of mind, self, and emotion.* New York: Cambridge University Press.

Rose, H. M. (1978). The geography of despair. *Annals of the Association of American Geographers, 68,* 453—464.

Rosen, B. M., Bahn, A. K., Shellow, R., & Bower, E. M. (1965). Adolescent patients served in outpatient clinics. *American Journal of Public Health, 55,* 1563—1577.

Rosen, J., & Gross, J. (1987). Prevalence of weight reducing and weight gaining in adolescent girls and boys. *Health Psychology, 6,* 131—147.

Rosen, J. C., Gross, J., & Vara, L. (1987). Psychological adjustment of adolescents attempting to lose or gain weight. Special issue: Eating disorders. *Journal of Consulting and Clinical Psychology, 55* (5), 742—747.

Rosenbaum, E., & Kandel, D. (1990). Early onset of adolescent sexual behavior and drug involvement. *Journal of Marriage and the Family, 52* (8), 783—798.

Rosenbaum, J. E. (1991). Are adolescent problems caused by school or society? *Journal of Research on Adolescence, 1,* 301—322.

Rosenberg, G. S., & Anspach, D. F. (1973). Sibling solidarity in the working class. *Journal of Marriage and Family, 35,* 108—113.

Rosenberg, M. (1979). *Conceiving the self* New York: Basic Books.

Rosenberg, M. (1985). Self-concept and psychological well-being in adolescence. In Robert L. Leahy (Ed.), *The development of self* Orlando, FL: Academic Press.

Rosenberg, M. (1986). Self-concept from middle childhood through adolescence. In J. Suls & A. G. Greenwald (Eds.), *Review, psychological perspectives on the self.* Hillsdale, NJ: Erlbaum.

Rosenthal, D. A., & Feldman, S. S. (1991). The influence of perceived family and personal factors on self-reported school performance of Chinese and Western high school students. *Journal of Research on Adolescence, 1,* 135—154.

Rosenthal, D. A., Gurney, R. M., & Moore, S. M. (1981). From trust to intimacy: A new inventory for examining Erikson's stages of psychosocial development. *Journal of Youth and Adolescence, 10* (6), 525—537.

Rosenvinge, J.H., Borgen, J.S. & Boerrensen, R. (1999). The prevalence and psychological correlates of anorexia nervosa and binge eating among 15 year old students: A controlled epedimiological study. European Eating Disorders review, 7 (5), 382-391.

Ross, S. I., & Jackson, J. M. (1991). Teachers' expectations for black males' and black females' academic achievement. *Personality and Social Psychology Bulletin, 17,* 78—82.

Rosser, R. & Eccles, J. (1998). Adolescents' perceptions of middle school: Relation to longitudinal changes in academic and psychological adjustment. *Journal of Research on Adolescence, 8 (1), 123-158.*

Rotheram, M. J. (1987). *Evaluation of imminent danger for suicide among youth.* Annual meeting of the American Orthopsychiatric Association, Chicago, Illinois.

Rotheram-Borus, M. J. (1993). Suicidal behavior and risk factors among runaway youth. *American Journal of Psychiatry, 150,* 103—107.

Rotheram-Borus, M. J., & Koopman, C. (1991). Sexual risk behaviors, AIDS knowledge, and beliefs about AIDS among runaways. *American Journal of Public Health, 81,* 208—210.

Rotheram-Borus, M. J., Rosario, M., & Koopman, C. (1991). Minority youths at high risk: Gay males and runaways. In M. F. Colten & S. Gore (Eds.), *Adolescent stress: Causes and consequences* (pp. 81—200). New York: Aldine de Gruyter.

Rotter, J. (1971, June). External control and internal control. *Psychology Today, 5,* 37ff.

Rowe, D., & Rodgers, J. (1989). Behavioral genetics, adolescent deviance and 'd': Contributions and issues. In G. Adams, R. Montemayor, & T. Gullotta (Eds.), *Biology of adolescent behavior and development* (pp. 38—67). Newbury Park, CA: Sage.

Rowe, I., & Marcia, J. F. (1980). Ego identity status, formal operations, and moral development. *Journal of Youth and Adolescence, 9* (2), 87—99.

Roy, A. (1990). Family rituals: Functions and significance for clergy and psychotherapists. *Group, 14* (1), 59—64.

Roye, C. (1998). Condom use by Hispanic and African American adolescent girls who use hormonal contraception. *Journal of Adolescent Health, 23 (4), 205-211.*

Rubenstein, J. L., & Feldman, S. S. (1993). Conflict-resolution behavior in adolescent boys: Antecedents and adaptational correlates. *Journal of Research on Adolescence, 3,* 41—66.

Rucinski, A. (1989). Relationship of body image and dietary intake of competitive ice skaters. *Journal of the American Dietetic Association, 89,* 98—100.

Ruebush, K. W. (1994). The mother-daughter relationship and psychological separation in adolescence. *Journal of Research on Adolescence, 4,* 439—451.

Ruggiero, V. (1994). *Warning: Nonsense is destroying America.* British Columbia: Thomas Nelson.

Russell, D. (1995). The prevalence of trauma and sociocultural causes of incestuous abuse of females: A human rights issue. In R.J. Kleber & C.R. Figley (Eds.), *Beyond Trauma: Cultural and Societal Dynamics* (171-186), NY, Plenum Press.

Rutledge, F. M. (1990). Suicide among black adolescents and young adults: A rising problem. In A. R. Stiffman & L. F. Davis (Eds.), *Ethnic issues in adolescent mental health* (pp. 339—351). Newbury Park, NJ: Sage.

Rutter, M. (1986). The developmental psychopathology of depression: Issues and perspectives. In M. Rutter, C. E. Izard, & P. B. Read (Eds.), *Depression in young people: Developmental and clinical perspectives.* New York: Guilford.

Rutter, M. (1987). Continuities and discontinuities from infancy. In J. Osofsky (Ed.), *Handbook of infant development.* New York:Wiley.

Rutter, M., Maughan, B., Mortimer, P., Ouston, J., & Smith, A. (1979). *Fifteen thousand hours: Secondary schools and their effects on children.* Cambridge, MA: Harvard University Press. Ryan, J. (1995). *Little girls in prettyboxes.* New York: Doubleday.

Ryan, R. M., & Lynch, J. H. (1989). Emotional autonomy versus detachment: Revisiting the vicissitudes of adolescence and young adulthood. *Child Development, 60,* 340-356.

Ryan, W. (1976). *Blaming the victim.* New York: Random House.

Rycek, R., Stuhr, S. McDermott, J., Benker, J., Schwartz, M. (1999). Adolescent egocentrism and cognitive functioning during late adolescence. Adolescence, 33 (132), 745-749.

Sadker, M., & Sadker, D. (1991). The gender gap in self esteem, achievement and instruction. *Communique, 4,* 27.

Sagrastano, L., McCormick, S., Paikoff, R., & Holmbeck, G. (1999). Pubertal development and parent-child conflict in low-income, urban, African American adolescents. *Journal of Research on Adolescence, 9 (1), 85-107.*

Santana, G. (1979). *Social and familial influences on substance use among youth.* Paper presented to the American Psychological Association, New York.

Santelli, J., Brener, N., Lowry, R., Bhatt, A., & Zabin, L. (1998). Multiple sexual partners among U.S. adolescents and young adults. *Family planning perspectives. 30 (6), 271-275.*

Santrock, J. W. (1987). The effects of divorce on adolescents: Needed research perspectives. *Family Therapy, 14* (2), 147—159.

Savin-Williams, R. C. (1990). *Gay and lesbian youth: Expressions of identity.* Washington, DC:Hemisphere.

Savin-Williams, R. C. (1994). Verbal and physical abuse as stressors in the lives of lesbian, gay male, and bisexual youths: Associations with school problems, running away, substance abuse, prostitution, and suicide. *Journal of Consulting and Clinical Psychology, 62 (2),* 261—269.

Savin-Williams, R. C. (1998). The disclosure to families of same-sex attractions by lesbian, gay and bisexual youth. *Journal of Research on Adolescence, 8 (1), 49-68.*

Savin-Williams, R. C., & Berndt, T. J. (1990). Friendship and peer relations. In S. Feldman and G. Elliot (Eds.), *At the threshold: The developing adolescent.* Cambridge, MA: Harvard University Press.

Savin-Williams, R. C., & Rodriguez, R. C. (1993). A developmental, clinical perspective on lesbian, gay male and bisexual youths. In T. P. Gullotta, G. R. Adams, & R. Montemayor (Eds.), *Adolescent sexuality, advances in adolescent development: An annual book series, Vol. 5* (pp. 77—101). Newbury Park, CA: Sage.

Savin-Williams, R. C., & Weisfeld, G. (1989). An ethological perspective on adolescence. In G. Adams, R. Montemayor, & T. Gullotta (Eds.), *Biology of adolescent behavior and development* (pp. 249—274). Newbury Park, CA: Sage.

Scaramella, L.V., Conger, R.D. & Simons, R.L. (1999). Parental protective influences and gender-specific increases in adolescent internalizing and externalizing problems. *Journal of Research on Adolescence, 9 (2), 111-141.*

Schaffer, B., & DeBlassie, R. R. (1984). Adolescent prostitution. *Adolescence 24,* 665—675.

Schichor, A., Bernstein, B., & King, 5. (1994). Self-reported depressive symptoms in inner- city adolescents seeking routine health care. *Adolescence, 29,* 379—388.

Schiller, K.S. (1999). Effects of feeder patterns on students' transition to high school. *Sociology of Education, 72 (4), 216-233.*

Schoem, D. (1991). *In separate worlds: Life stories of young Blacks, Jews, and Latinos.* Ann Arbor: University of MichiganPress.

Schubiner, H., Scott, R., & Tzelepis, A. (1993). Exposure to violence among inner-city youth. *Journal of Adolescent Health, 14,* 214-219.

Schwerin, M., & Corcoran, K. (1990). *A psychological model of compulsive steroid use.* Paper presented at the 98th Annual Convention of the American Psychological Association, August 10—14, Boston, MA.

Scott, D. (1988). *Anorexia and bulimia.* New York: New York University Press.

Searight, H. R., Binder, A., Manley, C., Krohn, E., Rogers, B., & Russo, R. (1990). *Autonomy and intimacy in the families of adolescent substance abusers.* Paper presented at the 98th Annual Convention of the American Psychological Association, August 10-14, Boston, MA.

Sebald, H. (1977). *Adolescence: A social psychological analysis* (2nd ed.). Englewood Cliffs, NJ: Prentice-Hall.

Sebald, H. (1992). *Adolescence: A social psychological analysis.* Englewood Cliffs, NJ: Prentice-Hall.

Secretary's Commission on Achieving Necessary Skills. (1991). *What work requires of schools: A SCANS report for America 2000.* Washington, DC: U.S. Department of Labor.

Seginer, R. (1998). Adolescents' perceptions of relationships with older siblings in the context of other close relationships. *Journal of Research on Adolescence, 8 (3), 287-308.*

Seginer, R., & Flum, H. (1987). Israeli adolescents' self-image profile. *Journal of Youth and Adolescence, 16 (5),* 455-472.

Seiden, R. H., & Freitas, R. P. (1980). Shifting patterns of deadly violence. *Suicide and Life-Threatening Behavior, 10,* 195—209.

Seidman, F. (1990, August). *Growing up the hard way: Pathways of urban adolescents.* Invited address, 98th annual convention of the American Psychological Association, Boston, MA. APA Audiotape 90-173. Aurora, CO: Sound Images.

Seidman, E., Allen, L., Aber, J. L., Mitchell, C., & Feinman, J. (1994). The impact of school transitions in early adolescence on the self-system and perceived social context of poor urban youth. *Child Development, 65,* 507—522.

Seiffge, K.I. (1999). Families with daughters, families with sons: Different challenges for family relationships and marital satisfaction? *Journal of Youth and Adolescence, 28 (3),* 325-342.

Seligman, M., & Petersen, C. (1986). A learned helplessness perspective on childhood depression: Theory and research. In M. Rutter, C. E. Izard, & P. B. Read (Eds.), *Depression in young people: Developmental and clinical perspectives.* New York: Guilford.

Selman, R. L. (1976). Social-cognitive understanding. In T. Lickona (Ed.), *Moral development and behavior.* New York: Holt, Rinehart & Winston.

Selman, R. L. (1980). *The growth of interpersonal understanding: Developmental and clinical analysis.* New York: Academic Press.

Selman, R. L. (1989). Fostering intimacy and autonomy. In W. Damon (Ed.), *Child development today and tomorrow* (pp. 409—435). San Francisco: Jossey-Bass.

Selman, R. L., & Schultz, L. H. (1990). *Making a friend in youth: Developmental theory and pair therapy.* Chicago: University of Chicago Press.

Selman, R. L., Schultz, L. H., Nakkula, M., Barr, D., Watts, C., & Richmond, J. B. (1992). Friendship and fighting: A developmental approach to the study of risk and prevention of violence. *Development and Psychopathology, 4,* 529—558.

Selman, R, Watts, C. & Schultz, L (Eds.) (1997). *Fostering friendship: Pair therapy for treatment and prevention.* NY: Aldine De Gruyter.

Selverstone, R. (1989). Adolescent sexuality: Developing self-esteem and mastering developmental tasks. *SIECUS Report, 18,* 1—3.

Selvini-Palazzoli, M., Boscolo, L., Cechin, G., & Prata, G. (1978). A ritualized prescription in family therapy: Odd days and even days. *Journal of Marriage and Family Counseling, 4,* 3—9.

Sessions, W. (1990). Gang violence and organized crime. *The Police Chief, 57,* 17.

Shaffer, D., & Fisher, P. (1981). The epidemiology of suicide in children and young adolescents. *Journal of the American Academy of Child Psychiatry, 20*, 545—565.

Shantall, T (1999). The experience of meaning in suffering among Holocaust survivors. *Journal of Humanistic Psychology, 39 (3)*, 96-124.

Shea, J. (1985). Studies of cognitive development in Papua, New Guinea. *Journal of Psychology*, 20(10), 33—61.

Shedler, J., & Block, J. (1990). Adolescent drug use and psychological health. *American Psychologist, 45* (5), 612—630.

Shelton, C. (1983). *Adolescent spirituality*. Chicago: Loyola University Press.

Shih, T. (1998). Finding the niche: Friendship formation of immigrant adolescents. *Youth and Society, 30* (2), 209-240.

Shirk, S. R. (1987). Self-doubt in late childhood and early adolescence. *Journal of Youth and Adolescence, 16(1)*, 59—68.

Short, J. (1990). *Delinquency and society*. Englewood Cliffs, NJ: Prentice-Hall.

Shulman, S. & Scharf, M. (2000). Adolescent romantic behaviors and perceptions: Age and gender related differences and links with family and peer relationships. *Journal of Research on Adolescence, 10 (1)*, 99-118.

Shure, M. B. (1992). *I can problem solve: An interpersonal cognitive problem-solving program*. Champaign, IL: Research Press.

Sidel, R. (1990). Mixed messages. *On her own: Growing up in the shadow of the American dream*. New York: Viking.

Siegel, D., Aten, M. & Roghmann, K. (1998). Self-reported honesty among middle and high school students responding to a sexual behavior questionnaire. *Journal of Adolescent Health, 23 (1)*, 20-28.

Siegel, J.M., Aneshensel, C.S., Taub, B., Cantwell, D. & Driscoll, A.K. (1998). Adolescent depressed mood in a multi-ethnic sample. Journal of Youth and Adolescents, 27 (4), 413-427.

Siegel, L. J., & Senna, J. J. (1981). *Juvenile delinquency: Theory, practice and law*. New York: West.

Silber, T. (1985). Some medical problems common in adolescence. *Medical Aspects of Human Sexuality, 19* (2), 79—85.

Silverberg, S. B., & Steinberg, L. (1987). Adolescent autonomy, parent-adolescent conflict, and parental well-being. *Journal of Youth and Adolescence, 16*, 293—312.

Sim, T. (2000). Adolescent psychosocial competence: The importance and role of regard for parents. *Journal of Research on Adolescence, 10 (1)*, 49-64.

Simmons, R. (1987). Social transition and adolescent development. In C. E. Irwin, Jr. (Ed.), *Adolescent social behavior and health* (pp. 33—53). New Directions for Child Development, no. 37. San Francisco: Jossey-Bass.

Simmons, R., & Blyth, D. (1987). *Moving into adolescence: The impact of pubertal change and school context*. New York: Aldine de Groyter.

Simonian, S., Gibbs, J., & Tarnowski, K. (1990). *Social skills and antisocial conduct of delinquents*. Paper presented at the 98th Annual Convention of the American Psychological Association, August 10-14, Boston, MA.

Simons, R. L., & Whitbeck, L. B. (1991). Sexual abuse as a precursor to prostitution and victimization among adolescent and adult homeless women. *Journal of Family Issues, 12* (3), 361—379.

Sizer, T. (1984). *Horace's compromise: The dilemma of the American High School*. Boston: Houghton Mifflin. Sizer, T. (1992). *Horace's school: Redesigning the American high school*. Boston: Houghton Mifflin.

Sizer, T. (1996). Dreams, interests and aspirations. *Journal of Research in Rural Education, 12 (3), 125-126.*

Sizer, T. (1997). National tests: Interesting theory, bad practice. *Our Children, 23 (2), 7-9.*

Skipper, J. K., & Nass, G. (1966). Dating behavior: A framework for analysis and an illustration. *Journal of Marriage and the Family, 29*, 412—420.

Skoe, E. (1998). The ethic of care: Issues in moral development. In Skoe, E., & von der Lippe, A. (Eds.) *Personality development on adolescence: A cross-national and lifesapn perspective, (143-171).* NY: Routledge.

Skoe, E., & Gooden, A. (1993). Ethic of care and real-life moral dilemma content in male and female early adolescents. *Journal of Early Adolescence, 13* (2), 154-167.

Skolnick, A. (1991). *Embattled paradise*. New York: Basic Books.

Skovholt, T. M., & Morgan, J. I. (1981). Career development: An outline of issues for men. *The Personnel and Guidance Journal*, 231—237. Slaby, A. E. (1990). Suicide among youth. *Suicide and Life-Threatening Behavior, 20*, 193—194.

Slaby, R. G., & Roedell, W. C. (1982). Development and regulation of aggression in young children. In J. Worrell (Ed.), *Psychological development in the elementary years* (pp. 97—149). San Diego: Academic Press. Slavin, R. E. (1983). *Cooperative learning*. New York: Longman.

Smetana, J. (1993). Conceptions on parental authority in divorced and married mothers and their adolescents. *Journal of Research on Adolescence, 3,* 19—39.

Smetana, J., Yau, J., Restrepo, A., & Braeges, J. (1991). Conflict and adaptation in adolescence: Adolescent-parent conflict. In M. F. Colten & S. Gore (Eds.), *Adolescent stress: Causes and consequences.* New York: Aldine de Gruyter.

Smith, C. L., & Rojewski, J. W. (1993). School-to-work transition: Alternatives for educational reform. *Youth and Society, 25,* 222—250.

Smith, E. (1999). The effects of the investment in the social capital of youth on political and civic behavior in young adulthood: A longitudinal analysis. *Political Psychology, 20 (3), 553-580.*

Smith, F. (1989). A biosocial model of adolescent sexual behavior. In G. Adams, R. Montemayor, & T. Gullotta (Eds.), *Biology of adolescent behavior and development* (pp. 143—167). Newbury Park, CA: Sage.

Smith, F., & Udry, J. R. (1985). Coital and non-coital sexual behaviors of White and Black adolescents. *American Journal of Public Health, 75,* 1200-1203.

Smith, E., Udry, J. R., & Morris, N. (1985, September). Pubertal development and friends: A biosocial explanation of adolescent sexual behavior. *Journal of Health and Social Behavior, 26,* 183—192.

Smith, R. (1990). *A theoretical framework for explaining the abuse of hyperactive children.* Unpublished doctoral dissertation, Boston College, Chestnut Hill, MA.

Smollar, J., & Youniss, J. (1985). Adolescent self-concept development. In R. L. Leahy (Ed.), *The development of self* Orlando, FL: Academic Press.

Smollar, J., & Youniss, J. (1989). Transformations in adolescents perceptions of parents. *International Journal of Behavioural Development, 12,* 71—84.

Snarey, J. R. (1985). Cross-cultural universality of social-moral development: A critical view of Kohlbergian research. *Psychological Bulletin, 97,* 202—232.

Snodgrass, S. E. (1992). Further effects of role versus gender on interpersonal sensitivity. *Journal of Personality and Social Psychology, 62,* 154-158.

Snyder, 5. (1991). Movies and juvenile delinquency: An overview. *Adolescence, 26* (101), 121—132.

Solow, R. A., & Solow, B. K. (1986). Mind-altering drugs: Effects on adolescent sexual functioning. *Medical Aspects of Human Sexuality, 20* (1), 64-74.

Sorenson, S. B., & Bowie, P.(1994). Girls and young women. In L. D. Eron, J. Gentry, & P. Schlegel (Eds.), *Reason to hope: A psychosocial perspective on violence and youth* (pp. 167—176). Washington, DC: American Psychological Association.

Soriano, F. I. (1994). U.S. Latinos. In L. D. Eron, J. Gentry, & P. Schlegel (Eds.), *Reason to hope: A psychosocial perspective on violence and youth* (pp. 119—132). Washington, DC: American Psychological Association.

Sousa, C. (1999). Teen dating violence: The hidden epidemic. *Family and Conciliation Courts review, 37 (3), 356-374.*

Sparks, E. (1994). Human rights violations in the inner city: Implications for moral educators. *Journal of Moral Education, 23,* 315-332.

Special Action Office for Drug Abuse Prevention. (1976). *Q and A* GPO: 1975, 0-576-576. Washington, DC: Executive Office of the President.

Spencer, M. B., & Markstrom Adams, C. (1990). Identity processes among racial and ethnic minority children in America. *Child Development, 61,* 290-310.

Spergel, I. A., Ross, R. F., Curry, G. D., & Chance, R. (1989). *Youth gangs: Problem and response.* Washington, DC: Office of Juvenile Justice and Delinquency Prevention. Spirito, A., Stark, L., Fristad, M., & Hart, K. (1987). Adolescent suicide attempters hospitalized on a pediatric unit. *Journal of Pediatric Psychology, 12* (2), 171—189.

Spirito, A., Stark, L., Grace, N., & Stamoulis, D. (1991). Common problems and coping strategies reported in childhood and early adolescence. *Journal of Youth and Adolescence, 20(5),* 531.

Spirito, A., Williams, C. A., Stark, L. J., & Hart, K. J. (1988). The Hopelessness Scale for Children: Psychometric properties with normal and emotionally disturbed adolescents. *Journal of Abnormal Child Psychology, 16,* 445—458.

Spivak, G., & Shure, M. B. (1974). *Social adjustment of young children: A cognitive approach to solving real-life problems.* San Francisco: Jossey-Bass.

Spivak, H., Hausman, A. J., & Prothrow-Stith, D. (1989). Practioners' forum: Public health and the primary prevention of adolescent violence—the Violence Prevention Project. *Violence and Victims, 4,* 203—212.

Sroufe, L. A., & Cooper, R. G. (1988). *Child development. Its nature and course.* New York: McGraw-Hill.

Stark, L. J., Spirito, A., Williams, C. A., & Guevremont, D. C. (1989). Common problems and coping strategies I: Findings with normal adolescents. *Journal of Abnormal Child Psychology, 17,* 203—212.

Statistical abstracts of the U.S. (1991). Washington, DC: Bureau of the Census.

Statistical abstracts of the U.S. (1998). Washington, DC: Bureau of the Census.

Steele, C & Aronson, J. (1998). Stereotype threat and the test performance of academically successful African Americans. In Jencks, C & Phillips, M (Eds.) *The Black-White test score gap (pp. 401-427), Washington DC: Brookings Institution.*

Steele, C., & Josephs, R. (1990). Alcohol myopia: Its prized and dangerous effects. *American Psychologist, 45* (8), 921—933.

Stein, D., & Reichert, P. (1990). Extreme dieting in early adolescence. *Journal of Early Adolescence, 10* (2), 108—121.

Steinberg, L. (1981). Transformations in family relations at puberty. *Developmental Psychology, 17* (6), 833—840.

Steinberg, L. (1987). Impact of puberty on family relations: Effects of pubertal status and pubertal timing. *Developmental Psychology, 23,* 451—460.

Steinberg, L. (1988). Reciprocal relation between parent-child distance and pubertal maturation. *Developmental Psychology, 24,* 122—128.

Steinberg, L. (1990). Autonomy, conflict, and harmony in the family relationship. In S. Feldman and G. Elliot (Eds.), *At the threshold: The developing adolescent.* Cambridge, MA: Harvard University.

Steinberg, L., & Dorubusch, S. (1991). Negative correlates of part-time employment during adolescence: Replication and elaboration. *Developmental Psychology, 27,* 304-313.

Steinberg, L., Greenberger, F., Garduque, L., Ruggiero, M., & Vaux, A. (1982). Effects of working on adolescent development. *Developmental Psychology, 18,* 385—395.

Steinberg, L., Mounts, N. S., Lamborn, S., & Dornbusch, S. (1991). Authoritative parenting and adolescent adjustment across varied ecological niches. *Journal of Research on Adolescence, 1,* 19—36.

Steiner-Adair, C. (1990). The body politic: Normal female adolescent development and the development of eating disorders. In C. Gilligan, N. Lyons, and T. Hanmer (Eds.), *Making connections: The relational worlds of adolescent girls at Emma Willard School.* Cambridge, MA: Harvard University Press.

Stern, L. (1990). Conceptions of separation and connection in female adolescents. In C. Gilligan, N. P. Lyons, & T. J. Hanmer (Eds.), *Making connections: The relational worlds of adolescent girls at Emma Willard School.* Cambridge, MA: Harvard University Press.

Stern, M., & Zevon, M. A. (1990). Stress, coping and family environment: The adolescent's response to naturally occurring stressors. *Journal of Adolescent Research, 5,* 290-305.

Stice, E., & Barrera, M. (1995). A longitudinal examination of reciprocal relations between perceived parenting and adolescents' substance abuse and externalizing behaviors. *Developmental Psychology, 31,* 322—334.

Stiffman, A., Earls, F., Robins, L., & Jung, K. (1987). Adolescent sexual activity and pregnancy: Socioenvironmental problems, physical health and mental health. *Journal of Youth and Adolescence, 16* (5), 497—569.

Stiles, D., Gibbons, J., Hardardottir, S., & Schnellmann, J. (1987). The ideal man or woman as described by young adolescents in Iceland and the United States. *Sex Roles, 17(5/6),* 313—320.

Stiles, D., Gibbons, J. L., & Schnellmann, J. (1987). The smiling sunbather and the chivalrous football player: Young adolescents' images of the ideal woman and man. *Journal of Youth and Adolescence, 7* (4), 411-427.

Stiles, D. A., Gibbons, J. L., & Peters, E. (1993). Adolescents' views of work and leisure in the Netherlands and the United States. *Adolescence, 28* (106).

Stiles, D. A., Gibbons, J. L., & Schnellmann, J. (1990). Opposite-sex ideal in the U.S.A. and Mexico as perceived by young adolescents. *Journal of Cross-Cultural Psychology, 21,* 180-199.

Straus, M. (1991). Discipline and deviance: Physical punishment of children and violence and other crime in adulthood. *Social Problems, 38,* 133—154.

Stricof, R., Novick, L. F., Kennedy, J., & Weisfuse, I. (1988). Seroprevalence of adolescents at a homeless facility. Paper presented at the American Public Health Association Conference, Boston, MA.

Striegel_moore, R. & Cachelin, F. (1999). Body image concerns and disordered eating in adolescent girls: Risk and protective factors. In Johnson, N., Roberts, M. & Worell, J. (Eds.) *Beyond appearance: A new look at adolescent girls., 85-108.* Washington, DC: APA.

Striegel-Moore, R., Silberstein, L., & Rodin, J. (1986). Toward an understanding of risk factors for bulimia. *American Psychologist, 41,* 246-263.

Subcommittee on Health and the Environment. (1987). *Incidence and control of chlamydia.* Washington, DC: U.S. Government.

Subrahmanyam, K., & Greenfield, P. M. (1994). Effect of video game practice on spatial skills in girls and boys. *Journal of Applied Developmental Psychology, 15,* (1), 13—32.

Sullivan, H. 5. (1953). *The interpersonal theory of psychiatry.* New York: W. W. Norton.

Sullivan, M. L. (1993). Culture and class as determinants of out-of-wedlock childbearing, and poverty during late adolescence. *Journal of Research on Adolescence, 3,* 295—316.

Sundal-Hansen, L. 5. (1984). Interrelationship of gender and career. In N. C. Gybers & Assoc., *Designing careers.* San Francisco, CA: Jossey-Bass. Sundberg, N. D., & Tyler, L. E. (1970). Awareness of action possibilities of Indian, Dutch, and American adolescents. *Journal of Cross-Cultural Psychology, 1,*153—170.

Super, D. F. (1957). *The psychology of careers.* New York: Harper & Row.

Super, D. E. (1983). Assessment in career guidance: Toward truly developmental counseling. *Personnel and Guidance Journal, 61,* 555—562.

Super, D. E. (1990). A life-span, life-space approach to career development. In D. Brown, L. Brooks, et al. (Eds.), *Career choice and development.* San Francisco: Jossey-Bass. 542 References

Super, D. E., Thompson, A., Lindeman, R., Jordaan, J. P., & Myers, R. A., and others. (1981). *Career development inventory.* Palo Alto, CA: Consulting Psychologists Press.

Super, D. E., & Thompson, A. S. (1981). *The adult career concerns inventory.* New York: Teachers College, Columbia University.

Sussman, S., Dent, C., Flay, B., Burton, D., Craig, S., MestelRauch, J., & Holden, 5. (1989). Media manipulation of adolescents' personal level judgments regarding consequence of smokeless tobacco use. *Journal of Drug Education, 19* (1), 43—57.

Swadi, H. (1999). Individual risk factors for adolescent substance use. Drug and Alcohol Dependence, 55 (3), 209-224.

Symonds, A. (1991). Gender issues and Horney theory. *The American Journal of Psychoanalysis, 51,* 301—312.

Takata, S., & Zevitz, R. (1990). Divergent perceptions of group delinquency in a midwestern community: Racine's gang problem. *Youth and Society, 21* (3), 282—305.

Takei, Y., & Dubas, J. 5. (1993). Academic achievement among early adolescents: Social and cultural diversity. In R. M. Leruer (Ed.), *Early adolescence: Perspectives on research, policy and intervention* (pp. 175—190). Hillsdale, NJ: Erlbaum.

Talwar, R., Nitz, K., & Leruer, R. (1990). Relations among early adolescent temperament, parent and peer demands, and adjustment: A test of the goodness of fit model. *Journal of Adolescence, 13* (3), 279—298.

Tanner, J. (1990). *Foetus into man.* Cambridge, MA: Harvard.

Tape, G. (1990). The African environment and cognitive development. Cited in M. H. Segall, P. Dasen, J. W. Berry, & Y. Poortinda (Eds.), *Human behavior in global perspective,* 154-155. New York: Pergamon Press.

Taylor, C. (1990). Gang imperialism. In C. R. Huff (Ed.), *Gangs in America.* Newbury Park, CA: Sage.

Taylor, R. D., Casten, R., Flickinger, S. M., Roberts, D., & Fulmore, D. N. (1994). Explaining the school performance of African-American adolescents. *Journal of Research on Adolescence, 4,* 21—44.

Tedesco, L., & Gaier, E. (1988). Friendship bonds in adolescence. *Adolescence, 89,* 127—136.

Teicher, J. (1973). A solution to the chronic problem of living: Adolescent attempted suicide. In J. C. Schoolar, *Current Issues in Adolescent Psychiatry.* New York: Bruner/Mazel.

Terman, L. M. (1925). *Genetic studies of genius.* Stanford, CA: Stanford University Press.

Tessler, D. J. (1980). *Drugs, kids, and schools.* Santa Monica, CA: Goodyear.

Thies, K. & Walsh, M.E. (1999). A developmental analysis of stress in children and adolescents with chronic disease. *Children's Health Care, 28* (1), 15-32.

Thoma, S., Narvarz, D., Rest, J. & Derryberry, P. (1999). Does moral judgment development reduce to political attitudes or verbal ability? Evidence using the Defining Issues Test. *Educational Psychology review, 11 (4),* 325-341.

Thompson, K. (1980). A comparison of black and white adolescents' beliefs about having children. *Journal of Marriage and the Family, 8* (2), 133—139.

Thornburg, H., & Aras, Z. (1986). Physical characteristics of developing adolescents. *Journal of Adolescent Research,] (1),* 47—78.

Timko, C., Stovel, K. W., Baumgartner, & Moos, R. H. (1995). Acute and chronic stressors, social resources, and functioning among adolescents with juvenile rheumatic disease. *Journal of Research on Adolescence, 5* (3), 361—385.

Tishler, C. (1981). Adolescent suicide attempts: Some significant factors. *Suicide and Life-Threatening Behavior, 11* (2), 86-92.

Tishler, C., & McKenry, P. (1983). Intrapsychic symptom dimension of adolescent attempted suicide. *The Journal of Family Practice, 16* (4), 731—734.

Toby, J. (1989). Of dropouts and stay-ins: The Gerschwin approach. *Public Interest, 95,* 3—13.

Toffler, A. (1970). *Future shock.* New York: Bantam.

Tolan, P. H., & Loeber, R. (1993). Antisocial behavior. In P. H. Tolan & B. J. Cohler (Eds.), *Handbook of clinical research and practice with adolescents* (pp. 307—331). New York: John Wiley.

Tolman, D. (1999). Female adolescent sexuality in relational context: beyond sexual decision making. In Johnson, N., Roberts, M. & Worell, J. (Eds.) *Beyond appearance: A new look at adolescent girls,* 227-246. Washington, DC: APA.

Tolone, W., & Tieman, C. (1990). Drugs, delinquency, and "nerds": Are loners deviant? *Journal of Drug Education, 20* (2), 153—162.

Tomasello, M., Krueger, A. C., & Ratner, H. H. (1993). Cultural learning. *Behavioral and Brain Sciences, 16* (3), 495—552.

Toolan, J. (1975). Depression in adolescents. In J. Howell (Ed.), *Modern perspectives in adolescent psychiatry.* New York: Bruner/Mazel.

Topol, P., & Reznikoff, M. (1984). Locus of control as factors in adolescent suicide attempts. *Suicide and Life-Threatening Behavior, 12* (3), 141—150.

Toro, J., Salamero, M., & Martinez, F. (1994). Assessment of sociocultural influences on the aesthetic body shape model in anorexia nervosa. *Acta Psychiatrica Scandinavica, 89,* (3), 147—151.

Towbes, L. C., Cohen, L. H., & Glyshaw, K. (1989). Instrumentality as a life-stress moderator for early versus middle adolescents. *Journal of Personality and Social Psychology, 57,* 109—119.

Treboux, D., & Busch-Rossnagel, N. (1990). Social network influences on adolescent sexual attitudes and behaviors. *Journal of Adolescent Research, 5* (2), 175—189.

Treffinger, D. J., Isaksen, S. G., & Firestein, R. (1983). Theoretical perspectives on creative learning and facilitation: An overview. *Journal of Creative Behavior, 17* (1), 9—17.

Trotter, J. (1999). Lesbian and gay issues in work with young people: Are schools "out" this summer? British Journal of Social Work, 29 (6), 955-961.

Trotter, A. (1991). Rites of passage. *The Executive Educator, 13* (9), 48—49.

Turiel, E. (1974). Conflict and transition in adolescent moral development. *Child Development, 45,* 677—691.

Turner, R. A., Irwin, C. F., & Millstein, S. G. (1991). Family structure, family processes, and experimenting with substances during adolescence. *Journal of Research on Adolescence, 1,* 93—105.

Tygart, C. (1991). Juvenile delinquency and number of children in a family: Some empirical and theoretical updates. *Youth and Society, 22* (4), 525—536.

Umbreit, M. (1991). Mediation of youth conflict: A multi-system perspective. *Child and Adolescent Social Work, 8* (2), 141—153.

Unger, D. U., & Wandersman, L. P. (1988). The relation of family and partner support to the adjustment of adolescent mothers. *Child Development, 59,* 1056—1060.

University of Michigan Institute for Social Research (1994, December). *Monitoring the future study.* Ann Arbor, Michigan: Unpublished data.

Urberg, K. (1990). *Social crowd influence in adolescent cigarette smoking.* Paper presented at the 98th Annual Convention of the American Psychological Association, August 10-14, Boston, MA.

U.S. Bureau of the Census. (1986). *Statistical abstracts of the United States, 1986.* Washington, DC: U.S. Government Printing Office.

U.S. Bureau of the Census. (1994, March). *The Black population in the United States.* Washington, DC: U.S. Government Printing Office.

U.S. Department of Education, National Center for Education Statistics. (1994). *The Condition of Education 1994.* Washington, DC.

U.S. Department of Labor, Bureau of Labor Statistics. (1992). *Employment and earnings.* Washington, DC: U.S. Department of Labor.

Useem, E. L. (1991). Student selection into courses in mathematics: The impact of parental involvement and school policies. *Journal of Research on Adolescence, 1,* 231—250.

Van Gennep, A. (1909/1960). *The rites of passage.* M. Vizedom & G. Caffee, translators. Chicago: The University of Chicago Press.

van Hoof , A. (2000). The identity status approach: In need of fundamental revision and qualitative change. *Developmental Review. 19 (4), 622-647.*

Vasquez, M. & De Las Fuentes, C. (1999). American-born Asian, African, Latina and American Indian adolescent girls: Challenges and strengths. In Johnson, N., Roberts, M. & Worell, J. (Eds.) *Beyond appearance: A new look at adolescent girls.* Washington, DC: APA.

Vasquez, R. (1982). The relationship of traditional Mexican-American culture to adjustment and delinquency among three generations of Mexican-American male adolescents. *Hispanic Journal of Behavioral Sciences, 13* (1), 41—55.

Vartanian, L. (1997). Separation-Individuation, social support, and adolescent egocentrism: An exploratory study. *Journal of Early Adolescence, 17 (3), 245-270.*

Vaughan, V., & Litt, I. (1990). *Child and adolescent development: Clinical implications.* Philadelphia: W. B. Saunders.

Ventura, M. (1989, Winter). The age of endarkenment. *Whole Earth Review,* 44—49.

Vernberg, E. (1990a). Experiences with peers following relocation during early adolescence. *American Journal of Orthopsychiatry, 60* (3), 466—472.

Vernberg, E. (1990b). Psychological adjustment and experiences with peers during early adolescence: Reciprocal, incidental, or unidirectional relationships? *Journal of Abnormal Child Psychology, 18* (2), 187—198.

Vigil, J. (1988). *Barrio gangs: Street life and identity in southern California.* Austin, TX: University of Texas Press.

Vitz, P. (1977). *Psychology as religion: The cult of self-worship.* Grand Rapids, MI: William B. Eerdmans. Vondracek, F. W. (1993). Promoting vocational development in early adolescence. In R. M. Lerner (Ed.), *Early adolescence: Perspectives on research, policy & intervention* (pp. 277—292). Hillsdale, NJ: Erlbaum.

Vygotsky, L. 5. (1962). *Language and thought.* Cambridge, MA: M.I.T. University Press. Wade, N. L. (1987). Suicide as a resolution of separation— individuation among adolescent girls. *Adolescence, 22* (85), 169—177.

Wagner, B. M., Compas, B. F., & Howell, D. C. (1988). Daily and major life events: A test of the integrative model of psychosocial stress. *American Journal of Community Psychology,* 16,189—205.

Waithe, M. (1989). Twenty-three hundred years of women philosophers: Toward a gender undifferentiated moral theory. In M. Brabeck (Ed.), pp. 3—18, *Who cares? Theory, research and educational implications of the ethic of care.* New York, NY: Praeger.

Walch, S. (1976). Adolescent- attempted suicide: Analysis of the differences in male and female behavior. *Dissertation Abstracts International, 38,* (6-b), V. 2892B.

Walker, A. (1991, April 28). Of violence and the young: Search for safe haven in gang turns tragic in fatal gunfire. *The Boston Sunday Globe,* pp. 1, 14.

Walker, A. K. (1995, August 9). As legions of youth smoke, Clinton plans a crackdown. *The Boston Globe,* pp. 13.

Walker, L. (1984). Sex differences in the development of moral reasoning: A critical review. *Child Development, 55,* 677—691.

Walker, L. (1989). A longitudinal study of moral reasoning. *Child Development, 60,* 157—166.

Wall, B. (1991, April 28). Bringing hope to violent streets. *The Boston Sunday Globe,* pp. A25—A27.

Wallace-Broscious, A., Serafica, F. C., & Osipow, 5. (1994). Adolescent career development relationships to self-concept and identity status. *Journal of Research on Adolescence, 4,* 127—149.

Wallach, M. A., & Kogan, N. (1965). *Modes of thinking in young children.* New York: Holt, Rinehart, & Winston.

Wallerstein, J. S. (1987). Children of divorce: Report of a ten-year follow-up of early latency-age children. *American Journal of Orthopsychiatry, 57(2),* 199—211.

Wallerstein, J. S & Lewis, J. (1998). The long-term impact of divorce on children: A first report from a 25-year study. *Family and Conciliation Courts Review, 36 (3),* 368-383.

Walsh, M. F. (1992). *'Moving to Nowhere": Children's stories of homelessness.* Westport, CT: Auburn House.

Waltz, G., & Benjamin, L. (1980). *Adolescent pregnancy and parenthood.* ERIC document 184528.

Ward, J. V. (1990). Racial identity formation and transformation. In C. Gilligan, N. P. Lyons, & T. J. Hanmer (Eds.), *Making connections: The relational worlds of adolescent girls at Emma Willard School* (pp. 215—231). Cambridge: Harvard University Press.

Washington, A. E., Arno, P. 5., & Brooks, M. A. (1986). The economic cost of pelvic inflammatory disease. *Journal of the American Medical Association, 225* (13), 1021—1033.

Waterman, A. (1999). Identity, the identity statuses and identity status development: A contemporary statement. *Developmental Review. 19 (4), 591-621.*

Watson, F. I., & Kelly, M. J. (1989). Targeting the at-risk male: A strategy for adolescent pregnancy prevention. *Journal of the National Medical Association, 81,* 453—456.

Watts, W. D., & Wright, L. (1990). The relationship of alcohol, tobacco, marijuana, and other illegal drug use to delinquency among Mexican-American, black, and white adolescent males. *Adolescence, 25* (97), 171—181.

Wechsler, D., & Isaacs, N. (1991). Binge drinking at Massachusetts colleges. *Journal of the American Medical Association, 267(21),* 2929—2931.

Wehlage, G., Rutter, R., Smith, G., Lesko, N., & Fernandez, R. (1989). *Reducing the risk: Schools as communities of support.* New York: Falmer Press.

Wehlage, G. G., & Rutter, R. A. (1986). Dropping out: How much do schools contribute to the problem? *Teachers College Record, 87,* 374-392.

Weiner, I. B. (1970). *Psychological disturbances in adolescence.* New York: Wiley-Interscience.

Weiner, I. B. (1992). *Psychological disturbances in adolescence* (2nd ed.). New York: Wiley-Interscience.

Weiner, I. B. (1993). *Psychological Disturbance in Adolescence.* New York: Wiley.

Wellman, D. (1977). *Portraits of white racism.* New York: Cambridge University Press. Wenar, C. (1994). *Developmental psychopathology from infancy through adolescence.* New York: McGraw Hill.

Wenz, F. (1979). Economic status, family anomie, and adolescent suicide potential. *Journal of Psychology, 98* (1), 457.

Werner, E. E. (1990). Protective factors and individual resilience. In S. Meisels & J. Shonkoff (Eds.), *Handbook of early intervention.* New York: Cambridge University Press. Westney, 0., Jenkins, R., Butts, J., & Williams, I. (1984). Sexual development and behavior in black preadolescents. *Adolescence, 99* (75), 557—568.

Werner, E. E. (1995). Resilience in development. *Current Directions in Psychological Science, 4 (3), 81-85.*

Werner, E. E. & Johnson, J.L. (1999). Can we apply, resilience? In *Resilience and development, positive life adaptation: Longitudinal research in the social and behavioral sciences* (pp. 259-268). NY: Kluwer Academic?Plenum Publishers.

Wetzel, J. R. (1987). *American youth: A statistical snapshot.* Washington, DC: William T. Grant Commission on Work, Family and Citizenship.

Whelan, R. (1982). Presidential message. *Society for Learning Disabilities and Remedial Education Newsletter, 2,* 1—3.

Whitaker, D. & Miller, K. (2000). Parent-Adolescent discussions about sex and condoms: Impact of peer influences of sexual risk behavior. *Journal of Adolescent Research, 15,* 251-273.

White, H. C. (1974). Self-poisoning in adolescents. *British Journal of Psychiatry, 124,* 24-35.

White, J. L. (1989). *The troubled adolescent.* New York: Pergamon.

Whitehouse, A. M., & Button, E. J. (1988). The prevalence of eating disorders in the U.K. college population: A reclassification of an earlier study. *International Journal of Eating Disorders, 7,* 393—397.

Whitely, B. F. (1983). Sex-role orientation and self-esteem: A critical meta-analytic review. *Journal of Personality and Social Psychology, 44,* 765—778.

Whitely, B. F. (1984). Sex-role orientation and psychological well-being: Two meta-analyses. *Sex Roles, 12,* 207—225.

Whiting, B. B., & Edwards, C. P. (1988). *Children of different worlds: The formation of social behavior.* Cambridge, MA: Harvard University Press.

Widmer, E. & Weiss, C. (2000). Do older siblings make a difference? The effect of older sibling support and older sibling adjustment on the adjustment of socially disadvantaged adolescents. *Journal of Research on Adolescence, 10 (1), 1-27.*

Widom, C. 5. (1991). Avoidance of criminality in abused and neglected children. *Psychiatry, 54,* 162—174.

Widom, C. 5. (1994). Childhood victimization and adolescent problem behaviors. In R. D. Ketterlinus & M. E. Lamb (Eds.), *Adolescent problem behaviors* (pp. 127—164). Hillsdale, NJ: Erlbaum.

Wiegman, O. & Van Schie, E. (1998). Video game playing and its relations with aggressive and prosocial behaviour. *British Journal of Social Psychology, 37 (3), 367-378.*

Wiland, L. J. (1986). Vision quest: Rites of adolescent passage. *Camping Magazine, 58* (7), 30-33.

William T. Grant Commission on Work, Family and Citizenship. (1988). *The forgotten half Pathways to success for America's youth and young families.* Washington, DC: Youth and America's Future. The William T. Grant Commission on Work, Family and Citizenships.

Williams, L. (1989). *Teens feel having sex is their own right.* New York: The New York Times.

Williams, T. (1990). Cocaine kids: The underground American dream. *Drugs, Society, and Behavior, 90/91.* Guilford, CT: The Dushkin Publishing Group.

Williams, J.M. & Dunlop, L.C. (1999). Pubertal timing and self reported delinquency among male adolescents. Journal of Adolescence, 22 (1), 157-171.

Williams, T., & Korublum, W. (1985). *Growing up poor.* New York: Free Press.

Williams, C. & Vines, S.W. (1999). Broken past, fragile future: Personal stories if high risk adolescent mothers. *Journal of the Society of Pediatric Nurses, 4, 15-23.*

Willits, F., & Crider, D. (1989). Church attendance and traditional religious beliefs in adolescence and adulthood: A panel study. *Review of Religious Research, 31(1),* 68—81.

Wilson, C., & Keye, W. (1989). A survey of adolescent dysmenorrhea and premenstrual symptom frequency. *Journal of Adolescent Health Care, 10,* 317—322.

Wilson, M. N. (1989). Child development in the context of the black extended family. *American Psychologist, 44* (2), 380—395.

Winnicott, D. W. (1965). *The maturational process and the facilitating environment.* New York: International Universities Press.

Winship, E. (1991, March 7). Sense about sex: Ask Beth: Nothing wrong with being flat-chested. *The Boston Globe,* 1991, p. 82.

Wirtz, P., & Harrell, A. (1990). *Adolescent problem drinking: Comparative profiles of problem domains.* Paper presented at the 98th Annual Convention of the American Psychological Association, August 10—14, Boston, MA. Wodarski, J. (1990). Adolescent substance abuse: Practice implications. *Adolescence, 25* (99), 667—688.

Wodarski, J. S., & Hedrick. M. (1989). Violent children: A practice paradigm. In J. S. Wodarski, *Preventive health services for adolescents* (pp. 135—148). Springfield, IL: Charles C Thomas. Wong, D. S., & Hart, J. (1994, May 20). Answers leave a confounding picture in poll of Mass. high school students. *The Boston Globe,* pp. 1, 33.

Woodruff-Pak, D. (1988). *Psychology and aging.* Englewood Cliffs, NJ: Prentice-Hall.

Wooley, S. C., & Keamey-Cooke, A. (1986). Intensive treatment of bulimia and body image disturbance. In K. D. Brownell & J. P. Foreyt (Eds.), *Handbook of eating disorders* (pp. 476-502). New York: Basic Books.

Work, H. (1989). Contemporary youth: Their psychological needs and beliefs. In M. Galanter (Ed.), *Cults and new religious movements: A report of the American Psychiatric Association.* Washington, DC: American Psychiatric Association.

Workman, J. & Freeburg, E. (1999). An examination of date rape, victim dress and perceiver variables within the context of attribution theory. *Sex Roles, 41 (3-4), 261-277.*

Wright, M. (1989). Body image satisfaction in adolescent girls and boys. *Journal of Youth and Adolescence, 18,* 71—84.

Wulff, D. (1991). *Psychology of religion: Classic and contemporary views.* New York: John Wiley & Sons. Wyatt, G. (1989). Reexamining factors predicting Afro-American and white American women's age at first coitus. *Archives of Sexual Behavior, 18* (4), 271—297.

Wyche, K. F., & Rotheram-Borus, M. J. (1990) Suicidal behavior among minority youth in the United States. In A. R. Stiffman & L. E. Davis (Eds.), *Ethnic issues in adolescent mental health* (pp. 323—338). Newbury Park, NJ: Sage.

Yamoor, C. M., & Mortimer, J. T. (1990). Age and gender differences in the effects of employment on adolescent achievement and well-being. *Youth and Society, 22,* 225—240.

Yancey, C. (1991, April 29). Kids killing kids: What can be done? Anti-crime partnership. *The Boston Globe.*

Yates, M. (1995, March). *Community service in adolescence: Implications for moral-political awareness.* Paper presented at the biennial meeting of the Society for Research on Child Development, Indianapolis, IN.

Yeaworth, R. C., York, J., Hassey, M. A., Ingle, M. F., & Goodwin, T. (1980). The development of an adolescent life change event scale. *Adolescence, 15,* 93.

Yesalis, C., Streit, A., Vicary, J.,Friedl, K., Brannon, D., & Buckley, W. (1989). Anabolic steroid use: Indications of habituation among adolescents. *Journal of Drug Education, 19* (2), 103—116.

Yoder, K.A. (1999). Comparing suicide attempters, suicide ideators and non-suicidal homeless and runaway adolescents. Suicide and Life Threatening Behavior, 29 (1), 25-36.

Youngs, G., Rathge, R., Mullis, R., & Mullis, A. (1990). Adolescent stress and self-esteem. *Adolescence, 25* (98), 333—341.

Youniss, J. (1989). Parent-adolescent relationships. In W. Damon (Ed.), *Child development today and tomorrow* (pp. 379—392). San Francisco: Jossey-Bass.

Zabin, L. S., Hardy, J. B., Smith, E. A., & Hirsch, M. B. (1986). Substance use and its relation to sexual activity among inner-city adolescents. *Journal of Adolescent Health Care, 7* (5), 320-331.

Zakin, D., Blyth, D., & Simmons, R. (1984). Physical attractiveness as a mediator of the impact of early pubertal changes for girls. *Journal of Youth and Adolescence, 13* (5), 439—450.

Zelnick, M., & Kantner, J. F. (1980). Sexual activity, contraceptive use and pregnancy among metropolitan teenagers: 1971—1979. *Family Planning Perspectives, 12* (57), 30-36.

Zelnick, M., & Shah, F. K. (1983). First intercourse among young Americans. *Family Planning Perspectives, 15* (2), 64-70.

Zhang, L. (1999). A comparison of U.S. and Chinese University students' cognitive development: The cross cultural applicability of Perry's theory. *Journal of Psychology, 133 (4),* 425-439.

Zoja, L. (1984). Sucht als unbewusster Versuch zur initiation, Teil I. (Addiction as an unconscious attempt towards initiation: I.) *Analyische Psychologie, 15(2),* 110—115.

CREDITS

PHOTOS

Chapter 1
All photos Digital Images © 2000 Artville

Chapter 2
All photos Digital Images © 2000 Artville

Chapter 3
Page 112: Corbis© 2000. All other photos Digital Images © 2000 Artville

Chapter 4
All photos Digital Images © 2000 Artville

Chapter 5
All photos Digital Images © 2000 Artville

Chapter 6
All photos Digital Images © 2000 Artville

Chapter 7
All photos Digital Images © 2000 Artville

Chapter 8
All photos Digital Images © 2000 Artville

Chapter 9
All photos Digital Images © 2000 Artville

Chapter 10
All photos Digital Images © 2000 Artville

Chapter 11
All photos Digital Images © 2000 Artville

Chapter 12
All photos Digital Images © 2000 Artville

Chapter 13
All photos Digital Images © 2000 Artville

Chapter 14
All photos Digital Images © 2000 Artville

LINE ART AND TEXT

Chapter 1
Figure 1.2: John F. Travers, "The Classic Experiment" in *The Growing Child*, © 1982 Scott, Foresman and Company, Glenview, IL. Reprinted by permission of the author. Table 1.4: A. Spirito et al *Journal of Youth and Adolescence*, 20(5):531 © 1991 Kluwer Academic/Plenum Publishers. Figure 1.5: John S. Dacey, *Adolescents Today*, 3d edition © 1986 Scott, Foresman and Company, Glenview, IL. Reprinted by permission of the author.

Chapter 2
Table 2.1: John S. Dacey, *Adolescents Today*, 3d edition © 1986 Scott, Foresman and Company, Glenview, IL. Reprinted by permission of the author.

Chapter 3
Table 3.3. From A.C. Petersen, et al, *Journal of Youth and Adolescence*, 17:117-134, 1986. Figures 3.5 and 3.6: From J. M. Tanner, in A. C. Petersen and B. Taylor "The Biological Approach to Adolescence," in *Handbook of Adolescent Psychology*, edited by J. Adelson, 1980. Copyright © 1980 Reprinted by permission of John Wiley & Sons, Inc., New York, NY. Applied View Boxes pp. 83, 85, 95, 96: John S. Dacey, *Adolescents Today*, 3d edition © 1986 Scott, Foresman and Company, Glenview, IL. Reprinted by permission of the author.

Chapter 4
Applied View Box p. 116: From J. B. Funk, Reevaluating the impact of video games, in *Clinical Pediatrics*, 32 (2):86–90. Copyright © 1993 Westminster Publications, Inc., Glen Head, NY. Reprinted with permission. Applied View Box p. 124: John S. Dacey, *Adolescents Today*, 3d edition © 1986 Scott, Foresman and Company, Glenview, IL. Reprinted by permission of the author.

Chapter 6
Applied View Box p. 186: From D.A. Rosenthal, et al. *The Journal of Youth and Adolescence*, 10(6):525-37 © 1981 Kluwer Academic/Plenum Publishers, New York, NY. Reprinted with permission. Figures 6.2 and 6.3: From D. Stiles et al (1987), *Sex Roles*, 17(5/6), 313–320 © 1987 Plenum Publishing Corp. New York, NY.

Chapter 8
Sociocultural View Box p. 251: From *Black Ice* by Lorene Cary© 1991. Reprinted with permission of Alfred A. Knopf, a Division of Random House, Inc.

Figure 8.2: John S. Dacey, *Adolescents Today*, 3d edition © 1986 Scott, Foresman and Company, Glenview, IL. Reprinted by permission of the author. Applied View Box p. 272: John S. Dacey, *Adolescents Today*, 3d edition © 1986 Scott, Foresman and Company, Glenview, IL. Reprinted by permission of the author.

Chapter 9
Figure 9.3 and 9.4: J. Jones and D. Barlow, Archives of Sexual Behavior, 19(3): 269-279, © 1990 Kluwer Academic/Plenum Publishing Corp., New York, NY.

NAME INDEX

SUBJECT INDEX

A

AA (Alcoholics Anonymous), 157, 425
Abstinence, 280, 297
  from drugs, 424–25
Abstract concepts, 111
Abstract thinking, 105, 107, 148
  practicing, 106
  self-concept and, 167–68
Academic competitions, sports and, 267
Academic quality, of secondary schools, 317–18
Academic tracking, disadvantages of, 323–24
Accommodation, 103
Achieved identities, 189–90, 481
  cultural identity, 191
Achievement (identity status), 189
Acquaintance rape, 302–5
  myths and realities of, 304–5
  substance abuse and, 407
Acquired immune deficiency syndrome (AIDS)
  growing concern about, 281
  risks for, 410, 460
Adaptation, 103
Addictive drug use, 401
Adolescence, 2–32
  beginning and end of, 22–24, 32
  methods of studying, 24–31, 32
    data-collection techniques, 25–28
    time-variable designs, 28–31
  in past cultures, 14–22, 32
    Age of Enlightenment, 16–17
    ancient times, 14–15
    early twentieth century, 17–18
    middle ages, 15–16
    retrospective accounts of, 18–22
  in today's culture, 4–14, 32
    adolescent spending habits, 13–14
    contributions of adolescents, 12–13
    demographics of, 4–6
    problems of, 8–12
    sociocultural diversity of, 6–7
Adolescence, 38
Adolescent Life Change Event Scale (ALCES), 358, 360
Adolescent moratorium, 187–88, 478
Adolescent Perceived Life Events Scale (APES), 357
Adolescent subculture, 265–70, 273
  adult anxieties about, 265–67
  elements of, 269–70
  origins of, 267–69
  sports *versus* academic competition, 266
Adult supervision
  avoidance by peer groups, 255
  effect of work roles on, 233–34
  negative effects on behavior, 250
Aesthetic needs, 49
Affection, sexual behavior and, 295

Affirmative action, 330
African Americans
  academic success and, 329–31, 345
  in aggressive gangs, 448
  bias of schools against, 325–26
  college programs for, 329
  cultural opposition to violence, 443
  dropouts, 439
  effect of work roles on family, 233
  ethnic socialization by parents, 193
  family factors in substance abuse, 404–5
  family relations of, 221–22
  incomes of, 6
  lack of studies of, 59
  myths about male teens, 443
  peer groups of, 263–64
  secondary school attendance of, 318–19
  self-esteem and, 180–82
  stress in female adolescents, 360–61
  substance abuse by, 410–11
  suicide rate for males, 380
  teenage parenthood and, 236, 238–40
  unemployment and underemployment, 334, 344
  violent delinquency and, 441
  vulnerability to depression, 376
Age cohort, 28, 29
Age of Enlightenment, 16–17
Aggressive behavior
  control by peer groups, 256
  as depressive equivalent, 378
  violent delinquency and, 435
Aggressive gangs, 431, 446–54, 465
  as family substitute, 447
AIDS. *See* Acquired immune deficiency syndrome (AIDS)
AIDS education, 298–99
Aid to Dependent Children, 237
Alateen, 157
ALCES (Adolescent Life Change Event Scale), 358, 360
Alcohol, 395–426
  abuse of, 407
  crime and, 405–6
  decline in use of, 412–13
  dual use with illegal drugs, 413, 416
  sexual assault and, 446
  sexual functioning and, 408
Alcoholics Anonymous (AA), 157, 425
Alice in Wonderland, 3
Alienation, prostitution and, 462
Amenorrhea, 76
American Council on Education, 326
American Psychiatric Association, 285, 286–87
America's Choice, 334
Amphetamines, sexual functioning and, 408
Anabolic steroids, 413, 416
Anal stage of development, 43
Ancestral worldview, 221–22

Ancient times, adolescence in, 14–15, 31, 32
Androgens, 23
Androgyny model, gender roles and, 197–99
Anger
  depression and, 378–79
  sexual behavior and, 295
Announcing function of initiation rites, 475
Anorexia nervosa, 76, 77, 368–69
  athletics and, 93–94
Anthropology, contributions of, 45–46
Antisocial behavior, 430–66
  aggressive gangs, 431, 446–54, 465
    attempts to eliminate, 452–54
    characteristics of joiners, 452
    violence and, 449–51
  delinquency. *See* Delinquency
  nonaggressive offenders, 431, 454–66
    prostitution, 462–64
    runaways. *See* Running away
    treatment of, 464
  types of, 431–32
  violence. *See* Violence
Anxiety
  of adults about adolescent subculture, 265–67
  changing self-concept and, 167
  divorce and, 172
  about relationships, 172
  about sexuality, 44
APES (Adolescent Perceived Life Events Scale), 357
Appraisal-focused coping, 362
Approach-oriented coping strategies, 362
Are You There, God? It's Me, Margaret, 66
Aristotle, 145
Art, Mind, and Brain: A Cognitive Approach to Creativity, 122
Artistic personality type, 338
Asceticism, 44
Ascribed identity, 480–81
Asian Americans
  cultural clashes with American society, 224
  family relations of, 222–23
  in gangs, 448, 451
  peer groups of, 263–64
Assigned identities, 191
Assimilation, 103
Attachment, 143, 144
Attachment theory, 215
Attempted suicide, 379
  history of, 387–88
  meanings of, 382–83
  rates of, 381
  warning signs of, 389
Attention control treatment, 454
Attention-deficit hyperactivity disorder, 402
Authoritarian parenting style, 217–18
Authoritative parenting style, 218–19
  drug abuse prevention and, 404